AF802976

Satellite Thermal Control Handbook

Satellite Thermal Control Handbook

David G. Gilmore
editor

Mel Bello
executive editor

The Aerospace Corporation Press • El Segundo, California

The Aerospace Corporation Press
2350 East El Segundo Boulevard
El Segundo, California 90245-4691

Library of Congress Catalog Card Number: 94-70360

ISBN: 1-884989-00-4

Copyright © 1994 by The Aerospace Corporation
Unless otherwise indicated, this material may be reproduced by or for the United States Government pursuant to the copyright license under the clause of DFARS 252.227.7013.

Printed on recycled paper

Contents

Preface

As we approach the 21st Century, one cannot help but reflect upon the many accomplishments and lessons learned in the field of spacecraft thermal control. From the early days of space experimentation and development in the 50s and 60s, through the challenges of routine manned space flight in the 70s and the maturing satellite programs of the 80s, to the downsizing of space programs in the 90s, the field of satellite thermal control has continued to evolve into a discipline unto itself.

In keeping with its goal of strengthening its relationship with its customers and industry, The Aerospace Corporation has prepared this handbook, which represents a compendium of corporate knowledge and heritage in the field of thermal control of unmanned Earth-orbiting satellites. The objective of this effort was to develop a *practical* handbook that provides the reader with enough background and specific information to begin conducting thermal analysis and to participate in the thermal design of satellite systems. It is assumed that the reader has had at least one introductory heat-transfer class and understands the fundamental principles of conductive, radiative, and convective heat transfer.

The handbook is written in such a way as to be useful to thermal engineers of all experience levels. The first two chapters provide a general overview of satellite systems and space flight thermal environments. Chapter III describes a number of actual satellite and component thermal designs to familiarize those new to the field with some of the current design approaches. Subsequent chapters discuss in detail thermal control hardware and the thermal design and testing process.

This work is published with the understanding that The Aerospace Corporation and its authors are supplying information but are not attempting to render engineering or other professional services. If such services are required, the assistance of an experienced professional should be sought. Any references to commercially supplied hardware or software are, furthermore, included as information only, and no product endorsment on the part of The Aerospace Corporation is intended or implied.

The Aerospace Corporation

The Aerospace Corporation operates as a Federally Funded Research and Development Center, performing space systems architecture, engineering, planning, and analysis and research, and supports space and launch systems acquisition. Its primary function is to assist government organizations in applying the resources of science and technology to the advancement of space systems. Progress is monitored, alternatives and tests are analyzed, and problems are resolved in cooperation with industrial contractors and government laboratories.

Supporting these tasks, laboratory research is directed toward advancements in space systems, engineering feasibility studies of new system concepts, and analyses and forecasts of both U.S. domestic and international technology. In fulfilling this architect-engineering role, Aerospace helps to minimize system development and operational risks, which in turn leads to acquisition schedule protection and cost containment.

Aerospace neither manufactures a product or contracts for systems, nor does it select or direct industrial contractors. Objectivity, excellence, corporate memory, rigorous security, confidentiality, and integrity are hallmarks of the corporation.

A private, nonprofit corporation, Aerospace was established in 1960 by an act of the United States Congress. The company operates in the public interest under a Board of Trustees composed of nationally recognized leaders in industry, government, and education.

The company has no stockholders and does not distribute dividends. Earnings are used to further the corporate mission through investments in research, equipment, and facilities, and to satisfy the company's citizenship obligations to its resident and professional communities.

A staff of 3,800 includes 2,400 engineers and scientists, two-thirds of whom hold advanced degrees; fully one-fourth hold doctorates. Highly qualified, motivated, innovative people, offering skills in a broad range of technical disciplines and extensive experience with a variety of space programs, are the company's most valued asset.

Scientific and engineering research centers, including the Ivan A. Getting and Allen F. Donovan Laboratories, house facilities for research in electronics and optics, aerophysics, materials sciences, information and computer science, space science, chemistry, and physics. A large-scale computer facility contains several major mainframes interconnected by a fiber-optic network with computers and workstations throughout the company.

Acknowledgments

This handbook was prepared by The Aerospace Corporation and includes text contributed by The American Institute of Aeronautics and Astronautics, Fairchild Space Company, Hughes Aircraft Company, Lockheed Missiles and Space Company, and TRW. Material attributed to employees of The Aerospace Corporation may be reproduced by or for the U. S. Government. Material attributed to other individuals may be protected by applicable copyright laws and is reprinted here with permission of the copyright owner.

The editors wish to acknowledge and thank the following individuals for their respective contributions, with special thanks to James Gilchrist for his encouragement and guidance and to Diane Rader for her outstanding effort in preparing the document for publication.

Leon Bledjian, Engineering Specialist, Spacecraft Thermal Department, The Aerospace Corporation

Dr. Robert L. Collins, Manager, Thermal Management Technology Section, Spacecraft Thermal Department, The Aerospace Corporation

Martin Donabedian, Manager, Cryogenics Section, Spacecraft Thermal Department, The Aerospace Corporation

Robert J. Eby, Robert D. Karam, Fairchild Space Company

C. Finch and B. Turner, British Aerospace

A.S. Friedman, Space and Communications Group, Hughes Aircraft Company

Dr. Donald F. Gluck, Project Engineer, Space Technology Directorate, The Aerospace Corporation

Gwynne E. Gurevich, Project Engineer, Space Flight Programs, The Aerospace Corporation

D. V. Hale, M. J. Hoover, M. J. O'Neill, Lockheed Missiles and Space Company

Brian E. Hardt, Member of the Technical Staff, Spacecraft Thermal Department, The Aerospace Corporation

Lee A. Hennis, Space and Communications Group, Hughes Aircraft Company

Dr. Tung T. Lam, Manager, Satellite Heat Transfer Section, Spacecraft Thermal Department, The Aerospace Corporation

Robert R. McMurchy, TRW

Robert Prager, Engineering Specialist, Spacecraft Thermal Department, The Aerospace Corporation

Dr. Wayne Stoll, R. Bettini, Hughes Danbury Optical Systems

Dr. Wayne Stuckey, Research Scientist, Mechanics and Materials Technology Center, The Aerospace Corporation

John W. Welch, Senior Member of the Technical Staff, Spacecraft Thermal Department, The Aerospace Corporation

Y. Yoshikawa, Lockheed Missiles and Space Company

Chapter 1

Satellite Systems Overview

David G. Gilmore
The Aerospace Corporation

INTRODUCTION

Over the past 30 years, hundreds of satellites have been built in support of scientific, military, and commercial missions. Most of these satellites, however, can be broadly categorized as one of three different types: spin stabilized, 3-axis stabilized, or pallets. Each of these types has a rather distinctive set of characteristics as to the satellite configuration, internal equipment, and approach to thermal control. The purpose of this chapter is to give a brief overview of the characteristics of each of these different types of spacecraft and the missions they support. Representative thermal designs for each type are discussed in more detail in Chapter III.

SATELLITE CONFIGURATIONS

The most common spacecraft configuration today is 3-axis stabilized. This type of spacecraft is characterized by a roughly box-shaped body and deployable solar array panels as typified by the Defense Meteorological Satellite Program (DMSP), the European Communications Satellite (ECS), and NASA's Tracking and Data Relay Satellite System (TDRSS), which are shown in Figure 1. The bodies of these spacecraft are kept inertially stable except for a slow rotation induced about one axis to keep the payload antennas or sensors continuously pointed toward the Earth as the satellite orbits. The solar array panels are then counter-rotated relative to the spacecraft body to keep them inertially fixed on the sun.

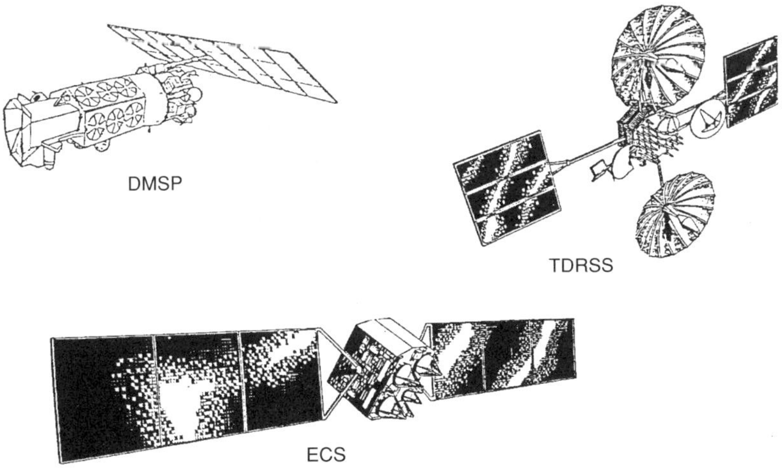

Figure 1. 3-axis stabilized satellites (Martin Marietta, TRW, British Aerospace)

A typical internal equipment complement for a 3-axis stabilized spacecraft is shown in the exploded view of a Fltsatcom satellite in Figure 2. It is common to refer to the spacecraft in terms of a "payload" and a "bus," or "platform." The payload is the equipment that services the primary mission, e.g., a cloud-cover camera for a weather satellite or an infrared sensor for a missile early warning system. Since Fltsatcom is a communications satellite, the "payload" is the communications subsystem, which consists of the antennas on the Earth-facing side of the vehicle and the communications electronics boxes mounted in the upper hexagonal compartment as shown in Figure 2. The "bus" consists of all other spacecraft subsystems that support the payload. These other subsystems typically include:

Structures Subsystem. The physical structure of the spacecraft to which all electronics boxes, thrusters, sensors, propellant tanks, etc. are mounted.

Electrical Power/Distribution Subsystem (EPS or EPDS). The equipment used to generate and distribute electrical power to the spacecraft, including solar arrays, batteries, solar array controllers, power converters, electrical harnesses, battery charge control electronics, etc.

Telemetry, Tracking, and Command Subsystem (TT&C). The electronics used to track, communicate with, and monitor the spacecraft from the ground. TT&C equipment generally includes receivers, transmitters, antennas, tape recorders, and state-of-health sensors for parameters such as temperature, electrical current, voltage, enable/disable status for various components, propellant tank pressure, etc.

Attitude/Velocity Control Subsystem (ACS or AVCS). The devices used to sense and control the vehicle attitude (orientation relative to an inertial coordinate system) and velocity. Typical components of the ACS system include sun and Earth sensors, star sensors if high-precision pointing is required, reaction or momentum wheels, Inertial Measurement Units (IMUs), Inertial Reference Units (IRUs), and the electronics required to process signals from the above devices and control satellite attitude.

Propulsion Subsystem. Liquid and solid rockets or compressed-gas jets and associated hardware used for changing satellite attitude, velocity, or spin rate. Solid rockets are usually used for placing a satellite in its final orbit after separation from the launch vehicle. The liquid engines (along with associated plumbing lines, valves, and tanks) may be used for attitude control and orbit adjustments as well as final orbit insertion after launch.

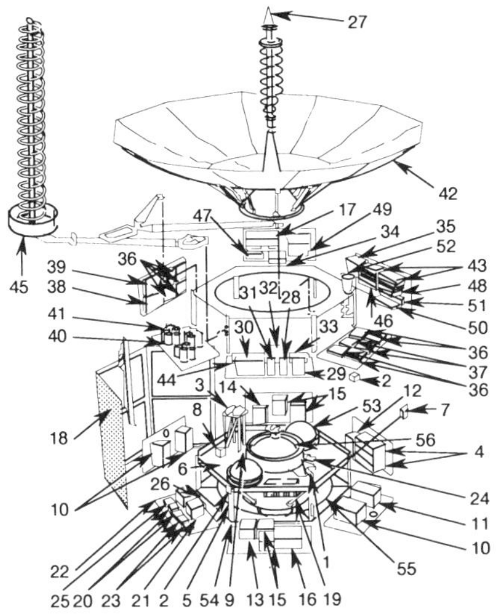

No.	Nomenclature	No.	Nomenclature
	ATTITUDE AND VELOCITY CONTROL		COMMUNICATION SYSTEM UHF TRANSPONDER
1.	Solar array drive assembly	28.	Preamp/downconverter/IF limiter no. 1
2.	Sun sensor assembly	29.	IF filter limiter no. 2
3.	Earth sensor assembly	30.	Processor receiver/synthesizer
4.	Control and auxiliary electronics	31.	Repeater receiver
5.	Spinning earth sensor assembly	32.	Command receiver/synthesizer
6.	Reaction wheel assembly	33.	Oven controlled crystal oscillator (2)
7.	Coarse sun sensor assembly	34.	AF processor
8.	Earth sensor electronics	35.	UHF command decoder
9.	Nutation damper assembly	36.	UHF transmitter Navy low power
	ELECTRICAL POWER/ DISTRIBUTION SYSTEM	37.	UHF transmitter Navy high power
10.	Battery assembly	38.	UHF transmitter (DODWB)
11.	Power control unit	39.	UHF transmitter (AFNB)
12.	Converter, spacecraft equipment	40.	UHF transmit filter
13.	Converter, communications no. 1	41.	UHF multicoupler filter assembly
14.	Converter, communications no. 2	42.	Transmit antenna assembly
15.	Converter, transmitter	43.	Frequency generator
16.	Payload switching unit no. 1	44.	Receive filter
17.	Payload switching unit no. 2	45.	UHF receive antenna assembly
18.	Solar panel assembly	46.	Signal distribution unit no. 1
19.	Electrical integration assembly	47.	Signal distribution unit no. 2
	TELEMETRY, TRACKING, AND COMMAND SYSTEM	48.	Passive hybrid
	S-BAND COMMAND GROUP		SHF TRANSPONDER
20.	S-band receiver	49.	FB processor
21.	Decrypter KIR23 (2 required)	50.	SHF receiver
22.	Command unit	51.	SHF transmitter
	S-BAND TELEMETRY GROUP	52.	SHF antenna
23.	S-band telemetry transmitter		PROPULSION SYSTEM
24.	PCM encoder	53.	Propellant tank
	S-BAND ANTENNA GROUP	54.	Fill and drain valve
25.	S-band diplexer	55.	Thruster assembly
26.	RF coaxial switch	56.	Apogee kick motor
27.	S-band antenna		

Figure 2. Fltsatcom (TRW)

Thermal Control Subsystem (TCS). Last but not least, the hardware used to control temperatures of all vehicle components. Typical TCS elements include surface finishes, insulation blankets, heaters, and refrigerators.

All of these subsystem components are shown in the sketch of Fltsatcom in Figure 2.

The second category of satellites is the spin-stabilized variety. These are less common than the 3-axis type and have historically been used mostly for relatively high-altitude missions in geosynchronous or Molniya orbits, although low-altitude spinning satellites do exist. A typical "spinner," INTELSAT VI, is shown in Figure 3. As the name implies, attitude stability is achieved by spinning the satellite like a top at approximately 15 rpm about the axis of the cylindrical solar array. In the case of INTELSAT VI, there is a large shelf on which the communications payload is mounted and which is despun relative to the rest of the spacecraft so that it points continuously at the Earth.

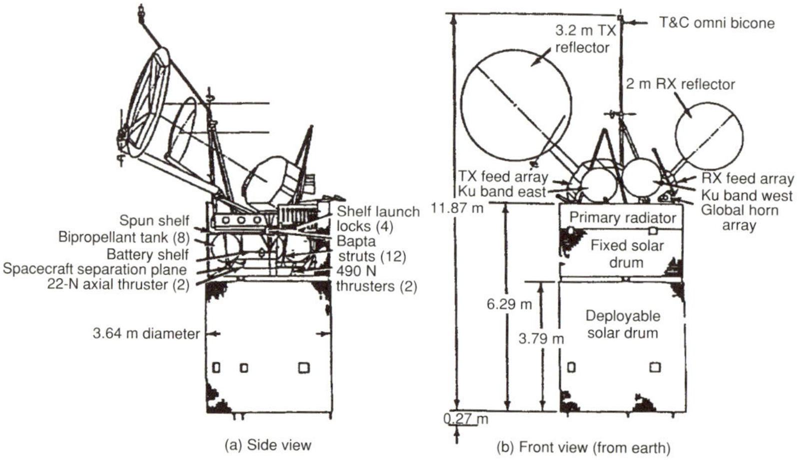

Figure 3. INTELSAT VI satellite (Hughes)

A spinner has all of the same basic subsystems as a 3-axis satellite: Structures, EPS, TT&C, ACS, TCS, and Propulsion. The payload is usually contained entirely on the despun section while most of the other subsystems are on the spinning side. It should be noted, however, that there are other types of spinners, such as the Defense Support Program satellites (DSP see Figure 4), that do not have a despun shelf. In the case of DSP, the payload, an infrared (IR) telescope/sensor, spins with the rest of the satellite, with the rotation of the vehicle being used to provide a scanning motion for the sensor.

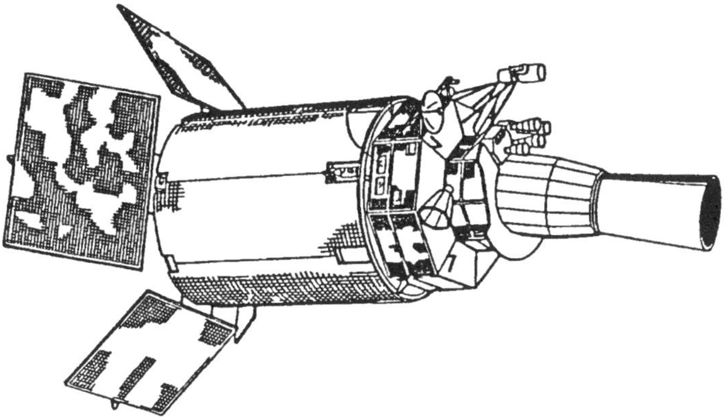

Figure 4. DSP satellite (TRW, Aerojet)

The third general type of spacecraft is the "pallet," which is not a complete satellite but a collection of one or more payloads plus some limited support services, such as power distribution, data recording, or telemetry sensors. These pallets may be anything from a small experiment mounted to the side of a host spacecraft to a large structure containing many instruments and mounted in the payload bay of the space shuttle. The principle difference between a pallet and a spacecraft is that the pallet is not able to function autonomously, but instead relies on the host vehicle for attitude control, electrical power, and TT&C support.

The Experiment Support System is a typical pallet system and is shown in Figure 5. It consists of a rather large structure that supports a half dozen experiments and an equipment compartment containing power distribution, command processing, and data recording equipment. The pallet is mounted in the space shuttle payload bay with electrical power, attitude control, and TT&C functions provided by the shuttle. In addition to the pallet itself, there is a command monitor panel mounted in the crew compartment to allow the astronauts to control the operation of the experiment on the pallet. Because of the support provided by the shuttle, the pallet does not have propulsion, attitude control, electrical power generation, or telemetry subsystems, and is incapable of operating in space on its own.

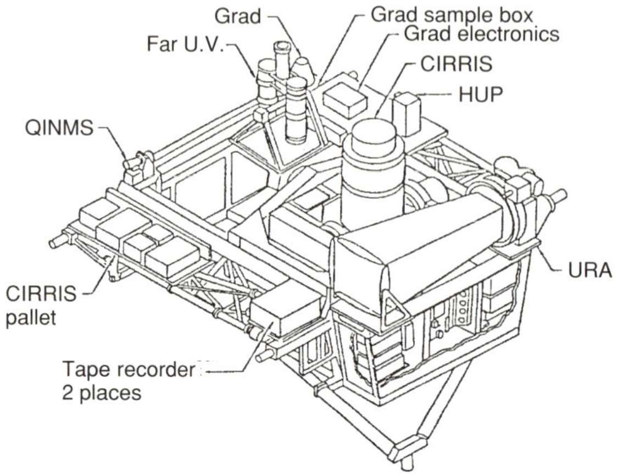

Figure 5. Experiment support system (Lockheed)

The final configuration to be discussed here is that of upper stages. Although they are not satellites per se, they sometimes share a similar level of complexity with and have some of the same subsystems as satellites. They are included in this handbook because upper stage thermal control after separation from the booster is quite similar to that of satellite systems.

Upper stages are generally used to raise a spacecraft from the relatively low orbit to which the booster delivers it to a higher operational orbit. Upper stages can use solid, liquid, or cryogenic propellants and have mission durations varying from a few hours to several days. The Inertial Upper Stage (IUS, see Figure 6) is an example of a solid-propellant upper stage that can be used in conjunction with either the space shuttle or expendable boosters. The IUS itself has two stages, the first generally being used to put the spacecraft into a highly elliptical transfer orbit, with the second stage fired at transfer-orbit apogee to circularize the orbit at the higher altitudes. Like a satellite, the IUS has Electrical Power, TT&C, Structures, Attitude Control, Thermal Control, and Propulsion subsystems.

ORBITS

There is a variety of different orbit types, which are used for different types of missions. The most common types of orbits, in order of increasing altitude, are known as low Earth orbit (LEO), Molniya, and geosynchronous (GEO), and are drawn to scale in Figure 7. The following paragraphs will briefly describe these types of orbits, and a more detailed discussion of orbit parameters can be found in Chapter II.

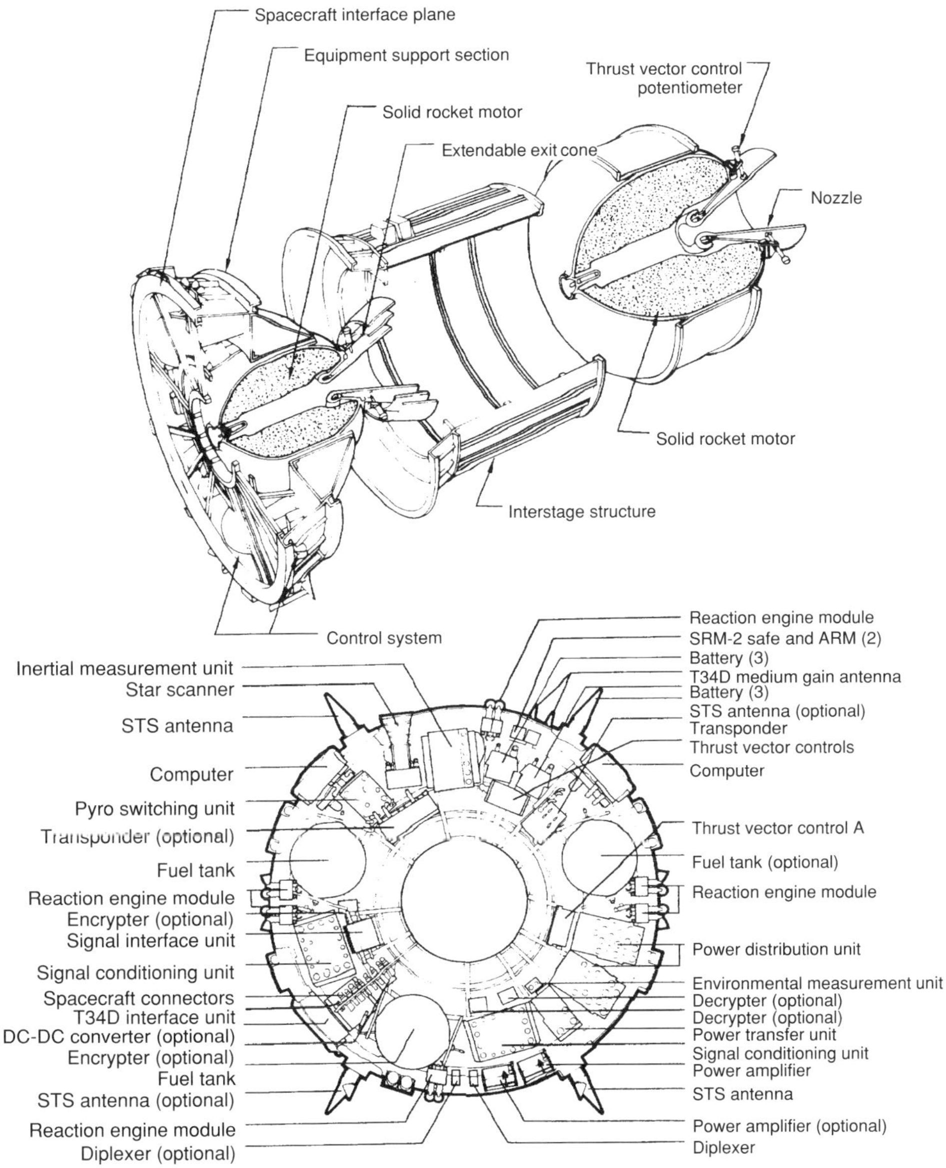

Figure 6. Inertial upper stage (Boeing)

Orbits whose maximum altitudes are less than approximately 1,000 nautical miles are generally considered low Earth orbits and have the shortest periods, on the order of an hour and a half. Some of these orbits are circular while others may be somewhat elliptical. The degree of eccentricity, however, is limited by the fact that the orbit (< 1,000 n. mi. altitude) is not much larger than the Earth, whose diameter is approximately 6,660 n. mi. (see Figure 7). The inclinations of these orbits, which is the angle between the plane of the equator and the plane of the orbit, can also vary from 0 deg to greater than 90 deg. Inclinations greater than 90 deg cause the satellite to orbit in a direction opposite to the Earth's rotation. Low Earth orbits are very often given high inclinations so that the satellite can pass over the entire surface of the Earth from pole to pole as it orbits. This coverage is important for weather and surveillance missions.

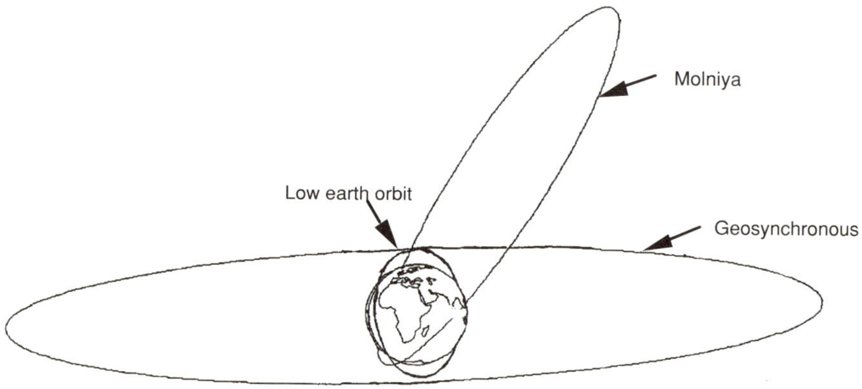

Figure 7. Orbit types

One particular type of low Earth orbit has the characteristic of maintaining the orbit plane at a nearly fixed angle relative to the sun (see Figure 8). The result of this is that, on every orbit, the satellite passes over points on the Earth that have the same local time, i.e., the same local sun elevation angle. Because the Earth rotates beneath the orbit, the satellite sees a different swatch of the Earth's surface on each revolution and can cover the entire globe over the course of a day. This characteristic of seeing the entire surface of the Earth at the same local sun angle is important for weather observation and visual surveillance missions. This type of orbit is known as sun synchronous and is discussed in more detail in Chapter II. Sun synchronous orbits may be placed so as to always see points on the Earth at anywhere from local sunrise/sunset to local noon, and are often referred to as "noon" or "morning" orbits.

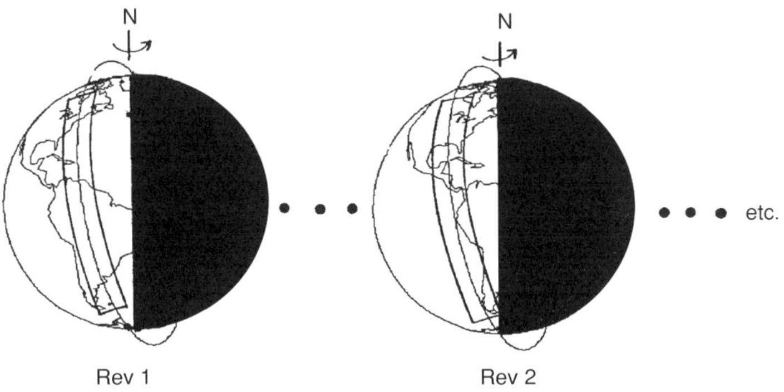

Figure 8. Sun synchronous orbit

The next higher type of common orbit is known as Molniya. These orbits are highly elliptical (apogee 21,000 n. mi., perigee 300 n. mi.), highly inclined (62 deg), and have the characteristic of providing good views of the north polar region for a large portion of the orbit (see Figure 9). Since the satellite travels very slowly near apogee it has a good view of the polar region for up to 8 hours

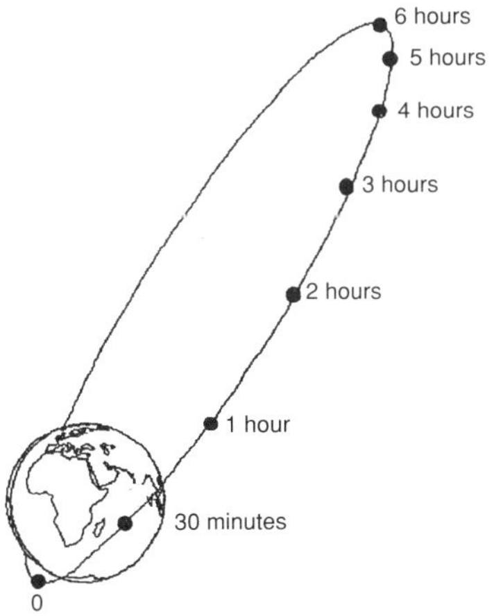

Figure 9. Molniya orbit

out of its 12-hour period. A constellation of three satellites can provide continuous coverage of the northern hemisphere for missions such as communications with aircraft flying over the polar region.

The highest common orbit type is known as geosynchronous. These orbits are circular, have very low inclinations (< 10 deg), and have an altitude of 19,360 n. mi. Thc distinguishing characteristic of this orbit is that it has a period matching the Earth's rotation and therefore remains over the same spot on the Earth at all times. This quality is valuable for a wide variety of missions including weather observation, communications, and surveillance.

One final useful observation is that most spacecraft go around their orbits in a counterclockwise motion as seen from above the north pole. This is done to take advantage of the initial eastward velocity given to the satellite due to the Earth's rotation (approximately 800 n. mi./hour at the Kennedy Space Center). To go around the orbit in the opposite direction would require the booster to overcome the initial 800 n. mi./hour eastward velocity before starting to build up speed in a westerly direction. This significantly affects booster size or allowable payload weight.

MISSIONS

The three general types of satellite platforms discussed earlier are used to support a wide variety of missions. The type of mission will dictate the orbit, the type of payload the satellite carries and, in some cases, the type of platform. Typical missions include communications, scientific observation, weather monitoring, navigation, remote sensing, surveillance, and data relay. This section briefly describes each of these missions.

The most common mission for both commerical and military satellites is communications; there are currently 124 operating communications satellites in orbit. Three of the spacecraft discussed earlier (ECS, INTELSAT VI, and Fltsatcom) are communications satellites. Their purpose is to relay radio, telephone, television, or data from one point on the Earth to another. These satellites are usually, but not always, in high-altitude geosynchronous orbits, where they remain over the same point on the Earth at all times. Communications can be provided between any two points on the side of the Earth to which the satellite has a direct view. Communications between two points on opposite sides of the Earth, however, require the use of multiple satellites with crosslinks between them. Both Fltsatcom and INTELSAT VI, which were discussed in more detail earlier (see Figures 2 and 3), are typical communications satellites that do not have crosslink capability. TDRSS is an example of a satellite that has crosslinks and is able to provide communications between any two points on the Earth.

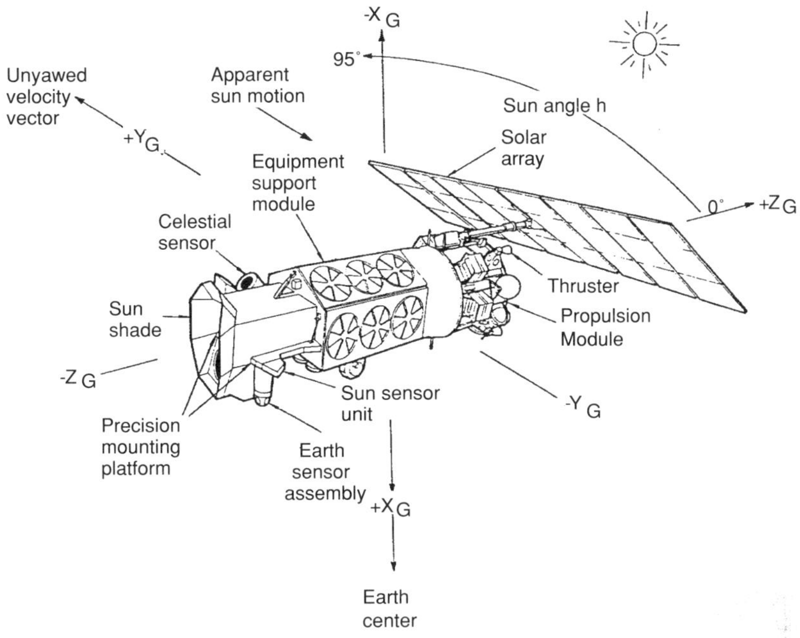

Figure 10. DMSP satellite (Martin Marietta)

Another mission that is common to both civilian and military space programs is weather monitoring. The DMSP spacecraft shown in Figure 10 is a typical low-altitude weather satellite. It carries both visual and infrared cameras for continuously photographing cloud patterns, as well as secondary sensors such as SSMI that can monitor such things as surface wind speeds, soil moisture content, and precipitation rates. Low-altitude weather satellites are usually in sun synchronous orbits. This allows them to scan the entire surface of the Earth at the same local sun angle over the course of a day. High-altitude weather satellites, such as NASA's GEOS, are usually in geosynchronous orbits that allow them to continuously photograph one entire hemisphere of the Earth.

Navigation constitutes a third type of spacecraft mission. For the U.S., this mission is currently fulfilled by one satellite program, NAVSTAR-GPS (Global Positioning System). The GPS system includes a constellation of 21 satellites in 12-hour circular orbits. Each satellite (shown in Figure 11) continuously broadcasts a signal that can be picked up by small receivers on the ground, in aircraft, or even in another satellite. If three or more GPS satellites are visible at any one time, the receiver can determine its own position and velocity to within 15 meters and 0.1 meter/second. Russia also operates a system of positioning satellites, known as GLONASS, that are located in similar orbits.

Surveillance is a general category for satellites whose mission is to monitor various activities on the Earth. This surveillance can be in the form of IR sensors to detect missile launches, radar to track aircraft or ships, visual observation of ground activities, or intercept of radio transmissions. Satellites designed to support each of these different missions have markedly different configurations.

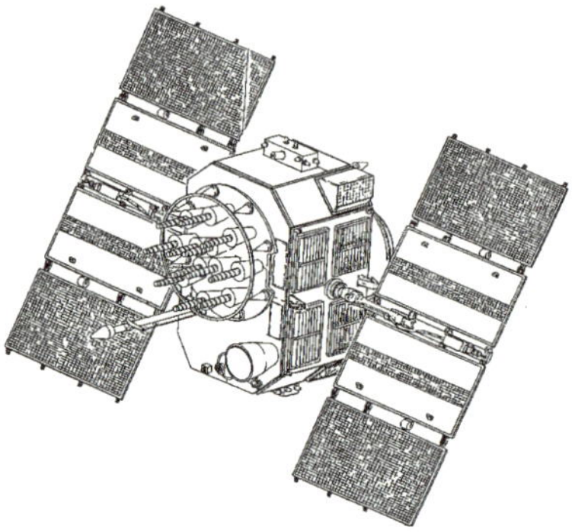

Figure 11. GPS (Rockwell)

The DSP spacecraft, shown in Figure 4, is an example of an IR surveillance satellite. The payload is an IR telescope that detects and tracks missiles by the heat emitted from their rocket plumes. The detectors in the telescope are cooled to approximately 150 deg K by a cryogenic radiator with a helium coolant loop. The entire satellite rotates at 6 rpm to provide a scanning motion that sweeps the linear detector array across the Earth's surface. Ground software reconstructs the sweep into an Earth image with all heat sources displayed. DSP provides the United States with its first warning of missile launches.

Space Based Radar (SBR) is an example of a radar surveillance satellite. A number of different configurations have been proposed for SBR, from flat phased arrays to a rotating reflector. These configurations, which are shown in Figure 12, are quite large, with antenna dimensions being on the order of 100 feet. The radars are being developed to track aircraft and ships, with some designs being proposed to track missiles and individual warheads for defense applications.

Relay satellites support another type of mission that is similar to that of communications satellites except that the communication link is between the ground and a second satellite as shown in Figure 13. Such links eliminate the need to have ground stations spaced all over the world and provide continuous contact with satellites in any orbit. An example of a relay satellite is NASA's Tracking and Data Relay Satellite System (TDRSS), shown in Figure 14. TDRSS is used to provide both ground-to-ground and ground-to-satellite links and is also used to communicate with shuttle astronauts.

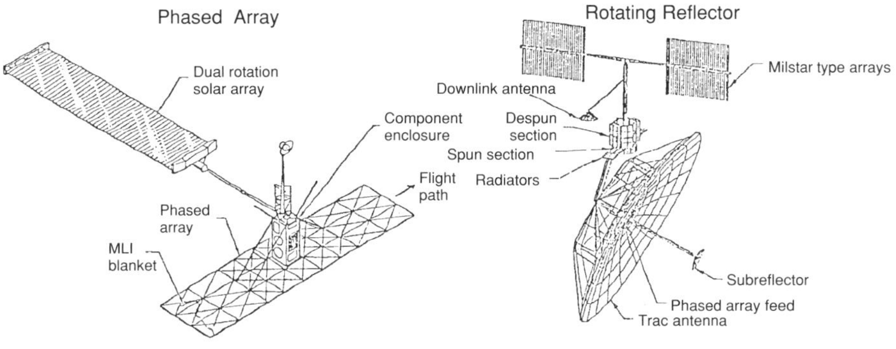

Figure 12. Space-based radar satellites

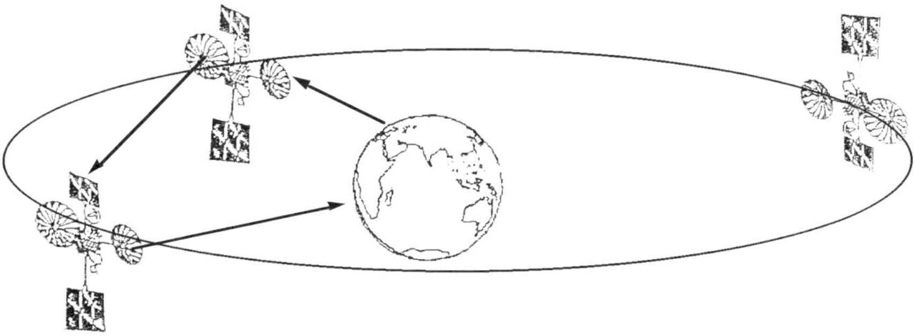

Figure 13. TDRSS relay (TRW)

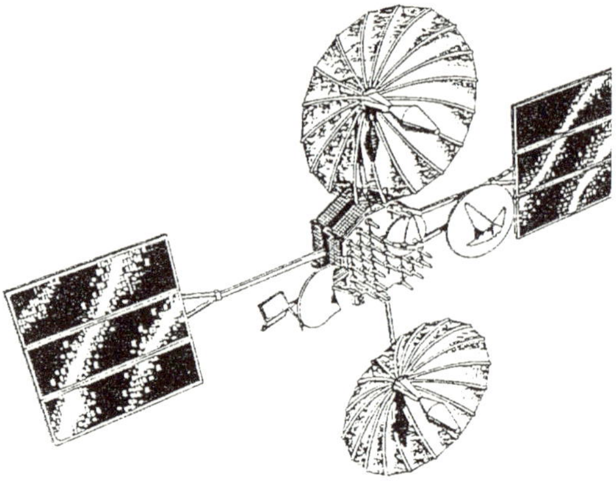

Figure 14. TDRSS satellite (TRW)

Most scientific satellites need go no higher than low Earth orbit to accomplish their missions. Astronomical satellites, such as the Infrared Astronomical Observatory (IRAS) and the Hubble Space Telescope (shown in Figure 15), need only get above the Earth's atmosphere to conduct their observations. For programs intended to study the Earth, such as NASA's Mission to Planet Earth, a low-altitude orbit is an advantage. There are, of course, interplanetary scientific spacecraft which leave Earth's orbit entirely, but these types of missions will not be covered in this handbook.

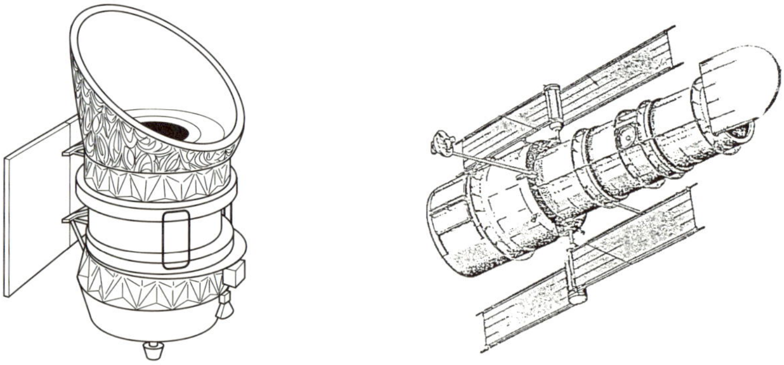

IRAS

Space Telescope

Figure 15. Scientific satellites (Ball, Lockheed)

Remote sensing missions are represented by the American Landsat and French SPOT (Systeme Pour L'Observation de la Terre) satellites (Figure 16). These vehicles gather images in a variety of wavelengths, which are used for managing crops and Earth resources and to support environmental and global change research. They are both placed in sun-synchronous polar orbits at approximately 450 n.mi. altitude.

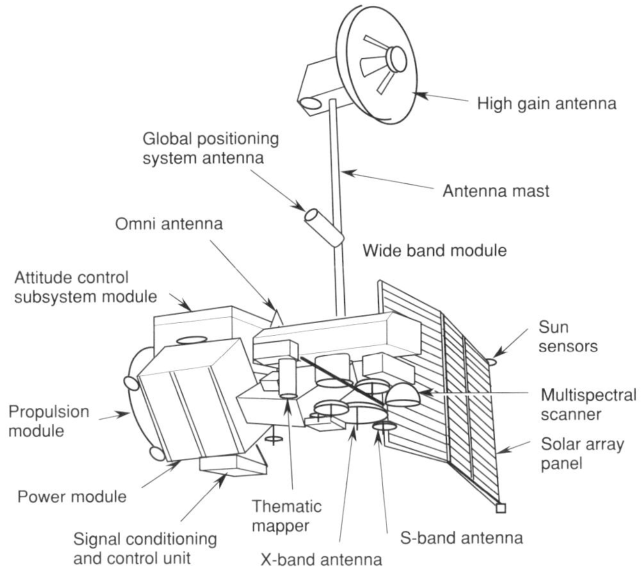

Landsat (Martin Marietta)

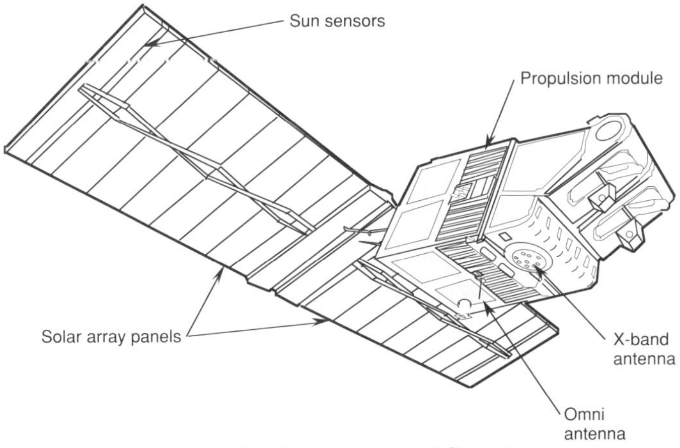

SPOT (Matra Marconi Space)

Figure 16. Remote sensing satellites

Chapter II

Satellite Thermal Environments

David G. Gilmore
The Aerospace Corporation

TYPES OF ENVIRONMENTAL LOADS

Due to the absence of atmospheric convection in space, overall thermal control of a satellite on orbit is usually achieved by balancing the energy emitted by the spacecraft as infrared radiation against the energy dissipated by internal electrical components plus the energy absorbed from the environment, as illustrated in Figure 1. Spacecraft thermal control is a process of energy management in which environmental heating plays a major role. The principle forms of environmental heating on orbit are sunlight, both direct and reflected off of the Earth, and IR energy emitted from the Earth itself. During launch or in exceptionally low orbits there is also a free molecular heating effect due to friction with the rarefied upper atmosphere. This chapter provides an overview of these different types of environmental heating.

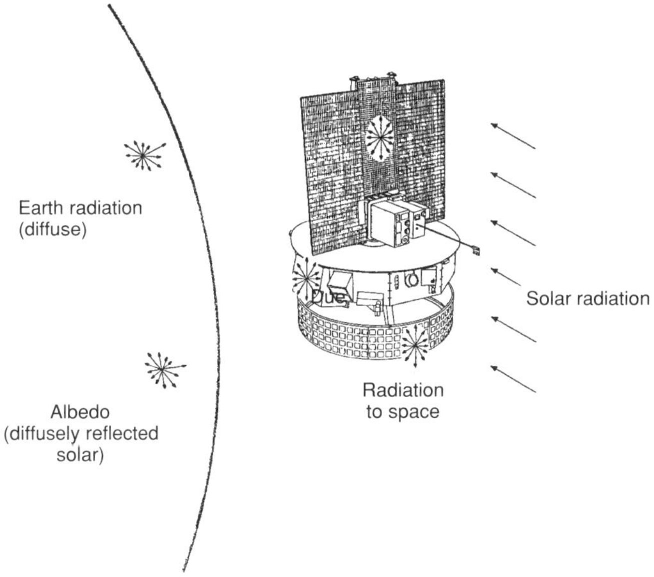

Figure 1. Satellite thermal environment

Direct Solar

Sunlight is the greatest source of environmental heating incident on most spacecraft. The emitted radiation from the sun is constant within a fraction of 1 percent at all times. However, due to the Earth's elliptical orbit, the intensity of sunlight reaching the Earth varies approximately $\pm$3.5 percent depending on the Earth's distance from the sun. At summer solstice (northern hemisphere) the intensity is at a minimum (415 Btu/hr ft^2) and at a maximum (444 Btu/hr ft^2) at winter solstice.

Solar intensity also varies as a function of wavelength, as shown in Figure 2. The energy distribution is approximately 7 percent ultraviolet, 46 percent visible, and 47 percent near (i.e., short wavelength) infrared with the total integrated energy being equal to the 415 to 444 Btu/hr ft^2 values mentioned above. It is important to note, however, that the infrared energy emitted by the sun is at a much shorter wavelength than the infrared energy emitted by a body near room

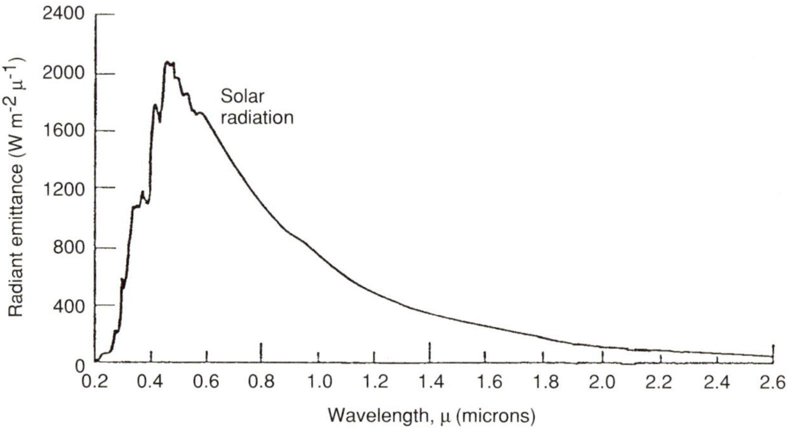

Figure 2. Solar spectral distribution

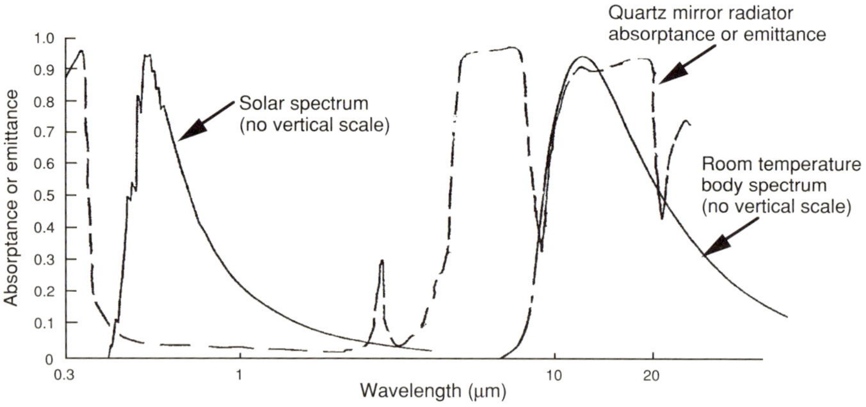

Figure 3. Solar and room temperature body spectral distributions

temperature. This distinction allows for the selection of thermal control finishes that are very reflective in the solar spectrum but highly emissive to room temperature (long wavelength) IR, as shown in Figure 3. This minimizes solar loads while maximizing the ability to reject waste heat from the spacecraft. These types of finishes are discussed in more detail in Chapter IV.

Albedo

Sunlight that is reflected off of a planet or moon is known as albedo. The Earth's albedo is usually expressed as a percentage of incident sunlight that is reflected back out to space and is highly variable. As a first approximation one can assume a value of about 30 percent; however, reflectivity is generally greater over continental as compared to oceanic regions and generally increases with decreasing local solar elevation angles and increasing cloud coverage. Due to the greater snow and ice coverage, decreasing solar elevation angle, and increasing cloud coverage, albedo also tends to increase with latitude. These variations make selection of the best values for albedo constants for a particular thermal analysis rather uncertain and it is not unusual to find variations throughout the industry.

Refs. 1 through 7 contain detailed discussions and a great deal of data pertaining to albedo and its variations. Table 1 (which was taken from Ref. 5) provides a good overview of how albedo varies with latitude and season. The data shown are monthly averages measured by satellite sensors and it should be noted that significant deviations from these values can occur for individual orbits. Since most spacecraft components are massive enough that short term (one orbit or less) variations in albedo will not cause large temperature changes, it is possible to use orbit average values for design purposes. Table 2 gives a set of reasonable worst-case orbit average values recommended in Ref. 1 for albedo as a function of orbit inclination. The "max" and "min" values are used for the "hot" and "cold" design cases, respectively.

Although the orbit average values in Table 2 are applicable to most satellites, there are situations that may require a more detailed evaluation of albedo effects. Any satellite component that is sensitive to albedo loads and has a low thermal mass may require that albedo variations around the orbit be considered. Similarly, any sensitive component (such as a cryogenic radiator) that only has a view to certain parts of the Earth may require an evaluation of albedo as a function of latitude. Table 3 (from Ref. 11) gives recommended albedo values, deduced from flight data, as a function of latitude. These values can be used to model albedo variations around individual orbits.

Table 1. Monthly Average IR and Albedo
(Stevens, Campbell, and Von der Haar)

Zonal Mean IR (W/M^2)

Latitude (deg)		Month J	F	M	A	M	J	J	A	S	O	N	D	Annual Min	Avg	Max
90	80	165	155	146	173	189	207	207	199	178	170	163	177	146	177	207
80	70	157	149	154	183	197	211	212	207	186	175	160	156	149	178	212
70	60	165	164	170	196	208	221	224	217	198	186	173	167	164	190	224
60	50	175	177	188	204	213	222	228	224	211	200	188	182	175	201	228
50	40	191	194	203	216	226	235	244	243	232	220	205	198	191	217	244
40	30	217	218	224	235	241	254	259	263	253	247	231	222	217	238	263
30	20	250	248	251	265	266	268	262	261	261	263	253	251	248	258	268
20	10	266	264	261	270	260	251	241	236	242	250	251	259	236	254	270
10	0	251	251	248	240	232	233	233	235	235	244	242	250	232	241	251
0	-10	240	240	240	243	257	261	261	261	260	257	248	246	240	251	261
-10	-20	248	247	250	264	270	273	272	276	271	266	257	253	247	262	276
-20	-30	261	256	254	263	258	260	260	264	259	258	256	262	254	259	264
-30	-40	253	251	244	239	233	229	231	232	233	238	239	249	229	239	253
-40	-50	232	232	225	217	213	209	205	207	211	219	220	229	205	218	232
-50	-60	217	217	208	204	199	195	188	187	192	201	208	216	187	202	217
-60	-70	209	204	193	186	177	172	164	161	165	180	197	209	161	184	209
-70	-80	196	184	165	153	146	146	131	124	128	151	183	220	124	160	220
-80	-90	187	171	148	121	105	110	104	94	94	126	170	190	94	135	190
	Min	157	149	146	121	105	110	104	94	94	126	160	156			
	Avg	215	212	209	215	216	219	218	216	211	213	213	218			
	Max	266	264	261	270	270	273	272	276	271	266	257	262			

Zonal Mean Albedo (Percent)

Latitude (deg)		Month J	F	M	A	M	J	J	A	S	O	N	D	Annual Min	Avg	Max
90	80	80*	78*	69	58	69	75	69	60	44	70*	75*	78*	44	69	80
80	70	81*	83	69	63	62	54	50	49	66	77	77*	80*	49	68	83
70	60	78	56	55	54	49	42	39	40	42	51	64	69	39	53	78
60	50	56	50	49	42	40	39	38	38	37	43	49	55	37	44	56
50	40	46	43	40	36	34	35	33	32	33	34	39	46	32	37	46
40	30	37	36	34	32	30	28	27	26	29	29	33	37	26	31	37
30	20	30	29	27	25	25	25	26	26	26	26	28	30	20	26	30
20	10	24	23	22	20	22	24	26	27	25	25	26	26	25	24	27
10	0	24	24	24	24	26	26	26	26	26	25	26	26	24	25	26
0	-10	25	24	24	24	21	22	23	23	22	23	24	25	21	23	25
-10	-20	24	23	23	22	21	21	22	22	22	23	24	24	21	22	24
-20	-30	23	24	24	23	24	24	25	25	25	25	24	23	23	24	25
-30	-40	27	28	29	29	30	30	30	30	29	29	28	27	27	28	30
-40	-50	33	34	34	37	38	39	39	36	35	34	34	33	33	35	39
-50	-60	41	41	40	42	44	47	48	45	43	44	43	42	40	43	48
-60	-70	46	47	46	54	62	72	77	65	56	56	52	49	46	56	77
-70	-80	61	62	61	86	86*	87*	87*	88	79	65	66	64	61	74	83
-80	-90	70	72	40	80*	81*	82*	82*	83*	80	67	75	75	40	74	83
	Min	23	23	22	20	21	21	22	22	22	23	24	23			
	Avg	45	43	39	41	42	42	42	41	39	41	43	44			
	Max	81	83	69	86	86	87	87	88	80	77	77	78			

* Estimated value

Table 2. Orbit average Earth IR and albedo (NASA)

Orbit Inclination (deg)	Emitted Radiation (Btu/h-ft^2)			Albedo (percent)		
	Min*	Ave	Max*	Min*	Ave	Max*
±90**	64	68	72	38	42	46
±80	66	70	74	34	38	42
±70	69	73	77	30	34	38
±60	71	75	79	26	30	34
±50	73	77	81	22	28	32
±40	76	80	84	19	25	29
±30	77	81	85	20	24	28
±20	76	80	84	20	24	28
±10	74	78	82	20	24	28

*Min/max based on uncertainties of ±4 Btu/h-ft^2 for Earth IR, +4 percent for albedo.

**Area averages are 74 Btu/h-ft^2 and 33 percent.

Table 3. Earth IR and albedo vs. latitude

Latitude		Min Earth IR* (Probable Cold Case)		Average		Max Earth IR** (Probable Hot Case)	
Max	Min	Earth IR (Btu/h-ft^2)	Albedo (percent)	Earth IR (Btu/h-ft^2)	Albedo (percent)	Earth IR (Btu/h-ft^2)	Albedo (percent)
90	80	46	69	56	69	66	75
80	70	47	83	56	68	67	50
70	60	52	56	60	53	71	39
60	50	56	56	64	44	72	38
50	40	61	46	69	37	77	33
40	30	69	37	76	31	83	26
30	20	79	29	82	26	85	25
20	10	75	27	81	24	86	20
10	0	74	26	76	25	80	24
0	-10	76	24	80	23	83	23
-20	-30	81	24	82	24	84	25
-30	-40	73	30	76	28	80	27
-40	-50	65	39	69	35	74	34
-50	-60	59	45	64	45	69	41
-60	-70	31	65	58	56	66	46
-70	-80	39	88	51	74	70	64
-80	-90	30	80	43	74	60	75

*Determined using lowest monthly average Earth IR with albedo value for the same month.

**Determined using highest monthly average Earth IR with albedo value for the same month.

In the rare situations in which a component might be so sensitive that the worst albedo at any time must be considered (not orbit or monthly average), it is recommended that the data contained in the references be reviewed and used as appropriate for the application in question. For most spacecraft, however, the values listed in Tables 2 and 3 are sufficient.

As a final clarification, it should be understood that the heat *flux* reaching a spacecraft will decrease as the spacecraft moves along its orbit and away from the subsolar point, even if the albedo constant remains the same. This is due to the reduction in solar energy per square foot incident on the Earth with the cosine of the angle from the subsolar point.

Earth Emitted IR

The Earth not only reflects sunlight, it also emits long-wave IR radiation. The Earth, like a satellite, achieves thermal equilibrium by balancing the energy received (absorbed) from the sun with the energy re-emitted as long-wavelength IR radiation. This balance is maintained fairly well on a global annual average basis. The intensity of IR energy emitted at any given time from a particular point on the Earth, however, can vary considerably depending on factors such as surface and air temperatures, atmospheric moisture content, and cloud coverage. The highest intensities are generally over clear tropical regions and tend to decrease with increasing latitude and increasing cloud cover. As a first approximation one can use a value of around 75 Btu/hr ft^2 emitted at the Earth's surface. Earth-emitted IR is also sometimes called "Earthshine."

The IR energy emitted by the Earth, which is around 0 deg F, is of approximately the same wavelength as that emitted by satellites, that is to say, it is of much longer wavelength than the IR energy emitted by the sun at 9500 deg F. Unlike short-wavelength solar IR, the Earth IR loads cannot be reflected away with special thermal control coatings since the same coating, would prevent the radiation of waste heat away from the spacecraft. Because of this, Earth-emitted IR energy can present a particularly heavy backload on spacecraft radiators in low-altitude orbits, which must emit energy at the same wavelength.

The concept of Earth-emitted IR can be confusing, since usually the spacecraft is warmer than the effective Earth temperature and the net heat transfer is from spacecraft to Earth. However, for analysis purposes, it is convenient to ignore the Earth when calculating view factors from the spacecraft to space and to assume that the Earth does not block the view to space. Then the difference in IR energy is added back in as an "incoming" heat rate called Earth-emitted IR, or Earthshine.

As with albedo, orbit average values of Earth IR can be used in most analyses. Recommended values for IR heating expressed in Btu/hr ft^2 units can be found in Table 2. For spacecraft with a particular sensitivity to variations in IR loads around the orbit, or for sensitive surfaces that only see the Earth at certain points, the values shown in Table 3 are recommended.

Free Molecular Heating

Another significant form of environmental heating is known as free molecular heating (FMH) and is due to bombardment of the vehicle by individual molecules in the outer reaches of the atmosphere. For most satellites this heating is only encountered during launch ascent just after the booster's payload fairing is ejected. It is desirable to drop the fairing as soon as possible after launch to minimize the amount of dead weight the booster must deliver to orbit. The point at which the fairing is separated is often determined by a trade-off between the desire to save weight and the need to protect the payload spacecraft from excessive atmospheric heating.

Fairing separation always occurs at altitudes that are high enough that the resultant heating is in the free or near-free molecular regime, i.e., the heating is modeled as collisions of the body with individual molecules rather than as a gas-flow heating problem. The heating rate is given by:

$$Q_{FMH} = \alpha \, (1/2) \, \rho \, V^3,$$

where ρ = atmospheric density
V = vehicle velocity
α = accommodation coefficient (around 0.6 to 0.8, but a value of 1.0 is recommended for conservatism).

The atmospheric density is a highly variable parameter that is governed by a number of factors that cause the upper atmosphere to expand or contract. These factors include; the level of solar electromagnetic activity ("F10.7," measured at a wavelength of 10.7 cm); the geomagnetic index (A_p); the longitude, latitude, and local hour of the point in question: altitude; day of year; etc. Atmospheric densities are calculated today using sophisticated atmospheric models with a dozen or more input parameters. The output of these models are atmospheric densities that will not be exceeded with a particular level of confidence (usually 97 percent confidence). The output is expressed probabilistically because the level of solar activity, which is a major factor, is not predictable precisely.

The velocity of the vehicle relative to the atmosphere can be calculated in a rather straightforward manner for a satellite in orbit. The velocity during launch ascent, however, must be calculated using sophisticated booster trajectory simulation programs that model and optimize the performance of the booster. Like the atmosphere, the trajectory and velocity of the booster is probabilistic, but to a lesser extent. These uncertainties are due to variations in rocket-motor performance, guidance-system accuracies, high-altitude wind effects, etc., and can result in the vehicle traveling at a different altitude or velocity than expected at any given time.

The atmospheric modeling and trajectory simulations are generally conducted by specialists in those areas, who then supply the thermal engineer with curves of worst-case heating versus time. With such a curve and a knowledge of the spacecraft attitude relative to the velocity vector, the thermal engineer may calculate the heat load on the satellite by simply multiplying the heating rate by the cross-sectional area of the surface in question and the cosine of the angle between the surface normal and velocity vector. A heating rate curve for one particular mission during launch ascent is shown in Figure 4.

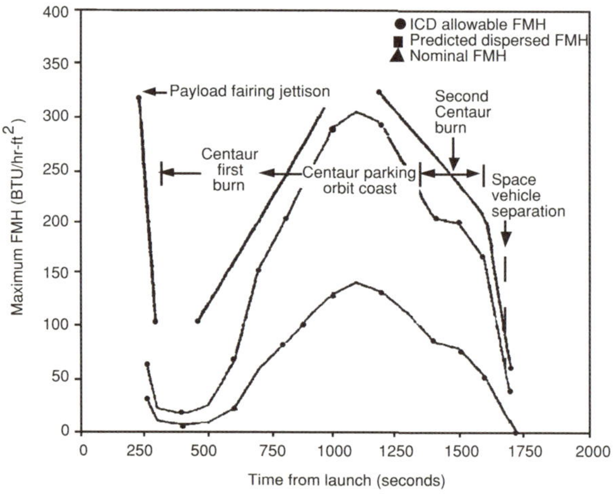

Figure 4. Example of free molecular heating profile (Martin Marietta)

As was mentioned earlier, most spacecraft see FMH only during launch. There are some spacecraft, however, that have orbits with very low perigee altitudes that can experience free molecular heating in their operational orbits. Generally speaking, operational-orbit free molecular heating rates should be assessed for any spacecraft with a perigee altitude below 100 n. mi.

Charged Particle Heating

In addition to the four principal environmental heating components discussed above, there is a fifth source; heating due to charged particles. Charge particle heating is very weak compared to the other heat sources and is generally not significant in the thermal design of room-temperature systems. At cryogenic temperatures, however, charge particle heating can become a significant factor in the thermal design due to the high sensitivity of such systems to environmental heat loads.

The near-Earth trapped charged particles, known as the Van Allen belts, lie about the plane of the geomagnetic equator and feature relativistic electrons and protons. The spatial characteristics of the Van Allen belts and the spectral properties of the trapped particles within the belts undergo both regular and irregular variations with time, accounted for by the solar activity level. The bulk of the Van Allen belts is approximately bounded between 1 to 8 Earth radii (Re). Van Allen (1958) discovered the inner proton belt peaking in intensity at an approximate altitude of 1.45 Re, while Fan et al. (1961), O'Brien et al. (1962), and Dessier and Karplus (1960) helped to establish the existence of other electron peaks. Vette (1966) developed a complete mapping of the Van Allen belt radiations.

Standard trapped particle environmental models include electron data for maximum solar activity periods AE6 and AE8 (Teague et al., 1976), solar minimum activity periods AE5 (Teague and Vette, 1976), an interim model AEI for outer zone electrons, and the solar maximum and minimum activity APS model for energetic trapped protons (Sawyer and Vette, 1976). These data represent omnidirectional integral intensities averaged over periods in excess of six months in orbit. However, over most regions of magnetospheric space, short-term excursions can vary from these averaged values by factors of 10^2 to 10^3, depending on the particle energies and on the type and intensity of the causative event (Stassinopoulos and Barth, 1982).

Data on trapped proton and electron fluxes as functions of energy for circular, geomagnetic equatorial orbits ranging in altitude from 1.5 Re to 6.6 Re (synchronous) are presented in Figs. 5 and 6, respectively. (The electron flux/energy data reported in the AE6 model is limited to altitudes up to 5.2 Re. The AES model includes flux levels for altitudes above 5.2 Re.) As illustrated, the concentration of relativistic (> Mev) protons is evident at lower altitudes (< Re), while at the synchronous altitude (6.6 Re) proton energies are less than 2 Mev. Conversely, electrons feature high flux levels and energies less than approximately 5 Mev over a wide spectrum of altitudes.

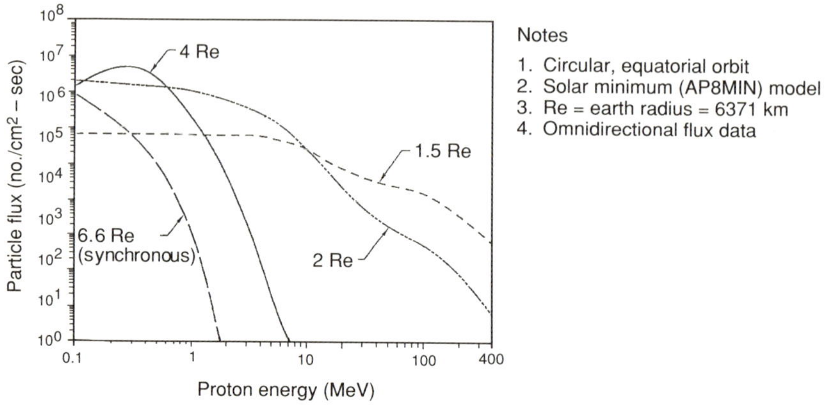

Figure 5. Proton flux and energy levels for several Earth equatorial orbits

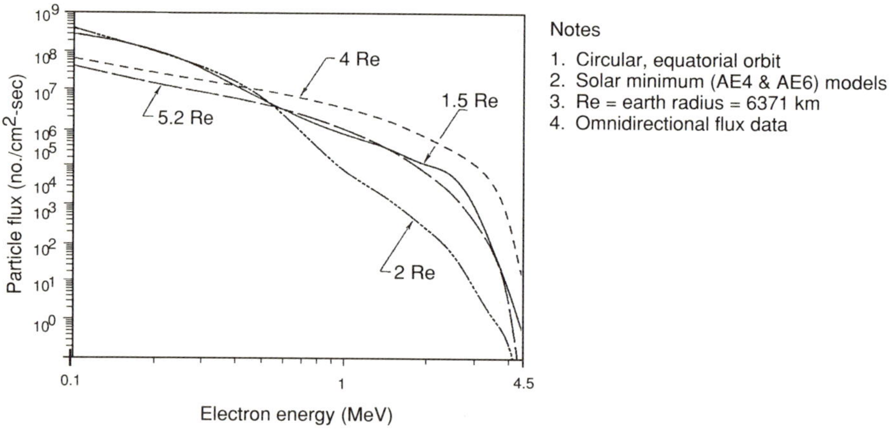

Figure 6. Electron flux and energy levels for several Earth equatorial orbits

The heating caused by these charged particles generally occurs in the first few hundredths of a centimeter of a material's thickness and is therefore essentially front-surface absorbed like solar, IR, or free molecular heating. The charged particle heating rates, while not significant at room temperature, can significantly raise the equilibrium temperature of a cryogenic radiator, as shown in Figure 7. A radiator design for steady-state operation at 70 K in circular equatorial Earth orbit will warm to approximately 72.9 K for the charged particle heating conditions at 1.5 Re, while warming to 74.7 K and 70.4 K for the conditions at 2.0 Re and 6.6 Re, respectively. In the theoretical limit where $T_{equiv} = 0$ K, the charged-particle-heating effect will warm the radiator to approximately 27.3 K for the synchronous (6.6 Re) circular equatorial Earth orbit condition. (It should be noted that the equilibrium temperature increase for the 4.0-Re altitude condition is nearly identical to the results for 1.5 Re, and therefore was not included in Figure 7. For such systems, charged particle heating must therefore be considered in the design and sizing of radiators. The reader is referred to Ref. 12 for a detailed discussion of this phenomenon.

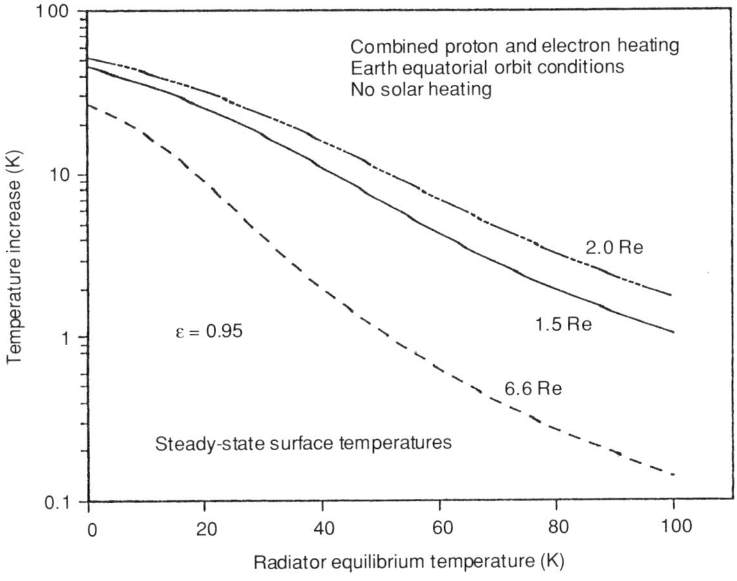

Figure 7. Equilibrium temperature increase of an aluminum radiator due to natural environment charged particle heating

ENVIRONMENTS IN TYPICAL ORBITS

In Chapter 1 the most common types of orbits were described: low Earth orbit (LEO), geosynchronous orbit (GEO), Molinya, and sun synchronous. In this section the thermal environments encountered in each of these orbits will be discussed. How one goes about calculating the actual heat loads that these environments impose on spacecraft surfaces, however, will be addressed in Chapter V.

To begin this discussion, some terminology definition is required. There are a number of orbital parameters that are commonly used in conducting analyses of

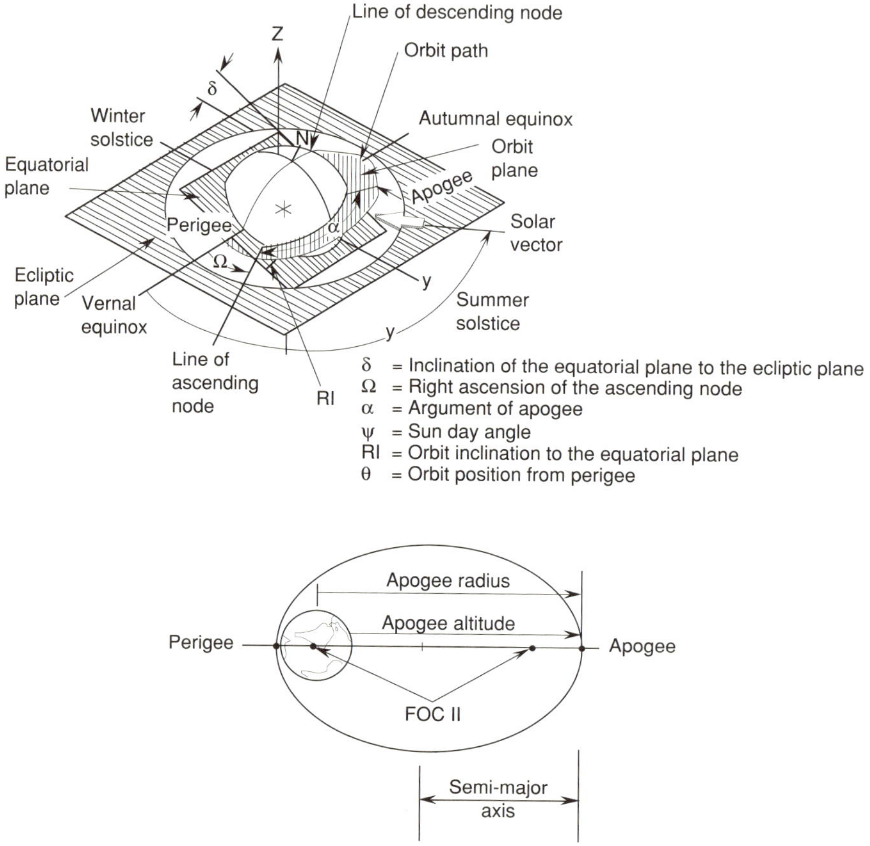

Figure 8. Orbit parameters (Turner Associates)

environmental heating. These are generally the same parameters used by orbit analysts to describe the spacecraft orbit, so it is usually a simple matter to get the inputs necessary to conduct the thermal analysis for any given program. The most important parameters are defined here and illustrated in Figures 8 and 9.

Equatorial Plane. The plane of the Earth's equator, which is, of course, perpendicular to the Earth's spin axis.

Ecliptic Plane. The plane of the Earth's orbit around the sun. From the point of view of the Earth, the sun always lies in the ecliptic plane. Over the course of a year the sun appears to move continuously around the Earth in this plane. Due to the tilt of the Earth's spin axis, the Equatorial Plane is inclined 23.4 deg from the ecliptic plane, which is shown in Figure 8 as the angle δ.

Sun Day Angle. The position angle of the sun in the ecliptic plane measured from vernal equinox. At vernal equinox this angle is 0 deg, at summer solstice 90 deg, 180 deg at autumnal equinox, and 270 deg at winter solstice. This angle is shown as ψ in Figure 8 and should not be confused with the "Right Ascension" of the sun, which is measured in the equatorial plane and is slightly different on most days of the year.

Orbit Inclination. The angle between the orbit plane and the equatorial plane, which is shown as RI in Figure 8. Orbit inclinations typically vary from 0 deg to 98 deg, although inclinations greater than 98 deg are possible. For inclinations less than 90 deg, the satellite appears to be going around its orbit in the same direction as the Earth's rotation. For inclinations greater than 90 deg, it appears to be going opposite the Earth's rotation and is known as a retrograde orbit.

Altitude. Distance of satellite above the Earth's surface.

Apogee/Perigee. Apogee is the point of highest altitude, perigee is the lowest.

Ascending Node. The point during orbit at which the spacecraft crosses the Earth's equator traveling from south to north (i.e., when it is "ascending"). The Descending Node is, of course, the crossing point during the southbound portion of the orbit.

Right Ascension and Declination. The position of an object in the celestial coordinate system (see Figure 9). Right ascension is the position angle in the equatorial plane measured from vernal equinox. Declination is the position angle above or below the equatorial plane.

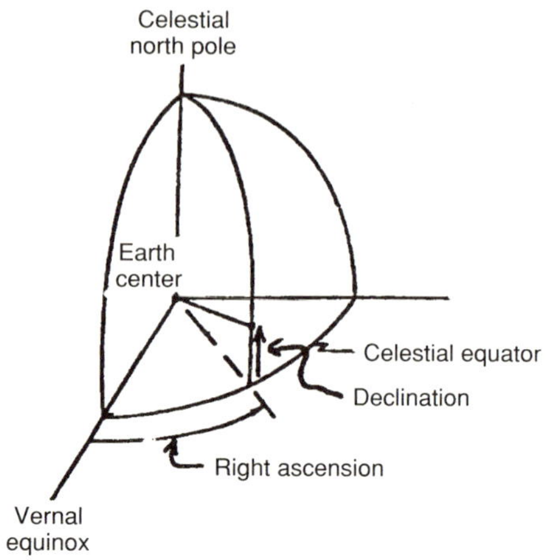

Figure 9. Celestial coordinates

Right Ascension of the Ascending Node (RAAN). The position angle of the ascending node measured from vernal equinox in the equatorial plane (Ω in Figure 8). The Earth's equatorial bulge (the Earth is not a true sphere) causes the ascending and descending nodes to drift slightly on each revolution about the Earth. This is known as "nodal regression." For most orbits the RAAN drifts continuously with time and varies from 0 deg to 360 deg.

Semi-Major Axis. The semi-major axis of the orbit ellipse.

$$a = \frac{r_a + r_p}{2}, \tag{1}$$

where a = orbit semi-major axis
r_a = orbit radius at apogee = Earth's radius + apogee altitude
r_p = orbit radius at perigee = Earth's radius + perigee altitude.

Period. The time required to make one revolution about the Earth. As orbit altitude increases, so does the period. The orbit period may be calculated using the relation shown in Eq. (2).

$$P = 2\,\pi\,(a^3/\mu)^{1/2}. \tag{2}$$

When P is the period, μ is the product of the universal gravitational constant and the mass of the planet (for Earth, $\mu = 8.1288\text{x}10^{11}\frac{\text{n. mi.}^3}{\text{hr}^2}$), and a is the semi-major axis of the orbit (for a circular orbit this is the orbit radius). The period of circular orbits versus orbit altitude is plotted in Figure 10.

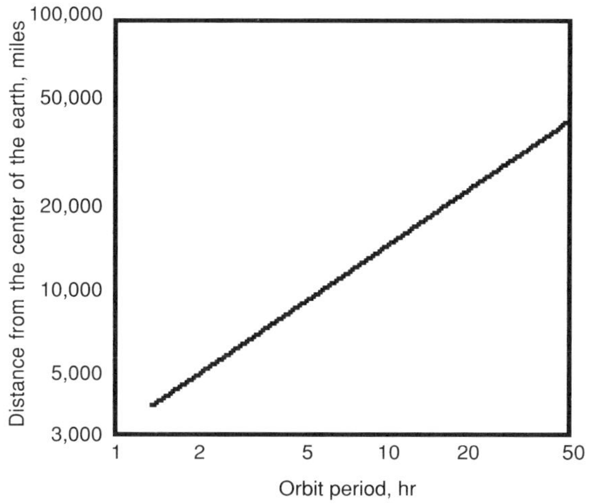

Figure 10. Total amount of time per orbit

Eccentricity. The degree of oblateness of the orbit, defined as the ratio of one-half the interfocal distance to the semi-major axis. For a circular orbit, the eccentricity is zero. As the orbit becomes more elliptical, the eccentricity increases. Eccentricity is related to the apogee and perigee radii and the semi-major axis by the following relationships:

$$r_a = a(1 + e), \text{ and} \tag{3}$$

$$r_p = a(1 - e), \tag{4}$$

when
r_a = orbit radius at apogee
r_p = orbit radius at perigee
a = orbit semi-major axis.

Argument of Apogee. For an elliptical orbit, the angle between the ascending node and apogee measured in the direction of satellite motion. This angle, which is shown as α in Figure 8, can vary from 0 deg to 360 deg.

Low Earth Orbits

Although the above parameters are used by orbit and thermal analysts to describe particular orbits in their analyses, there is another parameter, known as the orbit beta angle (β), which is more useful in visualizing the orbital thermal environment, particularly for low Earth orbits. The beta angle is defined as the minimum angle between the orbit plane and the solar vector, and can vary from -90 deg to +90 deg, as illustrated in Figure 11a. The beta angle is defined mathematically as

$$\beta = \sin^{-1} (\cos \delta_S \sin RI \sin(\Omega - \Omega_S) + \sin \delta_S \cos RI), \quad (5)$$

where
δ_S = declination of the sun
RI = orbit inclination
Ω = right ascension of the ascending node
Ω_S = right ascension of the sun
β = beta angle.

As viewed from the sun, a $\beta = 0$ deg orbit would appear edgewise, as shown in Figure 11b. A satellite in such an orbit would pass over the sub-solar point on the Earth (the point on the Earth where the sun is directly overhead) where albedo loads are the highest, but it would also have the longest eclipse time due to shadowing by the full diameter of the Earth. As the β angle increases, the satellite passes over areas of the Earth further from the sub-solar point, thereby reducing albedo loads; however, the satellite will also be in the sun for a larger percentage of each orbit due to decreasing eclipse times. At some point, which varies depending on the altitude of the orbit, eclipse time drops to zero. At a beta angle of 90 deg a circular orbit appears as a circle as seen from the sun, there are no eclipses no matter what the altitude, and albedo loads are near zero. Figure 11b shows how orbits of various beta angles appear as seen from the sun. It should also be noted here that beta angles are often expressed as positive or negative; positive if the satellite appears to be going counter-clockwise around the orbit as seen from the sun, negative if clockwise.

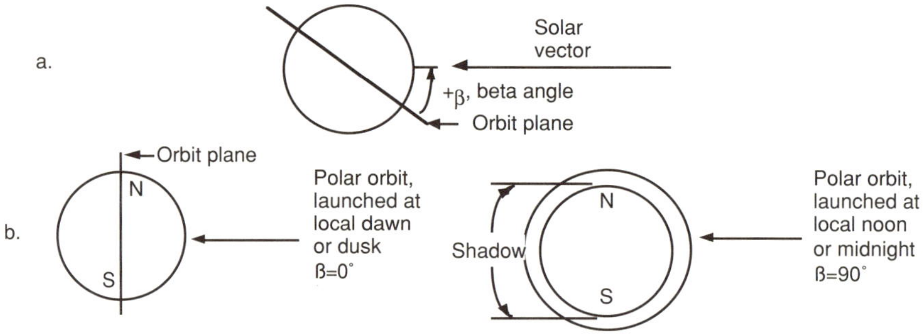

Figure 11. Orbit beta angle

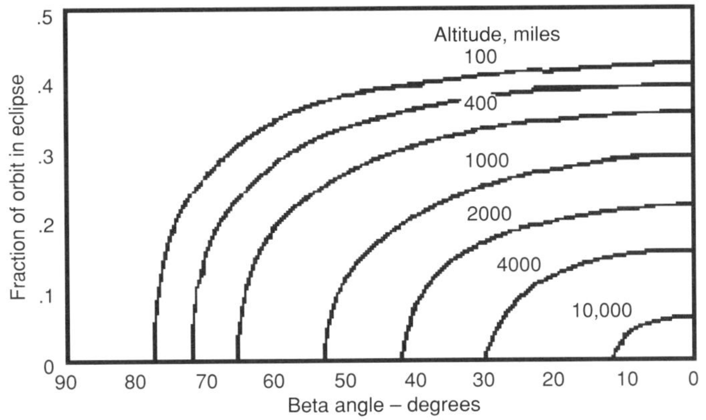

Figure 12. Eclipse durations

Figure 12 shows how eclipse times vary with beta angle for circular orbits of different altitudes. The eclipse fraction of a circular orbit can be calculated from Eq. (6).

$$f_E = \begin{cases} \dfrac{1}{180^\circ} \cos^{-1}\left([\dfrac{(h^2 + 2Rh)^{1/2}}{(R + h)\cos\beta} \right) & \text{if } |\beta| < \beta^* \\ 0. & \text{if } |\beta| \geq \beta^*, \end{cases} \tag{6}$$

where R = Earth's radius = 3444. n. mi.
h = orbit altitude
β = orbit beta angle
β^* – beta angle at which eclipses begin

β^* may be calculated using Eq. (7), as follows:

$$\beta^* = \sin^{-1}[R/(R + h)] \quad 0^\circ \leq \beta^* \leq 90^\circ. \tag{7}$$

Both Eqs. (6) and (7) assume that the Earth's shadow is cylindrical, which is valid for low orbits where there is no appreciable difference between the umbral and penumbral regions of total and partial eclipsing, respectively. For 12-hour and geosynchronous orbits, these equations may be slightly in error.

For any given satellite, the orbit β angle will vary continuously with time due to the orbit nodal regression and the change in the sun's right ascension and declination over the year. The regression rates as a function of inclination for circular orbits of different altitudes are shown in Figure 13. The sun's right

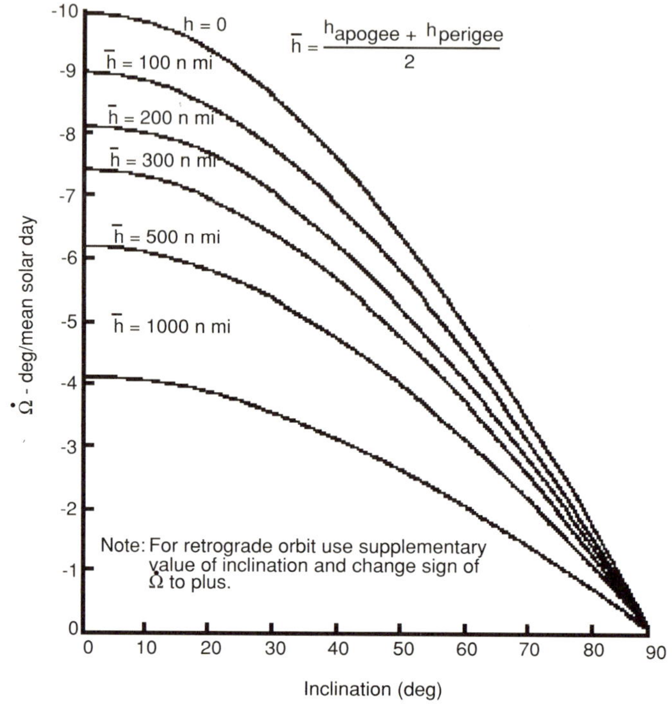

Figure 13. Regression rate due to oblateness vs inclination for various values of average altitude

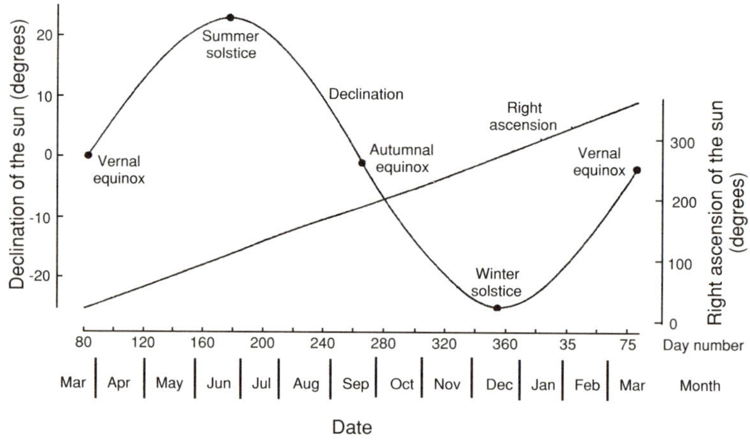

Figure 14. Solar declination and right ascension vs date

ascension and declination over the year are shown in Figure 14. The β angle history for one particular satellite in a 500-km altitude, circular orbit is shown in Figure 15. The absolute value of the β angle can vary from 0 to a maximum that equals the orbit inclination plus the maximum declination of the sun (i.e., inclination plus 23.4 deg).

If the nodal regression of an orbit proceeds eastward at exactly the rate at which the sun right ascension changes over the year, thereby "following" the sun, the orbit is called sun-synchronous. Since the mean sun moves uniformly eastward along the equator through 360 deg a year (about 365.242 mean solar days), the required rate of nodal regression is 360/365.242, or 0.985647 deg/day. For circular orbits, sun-synchronism is possible for retrograde orbits (i.e., inclination > 90 deg) up to an altitude of about 3226 n. mi.

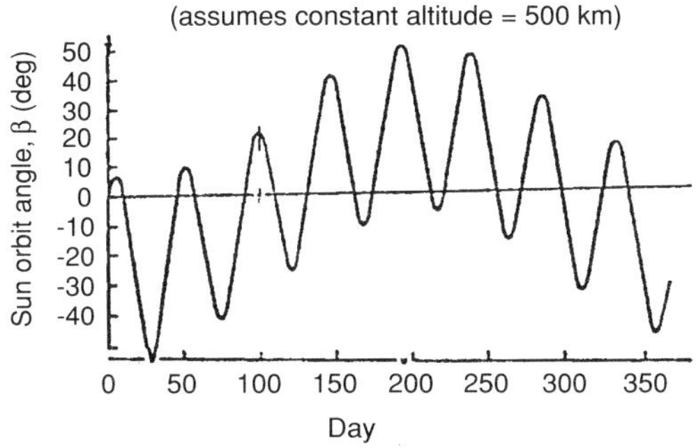

Figure 15. Sample beta angle drift (Lockheed)

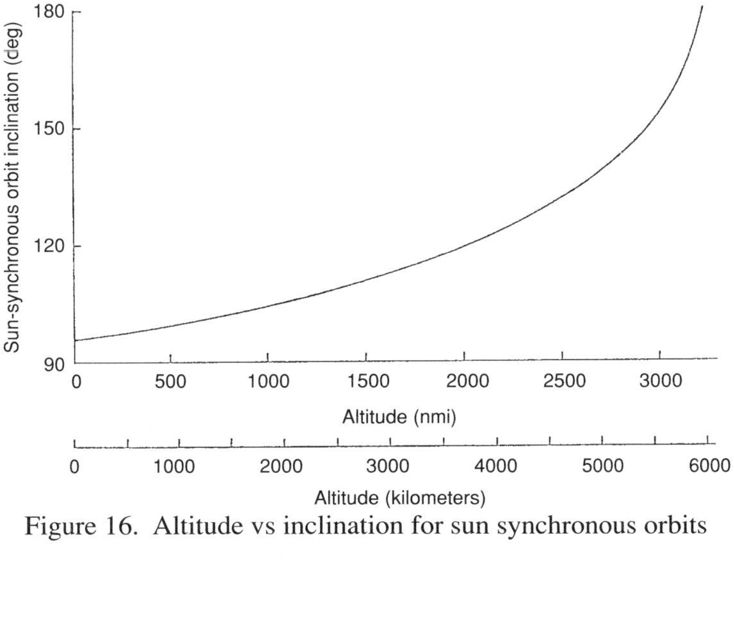

Figure 16. Altitude vs inclination for sun synchronous orbits

The orbit inclination required to achieve sun-synchronism in circular orbits is shown as a function of orbit altitude in Figure 16. It should be noted that, due to the change in the sun's declination over the year, the β angle is not constant but varies over a small range. The β-angle histories for 450 n. mi. sun-synchronous orbits with different initial Right Ascension of Ascending Nodes (RAAN) is shown in Figure 17.

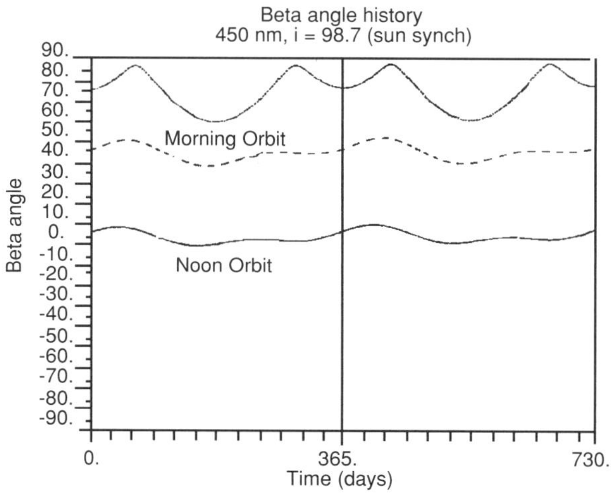

Figure 17. Beta angle histories for sun synchronous, 450-nm orbits

The chief advantage in thinking in terms of orbit β angles is that it simplifies the analysis of orbital thermal environments. By analyzing the environments at several discrete β angles, one can be confident that all possible combinations of orbit RAAN and sun day angles have been covered. Figure 18 shows such an analysis for a spinning cylindrical satellite in a 300-n. mi.-altitude low Earth orbit. Earth-emitted IR is usually considered constant over the Earth and therefore independent of orbit inclination, RAAN, or β angle. The IR load to the satellite therefore is constant with β angle. Since the eclipse time decreases with β, however, the satellite spends more time in the sun, thereby increasing the orbit average solar load, as shown in Figure 18. Also, as the β angle increases, the albedo loads decrease, as can be seen by comparing the "solar" and "solar plus albedo" curves in Figure 18. The net result for this particular satellite was that solar panel orbit average temperature (which provides a radiative heat sink for the internal components) was at a minimum at $\beta = 0$ and a maximum at $\beta = 65$ deg.

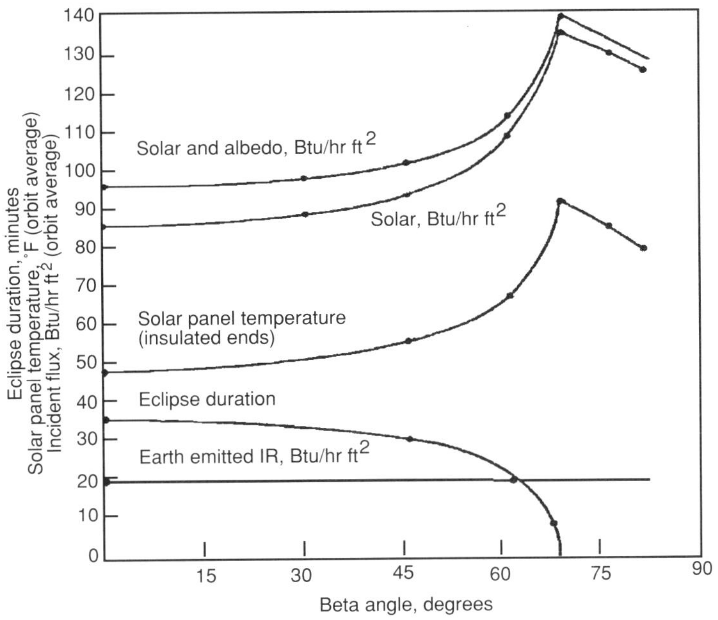

Figure 18. Cylinder in Low Earth Orbit

Geosynchronous Orbits

As orbit altitude increases, environmental loads from the Earth (IR and albedo) decrease rapidly. Figure 19 shows these loads on a black plate over the sub-solar point for various altitudes. By the time one reaches geosynchronous orbit these loads are insignificant for most thermal design analyses. The one exception to this is cryogenic systems, which operate at such low temperatures that even small environmental back loads from the Earth are significant to the thermal design.

With Earth loads being so small, the only significant environmental load for non-cryogenic systems in geosynchronous orbit is solar. At this altitude the spacecraft is in the sun most of the time, and the maximum possible eclipse duration is only 72 minutes out of the 24-hour orbit. Since most geosynchronous orbits have inclinations of less than 4 deg, eclipses occur only around vernal and autumnal equinox and are known as "eclipse seasons." During summer and winter the sun's declination causes the Earth's shadow to be cast above or below the satellite orbit, making eclipses impossible, as shown in Figure 20. For circular, 24-hour orbits inclined by more than a few degrees, eclipses could occur during seasons other than equinox, but such orbits are rather rare and the maximum eclipse duration would be the same.

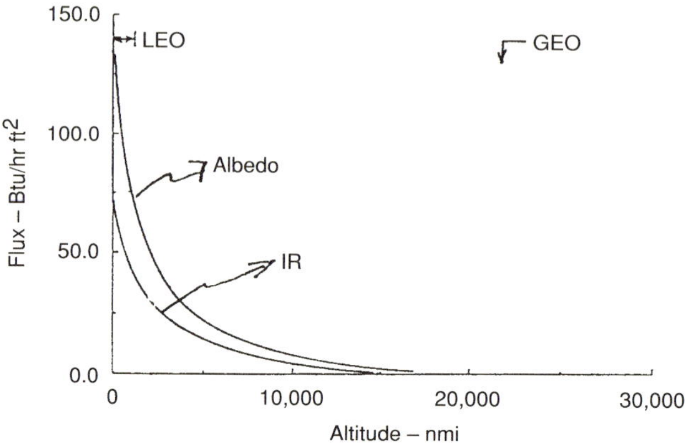

Figure 19. Earth heat loads vs altitude

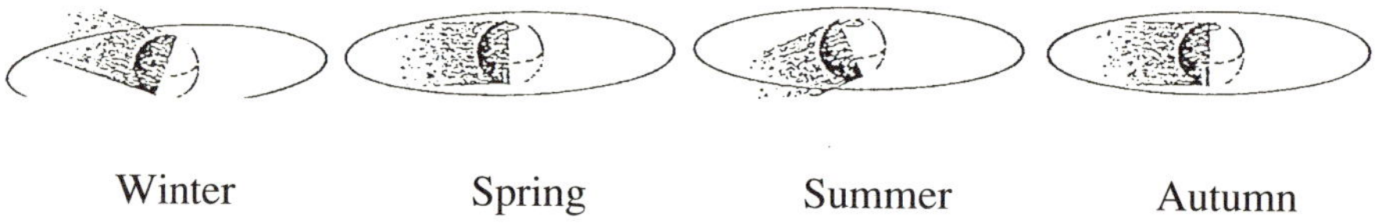

Figure 20. Geosynchronous orbit eclipse, once per orbit, spring and autumn only (TRW)

Many of the satellites in geosynchronous orbit are the 3-axis type, with one side of the vehicle constantly facing the Earth, as shown in Figure 21. For satellites such as this, the north and south faces receive the lowest peak solar flux, since the sun can only rise to a 23.4 deg angle above the surface (or maybe a little higher if the orbit has a slight inclination). As the spacecraft goes around the orbit, the sun cones about at a fixed elevation angle from these surfaces, as shown in Figure 21. This elevation angle changes from +23.4 deg in summer (sun on the north surface) to -23.4 deg in winter (sun on the south surface). The other four surfaces will see the sun circle around them during the orbit, with the result being a cosine variation in intensity from no sun to a full sun normal to the surface. Since the sun can only rise to an angle of 23.4 deg "above" the north/south faces, the maximum solar load on these surfaces is (sin 23.4 deg) (1.0 normal sun) = .4 suns, while the maximum load on all the other faces is 1.0 suns. It is therefore common practice to mount the highest-power dissipation components on the north and south faces where the reduced solar loads make it easier to reject heat from the spacecraft.

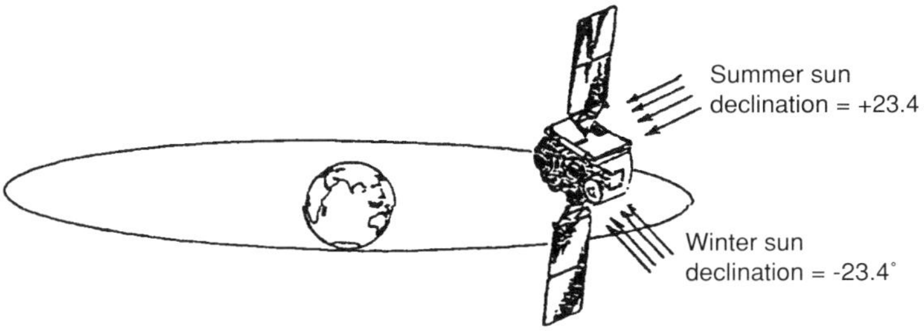

Figure 21. Solar illumination of GEO satellite

In addition to eclipses caused by the Earth there can also be eclipses caused by the moon. These are far less frequent than Earth eclipses and are of shorter duration, so they are not thermal design drivers for most spacecraft. Furthermore, while it is physically possible to have consecutive eclipses due to the Earth and the moon, the probability of this actually occurring is extremely remote and it is usually not considered in the thermal design of a spacecraft. Nonetheless, an assessment of the impact of consecutive eclipses on vehicle survival is a good idea if the spacecraft orbit could result in such a condition. There has been at least one spacecraft that has unexpectedly encountered consecutive eclipses and, although the vehicle survived, payload temperatures fell well below allowable limits.

12-Hour Circular Orbits

The thermal environment in 12-hour circular orbits is much like that in geosynchronous orbits. Earth loads (IR and albedo) are not significant unless cryogenic systems are involved, leaving solar as the only environmental load. At this time, these orbits are being used primarily by the Global Positioning System (GPS) and its Russian counterpart, known of GLONASS. For both of these programs there are a number of satellites in 12-hour circular orbits with many different inclinations.

Unlike most geosynchronous vehicles, the angles of solar illumination on the spacecraft can vary considerably due to the various orbit inclinations, but the maximum eclipse length is 56 minutes for all 12-hour circular orbits.

Molniya

Molniya orbits are unusual in that they have an extreme degree of eccentricity, i.e., they are very elliptical, and have a high inclination (62 deg). With perigee altitudes in the low Earth orbit range of around 300 n. mi., and apogee altitudes around geosynchronous altitude (21,400 n. mi.), a spacecraft in such an orbit goes through a wide swing in thermal environments. Around perigee there are high Earth loads, but at apogee only the solar loads are significant. Since the spacecraft velocity is much higher around perigee, it tends to spend most of the 12-hour orbit period at higher altitudes and relatively little time at low altitudes where Earth loads are significant. Figure 22 shows the position of a spacecraft in a Molniya orbit at one-hour intervals, and shows the Earth IR load versus time on a flat plate facing the Earth to illustrate the environmental changes that occur around the orbit.

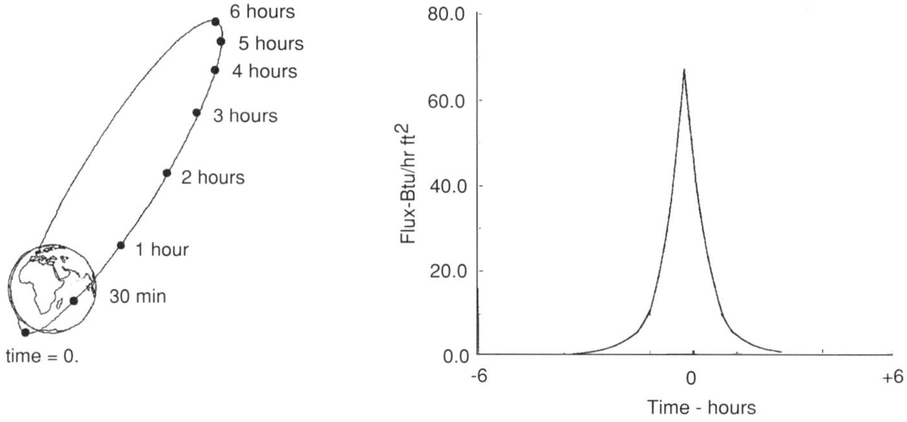

Figure 22. Earth IR heating in Molniya orbit, flat black plate facing Earth

Eclipse times for Molniya orbits vary considerably with season. During summer and equinox the Earth's shadow is cast on the southern portion of the orbit where the spacecraft is at low altitude and traveling very fast (see Figure 23). This results in relatively short eclipse times. During the winter the Earth's shadow is cast on more northerly portions of the orbit where the spacecraft is at higher altitude and lower velocity, the result is longer eclipse times. The range of eclipse times for Molniya orbits is 0 (for high beta angle orbits) to 72 minutes for certain winter eclipses.

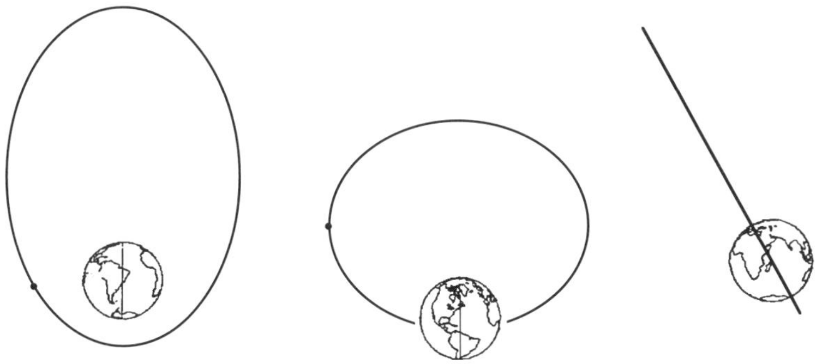

Figure 23. Molniya eclipses

LAUNCH/ASCENT ENVIRONMENT

Spacecraft thermal control systems are usually designed to the environment encountered on-orbit. Vehicle temperatures during transportation, prelaunch, and launch/ascent must be predicted, however, to insure temperature limits will not be exceeded during these initial phases of the mission. In some cases, thermal design changes or constraints on launch environments such as maximum eclipse duration or Free Molecular Heating (FMH) rates are necessary to prevent over/under temperatures from occurring on the spacecraft.

The transportation and prelaunch phases usually include the shipping of the spacecraft, preparations and testing in the clean room at the launch site, and the final countdown period with the spacecraft on the booster at the launch pad. A typical transportation sequence is shown in Figure 24. Thermal control during these phases is generally achieved by controlling the environment. For transportation, when the spacecraft is unpowered, ambient temperature and humidity limits are specified to keep all components within non-operating temperature limits and to prevent moisture condensation. During testing and storage at the launch site, room temperature conditions may be acceptable or constraints may be required on how long the vehicle may be powered up to prevent reaching operating temperature limits. If these "passive" approaches are not sufficient, special air-conditioning units may be required to blow cold air into or onto the spacecraft when it is powered on, although this is unusual.

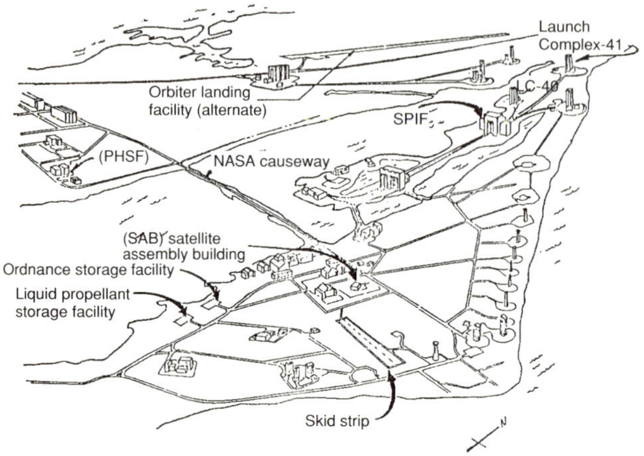

Launch Operations at
Eastern Range
(Cape Canaveral)

Overview

- C5-A arrives at skid strip
- Satellite shipping container convoys to Payload Hazardous Safe Facility (PHSF)
- Satellite processed to vertical in PHSF and convoys to LC-41
- Hoist satellite at LC-41 and mate to Booster

Figure 24. Launch site processing

Once the spacecraft is encapsulated in the booster fairing or placed in the Space Shuttle cargo bay on the pad, thermal control is achieved by blowing conditioned air or nitrogen through the fairing enclosure. This is generally a relatively low-flow-rate purge, which is like the flow in an air-conditioned room rather than a forced convection condition. The inlet temperature of this conditioned gas is usually specifiable over some nominal range such as 50 deg to 80 deg F for the Titan IV or 45 deg to 90 deg F for the Shuttle. The temperature of the gas may warm or cool significantly due to heat gain/loss to the payload fairing or Shuttle vehicle as the gas flows through the payload compartment. The electronic waste heat generated by most spacecraft, however, is usually not sufficient to cause a significant rise in purge-gas temperature.

With some spacecraft, the thermal analysis of prelaunch conditions may show that purge gas alone may not be sufficient to provide adequate cooling for all components. If this is the case, special air- or liquid-cooling ducts/loops may be required to provide extra cooling. However, since these cooling loops add significant cost and complexity to launch thermal control and may sometimes present reliability problems, other options such as intermittently turning off components should be investigated before special cooling provisions are implemented.

From lift-off through final orbit insertion the thermal environment becomes more severe and the approach is to predict spacecraft temperatures for the worst hot and cold conditions and, where necessary, implement constraints on such things as maximum eclipse time and maximum FMH. Changes to the thermal design or severe constraints on launch are usually implemented only as a last resort.

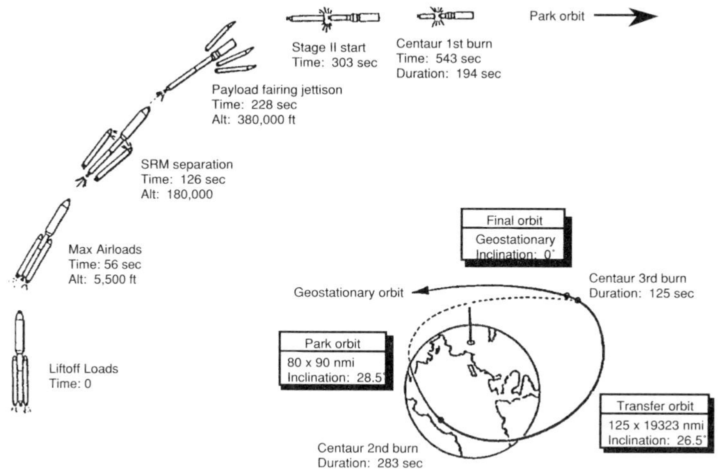

Figure 25. Representative geosynchronous mission profile (Martin Marietta)

A typical launch/ascent sequence for an expendable booster is shown in Figure 25. For the first few minutes the environment surrounding the spacecraft is driven by the payload fairing temperature, which rises rapidly to between two and four hundred deg F due to aerodynamic heating. Fairing temperatures for the Atlas II booster are shown in Figure 26. During this same period, there is a very slight cooling effect due to the depressurization of the gas in the payload compartment. This cooling effect, however, is very feeble, being noticeable for only a few minutes on very low-mass items such as the outer layer of an MLI blanket, and is usually ignored in launch thermal analysis. The effects of the payload fairing temperature rise are more significant, but will still only cause a temperature rise on relatively low-mass, exposed components such as solar arrays, insulation blankets, antennas, and very lightweight structures. This effect is further mitigated on some boosters by acoustic blankets inside the fairing that also provide an insulating effect.

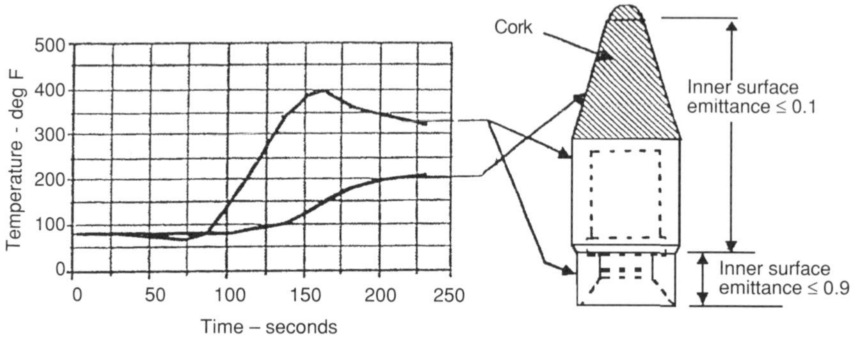

Figure 26. Atlas fairing temperatures (General Dynamics)

Within 2 to 5 minutes after liftoff the vehicle is high enough that aerodynamic effects are gone and FMH drops low enough that the fairing may be jettisoned to save weight and thereby increase payload capacity. Since it is desirable to drop the fairing as soon as possible, FMH rates are usually still very significant for up to 30 minutes after fairing separation. Curves of FMH versus time are usually generated by the booster contractor using sophisticated atmospheric and trajectory simulation codes, and are supplied to the spacecraft thermal engineers. These curves may be complex, rising and falling as the booster altitude and velocity change, as shown in the sample curve of Figure 4.

From the time of fairing separation onward, the spacecraft is exposed to a combination of FMH, solar, Earth IR, and albedo loads, and sometimes plume heating effects from the main rocket engines and attitude-control thrusters. During rocket firing, the attitude is set by guidance considerations. Between burns, however, the attitude may be changed for thermal or other reasons. It is not uncommon for the upper stage/spacecraft to go into a "barbeque roll" during these coast periods to maintain a moderate thermal environment for the payload. A thermal analysis is required to verify that spacecraft temperatures remain within limits under the combination of conditions discussed above. If temperature limits are exceeded, constraints on FMH, eclipse time, vehicle attitudes, or prelaunch purge temperatures are negotiated with the booster contractor to moderate the thermal environment. If such constraints are impractical, thermal design changes may be required to resolve the problem.

The ascent phase typically lasts 30 to 45 minutes and results in either direct insertion into the final mission orbit, a temporary parking orbit, or a transfer orbit. Direct insertion into the final orbit may occur for low Earth or highly elliptical (e.g., Molniya) orbits. Higher altitude circular orbits such as geosynchronous or 12-hour orbits require an elliptical transfer orbit to move the spacecraft to the higher altitude. An apogee-kick-motor burn at the apogee of the transfer orbit circularizes the orbit at the desired altitude (see Figure 25). During the parking or transfer orbits, the spacecraft will be exposed to the usual solar, IR, and albedo loads and is usually in a reduced power mode with appendages such as solar arrays stowed. Eclipses during transfer orbits to geosynchronous altitudes can be as long as 3 1/2 hours. This is almost three times longer than the maximum eclipse in geosynchronous orbits and can present thermal-control problems if eclipse times are not limited by launch constraints. Because of the reduced power dissipation and long eclipses, the most common concern during this period is unacceptably low temperatures on the spacecraft, although high temperatures can occur if the spacecraft is inertially stable with the sun shining continuously on a sensitive component.

Once the spacecraft reaches its final orbit, there is a period lasting anywhere from a few hours to several weeks during which the spacecraft attitude is stabilized, appendages such as solar arrays and antennas are deployed, and bus and payload electronics are powered up. The thermal control system must maintain acceptable temperatures during this period and survival heaters are sometimes required. The sequence of events are also sometimes driven by thermal considerations, e.g., when and which attitudes are achieved, when payloads are turned on, etc.

For launches on the Space Shuttle, the prelaunch, ascent, and transfer-orbit events are similar to those of the expendable booster. The Shuttle, however, has a far more complex park-orbit condition during which the spacecraft may be exposed to a wide range of thermal conditions for periods ranging from 6 hours to several days, with the longer duration typically due to contingency operations or multiple payload deployments. Unlike an expendable booster, which jettisons its fairing a few minutes after liftoff, the Shuttle doors may remain closed for up to 3 hours, limiting the payload spacecraft's ability to reject waste heat. Once the doors are open, the bay may be pointed toward Earth, which is fairly benign, or the more severe environments of deep space or the sun. Maneuvers are also required periodically for Shuttle guidance-system alignments, communications, etc. Since the bay liner is insulated, the overall spacecraft may be exposed to more extreme conditions than if it were on a conventional booster, where it would simultaneously see a combination of sun, Earth, and deep space. In addition to the complex on-orbit environment, abort reentry conditions must also be considered. This additional complexity, along with safety considerations, makes the thermal integration process an order of magnitude more difficult for a Shuttle launch than for a launch on a conventional booster.

For a more in-depth discussion of spacecraft to launch-vehicle thermal integration, the reader is referred to Refs. 9 and 10, which cover integration with the Titan IV and Space Shuttle launch vehicles, respectively.

REFERENCES

1. "Space and Planetary Environment Criteria Guidelines for Use in Space Vehicle Development, 1982 Revision (Volume 1)," NASA TM-82478, January 1983.

2. "Solar Electromagnetic Radiation," NASA SP-8005, May 1971.

3. "Earth Albedo and Emitted Radiation," NASA SP-8067, July 1971.

4. D. F. Gluck, "Space Vehicle Thermal Testing: Environments, Related Design and Analysis, Requirements, and Practice," The Aerospace Corporation, TOR-86A(2902-08)-1, 27 September 1987.

5. G. L. Stephens, G. G. Campbell, and T. H. Von der Haar, "Earth Radiation Budgets," J. Geophys. Res., Vol. 86, 20 October 1981.

6. G. L. Smith, D. Rutan, and T. D. Bess, "Atlas of Albedo and Absorbed Solar Radiation Derived from Nimbus 7 Earth Radiation Budget Data Set - November 1978 to October 1985," NASA Ref. Pub. 1231, 1990.

7. R. L. Collins, "Albedo and Earthshine," The Aerospace Corporation IOC 87.5462.4-25, 27 April 1987.

8. M. Donabedian, "Thermal Uncertainty Margins for Cryogenic Sensor Systems," The Aerospace Corporation, ATM 90(9975)-3, 21 November 1989.

9. R. Pleasant, W. Kelly, "Thermal Integration of Satellite Vehicles with Titan/Centaur," AIAA 89-1723, AIAA 24th Thermophysics Conferences, June 12-14, 1989.

10. "Shuttle/Payload Integration, Information Document for Thermal," NASA JSC-21000-INF, Sept. 1986.

11. D. S. Glaister, "Space Environment for Cryogenic Radiator," The Aerospace Corporation, ATM No. 90(9975)-35, 3 May 1990.

12. R. D. Jimenez, "Natural Environment Charged Particle Heating of Spacecraft Cryogenic Components," The Aerospace Corporation, TOR-0088(3062)-3, 19 September 1988.

Chapter III

Thermal Design Examples

David G. Gilmore
The Aerospace Corporation

Y. Yoshikawa
Lockheed Missiles and Space Company

R. Stoll and R. Bettini
Hughes Danbury Optical Systems

C. Finch and B. Turner
British Aerospace

INTRODUCTION

The purpose of the thermal-control system is to maintain all satellite components within allowable temperature limits for all operating modes of the vehicle when exposed to the thermal environments discussed in the last chapter. For the purposes of familiarizing the reader with how thermal control is achieved, typical thermal designs for various satellite components will be described here. While these are designs that are in wide use on current spacecraft, it should be emphasized that these are not the only thermal designs that are possible for these systems, and that creative alternative solutions to the thermal design of a satellite or component are always desirable. The designs described in the discussion that follows should therefore be considered as illustrative examples only.

Establishing a thermal design for a spacecraft is usually approached as a two-part process. The first is to select a thermal design for the body, or basic enclosures of the spacecraft, that will serve as a thermal sink for all of the internal components. The second step is to select thermal designs for various components located both within and outside of the spacecraft body. The following sections give a qualitative description of typical designs. A more detailed discussion of how these designs are selected and the thermal analyses required to verify them can be found in Chapter V.

SPIN-STABILIZED SATELLITES

Although there are a number of spin-stabilized satellite thermal designs, the most common is that typified by DSCS II, Satellite Data Systems, NATO II, and a host of commercial communications satellites. The approach is to use the spinning solar array as a heat sink for the internal components. A cylinder spinning with the sun normal to the spin axis will run around room temperature if the ratio of solar absorption to infrared emittance (α/ε) is near 1.0, as is approximately the case with the solar cells that cover the cylinder. Because of this, the spinning solar array makes a convenient heat sink for internal components.

Figure 1 illustrates the thermal balance in a typical spinning satellite. Electronic boxes are usually mounted on shelves and radiate their heat to the solar array, and sometimes also to the forward or aft ends if extra radiator area is required. The electronic boxes are typically painted black for high infrared emittance and are mounted in such a way as to insure good heat conduction to the shelf. For most boxes the combined surface area of the box and shelf is sufficient to radiate the waste heat to the solar array without a large temperature difference developing between the box and room temperature array. If a box is small and has a high heat dissipation, a thermal "doubler" (a sheet of high conductivity material such as aluminum, beryllium, or copper) may be placed under the box to help spread the heat out on the shelf and increase the effective radiating area of the box.

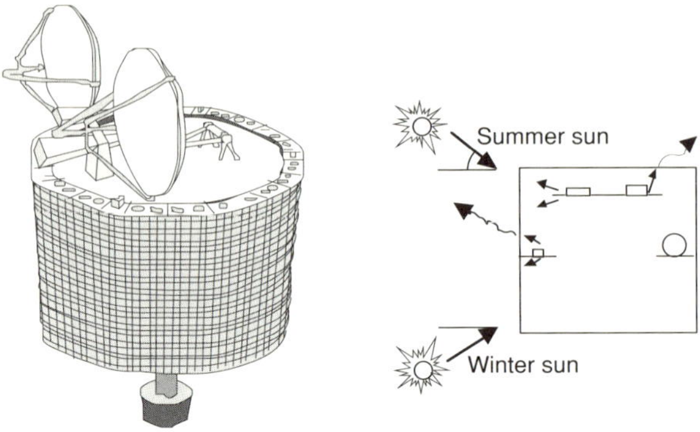

Figure 1. "Spinner" thermal balance

Because most spinners are used in high-altitude geosynchronous orbits, eclipses occur only once per day and last for a maximum of 72 minutes. During eclipse the solar array temperature drops dramatically, typically from room temperature to on the order of -100 deg F. During this period, the temperature of the electronics boxes and other components will also drop; however, due to their high thermal mass, they will not cool nearly as fast as the relatively lightweight solar array. The result is that it is often possible to "coast" through the eclipse without falling below the minimum allowable operating temperature of the electronics. However, if the thermal design analysis shows that some components get too cold, then either a lower emittance finish on the cold units or a heater may be required to reduce their radiative coupling to the solar array or provide extra heat during eclipse. The use of heaters during eclipse is minimized, however, since they drive up the size, and therefore the weight, of batteries.

3-AXIS-STABILIZED SATELLITES

The most common type of satellite today is the 3-axis-stabilized variety typified by TDRSS, SPOT, Fltsatcom, DMSP, and many others. Almost all of these satellites use the same basic approach to thermal control of the satellite body, e.g., insulating the spacecraft from the space environment using multilayer insulation (MLI) blankets and providing radiator areas with low solar absorptance and high infrared emittance to reject the satellite waste heat. The overall thermal balance of such a satellite is illustrated in Figure 2.

The high-power-dissipation boxes in a 3-axis satellite are usually mounted on the walls of the satellite, which provides them with a direct conduction path to the radiating areas on the outside surface. As with the spinner, some of the high-power boxes may require a doubler to spread out the heat over a wider area of the wall to which they are mounted. Boxes that are mounted on shelves, panels, and structures internal to the vehicle radiate their waste heat directly or indirectly to the outside walls of the spacecraft, where the heat is then rejected to space. Because this type of design is insulated and uses low solar absorptance radiators, it is less sensitive to sun position, albedo loads, and eclipses than is the spinner-type satellite discussed earlier.

- Insulate Main Body with Multilayer Insulation Blanket (MLI) Blanket
- Provide Low Solar Absorptance (α), High Infrared Emittance (ε) Radiators to Reject Waste Heat
- Use Heaters to Protect Equipment when Satellite is in Low Power Mode
- Use Surface Finishes and Insulation to Control Appendage Temperatures (Antennas and Solar Arrays Typically have very wide temperature ranges)

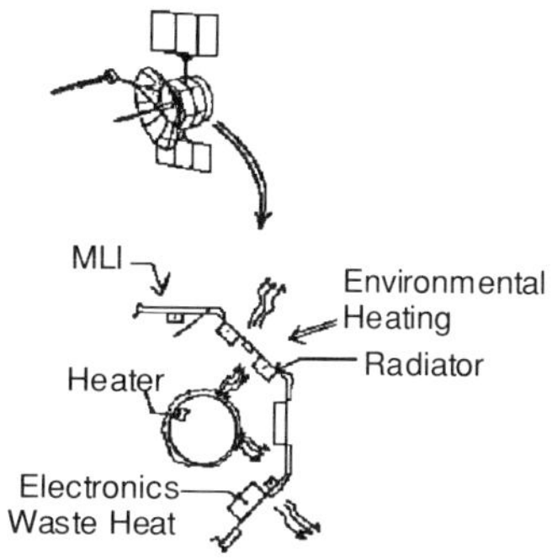

Figure 2. 3-Axis satellite thermal control

PROPULSION SYSTEMS

Almost all satellites have on-board propulsion systems for attitude control and/or small orbit corrections. The propulsion system typically consists of small (less than 5 pounds thrust) compressed gas or liquid-propellant thrusters, and all of the assorted tanks, lines, valves, etc., that are used to store propellants and feed the thrusters. In addition, some satellites may also have a solid rocket motor to provide the final boost from transfer orbit to operational orbit. Propulsion-system components have special thermal-control requirements to avoid freezing of liquid propellants, to prevent temperature gradients within solid propellants, and to limit temperature differences between fuel and oxidizer in liquid bi-propellant systems.

The most common propellant used for on-board propulsion systems today is hydrazine. In a hydrazine system a catalyst in each thruster is used to trigger a decomposition of the liquid hydrazine into a number of gases, including nitrogen, ammonia, and water, accompanied by the release of a large amount of heat. A schematic of a typical hydrazine propulsion system is shown in Figure 3 and includes tanks, lines, valves, thrusters, and filters.

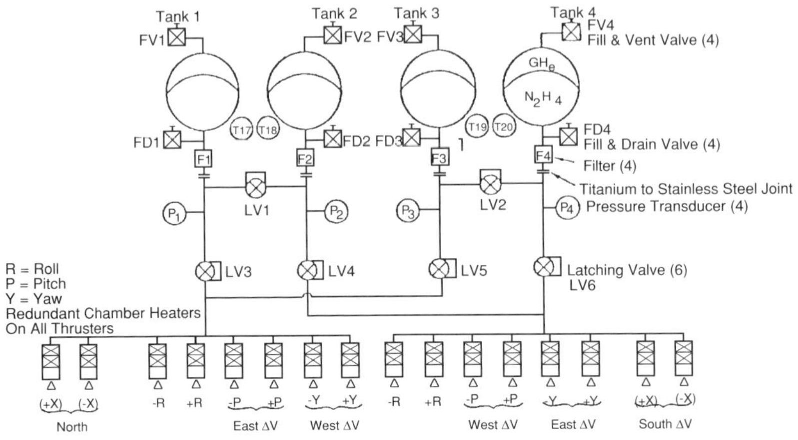

Figure 3. DSCS propulsion system schematic
(Martin Marietta)

A typical thermal design for a propulsion system is shown in Figure 4. The general approach is to conductively isolate all of the propulsion components from the vehicle structure using low-conductivity standoffs and attachment fittings, and to cover the components in a low-emittance finish or MLI to provide radiative isolation. Heaters are also often used, especially on low-mass items such as propellant lines, which may cool very quickly during eclipses or other short-term cold conditions. Heaters may be either hard wired (on all the time) or controlled to a fixed temperature using thermostats or solid-state controllers. The heater power density (watts/inch of line) may sometimes also have to be varied along the line to insure that acceptable temperatures are maintained as the line runs through "hot" and "cold" areas of the spacecraft. The heaters and isolation are required because the spacecraft may get quite cold during some launch or operational modes, and hydrazine freezes at 35 deg F.

Thermal control of thrusters is a bit more complicated. Not only must the thruster be kept above the freezing point of the hydrazine (or other propellants), but the vehicle must be protected against heating from the rocket plume and heat soak-back from the rocket engine body during and after the firing. Figure 5 shows the thermal design for a Milstar bi-propellant thruster, which is located on the exterior of the satellite. The entire thruster assembly is thermally isolated from the spacecraft using low-conductivity titanium standoffs. The thruster valves and injector are covered with MLI to minimize heat losses when the thruster is not operating; however, a total of 8 square inches of radiator area has been provided to

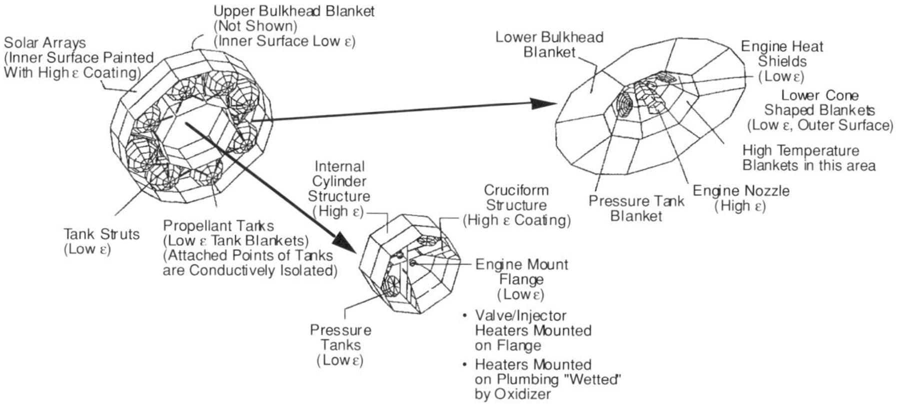

Figure 4. Integrated apogee boost system thermal design (Martin Marietta)

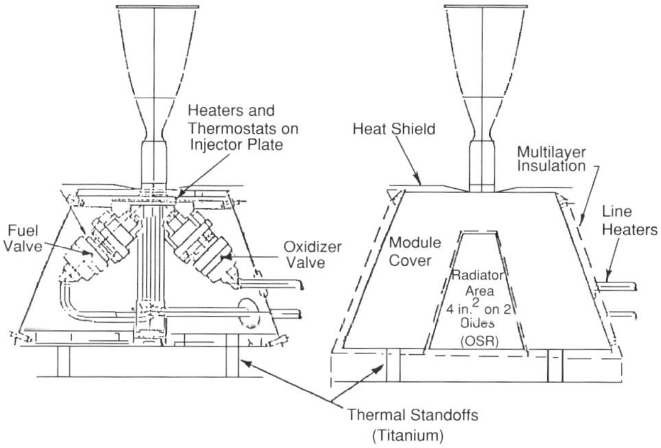

Figure 5. MILSTAR thruster thermal design (Lockheed)

help cool the thruster after firing. In order to keep the thruster warm during non-operating periods, thermostatically controlled heaters are provided on the injector plate. These heaters are sized to make up for heat lost by radiation from the exposed nozzle and the small radiator areas on the sides of the thruster enclosure. (The nozzle is not covered by insulation since it gets extremely hot during engine firing and must be able to radiate freely to space.) In addition, a single-layer low-emissivity heat shield is used to protect the enclosed elements from radiant heating from the nozzle as well as heating from the rocket plume.

Most liquid thrusters are designed to limit the conduction path between the combustion chamber/nozzle and the valve bodies. This isolation is more evident in the hydrazine thruster shown in Figure 6. Here isolation is achieved using a tubular support of low-conductivity stainless steel filled with holes. Fuel is fed to the thrust chamber through long, slender stainless-steel tubes. During and after a firing, the nozzle and combustion chamber become very hot, but the heat is primarily radiated to space rather than conducted back to the valves.

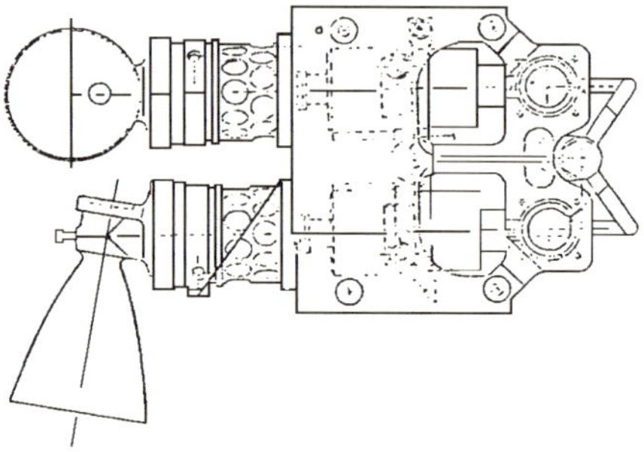

Figure 6. Hydrazine thruster module

Plume shields, such as those on the Milstar thruster discussed above, are often used to protect spacecraft hardware physically near to thrusters or large rocket motors. These heat shields are typically made of thin sheets of high-temperature, low-emissivity metals such as stainless steel or titanium. The metal can withstand the high temperatures to which the shield is driven and the low emissivity limits the heat reradiated from the shield back toward the spacecraft. (The space-facing side of such shields often have high-emissivity finishes to help reduce shield temperature.) A large heat shield used to protect the back end of a spacecraft from the plume of a large solid rocket motor is shown in Figure 7. It should be noted that the plumes from solid rockets produce much higher radiant heating rates than do liquid motors because the plume is full of solid particles, which have a much higher emissivity than the gases in a liquid motor plume.

Solid rocket motors are often used to transfer a spacecraft from the transfer orbit in which the launch vehicle has placed it to the final operational orbit. These solid motors usually require that the propellant be kept within a certain temperature range and that temperature gradients in the propellant be kept below a specified value. The most common approach to achieving this is to wrap the motor in MLI and provide conduction isolators at the mounting points, as shown in Figure 8. Sometimes it is also necessary to provide insulating shields or blankets on the nozzles and across the nozzle exit plane, since an exposed nozzle can cause a large heat leak and/or temperature gradient in the propellant. (The blanket across the nozzle exit plane is, of course, blown off when the motor ignites.) If the motor is to be used immediately after launch, insulation alone may be satisfactory since the motor is massive and will cool very slowly. If, however, several days may elapse before the motor is used, heaters may be required on the motor case to keep the propellant from getting too cold.

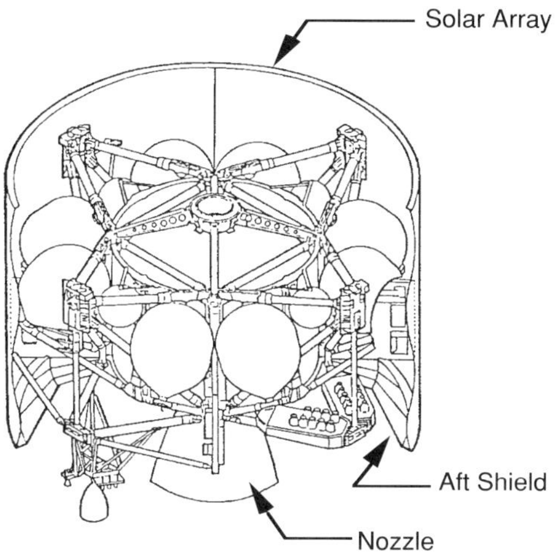

Figure 7. Plume shield

BATTERIES

Two different types of batteries are commonly used on spacecraft; nickel cadmium and nickel hydrogen. The thermal-control requirements and thermal design are somewhat different for each type.

The most common battery type used on older spacecraft power systems is nickel cadmium (NiCd). NiCd batteries are usually maintained around 0 deg to 10 deg C to maximize their life. As the temperature of these batteries rises above that range, their maximum useful life decreases significantly. Below this range, the electrolite may freeze and damage the battery. Another requirement, which is common to many types of batteries, is that all batteries on the satellite and all cells within a battery must be kept at the same temperature, plus or minus some specified value, such as ± 5 deg C. The isothermality requirement is necessary to ensure that all cells charge and discharge at the same rate.

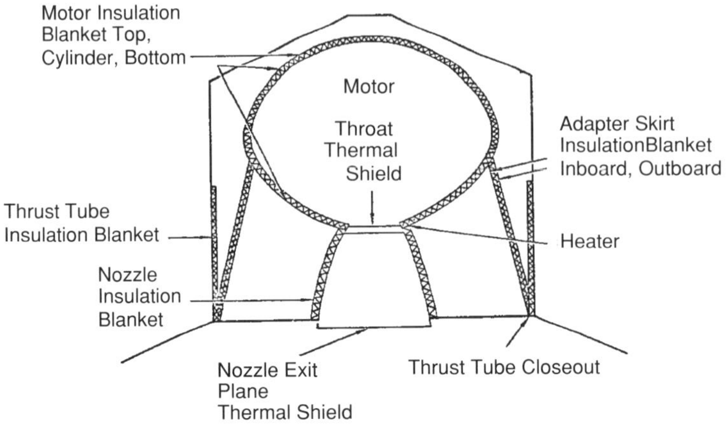

Figure 8. Solid rocket motor thermal design (Hughes)

Although some (usually small) NiCd batteries are mounted inside a spacecraft and simply painted black to radiate the waste heat from charge and discharge inefficiencies to the spacecraft interior, the most common thermal design uses radiators and thermostatically controlled heaters, as shown in Figure 9. Heat from individual rectangular battery cells is conducted down aluminum fins placed between cells to a baseplate, which in turn radiates off its other side directly to the space environment. The radiator is usually sized to keep the batteries somewhat below the maximum allowable temperature under worst hot-case conditions, and thermostatically controlled heaters are then used to maintain minimum allowable temperatures under cold-case conditions. This design insures that battery temperatures will be precisely controlled at all times.

At the time of this writing, nickel hydrogen batteries are beginning to appear on most new spacecraft programs, especially those requiring long life and minimum battery weight. Like NiCd, the NiH_2 battery requires a closely controlled isothermal operation in the 0 deg to 20 deg C range. Since NiH_2 batteries are high-pressure devices, however, they are manufactured as cylindrical pressure vessel cells and typically packaged together on "trays" to make batteries, as shown in Figure 10.

For the purposes of discussion, the thermal designs used for these batteries can generally be divided into four types. The simplest form, shown in Figure 10, involves direct conduction coupling between the cells and a baseplate/radiator with a heater used to control minimum temperature, as is used on the GPS II and APEX programs. The second category, also shown in Figure 10, introduces fixed-conductance heat pipes to isothermalize the batteries and couple them to remote radiators, as in a configuration like that found on MILSTAR. The third category

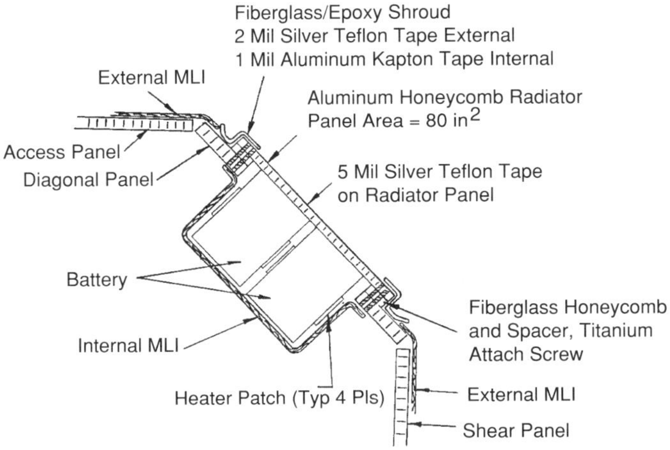

Figure 9. Nickel cadmium battery thermal design (TRW)

introduces variable-conductance heat pipes to minimize heater power and/or accommodate wide variations in environmental back loads that can occur in some applications. The fourth category is similar to the third except that louvers are used in place of variable-conductance heat pipes, as typified by the Hubble Space Telescope (see Figure 10).

The design approach used on these systems is to size the radiator for something between the orbit average and the peak heat load. Ideally, the radiator would be sized to the peak load occurring during discharge or overcharge (plus any environmental load) so that the battery could be kept around 0°C at all times. This, however, would result in a very large radiator and very high heater power during the charge and trickle charge periods when the battery is generating little or no waste heat. To reduce the radiator size and heater power, a radiator size closer to that required for the orbit average-heat dissipation is usually chosen and the cell temperatures are allowed to rise to around room temperature during discharge and overcharge. The minimum possible radiator size is that required for the orbit average power; however, the radiator is sized somewhat above orbit average-heat level so cell temperatures can be pulled back down below 5°C quickly after the discharge or overcharge heat pulses. This oversizing reduces the amount of time the battery is above the desired temperature range, but results in the need for heaters during the charge phases, even for the hot-design-case conditions.

It should be noted that the waste-heat rates of batteries are sometimes difficult to quantify. Thermal dissipation varies with state of charge, temperature, and charge rate, and may be different for different batteries of the same general type. Complex power thermal models are sometimes constructed to deal with these variables; however, close coordination between thermal and power-system engineers is usually sufficient to insure that conservative, but reasonable, heating rates are used in the thermal analysis.

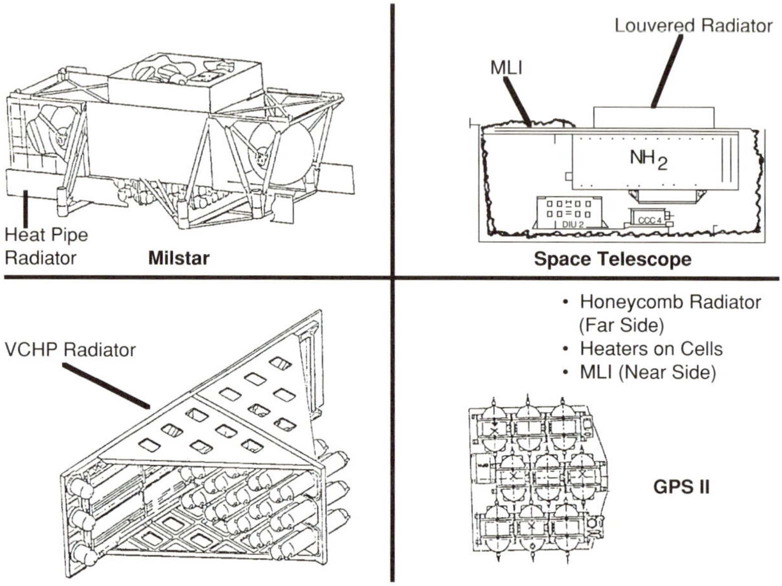

Figure 10. Battery thermal designs

ANTENNAS

There are many types of antennas used on spacecraft, including helixes, solid reflectors, mesh reflectors, horns, etc., as shown in Figure 11. The thermal-control requirements for antennas are usually to maintain temperatures within the allowable ranges for the materials of which they are made and, especially for reflectors, to keep thermally induced distortions within acceptable limits. For most antennas an acceptable design can be found using paints, insulation blankets, and/or low-coefficient-of-thermal-expansion structural materials.

Typical antenna thermal designs are shown in Figure 12. Horns, whether transmitting directly to the Earth or used in conjunction with a reflector, are often simply covered with MLI with an astroquartz or white-painted plastic film (such as Kapton) covering the aperture. Aluminized Kapton, which is typically used on other parts of the spacecraft, cannot be used to cover an antenna aperture because the conductive aluminum layer is not transparent to RF energy. Any material used in the path of an antenna beam must be close to 100 percent RF transparent so as not to attenuate the signal.

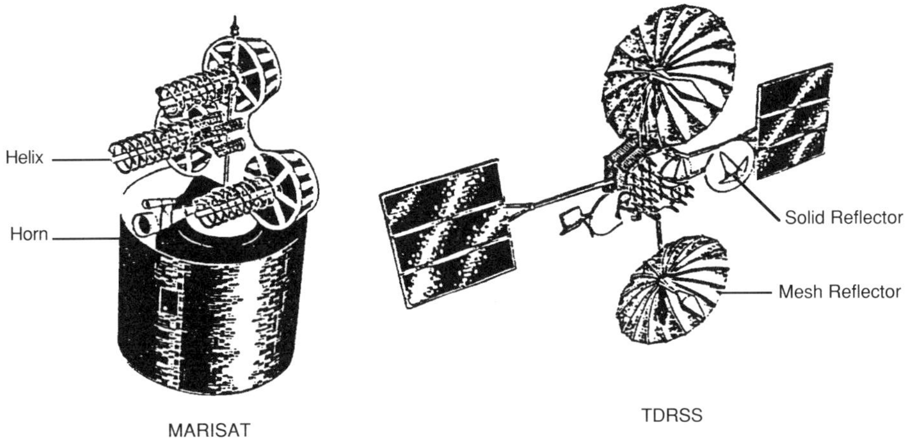

Figure 11. Satellite antennas

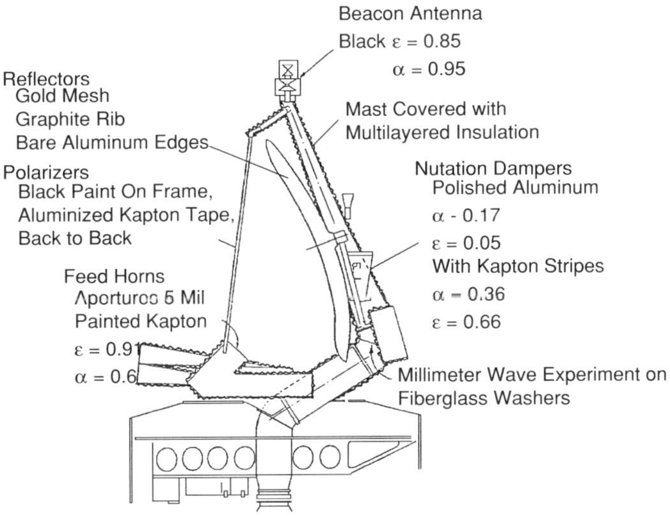

Figure 12. Representative antenna thermal design (Hughes)

Solid-dish antenna reflectors are generally painted white on the exposed (reflecting) side and covered with MLI on the back. The white paint limits the solar heating of the antenna and reduces temperature gradients in the dish that can be caused by shadowing. If support struts are present for a center reflector or center feed, the support structure is generally also painted white. Since uneven illumination of the dish or struts can cause thermal distortions that degrade RF beam quality, a thermo-structural analysis is generally required to verify the design.

Mesh antennas are generally more difficult to analyze than solid reflectors due to the complex shadowing and radiation interchange that occurs with a sparse open structure. As with other antennas, however, the principal thermal control-requirements are to keep all materials within allowable temperature ranges and to limit thermal distortion to acceptable levels. These requirements can usually be met by painting the low-coefficient-of-thermal-expansion antenna structural ribs with a low-absorptance, low-emittance paint, or covering them with MLI, as shown in Figure 12. Either of these approaches tends to minimize temperature gradients across the diameter of the tubular ribs, thereby limiting thermal bending and dish distortion. It should be noted that the use of paints, if possible, is preferable to MLI since it provides a much "cleaner" design from a mechanical-packaging and antenna-deployment standpoint. The antenna mesh (usually gold- or silver-coated stainless steel) is generally left bare and allowed to cycle between very high (+300 deg F) and very low (-200 deg F) temperatures, since it would be difficult to apply a thermal coating to the fine wire mesh.

Helix antennas are like the others in that the requirement is to maintain material temperature limits and minimize distortion; however the distortion problem is usually much less severe than it is with dish or mesh reflectors. Temperatures of helix antennas can generally be maintained with paints and bare metal finishes, and do not present a challenging thermal-design problem.

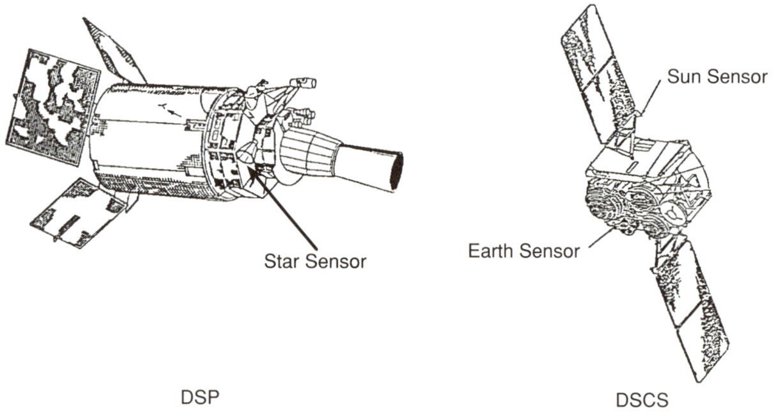

Figure 13. Attitude control sensors

SUN/EARTH/STAR SENSORS

All spacecraft have sun, Earth, or star sensors to determine attitude. These devices range in size from small sensors that could fit in the palm of the hand to rather large units up to two or three feet across. They may be mounted internal or external to the spacecraft, with sun sensors sometimes being mounted out on the solar-array structure (see Figure 13).

Attitude-sensor thermal designs are variable depending on the installation or temperature sensitivity of the device. Figure 14 shows the thermal design of a sun sensor mounted internal to a spin-stabilized satellite. The sensor is conductively isolated from the solar array to limit temperature drops during eclipse. The inside face of the unit is painted black to give good thermal coupling to the relatively stable temperature of the internal spacecraft hardware. The outside face is 80 percent polished aluminum and 20 percent black paint, which gives an α/ε of .34/.22. This α/ε ratio was tailored to produce some warming when in the sun while limiting heat loss during eclipse.

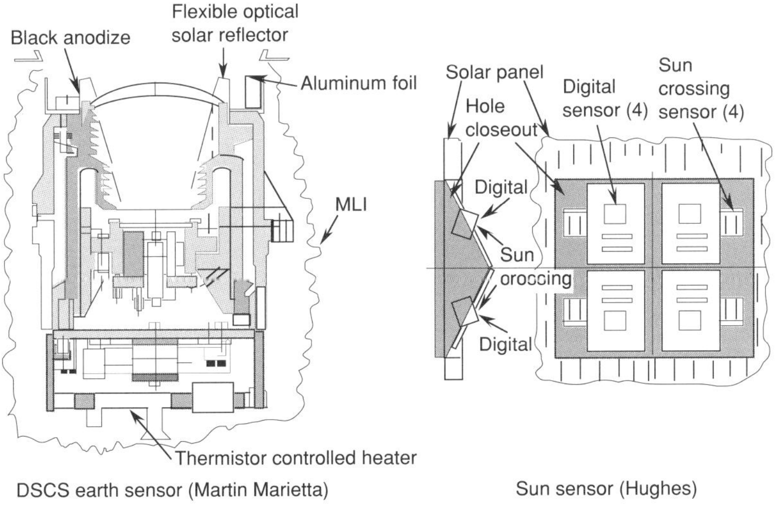

Figure 14. Sun and Earth sensor thermal designs

Figure 14 also shows an Earth sensor that has a very tight temperature-control requirement of ± 1 deg F on the sensing element. The approach used in this application is to conductively isolate the sensor from the spacecraft structure to which it is mounted, radiatively isolate the sensor from the external environment using MLI blankets, and provide a small radiator area and a proportionally controlled heater to maintain precise temperature control of the sensor element.

Figure 15 shows a large star-sensor assembly used for high-precision attitude control. Thermal-control requirements for this device include not only its maximum and minimum operating temperature (+10 to +40 deg C), but also limitations on temperature gradients, which could misalign the optical elements.

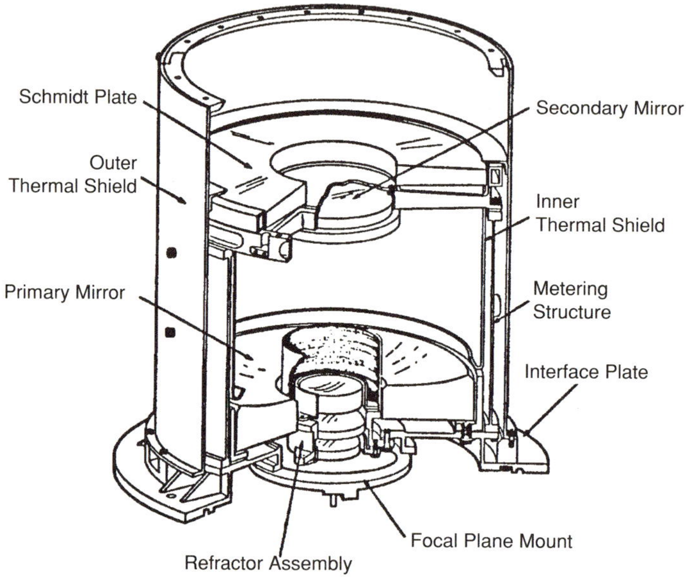

Figure 15. Star sensor thermal design (Kodak)

The thermal design used for this star sensor is shown in Figure 15. First of all, the entire device is thermally isolated from the spacecraft with plastic mounting blocks to reduce sensitivity to spacecraft temperature changes. A shutter is also provided to prevent sunlight from coming directly down the optical boresight. Thermal expansion of the "metering" structure slightly varies the separation of the optical elements to counteract temperature-induced changes in the mirror curvature and maintain focus over the range of operational temperatures. This structure, however, cannot have temperature gradients across its diameter, since this would cause the primary and secondary mirrors to rotate out of plane with one another. To prevent such temperature gradients, the metering structure is protected by thermal shields on both the inside and outside. These shields are made of high-thermal-conductivity aluminum, which is thick enough to conduct heat from hot areas where the sun may be shining to cold areas in the shade. The inner and outer shields are also thermally coupled with high conductivity "posts," which run through small holes cut in the metering structure. The shield surfaces facing the metering structure, and the metering structure itself, also have low-emissivity finishes to further reduce sensitivity to both temperature gradients and temperature fluctuations in the shields. Finally, the outer surface of the outer shield has a thermal finish that is tailored to reduce sensitivity to the asymmetric environments that the sensor will see on this particular spacecraft. All of these design features work together to insure a highly isothermal optical support structure.

COOLED DEVICES

Some spacecraft payloads require cooling to low temperatures. The most common types of cooled devices are IR-sensor focal planes and optics, and low-noise amplifiers for RF receivers. Several methods of cooling exist for such applications, including radiators, stored cryogen coolers, and refrigerators. The following paragraphs briefly describe specific designs utilizing coolers. For a more complete discussion of these technologies the reader is referred to Chapter VIII.

The DSP satellite uses a system of radiators to cool the optics and focal plane, as shown in Figure 16. The optical elements (mirrors) and the telescope enclosure and baffles are cooled passively by covering the telescope enclosure with low absorptance/high emittance quartz mirrors. Cooling the optics and enclosure reduces the amount of IR radiation emitted from those surfaces. Without this cooling, the sensors at the focal plane would not be able to see their targets over the IR "noise" created by the telescope itself. The focal-plane assembly is connected to a phase change material (PCM) heat sink and a passive radiator by a pumped-helium loop. The operating principal of this system (shown in Figure 16) is to transport heat from the focal plane and PCM to the radiator using a pumped-helium loop during that half of the orbit when there is no sun shining on the radiator. During the other half-orbit, solar illumination heats the radiator to temperatures well above those of the focal plane. To avoid a focal-plane temperature rise, the helium circulation is shut off, effectively decoupling the radiators, and the heat loads from the focal plane are stored in the PCM. When the sun moves behind the vehicle, the circulator is turned back on to reject the focal-plane heat and the excess heat stored in the PCM. Minimizing heat leaks into the forward-facing radiator by the use of MLI and low-conductance supports on the backside is critical to achieving low-temperature performance. Even small heat leaks into the radiator during the shadowed half-orbit can raise its temperature considerably from 173 deg K. (Due to the T^4 nature of radiation-heat transfer, it takes only one-fifth as much heat to raise radiator equilibrium temperatures one degree at 173 deg K than it does at room temperature. For lower temperature radiators the sensitivity is even greater, e.g., a factor of 50 greater sensitivity at 80-deg K than at room temperature. For this reason, low-temperature radiators are extremely sensitive to heat loads from the environment or heat leaks from the spacecraft.)

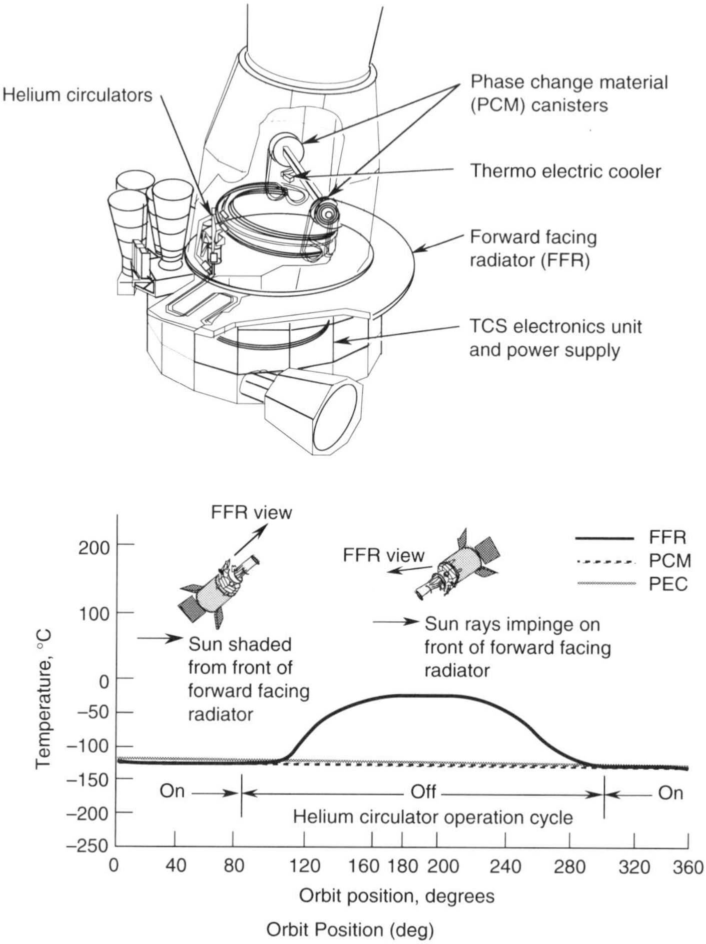

Concept

- Helium Circulator Transports Focal Plane/PCM Heat to Non-Solar Illuminated Forward Facing Radiator (FFR)
- Circulator Turned Off When FFR Solar Illuminated—Focal Plane Heat Melts PCM
- PCM Is Refrozen When Circulator Resumed

Figure 16. DSP sensor thermal control (Aerojet)

Devices requiring cooling to very low temperatures and having limited lifetime requirements (less than 1 or 2 years) usually employ stored cryogen cooling systems. Such designs use a cryogenic fluid or solid stored in a dewar as a heat sink to absorb waste heat from the device and maintain it at a low temperature. An example of such a system is the Infrared Astronomical Satellite (IRAS). The cryogen in this case is 154 pounds of helium stored at 1.85 deg K in a tank, which is itself wrapped around the telescope assembly, as can be seen in Figure 17. As the telescope is operated, it generates heat and this, along with the parasitic heat leaks through the tank insulation and supports, causes the helium to boil off. Rather than simply venting this vapor to space, it is routed through heat-exchange tubes mounted on thermal shields surrounding the tank in various stages. The thermal capacity of the vapor is thereby used to absorb some of the heat getting through the insulation and is eventually vented back out to space. Performance of the MLI insulation and the low-conductance tank-support struts is critical to reducing parasitic heat leaks and maintaining the lifetime of the system. Shielding of the instrument is also important. For this particular satellite, the dewar/telescope assembly is shadowed from the sun by the solar array and a sunshade/radiator system (shown in Figure 18), which, along with spacecraft attitude constraints, are used to block solar and Earth heat loads from entering the telescope aperture.

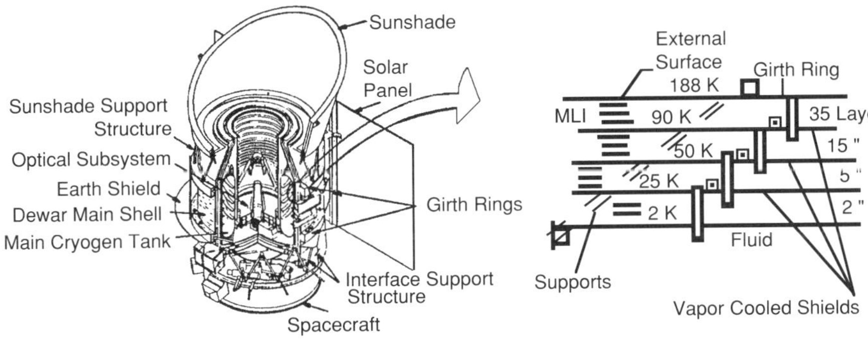

- 154 lb He Stored at 1.85 K
- 2.2 K Focal Plane
- Design Life = 1 Year
- Volume = 19 ft^3
- Dry Weight (Tankage) = 850 lb
- Instrument Heat Load = 12.2 MW
- Heat Leak Dominated by Supports (55 Percent)
- MLI Heat Leak Only 17 Percent of Total
- Parasitic Heat Leak:
  - Total = 33 MW
  - Q/A = 0.004 Btu/hr-ft^2

Figure 17. IRAS thermal design (Ball)

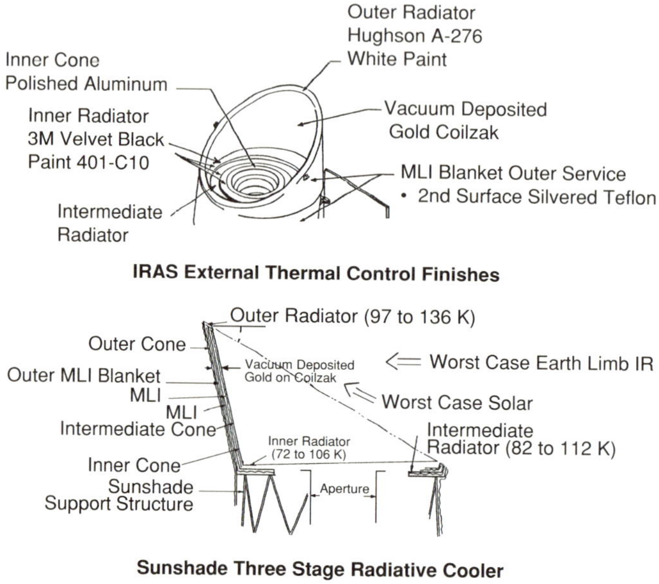

- Inner Shield Surface Specularly Reflects Earth Limb IR
- Black Painted Radiators Have High Emissivity, No Solar Exposure
- Inner Cone Self Emission Less Than 1×10^{-16} Watts at the Detector
- Twenty-Four Low Heat Leak Fiberglass Rods Support Sunshade Outer Stage Off of Outer Shell Girth Ring
- Outer MLI Blanket Protects Outer Radiator from Solar
- Inner Radiator Isolated from Other Radiator by Staged MLI Blankets
- Sunshade Protects Aperture from Ram Effect Molecular Contamination

Figure 18. IRAS sunshade (Ball)

For applications with moderate to large cooling requirements and a lifetime in excess of one or two years, refrigerators are normally employed, although they do have drawbacks, which will be discussed in Chapter VIII. An example of a refrigerator design is the DSP Third Color Experiment cryocooler shown in Figure 19. Here a refrigerator is mounted in the telescope assembly to provide additional cooling to a set of sensors that are mounted on, but conductively isolated from, the primary focal plane. A heat pipe is used to transfer heat from the sensor (TCE Segment V in the figure) to the refrigerator compressor cold heat pipe, to a radiator mounted on the side of the spacecraft. The temperature boost given by the refrigerator results in a much smaller radiator area due to the T^4 nature of radiation heat transfer. The reduced size and weight of the radiator more than compensates for the weight of the refrigerator and the extra electrical power system weight required to run it.

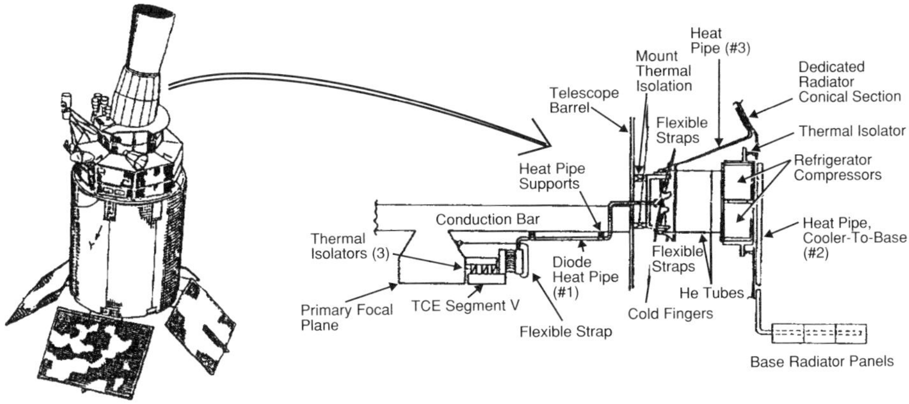

Figure 19. DSP sensor detail (Aerojet)

SOLAR ARRAYS

Thermal control of solar arrays is generally a fairly straightforward matter. The solar cells preclude the use of any thermal finishes on the sun-facing side of the array, so the thermal radiative properties are controlled by the high-absorptance/high-emittance solar cells themselves (see Figure 20). To keep array temperatures as low as possible (which increases electrical efficiency), the back of the array is usually painted with high-emittance black or white paint, with the white paint used primarily in low-altitude orbits where albedo loads from the Earth may illuminate the back side of the array. Due to their high absorptance, high emittance, large area, and low weight, solar arrays typically cycle over wide temperature ranges as they go from sunlight to eclipse; +150 deg F to -100 deg F in low Earth orbit, and +120 deg F to -230 deg F in geosynchronous orbit.

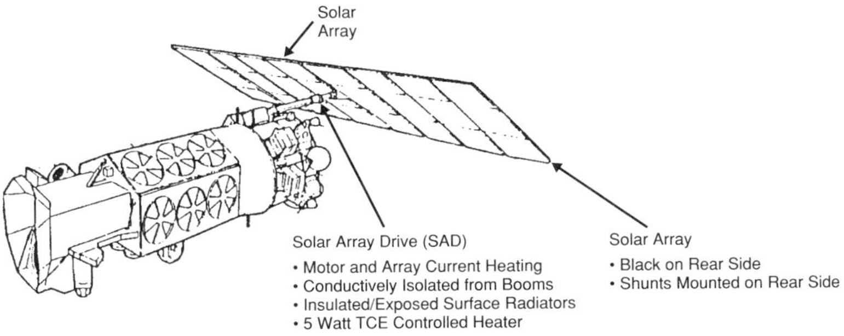

Figure 20. DMSP solar array and drive (Martin Marietta)

The support structure for the solar array is sometimes thermal isolated from the array drive motor on the spacecraft by low-conductance spacers. This is done primarily to prevent heat leaks out of the motor, since the structure temperatures themselves can usually be controlled to acceptable ranges with paint finishes. Occasionally, special thermal shields may also be required on the edges of arrays to protect them from rocket motor plumes or free molecular heating during launch. However, this is not very common.

SYSTEM OVERVIEW—THE HUBBLE SPACE TELESCOPE

To give the reader an appreciation for the extensive application of thermal control in the development of a typical satellite, the following top-level description of the thermal design of NASA's Hubble Space Telescope has been included in this chapter. The rather lengthy discussion is intended to illustrate the fact that the thermal engineer must consider the need for thermal control of all components of the vehicle. A great deal of analysis, parametric studies, design iterations, and test is required to arrive at the final thermal design. The total thermal effort can easily exceed 20 man-years for some satellites.

The Hubble Space Telescope is an example of a large optical imaging satellite. Despite the sophistication of the satellite and the complexity of its mission operations, thermal control has been accomplished using common thermal control hardware; thermal surface finishes, multilayer insulation (MLI), heaters, thermal isolators, and louvers. (The reader is referred to Chapter IV for detailed discussions of each of these elements.) It was designed and built by Lockheed Missiles and Space Company and Hughes Danbury Optical Systems (formerly Perkin-Elmer) under a NASA contract. For the purposes of discussion, the satellite has been broken down into the following sections, which are illustrated in Figure 21: the aperture door, light shield, forward shell, optical telescope assembly, support system module equipment section, optical telescope assembly equipment section, aft shroud, solar array, and external components (latches and drives, sun sensors, low-gain antennas, magnetic sensing systems, magnetic torque bars, and high-gain antennas).

The thermal control system maintains all component and structure temperatures within allowable limits under all required mission conditions, including normal operation, orbit maintenance while docked with the Space Shuttle, and safemode hold. The satellite is in a 28.5-deg inclination, circular orbit at altitudes that range from 398 to 593 km. The orbit beta angle varies between $\pm$52 deg and eclipse time ranges from 26 to 36 minutes, as shown in Figure 22. There are some attitude restrictions to prevent adverse solar illumination conditions on the telescope.

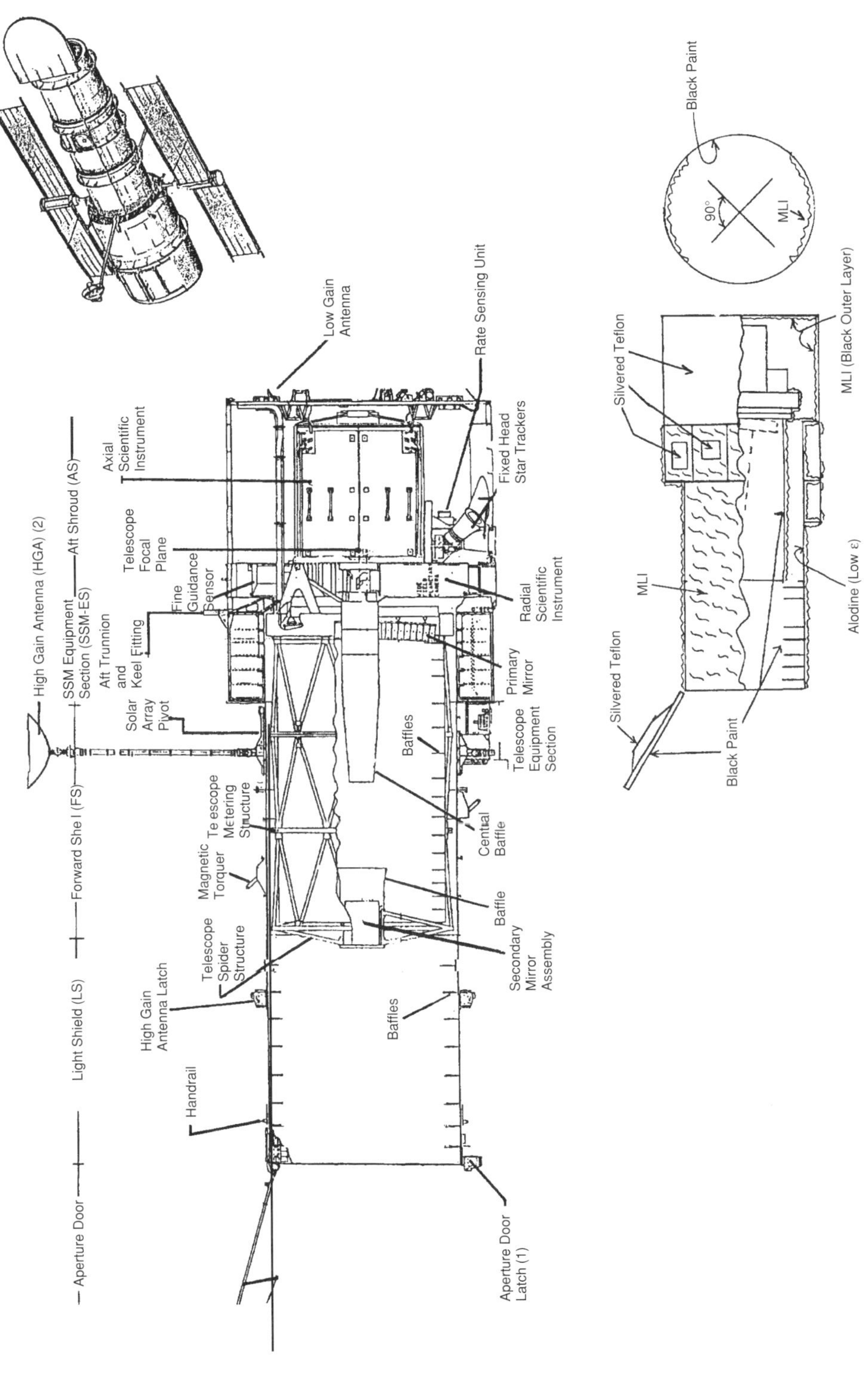

Figure 21. Hubble Space Telescope

The general approach was to keep the thermal design as passive as possible. To minimize sensitivity to the wide range of sun angles, the vehicle external surface is almost entirely covered with MLI with a low α/ε silvered Teflon or aluminized Teflon outer layer. Low contamination materials were used in constructing the vehicle, and venting paths were designed to prevent contamination of thermal and optical surfaces.

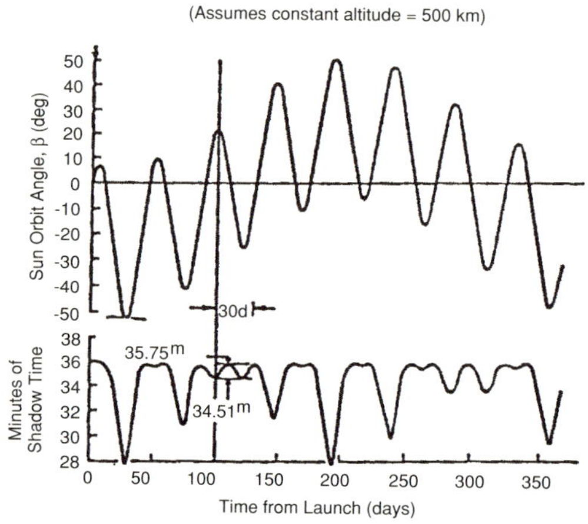

Figure 22. Space Telescope beta angle and eclipse time (Lockheed)

The Aperture Door

The aperture door is a honeycomb structure 1.5 in. thick. The core is aluminum honeycomb (3.8-in. cells, 1.6 lb/ft^3, and 1/4-in. cells, 3.4 lb/ft^3), and the facesheets are 0.012-in.-thick aluminum.

The surface finish of the door that faces the optics and never sees the sun is a glossy black paint (glossy black Chemglaze Z302) as required by the telescope straylight analysis. The outer surface is covered with aluminized Teflon tape to minimize temperatures and gradients with full solar heating. The aperture door has one flight-temperature sensor located at the center of the outer surface.

The aperture door has a radiative coupling with the telescope, and the orbit average temperature must be maintained below 33 deg C for the hot case and above -90 deg C for the cold case. The passive thermal design of the door has maintained its temperature within these limits.

The Light Shield

The light shield is the 10-foot diameter, 13-foot long forward portion of the barrel structure in front of the telescope. It has eight internal baffles for straylight control as well as a baffle at the forward end. The baffles and internal surface of the light shield are coated with an optical black paint (flat black Chemglaze Z306) as required by the telescope straylight analysis. The α/ε of the black paint is 0.95/0.92. The external surface of the light shield is covered with MLI blankets (outer layer aluminized Telfon, 15 layers of 1/3-mil embossed double aluminized Kapton, and an inner layer of 1-mil single aluminized Kapton). An effective emittance of 0.02 has been used for the MLI blankets. The MLI blankets are mounted on the structure to reduce the structural temperature variation, and also function as part of the meteoroid protection system. There are eight flight-temperature sensors on the light-shield structures. There is an orbit-average-temperature requirement of -33 deg to -59 deg C, which is met with this design.

The Forward Shell

The forward shell is the 10-foot diameter, 10-foot long cylinder that encloses the telescope assembly. The forward shell internal surface finish is alodine with an emittance of approximately 0.15. The external surfaces are covered with MLI blankets identical to the light-shield MLI blankets (outer layer of aluminized Teflon, 15 layers of 1/3-mil embossed double aluminized Kapton, and an inner layer of 1-mil single-aluminized Kapton). The MLI covers the external rings except for the structural ring at station 358, which is covered with aluminized Teflon. The forward shell has eight flight-temperature sensors. Temperature of the forward shell is maintained between -23 deg to -53 deg C on an orbit-average basis.

The Support System Module Equipment Section

The support system module equipment section (SSM-ES) consists of an annular ring of compartments surrounding part of the telescope, as shown in Figure 21. It has outside and inside diameters of approximately 14 feet and 10 feet, respectively, and is 5-feet long. The ring is broken into 12 compartments that house various electronics boxes, as shown in Figure 23.

The equipment section thermal control subsystem is designed to control temperatures of all its internal components, control temperatures and gradients of the structure that interfaces with the telescope and scientific instruments, and control conduction heat transfer through the attachments. The thermal design is primarily passive using insulation (MLI), low α/ε surface properties, component locations and mounting configurations, and augmented with thermostatically controlled heaters. Additionally, louver assemblies are used on the two battery-bay doors to conserve heater power.

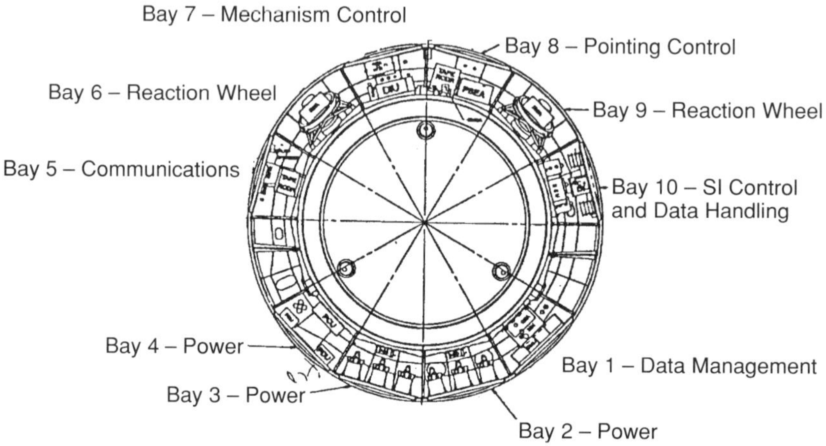

Figure 23. Support systems module equipment section

The design approach is to cover all surfaces of the equipment section with MLI except for some radiating areas on the "doors" of the equipment compartments, as shown in Figure 24. The radiator areas are covered with silvered Teflon for high emittance and low solar absorbtance. MLI also covers the equipment-section surfaces facing the telescope and the science instruments in the aft shroud area to limit thermal interactions with those components. In addition, some of the equipment-section compartments are thermally isolated from one another with MLI, as shown in Figure 24.

The majority of the electronic components are mounted on the honeycomb doors, except for Bays 6, 9, 11, and 12, which do not have honeycomb doors. The battery-bay doors (Bay 1 & 2) and the communications bay door (Bay 5) have additional aluminum doubler plates (0.12 in.) bonded on the internal door surfaces under the components for better heat distribution. All other components are mounted directly on the door or mounted on the structure at the bottom of the bay called the "tunnel structure." There are structural beams along the bottom of the bay with additional members to mount components. The reaction wheel assemblies (RWA) have separate mounting structures to provide the correct orientation.

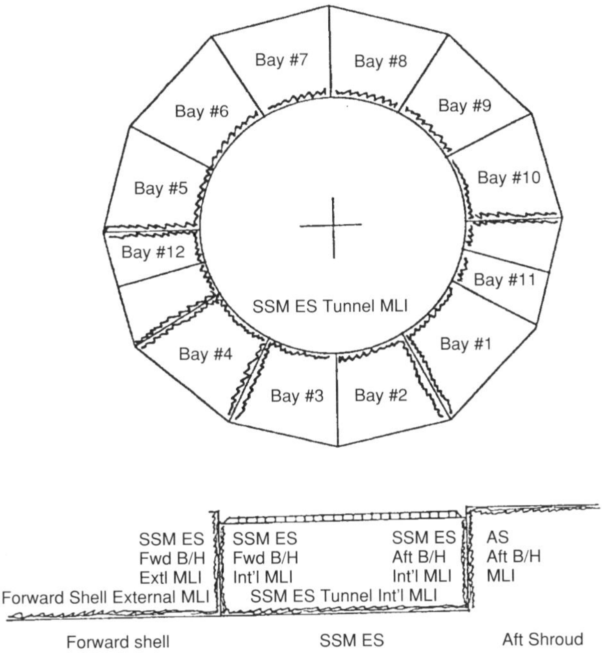

Figure 24. SSM equipment section MLI

Thermostatically controlled heaters are used if needed to maintain minimum temperatures during normal component operation, and survival temperatures when not operating. The batteries have integral internal heater systems. The RWAs have heaters mounted on them, and the remaining Equipment Section heaters are mounted on the doors or component mounting structures. There are sixteen primary heater circuits for the Equipment Section. They are as follows: one circuit for the computer in Bay 1; six circuits, one for each battery in Bay 2 and 3; one circuit for Bay 4; one circuit for the tape recorders in Bay 5; one circuit for the communication equipment in Bay 5; one circuit for each of the four RWAs in Bay 6 and 9; one circuit for the Bay 7 and Bay 8 door heaters, plus the tape recorder in Bay 8; and one circuit for the science-instrument electronics trays in Bay 10. There are also 16 secondary (backup) heater circuits. Several heater circuits control more than one heater system, i.e., the Bay 5 tape-recorder circuit has two tape-recorder heater systems; the communication circuit has two separate heater systems on the tray; each RWA circuit has separate heater systems for the inboard and outboard bearing; Bay 7 and Bay 8 are on one heater circuit, with heater systems on the Bay 7 door, Bay 8 door, and the tape-recorder mounting structure; and the science instrument electronics circuit has two heater systems on the tray.

Table 1. Space Telescope heaters

			Heaters								Thermostats		
LOCATION	TYPE	SIZE in x in	# of Htrs per Strip	# of Htr Strips	# of Htrs	Nom Resis Ω @25˚C ±3%	Each Htr Watts @ 28V	Total Watts @28V	Close (˚F)	Open (˚F)	Close (˚F)	Open (˚F)	# Req'd
Bay 1 DF224	Primary	0.5x8.0	1	2	2	50.7	15.46	30.9	18±2	15 to 21	-9±1	-9 to -6	2
Computer	Secondary		1	2	2		15.46	30.9	11±2	11 to 18	-12±1	-12 to -9	2
Bay 4 Door	Primary	1.0x10.0	2	4	4	152.0	5.16	20.6	-5±5	10±5	-21±3	-12±3	2
(PDU/PCU)	Secondary				4		5.16	20.6	-15±5	0±5	-26±3	-18±3	2
Bay 5	Primary	1.5x7.0	2	4	4	50.7	15.46	51.9	20±2	20 to 27	-7±1	-7 to -3	4
Comm tray	Secondary	2.0x5.0			4	50.7	15.46	51.9	15±2	16 to 21	-9 ±1	-9 to -6	4
Tape recorder	Primary	.75x10.0	2	6	6	101.4	7.73	46.4	25±2	25 to 30	-4±1	-4 to -1	6
1, 2, 3	Secondary				6		7.73	46.4	20±2	20 to 27	-7±1	-7 to -3	6
Bays 6 & 9	Primary	0.5x17.2	2	8	8	50.7	15.46	123.7	10±2	10 to 17	-12±1	-12 to -8	20
RWA (1-4)	Secondary	0.5x26.4			8	50.7	15.46	123.7	5±2	5 to 10	-15±1	-15 to -12	20
Bay 7 Door	Primary	1.0x10.0	2	2	2	76.0	10.32	20.6	7±2	7 to 12	-14±1	-14 to -11	2
	Secondary				2		10.32	20.6	3±2	3 to 8	-16±1	-16 to -13	2
Bay 8 Door	Primary	1.0x10.0	2	2	2	76.0	10.32	20.5	-5±5	10±5	-21±3	-12±3	2
	Secondary				2		10.32	20.5	-15±5	0±5	-26±3	-18±3	2
Bay 10	Primary	1.25x7.5	2	4	4	27.6	28.41	113.5	20±2	20 to 27	-7±1	-7 to -3	8
Si&DH Tray	Secondary				4		28.41	113.5					8
Magnetometer	Primary	1.0x4.0	2	2	4	209.1	3.75	15.0	-15±5	0±5	-26±3	-18±3	4
(MSS 1&2)	Secondary				4		3.75	15.0	-25±5	-10±5	-32±3	-23±3	4
Coarse Sun	Primary	1.0x1.0	2	5	5	192.0	4.08	20.42	-25±5	-10±5	-32±3	-23±3	10
Sensor (1 to 5)	Secondary				5		4.08	20.42	-30±5	-15±5	-34±3	-25±3	10
FHST Δ Plate	Primary	1.75x2.0	2	9	9	94.1	8.33	75.0	30±2	30 to 37	-1±1	-1±3	6
(1, 2, & 3)	Secondary				9		8.33	75.0	27±2	27 to 32	-3±1	-3±0	6
Mechanisms	Primary	.88x12.1	2	1	1	78.4	10.0	10.0	-15±5	0±5	-28±3	-18±3	2
AD Hinge Min	Secondary				1		10.0	10.0	-25±5	-10±5	-23±3	-23±3	2
AD Passive	Primary	.88x11.0	2	1	1	52.5	14.93	14.9	-60±5	-44±5	-51±3	-42±3	2

Table 1. Space Telescope heaters (continued)

		Heaters							Thermostats				
LOCATION	TYPE	SIZE in x in	# of Htrs per Strip	# of Htr Srips	# of Htrs	Nom Resis Ω @25˚C ±3%	Each Htr Watts @ 28V	Total Watts @28V	Close (˚F)	Open (˚F)	Close (˚F)	Open (˚F)	# Req'd
Hinge	Secondary				1		14.93	14.9	-73±5	-57±6	-58±3	-49±3	2
AD Active	Primary	.88x12.1	2	1	1	78.4	10.0	10.0					
Hinge	Secondary				1		10.0	10.0					
AD Latch	Primary	.88x12.1	1	1	1	78.4	10.0	10.0	-25±5	-10±5	-32±3	-23±3	2
Motor													
HGA Hinge	Primary	.88x12.1	1	2	2	78.4	10.0	20.0	-25±5	-10±5	-32±3	-23±3	4
Motor													
HGA Latch	Primary	.88x12.1	1	2	2	78.4	10.0	20.0	-25±6	-10±5	-32±3	-23±3	4
Motor													
HGA TAG	Primary	vendor	1	4	4	114.0	5.88	27.5	56±2	55 to 71	19±2	19 to 22	8
(Gimbals)	Secondary	supplied	1	4	4		5.88	27.5	61±2	61 to 66	16±2	16 to 19	8
S/A Latch	Primary	.88x12.1	1	4	4	78.4	10.0	40.0	-25±5	-10±5	-32±3	-23±3	8
Motor													
Misc:													
SADM Survival	Primary		1	2	2	118.0	6.64	13.3	-45.4	-20.2	-43.0	-29.0	2
(±Wing)	Secondary		1	2	2		6.64	13.3					
PDM Survival	Primary		1	4	4	240.0	3.27	13.1	-45.4	-20.2			2
(±Wing)	Secondary		1	4	4		3.27	13.1			-43.0	-29.0	
SDM Survival	Primary		1	2	2	88.7	8.84	17.7	-20.2	5.0	-29.0	-15.0	4
(±Wing)	Secondary		1	4	4	153.0	5.12	20.5					
Diode Tray	Plus Wing		1	10	10	14.9	2.10	21.0	-45.4	-20.2	-43.0	-29.0	4
2 strings of 5	Minus Wing		1	10	10	14.9	21.0	21.0			-43.0	-29.0	4
per (±) Wing													
Battery Bay 2	Primary				6		40.0	240.0	29 to 34	30 to 38	-2 to 1	-1 to 3	
Type 44	Secondary								18 to 23	19 to 27	- 8 to - 5	-7 to -3	
Battery Bay 3	Primary				6		40.0	240.0	29 to 34	30 to 38	-2 to 1	-3 to 3	
Type 44	Secondary								18 to 23	19 to 27	-8 to -5	-7 to -3	
Battery Bay 2	Primary				6		40.0	240.0	28 to 32	29 to 36	-2 to 0	-2 to 2	
NiH2	Secondary								23 to 27	24 to 31	-5 to -3	-4 to -1	
Battery Bay 3	Primary				6		40.0	240.0	28 to 32	29 to 36	-2 to 0	-2 to 2	
NiH2	Secondary								23 to 27	24 to 31	-5 to -3	-4 to -1	

(*) Heater strip has dual heaters (Primary + secondary)

Each heater system, both primary and secondary, has two thermostats wired in series with their heater elements. Each heater system is thereby protected against an open heater, thermostat, or wire. A failed closed thermostat in either the primary or secondary system is backed up by the second thermostat wired in series. It requires two independent failures to disable these heater systems. The primary and secondary heater elements can be on the same strip. If there is a second heater strip, the primary heaters are wired in parallel with the secondary heaters. A list of all Space Telescope heaters is given in Table 1. These heaters are all enabled prior to launch and their status was verified with the first available telemetry data received during deployment operations from the Space Shuttle.

Many of the electronic components in the Equipment Section have no internal temperature sensors. To provide temperature data for these components, there are a total of twenty temperature sensors placed near the interface of these components and their respective mounting structures.

The following sections give descriptions of the thermal design of selected bays in the equipment section. Bay 1 is representative of a typical electronics-box bay with fairly wide allowable temperatures, and uses radiator area on the door, MLI, and heaters to achieve thermal control. Bays 2 and 3 contain NiH_2 batteries, which must be controlled within a relatively narrow temperature range (-5 deg to +20 deg C). MLI, heaters, radiator areas, louvers, and aluminum doublers are used on these bays. Bays 7 and 8 have relatively low levels of electronic waste heat to dissipate and therefore have no radiators. These bays rely on MLI and heaters to keep components within limits.

Bay 1

The DF 224 computer, data management unit (DMU), and one of the two gimbal electronics assemblies (GEA) are located in Bay 1. The DMU is mounted on the door and both the computer and GEA are mounted on the tunnel structure. Figure 25 shows the location of components and monitors in Bay 1. A summary of the thermal characteristics of Bay 1 follows.

Percentage door radiator/MLI: 100 percent silvered Teflon radiator, 0 percent MLI.

Thermostats: There are two primary thermostats wired in series, as are the two secondary thermostats.

Set points	Open	Close
Primary	-9 deg C to -6 deg C	-9 deg C ± 1 deg C
Secondary	-12 deg C to -9 deg C	-12 deg C ± -1 deg C

Heater system: The two primary heaters are wired in parallel, as are the two secondary heaters. There are a total of 4 strips, each at 15.46 watts at 28 V; therefore the primary heaters, as well as the secondary heaters, can supply a total of 30.9 watts. These heaters are located at the computer mounting structure.

Temperature limits:	Operating	Non-Operating
DF 224 Computer	-18/49 deg C	-54/57 deg C
DMU	-40/35 deg C	-50/55 deg C
GEA	-29/60 deg C	-43/60 deg C

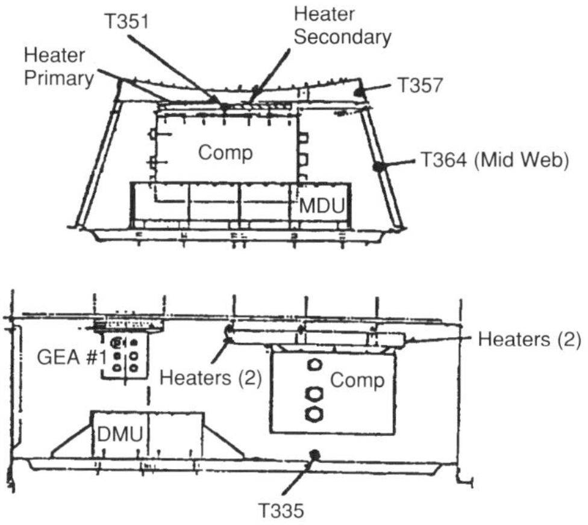

Figure 25. SSM equipment section Bay 1

Bays 2 and 3

Nickel hydrogen (NiH_2) batteries, charge current controllers (CCC), one of the four data interface units (DIU), and oscillators are located in Bays 2 and 3. One NiH_2 battery module, which contains three NiH_2 batteries, is located in Bay 2 and one module in Bay 3. Each module is mounted on the inner door surface, which has an aluminum plate (0.12 in.) bonded to the surface. There are three CCCs mounted on the tunnel structure in both Bays 2 and 3. The two Oscillators are mounted on the tunnel structure in Bay 2 and the DIU is mounted on the tunnel structure in Bay 3. Figure 26 presents the location of components and monitors in Bay 2. A summary of the thermal characteristics of the battery bays follows.

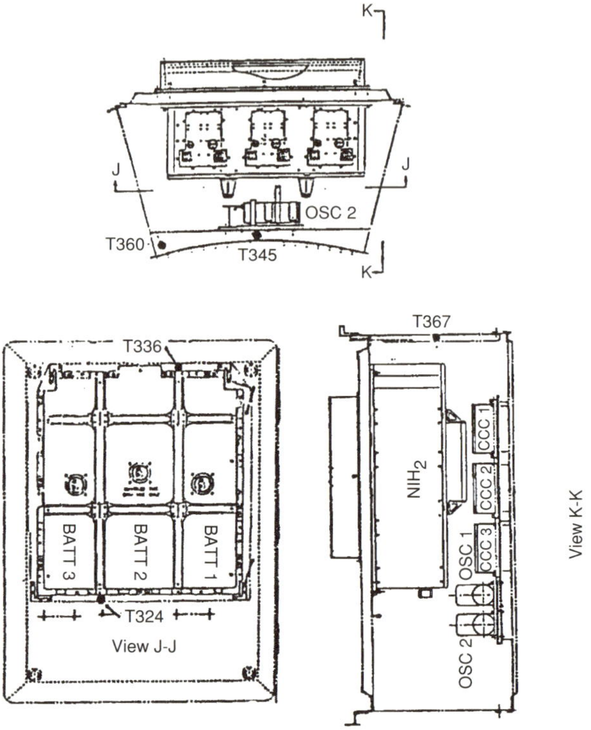

Figure 26. SSM equipment section Bay 2

Percentage door radiator/MLI:	Bay 2:	63 percent silvered Teflon radiator, 37 percent MLI.
	Bay 3:	69 percent silvered Teflon, 31 percent MLI.

Thermostats: Each battery has primary and secondary thermostats located on cells 8 and 10.

Set points	Open	Close
Primary	-2 deg C to 2 deg C	-2 deg C to 0 deg C
Secondary	-4 deg C to -1 deg C	-5 deg C to -3 deg C

Heater system: Individual primary and secondary heater patches are located on each individual battery cell sleeve, within each battery. Both heater sets are wired in parallel. Battery primary and secondary heaters are rated at 40 watts each. Total power for six battery primary heaters is 240 watts. Similarly, there is a total of 240 watts for the six secondary heaters. A schematic for these heaters is shown in Figure 27.

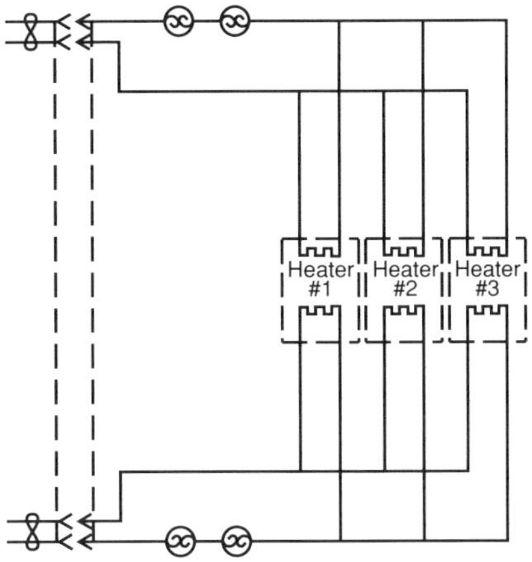

Figure 27. Battery heaters

Temperature limits:	Operating	Non-Operating
Batteries (NiH_2)	-5/20 deg C	-20/38 deg C
Clk Osc Inner	62.5/67 deg C	-60/45 deg C
DIU #2	-40/60 deg C	-40/60 deg C

Bays 7 and 8

Bay 7 contains the solar array drive electronics (SADE), deployment control electronics (DCE), one of the four data interface units (DIU), one of the two gimbal electronics assemblies (GEA), and the mechanism control unit (MCU). The two SADEs, DCE, and GEA are mounted on the inner door surface. The MCU and DIU are mounted on the tunnel structure. Bay 8 contains the instrument control unit (ICU), retrieval mode gyro assembly (RMGA), pointing and safemode electronics assembly (PSEA), magnet torque electronics (MTE), and one of the three tape recorders (T/R). The ICU is mounted on the inner door surface. The RMGA, PSEA, MTE (monitors located internally to the PSEA), and T/R are mounted on the tunnel structure. Figures 28 and 29 present the location of components and monitors in Bays 7 and 8. A summary of the thermal characteristics of Bays 7 and 8 follows.

Percentage door radiator/MLI:	Bay 7:	100 percent MLI.
	Bay 8:	100 percent MLI.

Thermostats: There are two primary thermostats wired in series as are the two secondary thermostats.

	Set points	Open	Close
Bay 7:	Primary (Door)	-14 deg C to -11 deg C	-14 deg C $\pm$ 1 deg C
Bay 7:	Secondary (Door)	-16 deg C to -13 deg C	-16 deg C $\pm$ 1 deg C
Bay 8:	Primary (Door)	-12 deg C $\pm$ 3 deg C	-21 deg C $\pm$ 3 deg C
Bay 8:	Secondary (Door)	-18 deg C $\pm$ 3 deg C	-26 deg C $\pm$ 3 deg C
Bay 8:	Primary (T/R)	-4 deg C to -1 deg C	-4 deg C $\pm$ 1 deg C
Bay 8:	Secondary (T/R)	-7 deg C to -3 deg C	-7 deg C $\pm$ 1 deg C

Heater system:

Bay 7

Two primary heaters are wired in parallel as are the two secondary heaters. Primary and secondary heaters are bonded onto a single strip. One heater strip is placed on the door between the two SADEs and the other strip between the DCE and GEA. Primary and secondary heaters are rated at 10.32 watts each. Total primary heater power is 20.6 watts, and the same is true for the secondary heaters.

Bay 8

Tray
Two primary heaters are wired in parallel, as are the two secondary heaters. Primary and secondary heaters are bonded onto a single strip. The two heater strips are placed on the door on opposite sides (along the V1 axis) of the ICU. Primary and secondary heaters are rated at 10.32 watts each. Total primary heater power is 20.56 watts, and the same is true for the secondary heaters.

Tape Recorder
Primary heaters are wired in parallel as are the secondary heaters. Primary and secondary heaters are bonded onto a single strip. A total of two strips are placed on the mounting bracket adjacent to the tape recorder. Primary and secondary heaters are rated at 7.73 watts each. Total Bay 8 tape recorder primary heater power is 15.5 watts, and the same is true for the secondary heaters.

Temperature limits:	Operating	Non-Operating
DIU #4 (Bay 7)	-40/60 deg C	-40/60 deg C
SADE (Bay 7)	-34/60 deg C	-34/60 deg C
DCE (Bay 7)	-34/60 deg C	-34/60 deg C
GEA (Bay 7)	-29/60 deg C	-43/60 deg C
MCU (Bay 7)	-40/60 deg C	-60/60 deg C
PSEA (Bay 8)	-12/54 deg C	-12/54 deg C
Tape Recorder (Bay 8)	-12/43 deg C	-40/43 deg C
ICU (Bay 8)	-30/60 deg C	-60/60 deg C

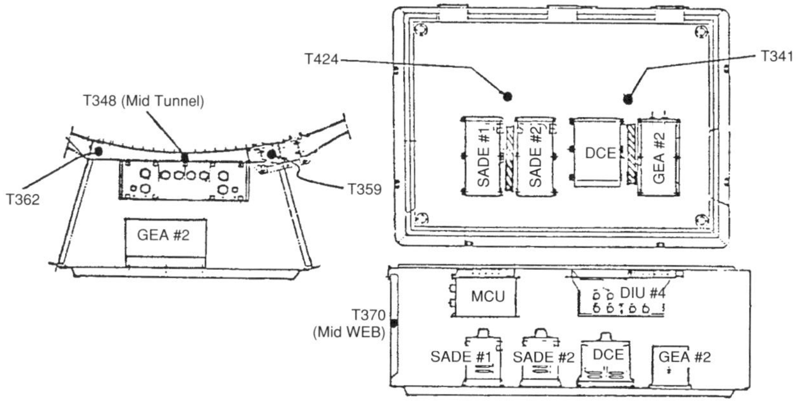

Figure 28. SSM equipment section Bay 7

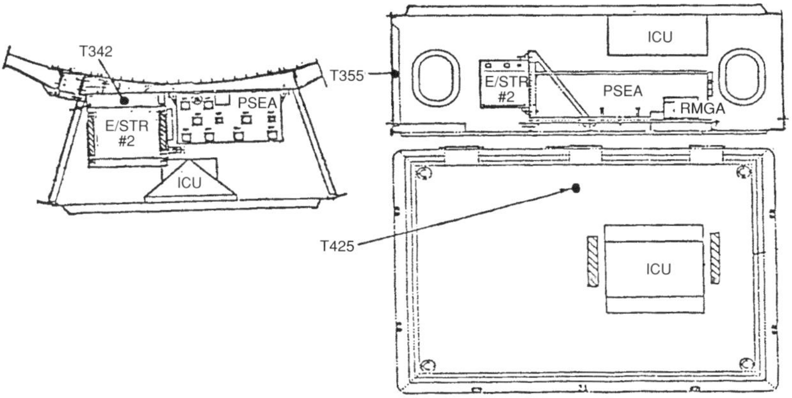

Figure 29. SSM equipment section Bay 8

Tunnel Structure

The six temperature sensors on the tunnel structure are isolated from the bays by the internal MLI blankets and will not respond rapidly to local power-level changes. Each of the sensors are placed approximately in the center of their respective panels. Figure 30 presents the location of the six thermistors located on the tunnel structure.

SSM Equipment Section Forward and Aft Bulkhead Structure

The equipment section forward bulkhead and aft bulkhead structures are covered with MLI on both forward and aft surfaces. The forward bulkhead external MLI is identical to the light-shield and forward-shell MLI, with an outer layer of aluminized Teflon except for the area covered by the optical telescope equipment section, which has an outer layer of DAK (double aluminized kapton). Both the forward and aft bulkhead MLI within the equipment section have outer layers of DAK. The aft bulkhead MLI outer surfaces facing the aft shroud components have an outer layer of black Kapton for straylight control. Figure 31 presents the locations of the six sensors on the forward bulkhead and the five sensors on the aft bulkhead. These are generally located close to the center of the bay. There are six additional sensors on the aft bulkhead to monitor the equipment section side of the telescope axial and tangential links.

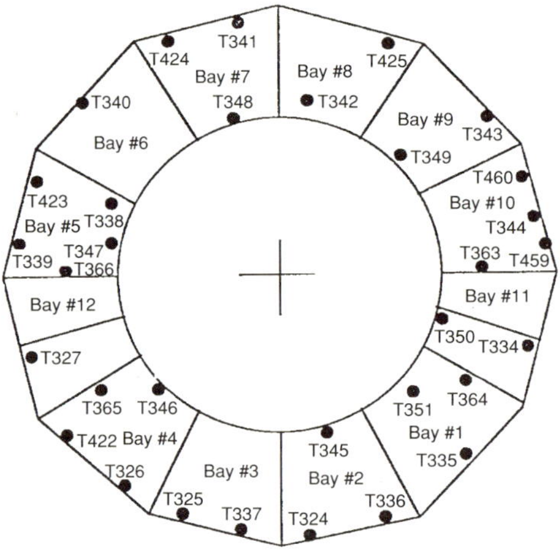

Figure 30. SSM equipment section tunnel structure thermistors

Optical Telescope Assembly Equipment Section

The optical telescope assembly has an equipment section in which its electronics are mounted. It consists of several compartments forming a section of a ring around the forward shell, just forward of the support system module equipment section, as shown in Figure 21. This OTA equipment section is shown in Figure 32 and consists of:

Bay A - (Empty)
Bay B - Data interface unit 1 (DIU 1)
Bay C - Optical control electronics (OCE)

Bay D - Fine guidance electronics 1 (FGE 1)
Bay E - Actuator control electronics (ACE)
Bay F - Fine guidance electronics 2 (FGE 2)
Bay G - Fine guidance electronics 3 (FGE 3)
Bay H - Electrical power/thermal control electronics (EP/TCE)
Bay J - (Empty)

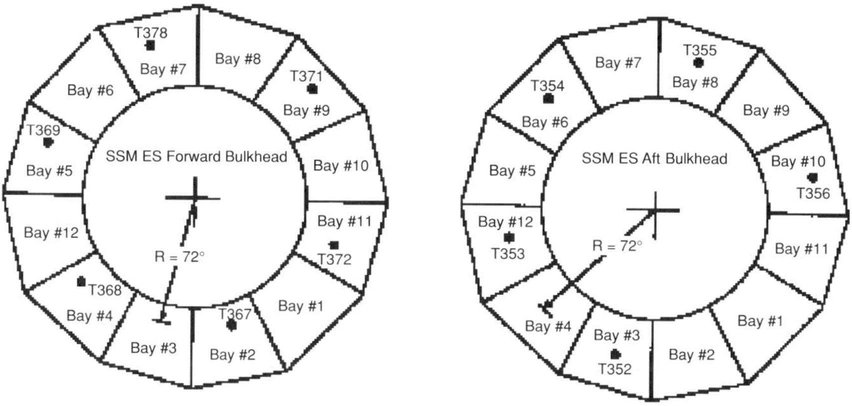

Figure 31. SSM equipment section bulkhead thermistors

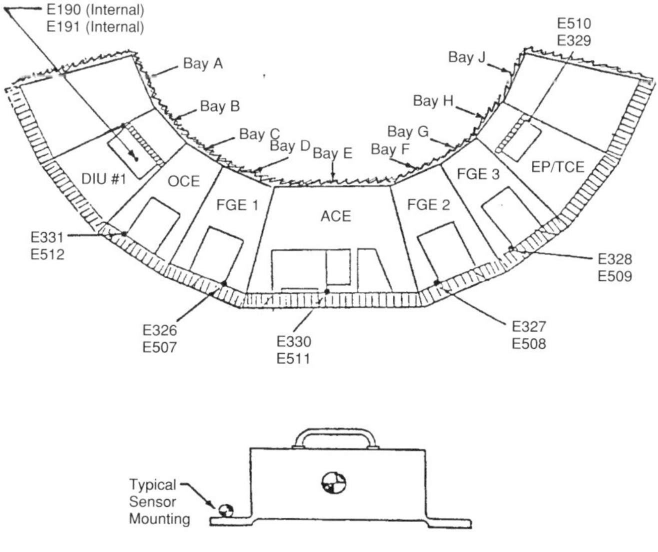

Figure 32. OTA equipment section

The OTA equipment section has a cold environment since it is located on the bottom of the vehicle and is therefore shielded from direct solar in all normal vehicle orientations.

The OTA equipment section has both active (heaters and thermostats) and passive (MLI and surface finish) thermal-control designs. Figure 33 presents the MLI pattern for each of the bay doors. The three FGEs and the OCE have heaters. The DIU and EP/TCE do not have heaters since they are always operating and do not drop below their minimum turn-on temperature of -40 deg C. The ACE also does not have heaters, since both the FGEs surrounding the ACE normally operate and also have their own heaters, to maintain the FGEs above their turn-on temperature. All heaters are located on the supporting rail. Each of the boxes have 2 thermistors and are located internally. The heaters are required for safemode operation to maintain acceptable turn-on temperatures.

All OTA equipment section electronic boxes are painted black except the EP/TCE, which is covered with MLI on three sides, to help retain heat. All boxes have the same temperature limits except for the DIU, which has different limits as listed below:

Surface properties: $\varepsilon > 0.85$ except EP/TCE has MLI on three sides.

Temperature Limits:	Operating	Non-Operating
DIU	-40/38 deg C	-40/60 deg C
All other	-23/35 deg C	-55/85 deg C

Minimum Turn-on: -40 deg C.

Thermostats: There are two primary thermostats in series and the two heaters are wired in parallel.

Set points	Open	Close
Primary	-28.9 deg C	-40 deg C
Secondary	-28.9 deg C	-40 deg C

Heater system: Two heater strips per box, with each strip containing both a primary and a secondary heater. The heater powers are as follows:

Box	Primary Heater @ 28 V	Secondary Heater @ 28 V
OCE	4.9 W	4.96 W
FGE 1	49.6 W	49.3 W
FGE 2	44.3 W	43.8 W
FGE 3	35.6 W	35.2 W

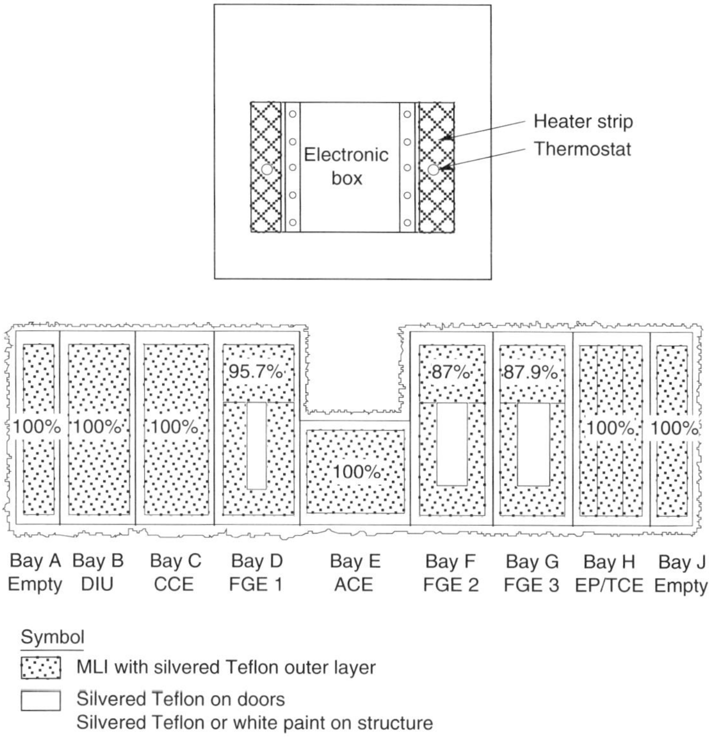

Figure 33. OTA equipment section thermal finishes

Aft Shroud

SSM aft shroud (AS) is a 14-foot diameter, 12-foot-long cylindrical section at the rear of the vehicle that encloses the three fine guidance sensors (FGS), the wide field planetary camera (WFPC), four axial scientific instruments (HSP, HRS, FOS, and FOC), the telescope focal plane structure (FPS), and a shelf with three RSUs and three FHSTs mounted on it (see Figures 21 and 34). The aft-shroud thermal design consists of the use of silvered Teflon on all external surfaces except for the FHST light shades and the astronaut handrails. All internal surfaces are black for straylight control. The internal top and bottom 90 deg cylindrical sections, as well as the entire internal aft bulkhead, are covered with black Kapton outer-layer MLI blankets. The side 90 deg surfaces are not covered with MLI and are used as radiators to reject heat from the internal instruments to space. These surfaces have black radiation shields on the inside face to control radiative couplings to the internal instruments.

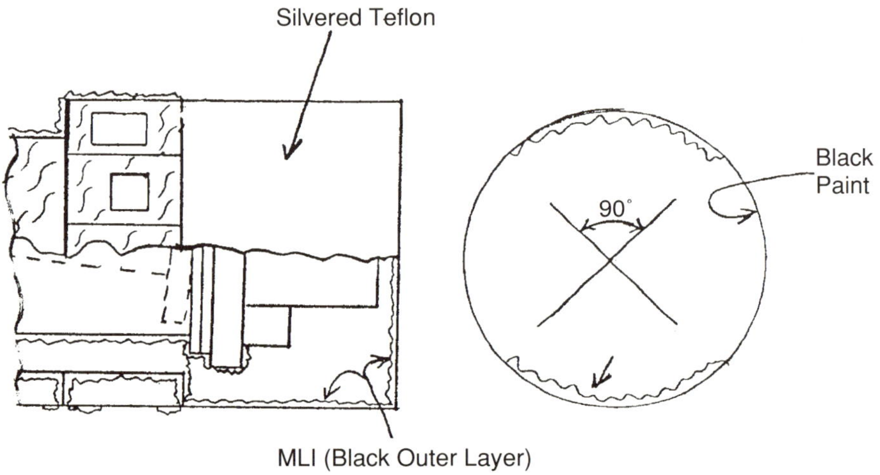

Figure 34. Aft shroud thermal finishes

The thermal interface between the aft shroud and the scientific instruments was difficult to establish and verify. Effective sink temperatures were established as the method for this thermal interface to allow the instrument contractors to perform their analyses, design, and test. These sink temperatures allowed the instrument to interface with the aft shroud by using only 3 temperatures instead or the actual radiation couplings to the hundreds of nodes in the shroud. The sink temperatures were calculated using the complete math-model radiation couplings and all the temperatures in the aft shroud. Unfortunately, flight-temperature monitors are not located on all the node points used for the sink-temperature calculations. Accordingly, an algorithm was constructed that weighted the node points that had flight sensors.

Optical Telescope Assembly

The telescope assembly is the primary payload and consists of a number of components including the optics, their support structure, baffles, electronics, and the science instruments at the focal plane, as shown in Figure 35. This entire assembly attaches to the SSM equipment section and is enclosed by the light shield, forward shell, aperture door, and aft shroud (Figure 21), which act as a thermal "cocoon" to isolate the telescope assembly from the thermal variations of the external environment.

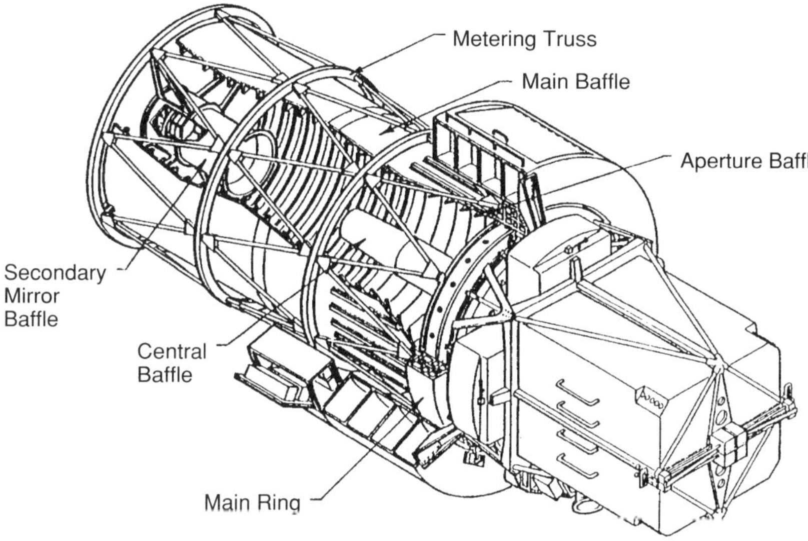

Figure 35. Optical telescope assembly

The dominant requirements that drove the thermal design of the telescope assembly were the .003 arc-sec (RMS) pointing stability over a 24-hour period, and maintaining optical wave front performance better than $\lambda/20$ (RMS). This optical requirement places strict limits on thermo-mechanical distortions of the optical mirrors and their supporting structure. These distortion limitations, in turn, call for very tight control of temperatures and temperature gradients.

The thermal design approach selected to meet these requirements was to insulate most structures with MLI or low-emittance surface finishes, provide conductive isolation at mechanical attachments, and use a large number of small heaters with very small dead bands to control temperatures at precisely 70 deg F. The following sections describe the thermal design of each of the telescope components shown in Figure 35.

The Main Ring

The main ring is the primary structural member to which all of the other telescope components are attached and which, in turn, attaches the telescope assembly to the SSM at the SSM equipment section through three tangential and three axial links. Ring temperatures are controlled using 36 heaters with a set point of 70°F and a control band of only $\pm$ 0.2°F, as shown in Figures 36 and 37. Heater powers and the effects of surrounding temperature variations are minimized by wrapping the entire ring and the ring-to-SSM attachment links in MLI with an $\varepsilon^* < .01$, and limiting the conductance at all of the attach points shown in Figure 36 to very small values, using low-conductivity materials where required. There are also a number of cables that pass through or are attached to the ring. Thermal interactions with these cables are minimized by wrapping them in MLI or low-emittance gold tape and attaching them to the ring with low conductance stand-offs. There are a total of 23 flight-temperature sensors on the ring.

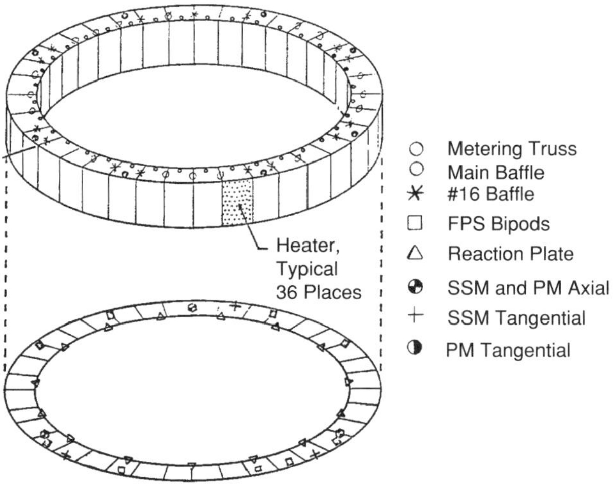

Figure 36. OTA main ring mounting points and heaters

The Primary Mirror Assembly

The primary mirror assembly (see Figure 38) consists of the primary mirror, the reaction plate, the mirror-to-reaction-plate mounts, and the mirror-figure control actuators. Mirror-figure distortion and mirror displacement relative to the main ring due to thermo-mechanical effects are the principle drivers in the thermal design. The design approach is to provide radiative isolation by wrapping the entire assembly in MLI ($\varepsilon^* < .01$) except, of course, for the front face of the mirror, which has a very low emittance of .01 to .03. (A value of .02 was used for design, .03 for heater sizing). The assembly is radiatively isolated from the baffle that passes through the central hole (see Figure 39) by gold tape or MLI and a guard heater. The guard heater system, shown in detail in Figure 40, reduces heat flow through the MLI by maintaining a very small ΔT between the mirror and the guard-heater plates. This effectively shields the mirror MLI from seeing the central baffle, which can get very cold due to its radiative view to space through the telescope aperture. There is also MLI between the guard-heater plate and the control baffle to reduce the heater power required to drive the guard-plate heaters. Conduction heat losses to the main ring and central baffle are controlled by designing low-conductance mountings between the reaction plate and the main ring, and between the reaction plate and the central baffle.

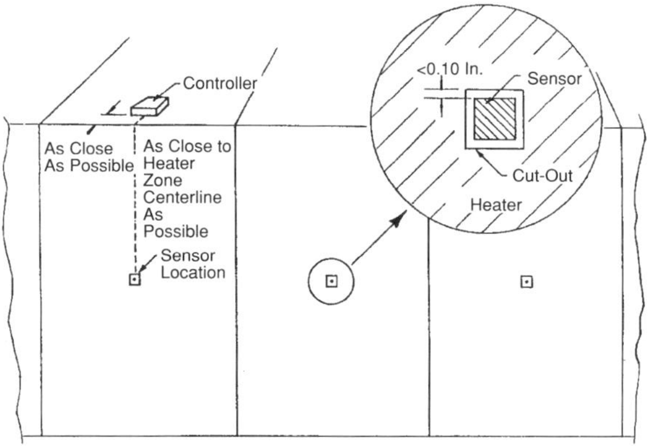

Figure 37. OTA main ring heater detail

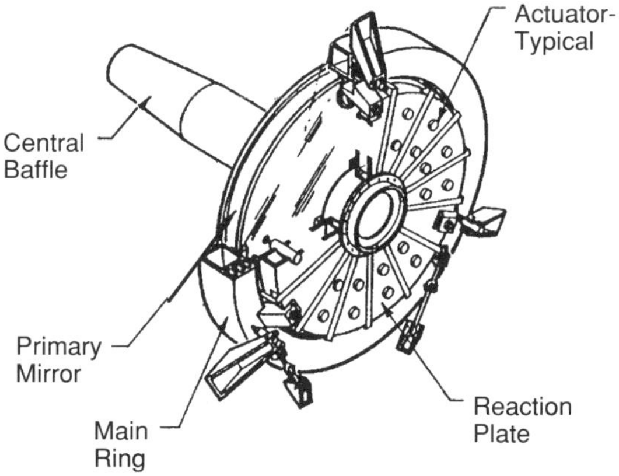

Figure 38. Primary mirror assembly

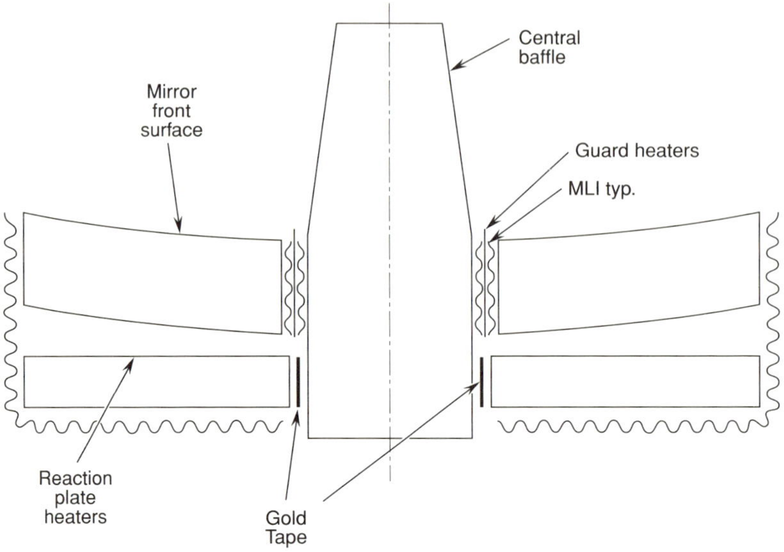

Figure 39. Primary mirror assembly central baffle

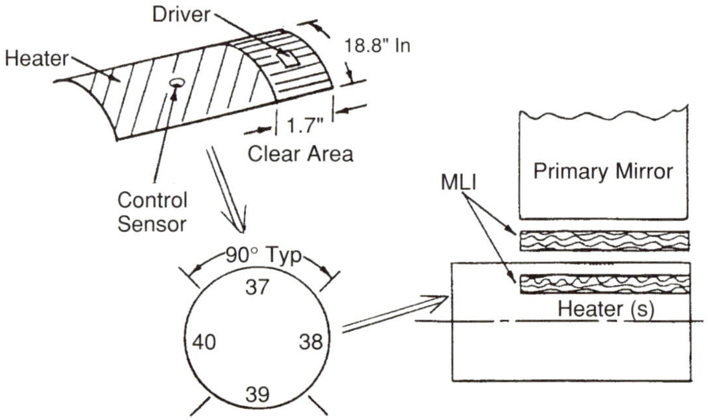

Figure 40. Central baffle guard heater

The front of the reaction plate and the rear of the mirror are both high emittance and form a radiant-interchange cavity. Temperatures of both are maintained by 36 precision-controlled heaters, with set points of 70 deg F and differentials of ± 0.2 deg F, mounted to the reaction plate, as shown in Figure 41. Heat from the reaction plate is then radiated to the mirror. Because of the high emittance of the mirror back face and very low emittance of its front face, its temperature follows that of the reaction plate and is not strongly influenced by the view to deep space or to the telescope enclosure. During extended non-operating periods, the 36 precision heaters are turned off and 18 back-up heaters on thermostats (50 deg F to 68 deg F dead band) are turned on. To minimize the radiative effects of the 36 figure-control actuators and the three axial links, they are covered with MLI or low-emittance finishes. Contact conductances between the mirror and the actuators and axial links are minimized.

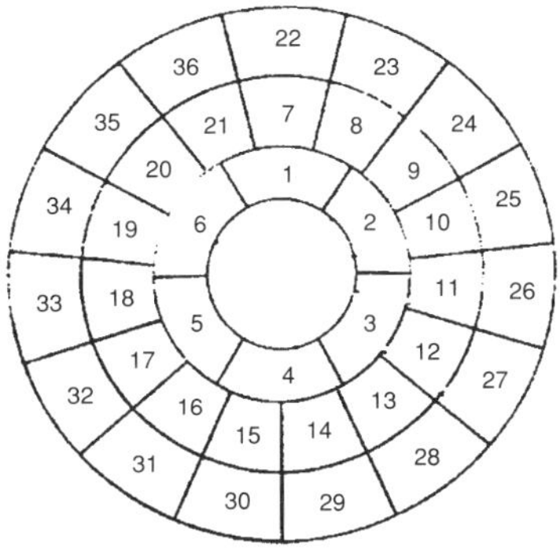

	Heating Sizing Power (Watts)	
Heater No.	Total	Clear Area (in.^{2})
1-6	2.0/Hr	7/Hr
7-21	1.2/Hr	18/Hr
22-36	1.2/Hr	18/Hr

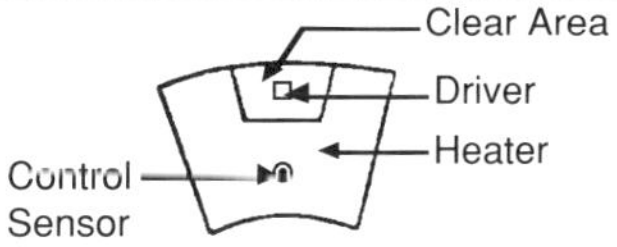

Typical Heater/Driver/
Control Sensor Relationship

Figure 41. Mirror reaction plate heaters

Baffle Assemblies

Three baffle assemblies are required for telescope straylight control, as shown in Figure 35. The main baffle assembly is the large cylinder extending forward from the main ring just inside the metering truss assembly. The central baffle extends from the mirror reaction plate forward through the hole in the center of the mirror. The secondary baffle extends rearward from the secondary mirror assembly in the front end of the telescope. The thermal designs of all of the baffles are passive. The principle thermal concern is to provide adequate conductive isolation between the baffles and the structure to which they are mounted so that they won't act as fins, carrying away energy from temperature-controlled structure. Also of concern is preventing baffle excursion into the telescope optical path due to thermal deformation and radiant sink temperature requirements of critical components viewing the baffles.

The interior surface of the main baffle must be painted black for optical straylight control. The exterior surface is covered with MLI to minimize the radiative influences of the surroundings. Conductive isolation is provided where the main baffle mounts to the main ring to avoid upsetting the thermal balance of the ring. The only significant thermal couplings for the main baffle are to the light shield and to the external environment by radiation out the telescope aperture. Main baffle temperatures are therefore driven by the external environments.

Both the interior and exterior surfaces of the central baffle must be painted black for straylight control. Low-conductance mounts are provided where the central baffle attaches to the mirror plate to avoid upsetting the reaction-plate thermal balance. The temperature of the central baffle is therefore driven by its radiative couplings to the main baffle and forward shield, and to the external environment through the telescope aperture.

The secondary baffles also must be painted black on both inside and outside surfaces for optical reasons. This baffle is not, however, conductively isolated from the secondary mirror housing. Its temperature, therefore, is the result of conductive coupling to the secondary mirror housing and radiative couplings to the main baffle, forward shield, aperture door, and the external environment.

Metering Truss Assembly

The metering truss must precisely maintain the position of the secondary mirror assembly with respect to the primary mirror during telescope operations. The truss is constructed of graphite epoxy ring and strut members, with four spider legs to hold the secondary mirror assembly, as shown in Figure 42a. Limits on truss temperatures and temperature gradients are based on despace, decenter, and tilt due to thermostructural distortions.

The entire metering truss, except for the spider mounts, is covered with MLI blankets, as shown in Figure 42b. In addition, the entire truss assembly sits between the MLI blankets on the outside of the main baffle and the low-emittance internal surface of the forward shell, which provides further isolation. The truss is bolted to the main ring at eight places without any special thermal isolation. The spider mounts have a high absorbtance for straylight control and a low emittance to minimize heat loss. All temperature and temperature gradient requirements are met with this passive design.

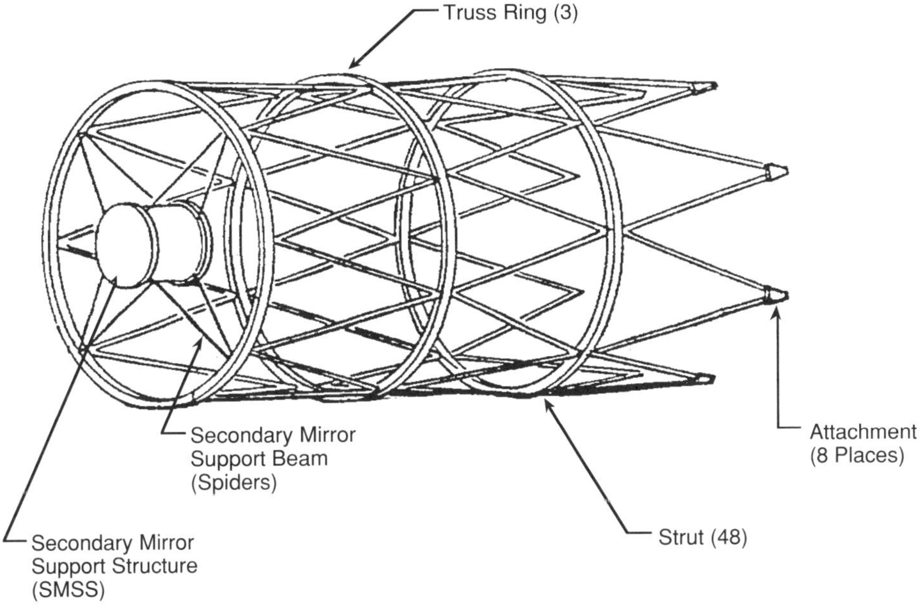

Figure 42a. OTA metering truss

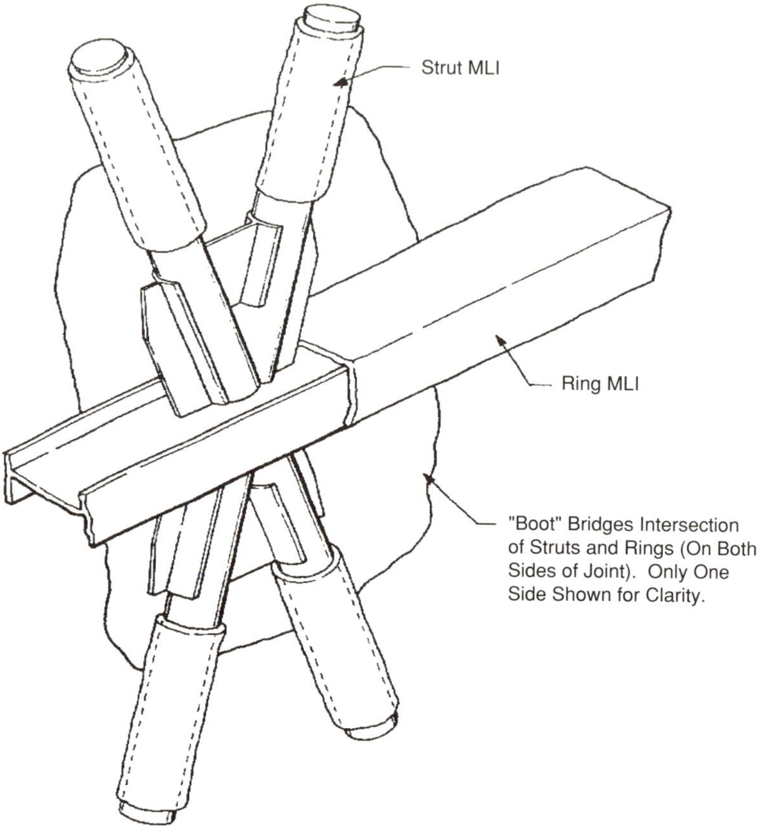

Figure 42b. Metering truss detail

Secondary Mirror Assembly

Thermostructural deformation of the secondary mirror and displacement of the mirror relative to the metering truss drive the temperature and temperature-gradient limits for the secondary-mirror assembly. The thermal design approach is to surround the secondary mirror with three precision-heater-controlled plates (shown in Figure 43), that act as a constant temperature (70° ± 0.2°F) enclosure for the mirror and as a mounting surface for the actuators that control mirror position. The inside surfaces of these plates have a high emittance to insure good radiation coupling to the back of the secondary mirror. The outside faces of these plates and the actuators mounted to them are covered with a low-emittance gold tape. Further isolation from the cylindrical hub is provided by MLI blankets and low-conductivity mounting pylons.

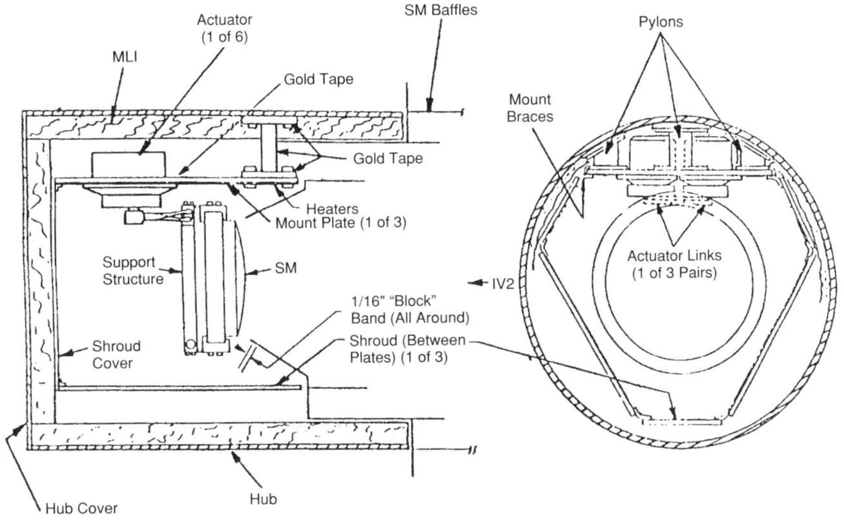

Figure 43. Secondary mirror assembly

The secondary mirror baffle, which is attached to the hub structure, extends down into the mirror cavity. The side of the baffle facing into the cavity is low emittance to minimize the effects of its wide temperature swings on cavity temperatures. The side facing the optical path is painted black for straylight control and is therefore high emittance. The low-emittance finish on the front of the secondary mirror, however, minimizes its radiative coupling to the baffle. The graphite epoxy/invar mirror support structure also has a low-emittance finish to decouple it somewhat from even the small temperature variation ($\pm$ 0.2 deg F) of the heater-controlled plates.

Focal Plane Assembly

The focal-plane assembly, shown in Figure 44, consists of the focal-plane structure, axial and radial science instruments, fine guidance-system sensors, and an equipment shelf to which the fixed-head star tracker and rate-sensor units are mounted. This entire assembly is located behind the primary mirror assembly and is attached to the telescope main ring.

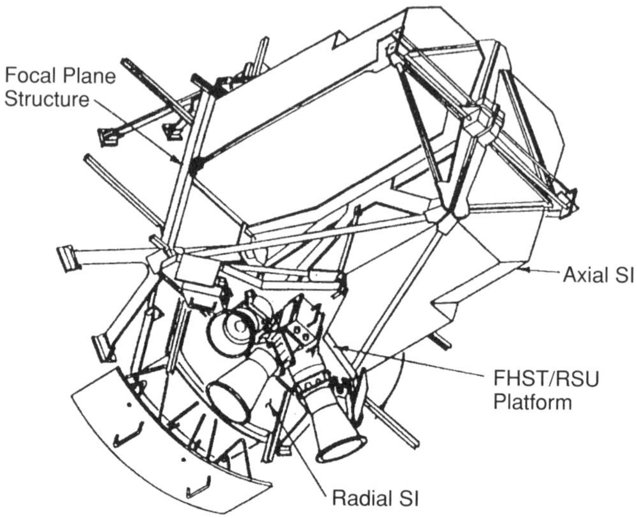

Figure 44. Focal plane assembly

The thermal design of the focal-plane assembly structure, shown in Figure 45, is the same as for the other telescope structural elements, i.e., place precision-controlled heaters on all structural members to control their temperature and wrap them in MLI to minimize heater power and temperature gradients. Conductive isolators are provided to limit heat loss to the science instruments, equipment shelf, and to the telescope aft ring. All cables leaving the focal plane assembly structure are wrapped in MLI, and guard heaters are installed a short distance from where the cable leaves the structure, as shown in Figure 45, to ensure that the cable temperature is the same as the structure so that no heat transfer will occur down the cable.

The equipment shelf (shown in Figure 46) is a dimensionally stable platform for mounting three fixed head star trackers and three rate sensing units. The platform is attached to the focal-plane assembly structure and is thermally controlled by a passive design that minimizes changes in temperature gradients in order to meet a 3-arc-sec alignment stability for the sensors. The thermal design approach is to cover the shelf in MLI and conductively isolate it from the focal plane assembly structure and the six sensors by the use of low-conductivity mounts. The thermal design of the shelf and sensors is shown in Figure 47.

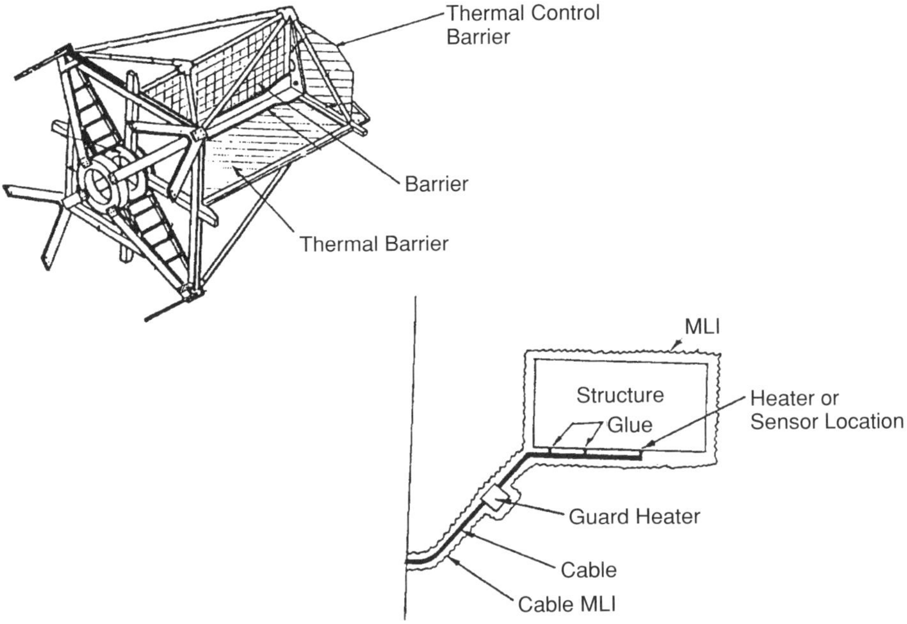

Figure 45. Focal plane assembly structure and cables

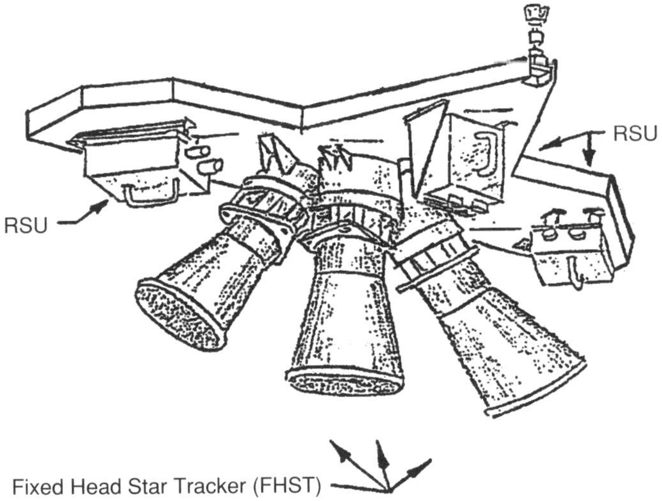

Figure 46. Focal plane assembly shelf

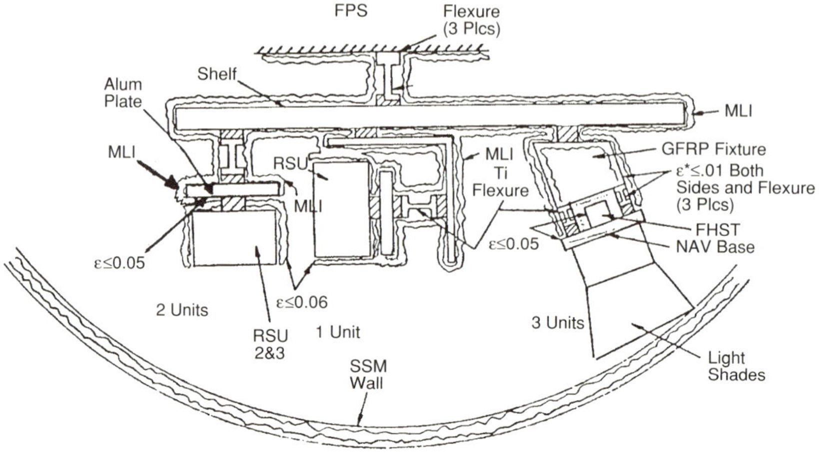

Figure 47. Equipment shelf thermal design

Solar Array Assembly

The Hubble Space Telescope solar array (SA) was designed by British Aerospace for the European Space Agency. Six major parts make up the SA; they are the primary deployment mechanism (PDM), solar array drive mechanism (SADM), secondary deployment mechanism (SDM), diode box assembly (DBA), the solar array blanket, and the solar array electronics (SADE and DCE) mounted in equipment section Bay 7. Figure 48 presents a sketch of an SA assembly and its major external parts. The SA thermal design is passive after array deployment (SA heaters are used prior to SA deployment) and uses a combination of three thermal-control tapes. The SA heater systems are left enabled after deployment to protect SA components, even though the cold-case thermal analyses have shown that heaters are not required. The types and properties of the three surface-finish tapes are as follows:

Aluminized Kapton	(α/ε = .12/.04)
Aluminized Teflon	(α/ε = .14/.62)
Silverized Teflon	(α/ε = .07/.82)

The general thermal-design approach for the SA components has been to use the lowest emittance possible consistent with maintaining acceptable maximum temperatures, allowing for any temperature increase during motor operation. The combination of the low solar absorptance and emittance results in minimizing the effect of changes in the environment while maintaining acceptable gradients.

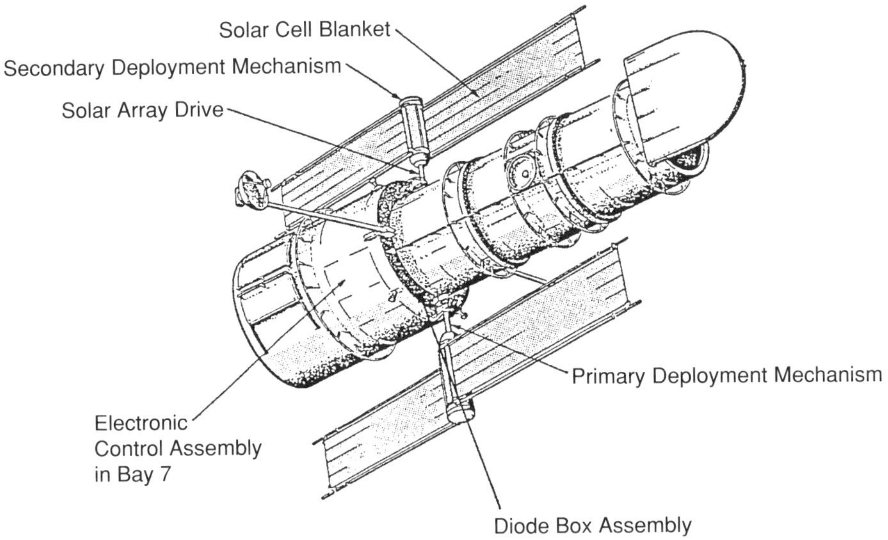

Figure 48. Solar array components

Primary Deployment Mechanism (PDM)

Both sides of the PDM are totally covered with aluminized Kapton. MLI is used on top of the mechanism and along the deployment arm as shown in Figure 49. The external surfaces of the MLI are 25 percent silvered Teflon and 75 percent aluminized Kapton. Temperature limits and PDM characteristics are listed below.

Surface properties: See Figure 49.

	Operating	Non-Operating
Temperature limits:	-43/55 deg C	-55/80 deg C

Thermostats: There are two primary thermostats in series and the two heaters are wired in parallel.

Set points	Open	Close
Primary	-29 deg C	-42 deg C
Secondary	-29 deg C	-42 deg C

Heater System: Four heater strips per PDM with each strip a primary and secondary heater @ 3.27 watts, therefore the primary heaters will supply a total of 13.1 watts, as will the secondary.

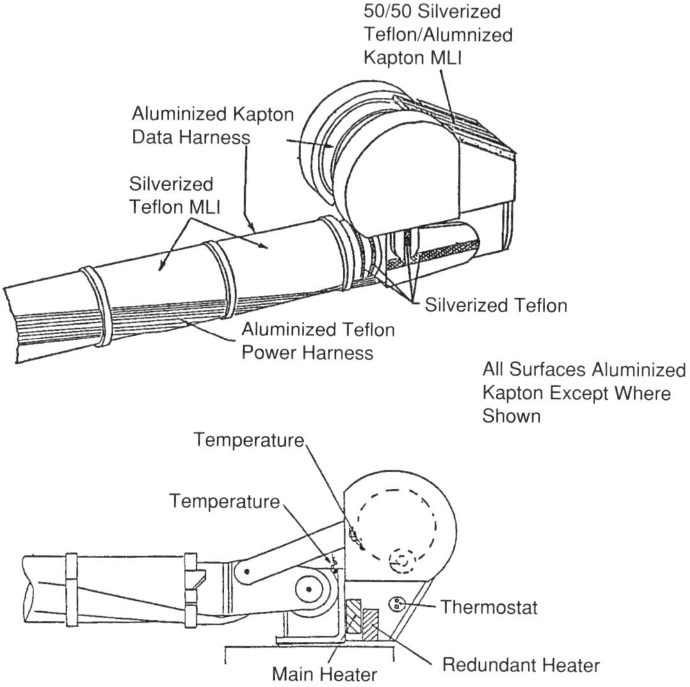

Figure 49. Solar array primary deployment mechanism

Secondary Deployment Mechanism (SDM)

The thermal design of the SDM is shown in Figure 50. Various combinations of aluminized Kapton, silvered Teflon, and aluminum Teflon are used to control various elements of the SDM. The SDM heater system is configured to allow bypass of the thermostats to directly power the heaters. The heaters are bypassed prior to secondary deployment of the SAs. During deployment, the SDMs were within their operating temperature limits of -10 deg C to 25 deg C. After deployment, the SDM thermostatically controlled heaters were reinstated and the SDMs have remained within their non-operational temperature limits of

Surface properties: See Figure 50.

Temperature limits:

	Operating	Non-Operating
SDM	-10/25 deg C	-55/80 deg C

Thermostats: There are two primary thermostats in series and the two heaters are wired in parallel.

Set points	Open	Close
Primary	-15 deg C	-29 deg C
Secondary	-15 deg C	-29 deg C

Heater system: Six heater strips per SDM, two primary and four secondary, 8.84 watts for each primary and 5.12 for each secondary. Therefore, the primary heaters will supply a total of 17.7 watts and the secondary will supply 20.5 watts.

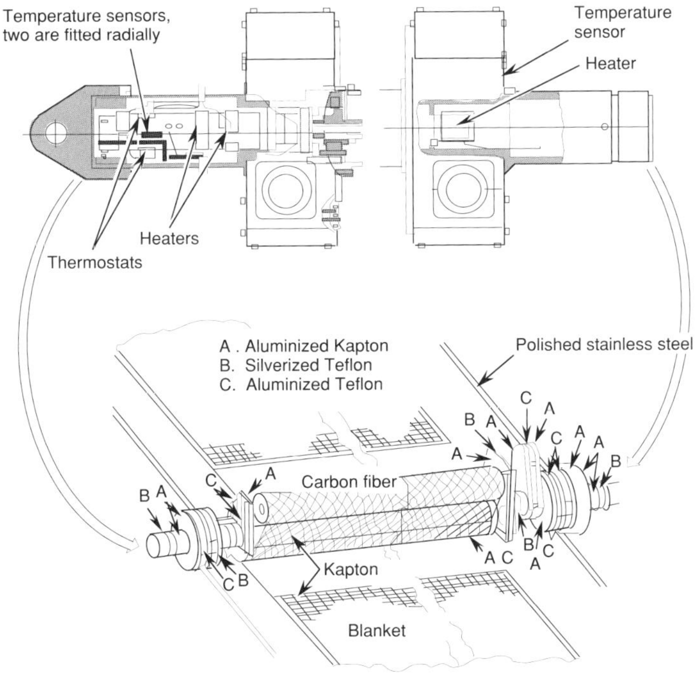

Figure 50. Solar array secondary deployment mechanism

Solar Array Drive Mechanism (SADM)

The SADM function is to slew the SA assemblies so that the sun's rays are normal to the blankets. The SADMs are located on the external skin of the forward shell. MLI is used over the cover around the motor. The outer surface of both the MLI and uninsulated areas are 25 percent Silvered Teflon and 75 percent aluminized Kapton. The SADM structure is thermally isolated from the support structure on the forward shell. Figure 51 presents a sketch of the SADM. Characteristics and information of the SADMS are as follows:

	Operating	Non-Operating
Temperature limits:	-43/55 deg C	-55/80 deg C

Thermostats: There are two primary thermostats in series and the two heaters are wired in parallel.

Set points	Open	Close
Primary	-29 deg C	-43 deg C
Secondary	-29 deg C	-43 deg C

Heater system: Two heater strips per SADM with each strip a primary and secondary heater @ 6.64 watts, therefore the primary heaters will supply a total of 13.3 watts as will the secondary.

Solar array blankets

Figure 51 presents a drawing of the BSFR (back surface field reflector) SA blanket. There are no heater systems on the SA blankets. The SA electrical conversion efficiency is about 11 percent, which would effectively reduce the (α) to 0.68 instead of the 0.76 shown below. Characteristics and information of the SA blankets are as follows:

Surface properties: Solar cell $\alpha/\varepsilon = .76/.83$ (BSFR).
Rear substrate $\alpha/\varepsilon = .54/.90$.

	Operating	Non-Operating
Temperature Limits:	-100/100 deg C	-105/105 deg C

Solar array diode box assembly (DBA)

The two DBAs are mounted externally on the forward bulkhead of the SSM equipment section approximately 9 deg from one another. The DBA brackets are conduction-isolated from the Equipment Section structure. The diode plates and box-surface finishes are shown in Figure 52. Thermostatically controlled heaters are mounted on the diode plates to maintain minimum temperatures prior to SA deployment. The prediction orbit-temperature range for the DBA is -20 deg C to 93 deg C. There are no temperature monitors on the DBA.

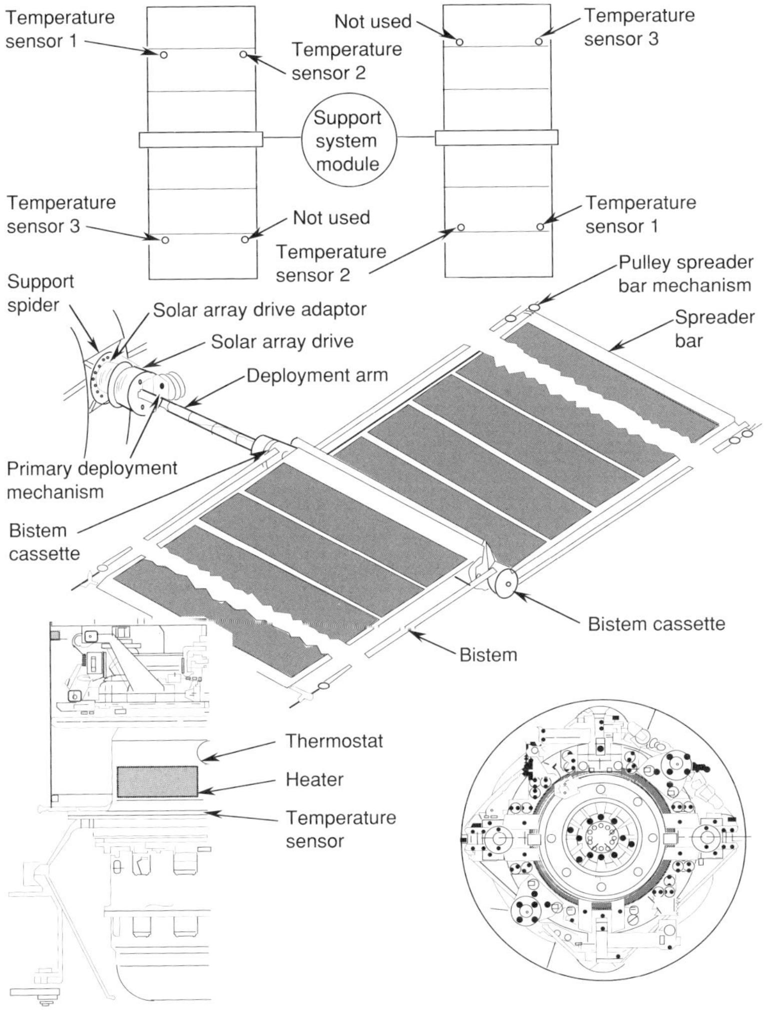

Figure 51. Deployed solar array

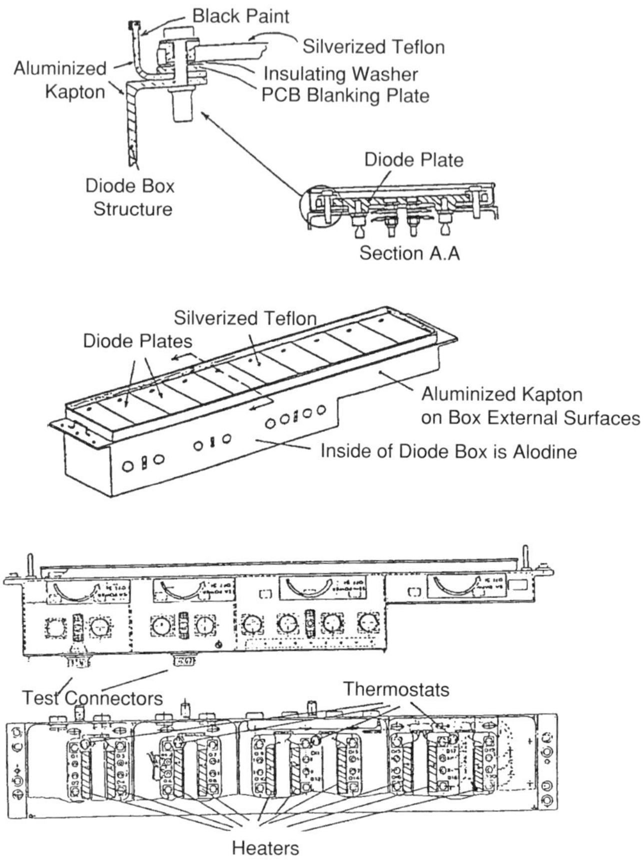

Figure 52. Solar array diode box assembly

External Components

The external components consist of the following:

- Latches and drives on the high gain antenna, solar arrays, and aperture door
- Course sun sensors (CSS)
- Low-gain antennas (LGA)
- Magnetic sensing systems (MSS)
- Magnetic torquer bars (MTE bars)
- High-gain antennas (HGA)

All of the external components have both active (heaters and thermostats) and passive (MLI, isolators, and surface finish) thermal-design features, except for the MTE bars, LGA, and HGA dish, which have only passive thermal control. The solar array mechanisms have heater systems, and the SA blankets and arm are passively controlled.

There are several heater circuits for the external components. The retrieval/deployment heater circuit enables or disables all the heaters on the latches and drives (except for the AD hinge) that are used for deployment and retrieval from the Space Shuttle. The LS/FS heater circuit controls the AD hinge, HGA two-axis gimbal (TAG) heaters, and the MSS heaters. All the CSS heaters are on a separate circuit. The latches and drives used only for deployment and retrieval have only a single-heater system, and all the other external-component heater systems have completely redundant heater systems. The solar arrays have heater circuits for the diode boxes, SADM/PDM, SDM, and SDM retrieval/survival heaters.

Latches and Drives

There are 2 HGA drives, 2 HGA latches, 2 forward SA latches, 2 SA aft latches, 1 AD drive, 1 AD hinge system (1 passive and 1 active), and 1 AD latch located on the external shell of the vehicle. Figures 53 and 54 show the location of these components along with the thermistors associated with them. During deployment, all of the latches and drives were maintained above their lower operational temperature limits by heaters (retrieval/deployment heater circuit). After deployment, this heater circuit was disabled and temperatures of the latches and drives were allowed to drop, except for the AD drive and hinges, which are on a different heater circuit. From temperature plots, each of the components dropped in temperature from ambient temperatures just after launch and started cycling on their heaters as expected.

The AD hinge heaters are always enabled, since the AD may be closed at any time. AD drive and AD hinge temperatures showed that the heaters are cycling properly. Characteristics and information of the latches and drives follows.

Surface properties: MLI $\alpha/\varepsilon = .12/.80$.

	Operating	Non-Operating
Temperature limits:	-40/40 deg C	-73/40 deg C

Thermostats: There are two thermostats in series with the heater.

Set points (Latch/Drive)	Open	Close
Primary	-23 ±3 deg C	-32 ±3 deg C

Setpoints (AD Drive)		
Primary	-18 ±3 deg C	-26 ±3 deg C
Secondary	-23 ±3 deg C	-32 ±3 deg C
Setpoints (AD Hinge)		
Primary	-42 ±3 deg C	-51 ±3 deg C
Secondary	-49 ±3 deg C	-58 ±3 deg C

Heater system: The AD drive and AD hinge system have both primary and secondary heater systems, whereas all latches and HGA drives have only primary heaters for retrieval and deployment.

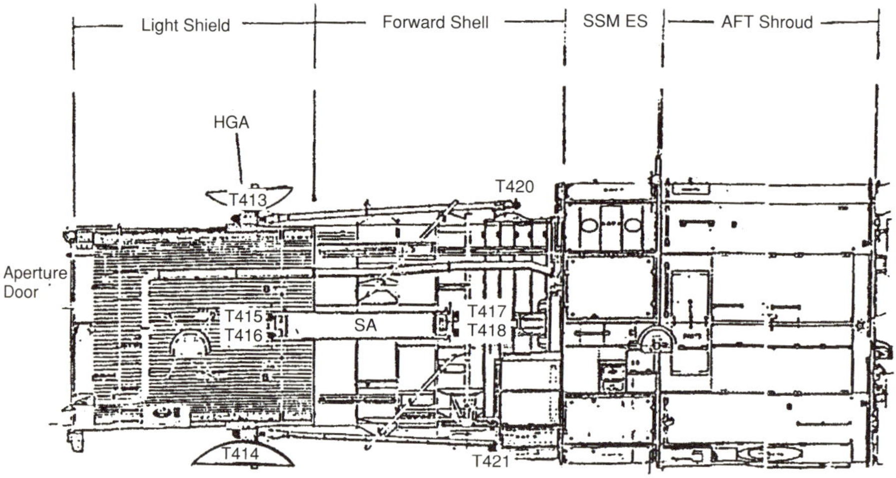

Flight Subsystem Thermistors

MSID	Mnemonic	Description
T413	T+HALCH	+HGA Latch Temp
T414	T-HGALCH	-HGA Latch Temp
T415	T+SAFLCH	+SA FWD Latch Temp
T416	T-SAALCH	-SA AFT Latch Temp
T417	T-SAFLCH	-SA FWD Latch Temp
T418	T-SAALCH	-SA AFT Latch Temp
T420	T+HGALDR	+HGA Hinge Dry Temp
T421	T-HGALDR	+HGA Hinge Dry Temp

Figure 53. Latch and drive thermistors

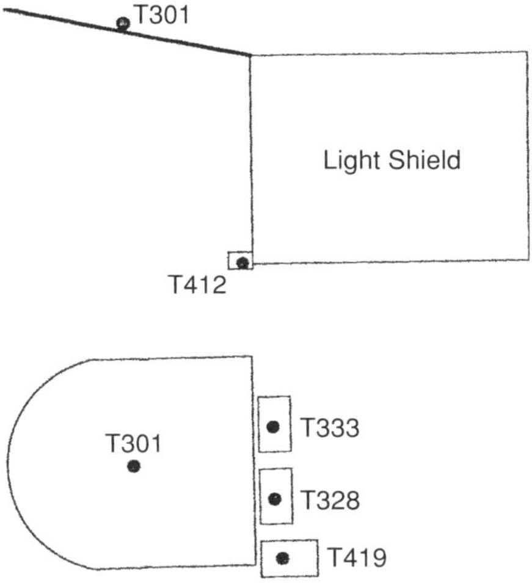

Figure 54. Aperature door and course sun sensor thermistors

Coarse Sun Sensor (CSS)

There are five coarse sun-sensor assemblies located on the vehicle; two at the forward end of the light shield and three on the aft bulkhead. Figure 55 presents the location and viewing directions of the CSSs along with a sketch of a CSS. Characteristics and information of the CSS follow.

Surface properties: MLI α/ε = .12/.80.

Temperature limits:	Operating	Non-Operating
	-40/38 deg C	-67/120 deg C

Thermostats: There are two primary thermostats in series with the heater.

Set points	Open	Close
Primary	-23.3 deg C ± 2.8 deg C	-31.7 deg C ± 2.8 deg C
Secondary	-26.1 deg C ± 2.8 deg C	-34.4 ± 2.8 deg C

Heater system: One heater strip per CSS with each strip containing both a primary and a secondary heater @ 4.08 watts. CSS 4 and CSS 5 are mounted on a common bracket and have only one heater system.

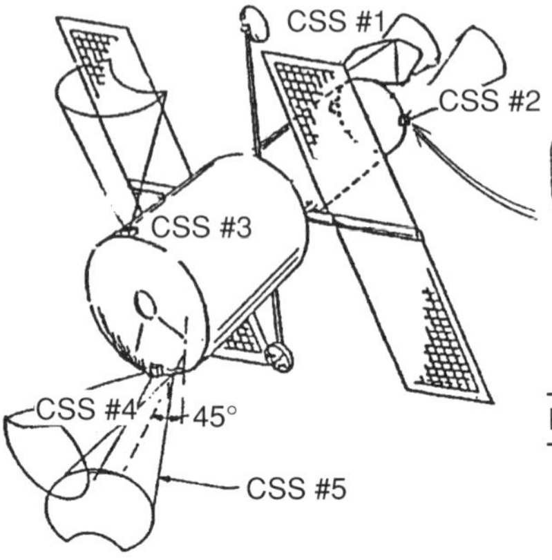

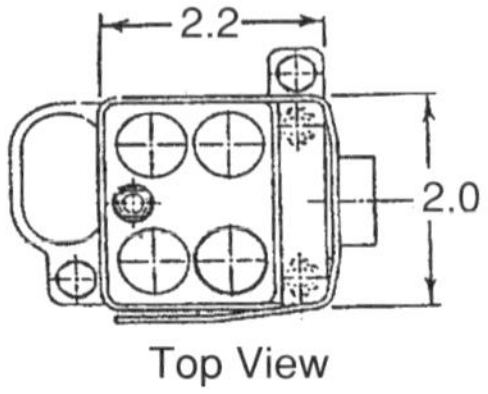

Flight Subsystem Monitors

MSID No.	MNEMONIC	Description
G360	GCSS5T	CSS 5 Temp
G361	GSS1T	CSS 1 Temp
G362	GCSS2T	CSS 2 Temp
G363	GCSS3T	CSS 3 Temp
G364	GCSS4T	CSS 4 Temp

Figure 55. Course sun sensor locations and thermistors

Low Gain Antenna (LGA)

There are two LGAs located on the vehicle; one is located on the aft bulkhead, the other on the forward end of the light shield. Figure 56 presents a sketch of a LGA. The predicted temperatures for the LGAs are -70.5 deg C for the cold case 41.1 deg C for the hot case. The temperature limits for the LGAs are -100 to +70 deg C. There are no flight thermistors located on the LGAs.

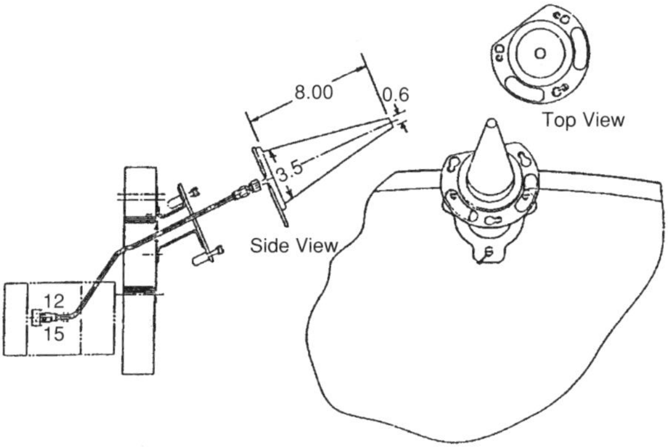

Figure 56. Low gain antenna

Magnetic Sensing System (MSS)

The two MSSs are located on the light shield near the forward end. Two small boxes make up each of the two MSSs; an electronic unit and a sensor unit, with the electronic unit containing the thermistor and heater systems. Figure 57 presents the location of the MSSs along with a sketch of a MSS. Characteristics and information of the MSSs follow.

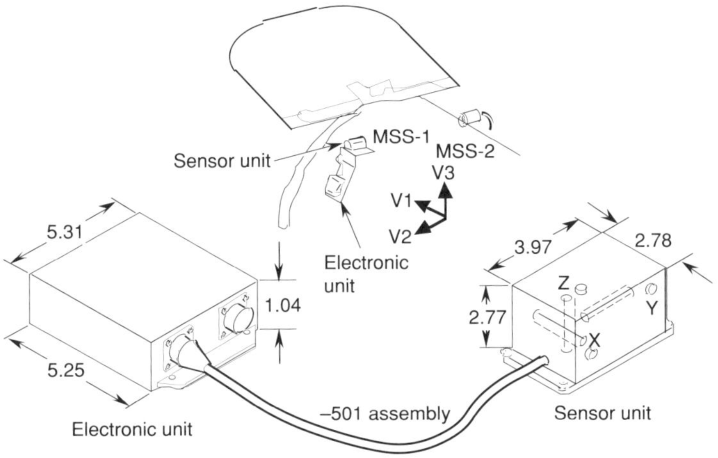

Flight subsystem monitors

MSID No.	MNEMONIC	Description
G314	GMS1T	MSS-1 temp
G318	GMS2T	MSS-2 temp

Figure 57. Magnetic sensing system

Surface properties: MLI $\alpha/\varepsilon = .12/.80$.

Temperature limits:	Operating	Non-Operating
Electronic	-40/72 deg C	-55/125 deg C
Sensor	-73/72 deg C	-100/+100 deg C

Thermostats: There are two primary thermostats in series and the two heaters are wired in parallel.

Set points	Open	Close
Primary	-17.8 deg C ± 2.8 deg C	-26.1 deg C ± 2.8 deg C
Secondary	-23.3 deg C ± 2.8 deg C	-31.7 deg C ± 2.8 deg C

Heater system: Two heater strips per MSS, with each strip containing both a primary and a secondary heater @ 3.75 watts. Therefore the primary heaters will supply a total of 7.5 watts, as will the secondary.

Magnetic Torquer Bars

There are four magnetic torquer (MT) bars mounted on the forward shell. Each bar is covered with MLI and is conduction-isolated from the forward shell by non-metallic spacers. There are no heater systems for the MT bars. Figure 58 presents a sketch of a MT bar and the locations of the bars relative to the vehicle. Bars 1 and 4 have greater temperature fluctuations than bars 2 and 3. Bars 1 and 4 are located on the half of the vehicle that receives direct solar heating, whereas bars 2 and 3 are located on the bottom of the vehicle and are shielded from the sun. Characteristics and information on the MT bars follows.

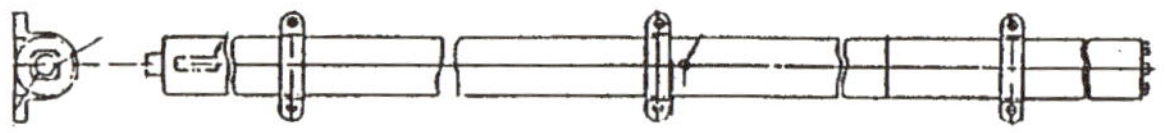

Flight Subsystem Monitors

MSID No.	MNEMONIC	Description
G771	GMT1ABT	MTE 1A Bar TMP
G772	GMT1BBT	MTE 1B Bar TMP
G773	GMT2ABT	MTE 2A Bar TMP
G774	GMT2BBT	MTE 2B Bar TMP
G775	GMT3ABT	MTE 3A Bar TMP
G776	GMT3BBT	MTE 3B Bar TMP
G777	GMT4ABT	MTE 4A Bar TMP
G778	GMT4BBT	MTE 4B Bar TMP

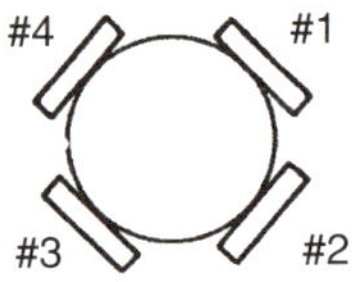

Looking Forward

Figure 58. Magnetic torquer bars

Surface properties: MLI α/ε = .12/.80.

Temperature limits:	Operating	Non-Operating
	-65/70 deg C	-65/70 deg C

HGA Two-Axis Gimbal (TAG)

The HGA two-axis gimbals are located between the HGA mast and the HGA dish. The TAGs point and track the HGA dishes to the TDRSS relay satellites. There are four thermistors for each TAG. One thermistor is located near each of the TAG's bearings. Figure 59 presents a sketch of a TAG. Characteristics and information of the TAGs follows.

Surface properties: Gold Alodine $\alpha/\varepsilon = .23/.05$.

Temperature limits:	Operating	Non-Operating
	-18/93 deg C	-18/93 deg C

Thermostats: There are two primary thermostats in series and the two heaters are wired in parallel.

Set points	Open	Close
Primary	18.9 deg C to 21.7 deg C	18.9 deg C $\pm$ 1.1 deg C
Secondary	16.1 deg C to 18.9 deg C	16.1 deg C $\pm$ 1.1 deg C

Heater system: Eight heater strips per TAG, with each strip a primary or secondary heater @ 6.88 watts; therefore the primary heaters will supply a total of 27.5 watts, as will the secondary.

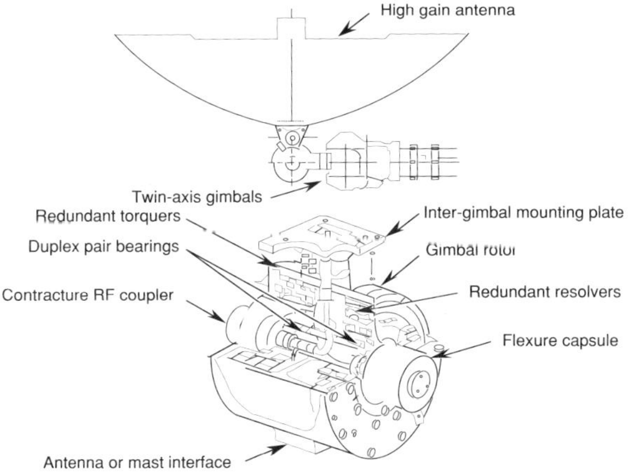

Flight subsystem monitors

MSID No.	MNEMONIC	Description
H526	HG1+GXT	GEA1 + GMBL X TMP
H527	HG1+GYT	GEA1 + GMBL Y
TMP		
H528	HG1–GXT	GEA1 – GMBL X TMP
H529	HG1–GYT	GEA1 – GMBL Y TMP
H530	HG2+GXT	GEA2 + GMBL X TMP
H531	HG2+GYT	GEA2 + GMBL Y TMP
H532	HG2–GXT	GEA2 – GMBL X TMP
H533	HG2–GYT	GEA2 – GMBL Y TMP

Figure 59. High-gain antenna two-axis gimbal

Chapter IV

Thermal Control Hardware

Section 1

Thermal Surface Finishes

David G. Gilmore
Wayne K. Stuckey
The Aerospace Corporation

INTRODUCTION

In spacecraft thermal designs, wavelength-dependent thermal control coatings are used for various purposes. Solar reflectors such as second-surface mirrors and white paints or silver- or aluminum-backed Teflon are used to minimize absorbed solar energy, yet emit energy almost like an ideal black body. To minimize both the absorbed solar energy and infrared emission, polished metal such as aluminum foil or gold plating is used. On the interior of the vehicle, if it is desired to exchange energy with the compartment and/or other equipment, black paint is commonly used. Thus, the existing state of the art uses a rather wide variety of wavelength-dependent coatings. The problems of in-space stability, outgassing, and mechanical adhesion to the substrate are all problems that have been resolved for most coatings. There are many fully qualified coatings, so development and qualification of a new coating for a new design is normally unnecessary.

The external surfaces of a spacecraft radiatively couple the spacecraft to space. Because these surfaces are also exposed to external sources of energy such as sunlight and Earth-emitted IR, their radiative properties must be selected to achieve an energy balance at the desired temperature between spacecraft internal dissipation, external sources of heat, and reradiation to space, as illustrated in Figure 1.

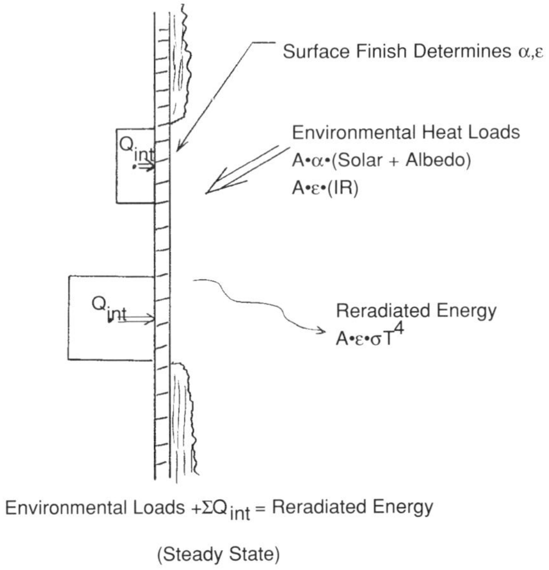

Figure 1. Radiator energy balance (no external blockage)

The two primary surface properties of importance are the IR emittance and the solar absorptance. Figure 2 indicates the range of properties available for different types of materials. Two or more coatings are sometimes combined in a checkerboard or striped pattern to obtain the desired characteristics.

There are four basic types of thermal control surfaces: solar reflector, solar absorber, flat reflector, and flat absorber (see Figure 3). The solar reflector reflects incident solar energy while absorbing and emitting infrared energy. Solar reflectors are characterized by a very low α/ε ratio. Solar absorbers absorb solar energy while emitting only a small percentage of the infrared energy. Polished metal surfaces approximate solar absorbers. Flat reflectors reflect energy throughout the spectral range, i.e., they reflect in both the solar and IR regions, while flat absorbers, such as black paint, absorb throughout the spectral range. Table 1 lists the most common thermal control finishes used on today's satellites, and their properties. Appendix A and reference 1 contain a much more extensive list of space-qualified finishes that have been used on actual satellites. It should be noted that the values given here are for "normal" temperature ranges. Substantial changes may occur at cryogenic or very high temperatures (see References 1 and 2).

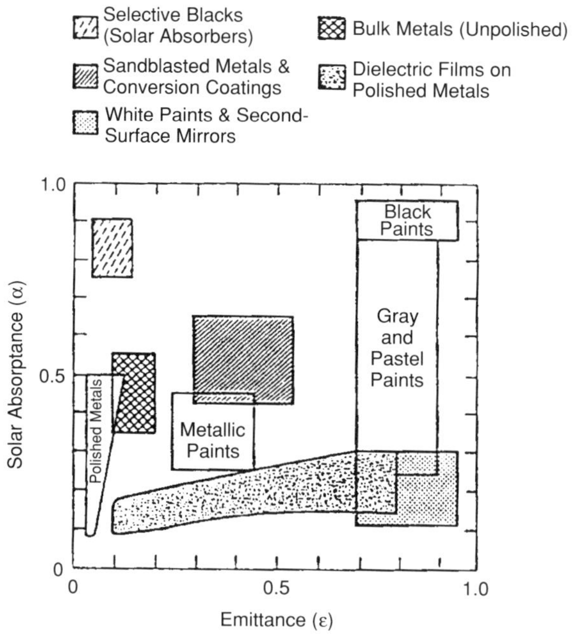

Figure 2. Surface properties by type of finish

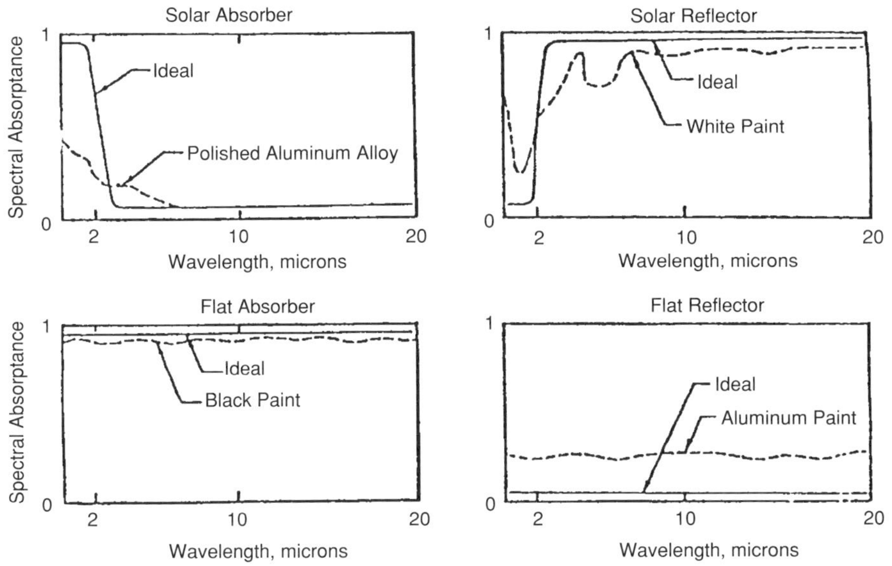

Figure 3. Ideal representation of four basic passive control surfaces

THERMAL SURFACE DEGRADATION

Thermal control finishes are affected in orbit by charged particles, ultraviolet radiation, high vacuum, and the contaminant films that deposit out on almost all spacecraft surfaces. The general result of these processes is an increase in solar absorptivity with little or no effect on infrared emittance. This is normally undesirable from a thermal control standpoint because spacecraft radiators must be sized to account for the substantial increase in absorbed solar energy that occurs due to degradation over the mission. These radiators, which are oversized to handle the high solar loads at "end-of-life," cause the spacecraft to run much cooler in the early years of the mission, sometimes necessitating the use of heaters to avoid under-temperatures of electronic components. The degradation is, therefore, a problem not only because of the solar load, but also because of the change in load over the course of the mission. The stability of coating properties is therefore important in order to minimize heater-power requirements.

Different materials are affected in different ways by the space exposure. White paints, such as S13G, are affected most strongly by ultraviolet (UV) radiation and charged particles (protons and electrons), and can degrade from $\alpha = .20$ to $\alpha = .70$ in just a few years, as shown in Figure 4. Kapton also shows substantial degradation and can turn almost black after several years in orbit (Figure 5).

Table 1. Properties of Common Surface Finishes

Surface Finish	α (Beginning-of-Life)	ε
Optical Solar Reflectors		
8 mil Quartz Mirrors	.05 to .08	.80
2 mil Silvered Teflon	.05 to .09	.66
5 mil Silvered Teflon	.05 to .09	.78
2 mil Aluminized Teflon	.10 to .16	.66
5 mil Aluminized Teflon	.10 to .16	.78
White Paints		
S13G-LO	.20 to .25	.85
Z93	.17 to .20	.92
ZOT	.18 to .20	.91
Chemglaze A276	.22 to .28	.88
Black Paints		
Chemglaze Z306	.92 to .98	.89
3M Black Velvet	~.97	.84
Aluminized Kapton		
1/2 mil	.34	.55
1 mil	.38	.67
2 mil	.41	.75
5 mil	.46	.86
Metallic		
Vapor Deposited Aluminum (VDA)	.08 to .17	.04
Bare Aluminum	.09 to .17	.03 to .10
Vapor Deposited Gold	.19 to .30	.03
Anodized Aluminum	.25 to .86 (see note)	.04 to .88 (see note)
Mylar		
1/4 mil Aluminized Mylar, Mylar side	(Material degrades in sunlight)	.34
Beta Cloth	.32	.86
Astro Quartz	~.22	.80
MAXORB	.9	.1

Note: Anodizing and similar surface treatments generally produce highly variable optical properties that are difficult to control and are very irregular in any given surface. The reader is referred to Ref. 3 for a description of a process that may produce controllable properties. This process has been used successfully to tailor the α and ε of structural elements on NASA's LDEF satellite.

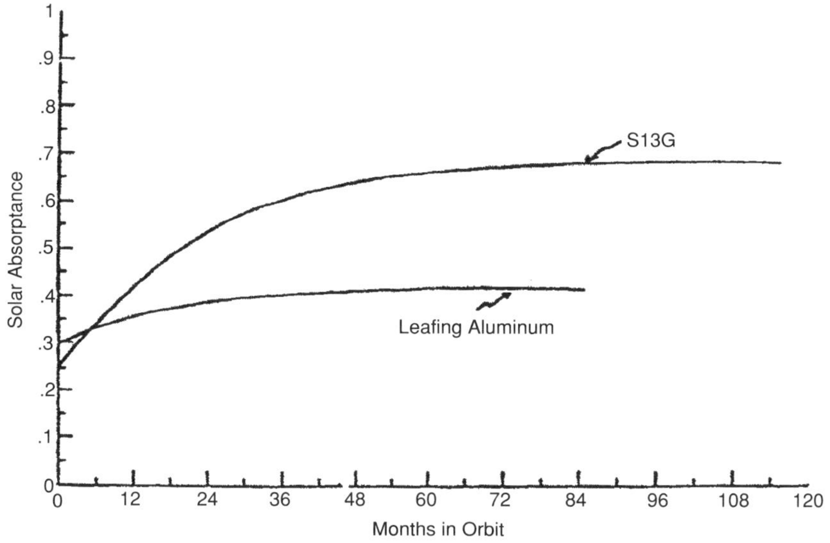

Figure 4. Degradation of selected paints in geosynchrous orbit

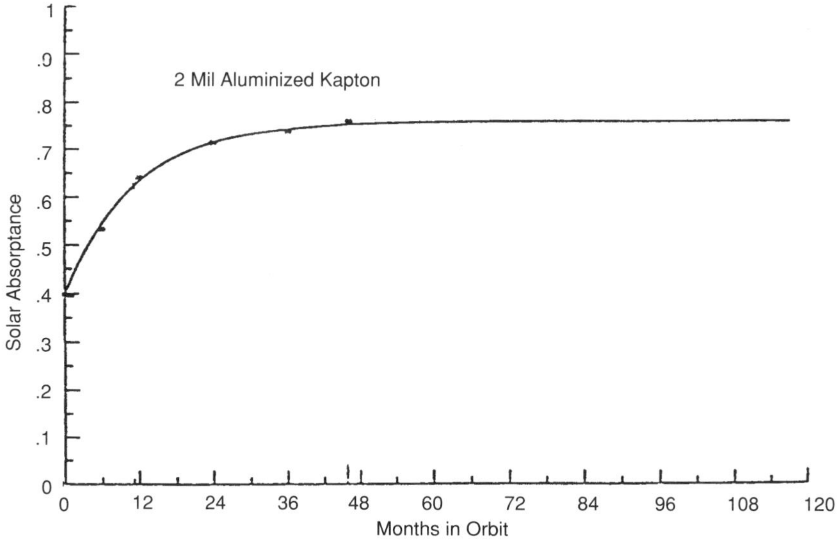

Figure 5. Degradation of aluminized Kapton in geosynchronous orbit

Quartz mirrors, on the other hand, experience essentially no damage due to the UV and charged particles, leaving only contamination as a source of increased absorptivity. Aluminized or silvered Teflon films show degradation due to both charged-particle damage to the Teflon and contaminant deposition. These degradation processes are illustrated in Figure 6.

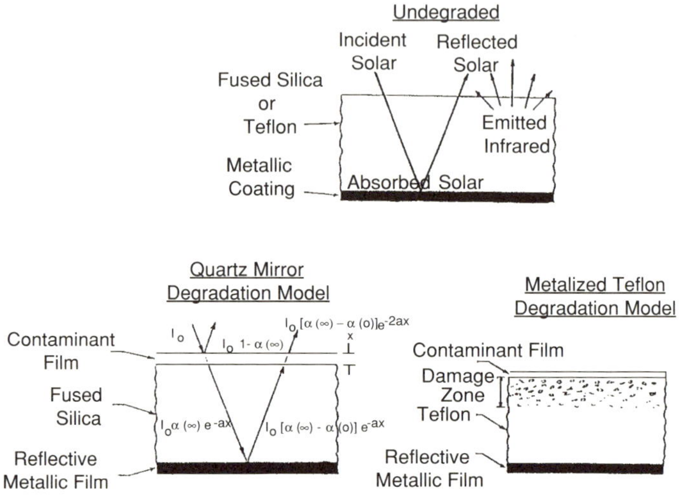

Figure 6. Degradation model

Contamination results when volatile condensable materials (VCMs) are outgassed from spacecraft components, primarily electronic boxes, and then deposit out on the thermal finishes. This outgassing is strongest early in the mission (the first few months to one year) and tapers off with time. Once the contaminants deposit on the thermal surfaces they are in turn darkened by the UV and charged-particle environment. The net result is a rather large increase in solar absorptance in the first few years of the mission, followed by a small steady increase until end-of-life. Figure 7 shows the observed rate of contamination induced $\Delta\alpha$ for quartz mirror radiators on several spacecraft. These variations are not completely understood, but are known to be strongly dependent upon such factors as the types of materials used in the spacecraft, the venting of outgassed materials across thermal surfaces as they leave the spacecraft, and the presence of sunlight, which enhances the deposition of contaminants onto surfaces. Because of these effects, many programs are switching to lower outgassing materials and redesigning vent paths to insure that these outgassed contaminants are directed out to space without impinging onto thermally sensitive surfaces. It is recommended that degradation rates-similar to those shown in Figure 7 be used for other low-absorptance surfaces that are not affected by UV or charged particles, such as polished metals.

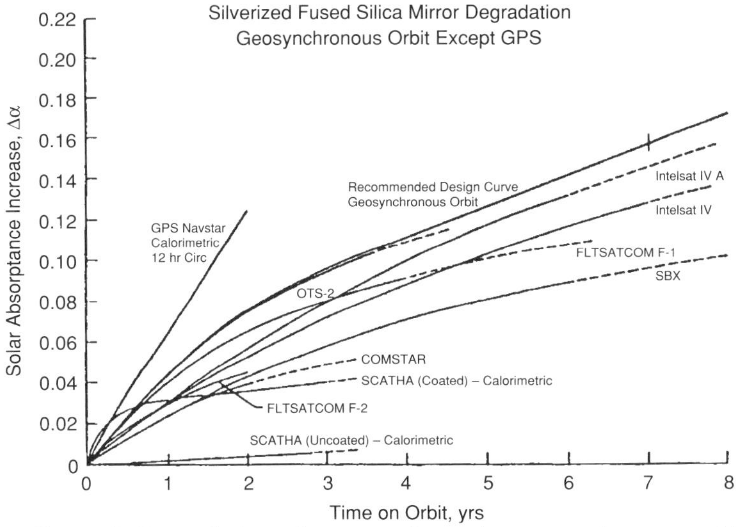

Figure 7. Degradation of quartz mirrors in geosynchronous orbit

As was mentioned earlier, aluminized and silvered Teflon films degrade due to both contamination and natural environmental effects. Degradation rates observed on a number of flight spacecraft are summarized in Figure 8. As can be seen from this data, the degradation is slightly greater than that for quartz mirrors, which experience only contamination-induced absorptance changes. High

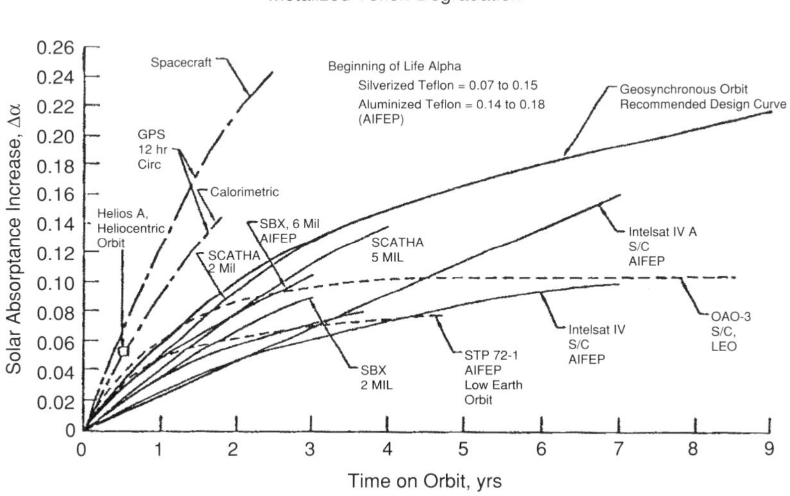

Figure 8. Metalized Teflon degradation (in geosynchronous orbit unless otherwise indicated)

absorptance surfaces, such as black paints, generally do not degrade. There may, however, be a slight (a few percentage points) reduction in absorptivity over time due to UV bleaching.

Most of the degradation data presented above comes from geosynchronous satellites. The change in solar absorptance is calculated by analytical correlation of thermal math models to flight temperature increases on the satellite. Because of the much shorter orbit period and the effects of highly variable Earth heating rates, it is much harder to separate out absorptance changes from other effects for low Earth orbit satellites. This may explain the shortage of data for LEO systems. The few studies done on these satellites, however, do show substantially less degradation than for the GEO vehicles, and qualitative evaluations of other LEO satellites seem to support this. The use of the GEO data presented here would therefore probably be conservative for low Earth orbit satellites.

THE LONG DURATION EXPOSURE FACILITY RESULTS

The Long Duration Exposure Facility (LDEF) was launched by the Space Shuttle in April 1984 into a 257-n.mi. orbit at 28.5-degrees inclination, and was recovered in January 1990 at 179 n.mi. There were 57 experiments containing over 10,000 specimens to test the effects of the low Earth orbit space environment on materials, components, and systems. The principal environmental factors affecting thermal finishes are solar UV radiation; bombardment by atomic oxygen atoms, which are present at very low densities in low orbits; electron and proton radiation; and micrometeoroids. Originally planned for one year, the exposure actually lasted almost six years. While many LDEF investigations are continuing, results to date have given valuable information on long-term performance in orbit.

The 69-month LDEF mission resulted in far longer space exposure of material surfaces than other hardware returned from orbit, such as from the short-duration Shuttle experiments or hardware returned from the Solar Maximum Repair Mission. LDEF was gravity-gradient stabilized, with one side of the vehicle continuously pointing down toward Earth center, and another side always facing the velocity vector, within one degree. The vehicle contained 86 experiment trays, which were oriented around the vehicle in 12 rows of 50-inch by 34-inch trays, with additional trays on the sides facing Earth and facing directly away from Earth. During the mission, the leading-edge materials, i.e., those facing into the velocity vector, were exposed to approximately 9×10^{21} oxygen atoms/cm^2, a level at which erosion of over 10 mils would be expected for many polymers. The trailing edge exposure was only about 10^4 oxygen atoms/cm^2, making atomic oxygen effects insignificant compared to solar UV and charged particles. Trailing edge samples are, therefore, more representative of higher-altitude orbits where atomic oxygen concentrations are insignificant. The solar exposure ranged from about 5,000 to 14,500 equivalent sun hours, depending on location on the LDEF, with 34,200 thermal cycles. The radiation environment on the surface was ~2.5×10^5 rads of electron and 1.6×10^3 rads of proton radiation.

The LDEF observations on thermal control materials are particularly significant for atomic-oxygen effects on the leading edge for low Earth orbits, while the trailing-edge samples show the effects of ultraviolet radiation. The Thermal Control Surfaces Experiment provided on-orbit leading-edge data on thermal properties of 25 materials during the first 18 months of the mission (Ref. 5). The inorganic binder paints, such as Z93 (zinc oxide in a potassium silicate binder) and YB-71 (zinc orthotitanate in a potassium silicate binder), were shown to be stable in the LEO environment. Some thermal control materials degraded more, others less, than predicted from ground tests. The thermal-control properties (alpha/epsilon) of organic binder paints, commonly used for their ease of application, were observed to degrade by as much as a factor of 3 on the trailing edge, but show much smaller changes on the leading edge. Data from paints flown on the M0003 experiment on LDEF are shown in Table 2 (Ref. 6).

Table 2. Solar Absorptance of Thermal Control Paints on LDEF M0003

Paint	Initial α	Leading Edge α	Trailing Edge α
YB-71	0.130	0.182	0.182
A276	0.282	0.228	0.552
S13GLO	0.147	0.232	0.458
D111	0.971	0.933	0.968

The polyurethane paint A276 on LDEF is interesting because the multiple locations on hardware completely around the vehicle allowed the effects of

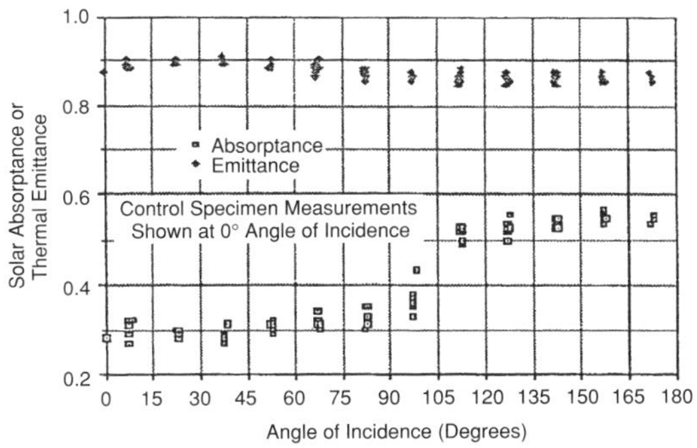

Figure 9. Optical properties from A276 white thermal control discs (Ref. 5)

orientation on performance of the paint to be clearly measured, as shown in Figure 9 (Ref. 7). The data from the trailing edge at or near 180° clearly shows the degradation of the paint by the solar ultraviolet (UV), while the degraded binder on the leading edge near 0° has been removed by the atomic oxygen erosion to maintain properties close to the initial values.

The Kapton and Mylar thermal blankets on LDEF were of particular interest because of the susceptibility of these kinds of materials to erosion from reaction with atomic oxygen on leading-edge surfaces. In fact, one of the initial observations of damage to LDEF materials on-orbit was the observation of the severely eroded Mylar blankets on the space-facing end of the LDEF structure. There were 5-mil Kapton blankets on the leading edge of LDEF where the Kapton had been completely removed and only the few thousand angstroms of metalization remained. The AO fluence of ~9 x 10^{21} oxygen atoms/cm^2 observed on LDEF leading-edge surfaces lead to a predicted loss of over 10 mils of Kapton, based on the reaction efficiency from earlier Shuttle flights of 3.0 x 10^{-24} cm^3/oxygen atom. The observed erosion for Kapton (and a number of other organic polymers) on LDEF was consistent with previously determined reaction efficiencies.

A variety of visible changes were observed on both the leading- and trailing-edge silvered FEP Teflon surfaces on LDEF. The 5-mil silvered Teflon blankets were visibly altered during the LDEF mission, but the thermal properties did not degrade significantly except in those areas that were contaminated. However, caution should be used in other applications depending on the thermal blanket thickness and the planned orbit. The cloudy, diffuse appearance of the Teflon on the leading edge was due to an unexpectedly high erosion of the Teflon layer. For short exposures in LEO, such as the prior Shuttle experiment to study atomic-oxygen effects, very low erosion had been observed, consistent with a recession rate of <0.1 x 10^{-24} cm^3/O atom. The LDEF has permitted the first orbital measurement of the erosion of the Teflon layer on the leading-edge from atomic oxygen; previous attempts could not measure the smaller-thickness decrease of the Teflon. The ~1 mil of erosion observed on LDEF is apparently due to synergistic effects of the vacuum ultraviolet (VUV) and atomic-oxygen environment (Ref. 8). Thermal measurements show the expected decrease in emissivity as the thickness is decreased. The diffuse reflectance increased for those areas toward the leading edge roughened by exposure to both atomic oxygen and solar UV, giving rise to the uniformly clouded appearance. LDEF data has shown that a value of 0.34 x 10^{-24} cm^3/O atom is clearly more appropriate for longer exposures. In practice, the known reaction efficiency and expected oxygen fluence are used to predict the expected life of a film with a given initial thickness. Most blanket areas from the trailing edge side, exposed only to solar UV, remained specular. The LDEF results for silvered Teflon indicate that the thermal performance shows minimal degradation from the solar UV exposures of up to

11,000 ESH. For the trailing-edge blankets, the ultraviolet exposure caused polymer chain scission at the surface and resulted in decreases of percent elongation to failure and ultimate tensile strength (Ref. 9). Another effect observed on LDEF silvered FEP Teflon blankets was the severe degradation associated with cracked silver-Inconel layers. Improper application, which produced cracking of the metallization, allowed migration of the Y966 adhesive through the metallization, and subsequent darkening by solar UV. This process led to increases in absorptance up to 0.25 in small areas. Lifetime predictions should also include consideration of the fraction of the blanket surface that will likely be darkened or destroyed by meteoroid and debris impacts, and potential absorptance increases due to contaminant films over a fraction of the surface. These considerations were minor for LDEF. Impacts darkened 2% or less of the surface area of each LDEF blanket, and delaminated <5% of the area on each blanket. Contaminant films caused absorptance changes as high as about 0.25, but only for relatively small surface areas.

The charged-particle level at the LEO altitudes for LDEF was too low to cause degradation, but at higher altitudes, up to geosynchronous altitudes, the Teflon is susceptible to degradation. Above geosynchronous altitudes, Teflon may have longer lifetimes due to the lower charged-particle environment. At the lower LEO altitudes, atomic-oxygen erosion may result in degraded properties, depending on total fluence levels. An estimate of the expected environmental degradation for a specified mission can be made from the mission profile, which establishes the orbit and required lifetime. End-of-life requirements for the optical properties must be established. At geosynchronous altitude, the SCATHA degradation curves could be used to estimate the performance life with exposure to the trapped-radiation charged-particle belt. In LEO orbits, the atomic-oxygen flux is strongly dependent upon altitude and solar activity. The atomic-oxygen and solar-UV fluences are determined based on the mission profile, and the total recession over the life of the mission is predicted. The minimum required thickness of the Ag/FEP material at end-of-life is based on the well-established values for emittance of FEP as a function of thickness. The actual recession rate used will depend on the expected duration of the mission. For short periods of time in LEO, recession rates of 0.15×10^{-24} cm^3/O atom will be sufficient to establish recession. For missions of greater length, the LDEF value 0.34×10^{-24} cm^3/O atom is clearly more appropriate. In practice, the known reaction efficiency and expected oxygen fluence are used to predict the expected life of a film with a given initial thickness.

REFERENCES

1. Y. S. Touloukian, "Thermophysical Properties of Matter," IFI/Plenum, 1972.

2. M. Donabedian, "Emittance of Selected Thermal Control Surfaces at Cryogenic Temperatures," The Aerospace Corporation, ATM 90(9975)-10, 15 December 1989).

3. R. J. Duckett and C. S. Gilliland, "Variable Anodic Thermal Control Coating on Aluminum," AIAA-83-1492, AIAA 18th Thermophysics Conference, June 1-3 1983.

4. Bland A. Stein, "LDEF Materials: An Overview of the Interim Findings," *Proceedings of LDEF Materials Workshop '91*, pp 1-56, eds. Bland A. Stein and Philip R. Young (NASA Conference Publication 3162) (1992).

5. D. L. Wilkes and L. H. Hummer, "Thermal Control Surfaces Experiment-Initial Flight Data Analysis-Final Report," AZ Technology Report No. 90-1-100-2, 1991.

6. M. J. Meshishnek, S. R. Gyetvay, and C. H. Jaggers, "Long Duration Exposure Facility Experiment Deintegration/Findings and Impacts," *LDEF-69 Months in Space First Post-Retrieval Symposium*, pp. 1073-1107, ed. Arlene S. Levine (NASA Conference Publication 3134) (1992).

7. J. L. Golden, "Results of Examination of the A276 White and Z306 Black Thermal Control Paint Discs Flown on LDEF," *LDEF-69 Months in Space First Post-Retrieval Symposium*, pp. 975-987, ed. Arlene S. Levine (NASA Conference Publication 3134) (1992).

8. C. S. Hemminger, W. K. Stuckey, and J. C. Uht, "Space Environmental Effects on Silvered Teflon Thermal Control Coatings," *LDEF-69 Months in Space First Post-Retrieval Symposium*, pp. 831-845, ed. Arlene S. Levine (NASA Conference Publication 3134) (1992).

9. G. Pippin, W. K. Stuckey, and C. S. Hemminger, "Performance of Silvered Teflon Thermal Control Blankets on Spacecraft," LDEF Materials Results for Spacecraft Application Conference, Huntsville, AL October 27, 1992.

Section 2

Mounting and Interfaces

Donald F. Gluck
The Aerospace Corporation

INTRODUCTION

This chapter deals with the thermal design considerations in mechanical interfaces. In some instances, such as mounting an electronics box to a radiator panel, one wishes to maximize the heat transfer by minimizing the thermal resistance in the contact region. In other cases, one would like to minimize heat transfer, such as using thermal isolators to limit conduction-heat soak-back from a firing thruster to the spacecraft structure. The bulk of this chapter addresses the surprisingly complex issue of contact conductance in bolted interfaces. Additional sections deal with the topics of thermal isolation and the highly uncertain conduction across bearings.

CONDUCTION COOLING

We deal here with the mounting of components (electronic units) to the spacecraft's structure. The mounting of piece parts and modules within a component is a related subject that is touched on but not directly dealt with here. Most electronic components are designed so that the power dissipated within is transported as heat to the component's mounting surface (baseplate). This heat is transferred by conduction to the spacecraft structure (here called the mounting plate) and thence by a variety of methods and paths to the space sink. A smaller number of components are cooled partially and, sometimes, largely by radiation. These components are designed so that heat can be radiated from various component surfaces, but usually not the mounting surface, to the surrounding space-vehicle enclosure or directly to space. We deal here with the predominant mounting interface and heat-transport method encountered with space vehicles: conduction cooling from the component's baseplate to the spacecraft's mounting plate.

It is important that the temperature rise across the mounting interface be small, as each part and device within the component is subject to this temperature rise. Reliability (mean time between failures) is adversely affected by high temperature. Functional performance can be adversely affected by high temperature. As a rule of thumb, the average temperature rise across the interface should be less than 5 to 10°F.

Space vehicle components have baseplates ranging in size from, say, 4-inches by 6 inches to 18 inches by 24 inches, with power levels up to 300 watts. Mounting is typically done by a bolt pattern along the perimeter of the base plate, such as that shown schematically in Figure 1. Flange mounting is convenient as bolts are accessible and can be torqued from above. When electronic boxes are built up from "slices" or modules (Figure 2), the bolts are arranged along two opposed sides of the baseplate. Where power dissipation or local power density (power per unit area) is large, additional screws in the inboard regions of the component are used. These secure from below, extending from the spacecraft mounting plate to the component baseplate (Figures 3 and 4). As these screws must be inserted from below, they complicate the assembly process. However, they increase the conductance from baseplate to mounting surface.

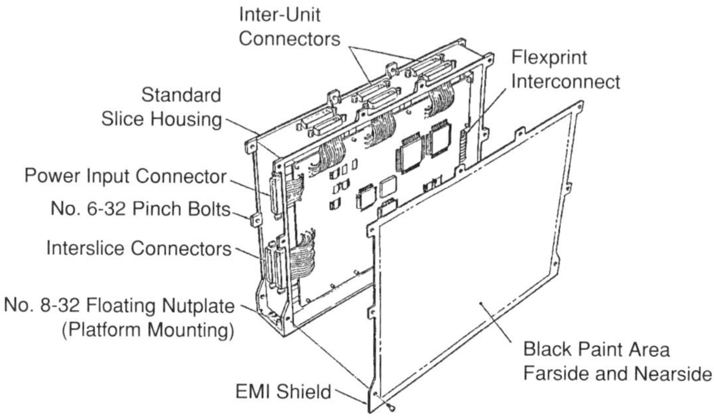

Figure 1. Typical digital module

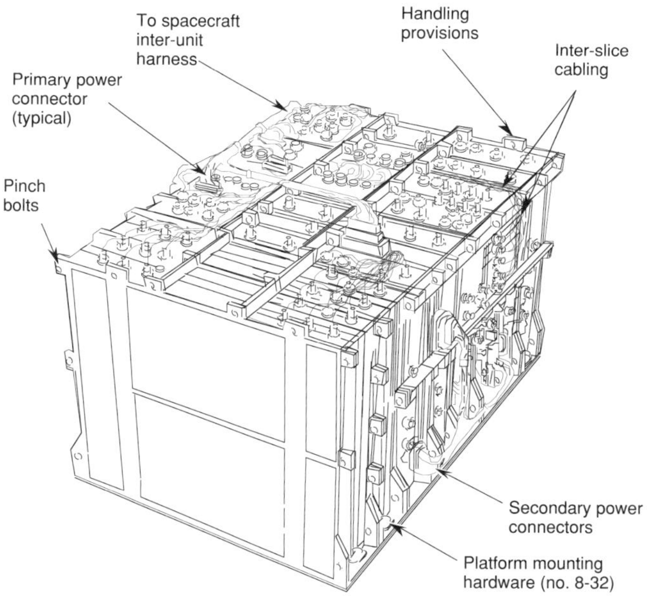

Figure 2. Opposed side perimeter mounting

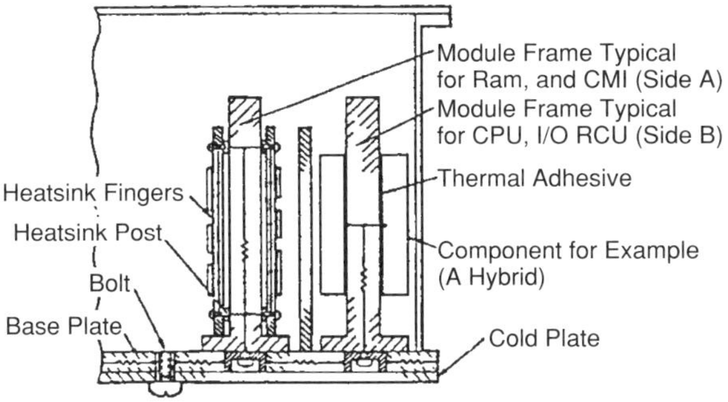

Figure 3. Schematic showing bolt inserted from mounting plate side

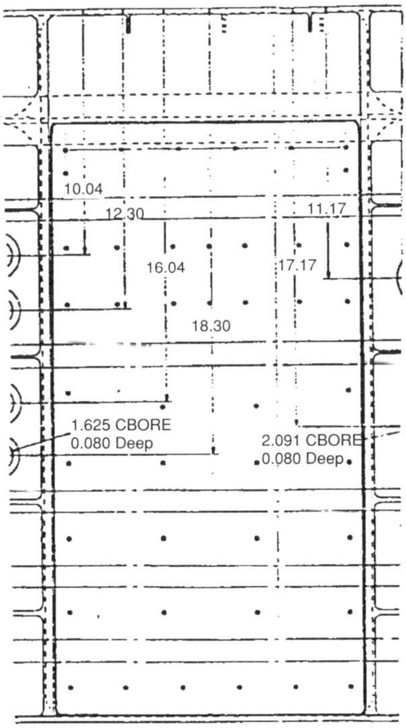

Figure 4. Component mounting footprint showing bolt holes

In this report we will treat a number of important areas under the general heading of component mounting. These are conductance of bare interfaces under uniform pressure; conductance of unfilled bolted joints—in the region of the bolt and overall, for a variety of bolt patterns, for special mounting configurations including honeycomb panels and thermal doublers; thermal/structural analysis; conductance of bolted joints with thermal enhancement fillers; interface control drawings and requirements; and interfaces during testing. Theory, design, and practice are covered.

CONDUCTANCE FOR SURFACES UNDER UNIFORM PRESSURE

Heat transfer from the component baseplate to space-vehicle mounting surface is a complex problem in contact-conductance heat transfer. Because of the vacuum condition of space, there is no convection at the interface. Simple calculations show that for the temperature levels of interest, say -60 to 180-deg F, radiation is negligible compared to conduction. As the conduction process takes place from one surface to another, it is called contact conductance. This problem has been studied widely for the case of two surfaces pressed together under uniform pressure. The problem of surfaces that are bolted together and therefore experience a non-uniform pressure profile is considerably more complex and will be discussed later.

Theory

Surfaces that are nominally smooth in a macroscopic sense are rough in a microscopic sense (Figure 5). Typically, actual contact occurs over only a small fraction of the apparent contact area. The theory of contact conductance has made important advances, particularly because of the work of Mikic (Refs. 1, 2 and 3). He and his co-workers assumed that the microasperity heights can be characterized by a random process that is stationary and for which the distribution of heights is Gaussian above a mean plane. The combined profile can then be

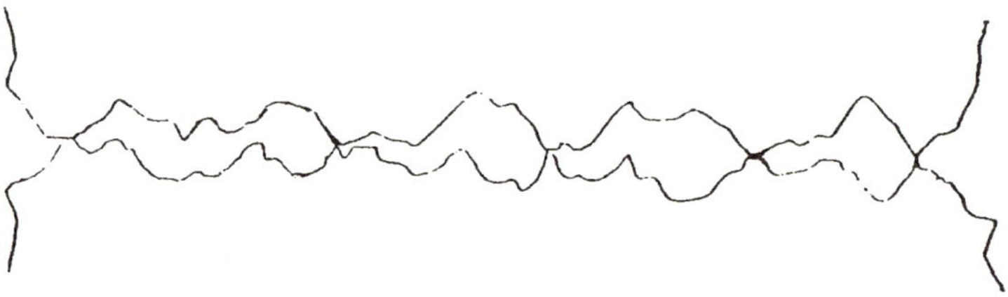

Figure 5. Interface contact, vertical scale expanded

characterized by the standard deviation of profile height, σ, and the mean of the absolute value of the slope, tan θ. Here $\sigma = \left(\sigma_1^2 + \sigma_2^2\right)^{1/2}$, where the subscripts denote the two surfaces. The term σ is also the root mean square (RMS) roughness. Typical values of σ_1 or σ_2 are 10 to 125 microinches, although much larger values are possible (Table 1). If both slopes are normally distributed, we have $\left|\overline{\tan\theta}\right| = \left(\left|\overline{\tan\theta_1}\right|^2 + \left|\overline{\tan\theta_2}\right|^2\right)^{1/2}$. The value $\left|\overline{\tan\theta}\right|$ or simply $\tan\theta$ is typically in the range 0.10 to 0.15.

Mikic distinguishes between the initial loading, during which deformation can be either elastic or elastic transitioning to plastic at higher loads, and subsequent loading up to the maximum pressure of the initial loading, during which deformation is elastic. For the first loading, deformation is predominantly elastic or elastic transitioning to plastic if the group $\gamma = \dfrac{H}{E' \tan\theta}$ is $\geq$ 3 or $\leq$ 0.33, respectively. Here H is the microhardness of the softer of the two materials and E' is given by

$$E' = \frac{E_1 E_2}{\left[E_2\left(1 - \nu_1^2\right) + E_1\left(1 - \nu_2^2\right)\right]} \tag{1}$$

where E is Young's modulus and ν is Poisson's ratio. Mikic develops predictive equations for heat transfer coefficients for plastic and elastic deformation:

$$h_p = 1.13 \frac{k \tan\theta}{\sigma} \left(\frac{P}{H}\right)^{0.94} \qquad \text{(plastic)} \tag{2}$$

$$h_e = 1.55 \frac{k \tan\theta}{\sigma} \left(\frac{P/\sqrt{2}}{E' \tan\theta}\right)^{0.94} \qquad \text{(elastic)} \tag{3}$$

where P is the apparent loading pressure, i.e., the pressure calculated by dividing force by the nominal flat surface area. The form of these equations is the same for plastic and elastic deformation. More recently, Sridhas and Yovanovich (Ref. 4) have critically reviewed elastic and plastic thermal contact-conductance models.

We do not deal here with fluid conductance within the gaps, as the vacuum condition of space is our central concern. Should one be interested in calculating conductance where fluids play a role, the approach of Song et. al. (Ref. 5) is recommended. Rarefaction effects can be important at pressures slightly below ambient because of the small characteristic dimension (RMS roughness).

Experimental Data

There is vast literature on contact conductance between two surfaces under uniform pressure. Some of this is summarized by Schneider (Ref. 6) and presented here in Figure 6. Some of the data from Refs. 7, 8, 9, and 10 include interface fillers. Data from Refs. 11, (circles), 12 (crosses), and 13 (points) in support of Mikic theoretical development are given in Figure 7. Swartz (Ref. 14) replotted data of Fried and Costello (Ref. 7), Fried and Kelley (Ref. 15), and Fried and Atkin (Ref. 16) (Figure 8) in a generalized form (Figure 9); these data are primarily applicable to vacuum conditions.

Table 1. Surface Roughness Produced by Common Production Methods

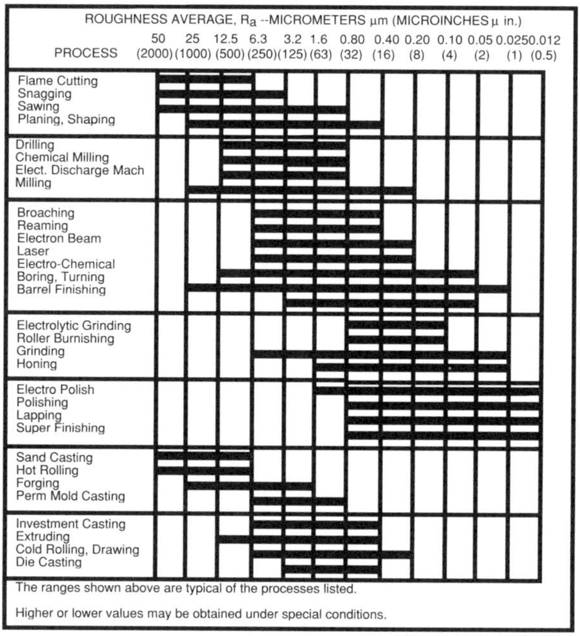

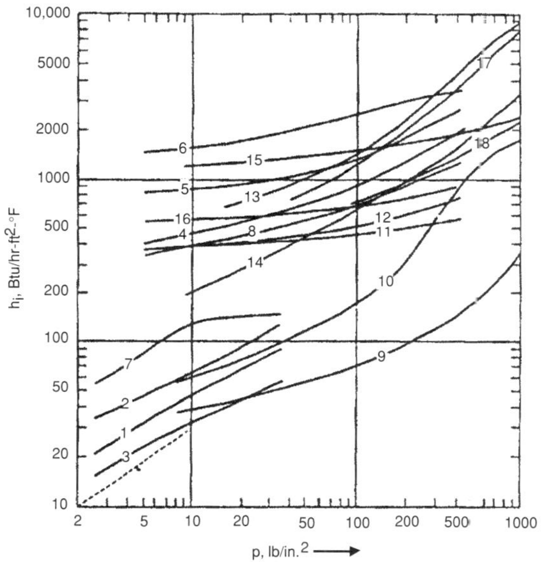

Thermal Interface Conductance Data

Handbook of Heat Transfer

Interface Conditions for Conductance Data in Fig.

Curve	Material Pair	RMS Surface Finish (m in.)	Gap Material	Mean Contact Temp. (°F)	Ref.
1	Aluminum (2024-T3)	48-65	Vacuum (10^{-4} mm Hg)	110	7
2	Aluminum (2024-T3)	8-18	Vacuum (10^{-4} mm Hg)	110	7
3	Aluminum (2024-T3)	6-18 (Not Flat)	Vacuum (10^{-4} mm Hg)	110	7
4	Aluminum (75S-T6)	120	Air	200	8
5	Aluminum (75S-T6)	65	Air	200	8
6	Aluminum (75S-T6)	10	Air	200	8
7	Aluminum (2024-T3)	6-8 (Not Flat)	Lead Foil (0.008 in.)	110	7
8	Aluminum (75S-T6)	120	Brass Foil (0.001 in.)	200	8
9	Stainless (304)	42-60	Vacuum (10^{-4} mm Hg)	85	9
10	Stainless (304)	10-15	Vacuum (10^{-4} mm Hg)	85	9
11	Stainless (416)	100	Air	200	8
12	Stainless (416)	100	Brass Foil (0.001 in.)	200	8
13	Magnesium (AZ-31B)	50-60 (Oxidized)	Vacuum (10^{-4} mm Hg)	85	9
14	Magnesium (AZ-31B)	8-16 (Oxidized)	Vacuum (10^{-4} mm Hg)	85	9
15	Copper (OFHC)	7-9	Vacuum (10^{-4} mm Hg)	115	9
16	Stainless/Aluminum	30/65	Air	200	8
17	Iron/Aluminum	—	Air	80	10
18	Tungsten/Graphite	—	Air	270	10

Aluminum pairs 1 and 2 show increased interface conductance with improved surface finish. Due to surface waviness, pair 3 exhibits lower conductance despite superior surface finish. Aluminum pairs 4, 5, and 6 although harder than pairs 1, 2, and 3, show higher h_i as a result of higher interface temperatures and because interstitial air improves overall conduction contribution. Aluminum pair 7 confirms the beneficial effect of inserting soft lead interface shim material to overcome waviness conductance loss in pair 3. Aluminum pair 8 illustrates, in comparison to pair 4, that use of shim material harder than aluminum degrades contact conductance. Stainless steel pairs 9, 10, and 11 indicate that sensitivity to high contact pressure increases as surface finishes improve.

In contrast to pair 8, the interface shim in pair 12 improves conductance (brass is softer than stainless steel). Magnesium pairs 13 and 14 suggest that oxidation of smooth surfaces can add thermal resistance sufficient to more than offset the usual effect of high surface finish; apparently oxide films have a greater effect on rough than on smooth surfaces. Copper pair 15 shows generally high conductance despite its vacuum interface. Joints of dissimilar materials, such as pairs 16, 17 and 18 are controlled in their contact-pressure sensitivity by the softer of the two materials. Another effect observed with dissimilar material interfaces is a dependence on direction of heat flow, this presumably arises from different temperature dependencies of both thermal and mechanical properties for the two materials and changes in interface warping when temperature pattern is inverted by reversing the direction of heat flow.

Figure 6. Thermal interface conductance data

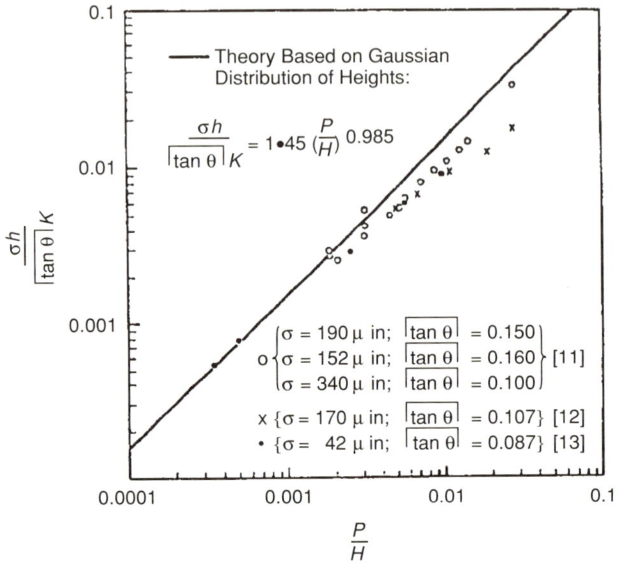

Figure 7. Data in support of Mikic theory

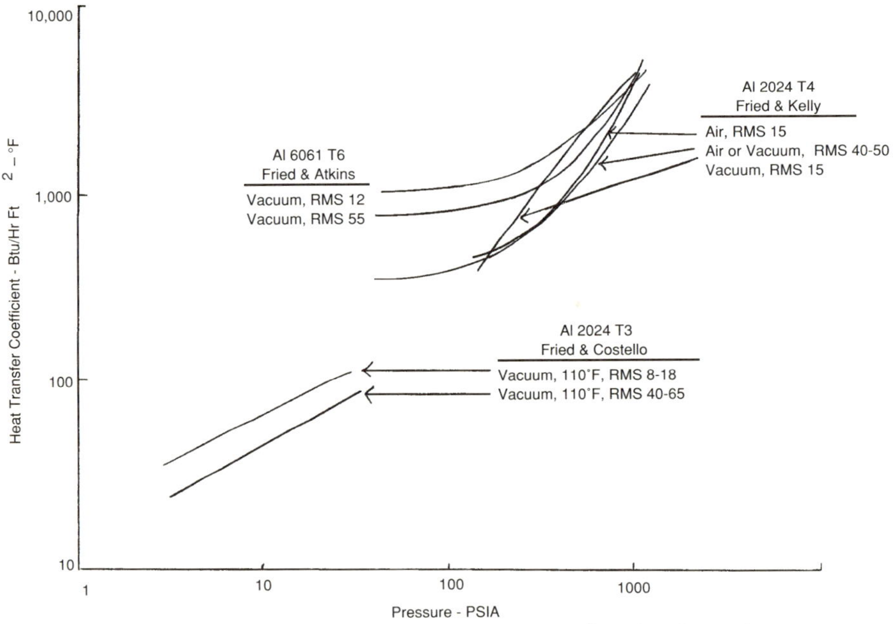

Figure 8. Heat transfer coefficient vs. pressure for aluminum in vacuum (RMS in microinches)

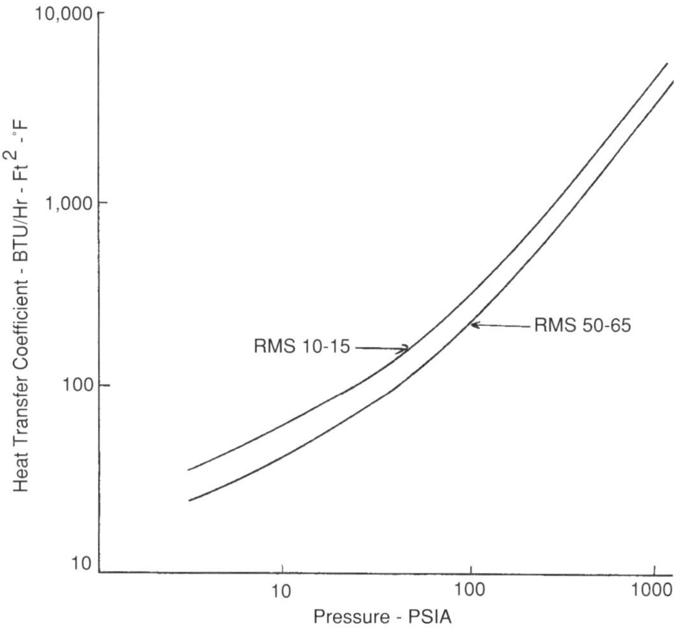

Figure 9. Generalized heat transfer coefficient vs. pressure for aluminum in vacuum

BOLTED JOINT CONDUCTANCE, NO INTERFACE FILLER

Theory

It is generally appreciated that at the macroscopic level, bolted plates deform elastically (Figure 10). Separation, though exaggerated in the figure, does occur, and at relatively small distances from the bolt. In its most basic statement, the bolted joint problem can be considered the contact-conductance problem for a non-uniform interface pressure. Figure 11a shows schematics of interfacial pressure profile with ($\sigma \neq 0$) and without ($\sigma = 0$) roughness.

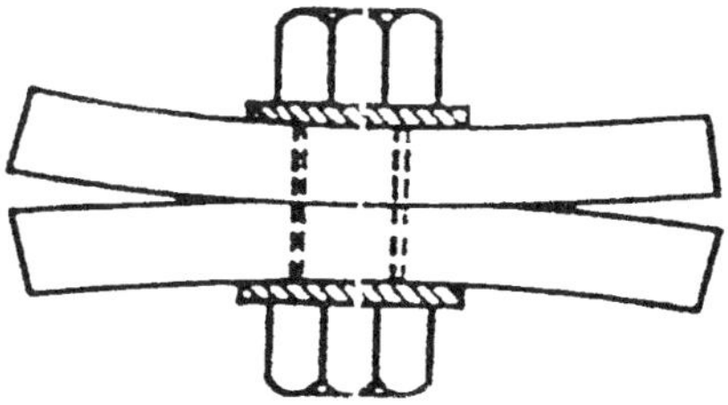

Figure 10. Bolted interface

An excellent theoretical treatment of this subject has been provided by Roca and Mikic (Refs. 17 and 18) for plates that when unstressed are nominally flat. A typical flatness specification for mounting plates calls for flatness less than 0.001 inch/inch and less than 0.01-inch overall for the footprint. Significant waviness or warpage cannot readily be analyzed and must be avoided in practice. Roca and Mikic extended the theory beyond the single plate mid-plane work of Fernlund (Ref. 19) and others to two plates with surface roughness. The biharmonic equation was used to characterize the elastic deformation of the plates. Their method assumed elastic deformation of the plates, asperity height above a mean plane is Gaussian, and asperity contact is normal with no tangential component. Both plastic and elastic asperity deformation were treated. The structural model used is shown in Figure 11b; Figure 11c shows typical calculated results obtained using an iterative method.

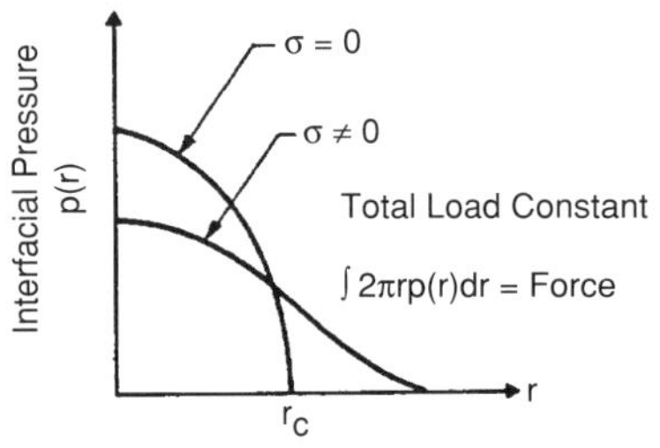

Figure 11a. Schematic showing interfacial pressure profiles with and without roughness

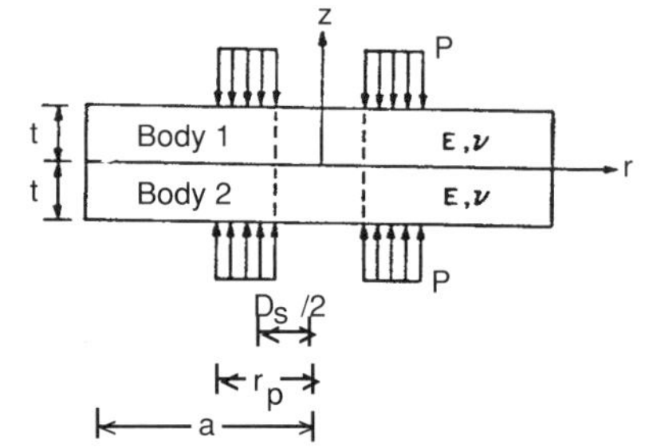

Figure 11b. Model used by Roca and Mikic for bolted joint

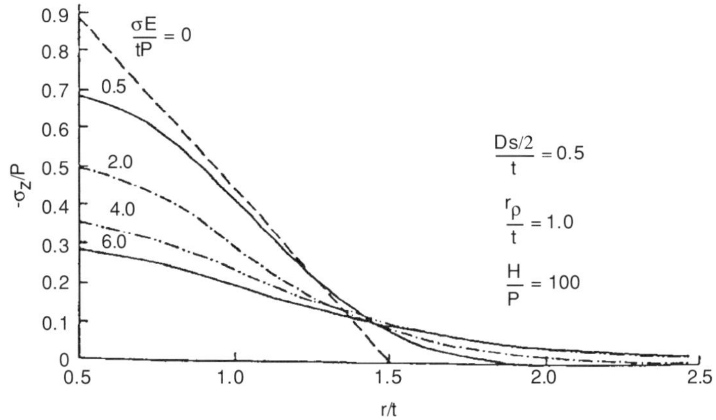

Figure 11c. Typical interfacial pressure profiles predicted by Roca and Mikic

The thermal model used by Roca and Mikic is shown in Figure 12a for the upper plate. Heat enters around the perimeter, flows radially inward and then passes from one plate to the other in the contact region. Boundary conditions are

$$k \frac{\partial T}{\partial Z} = h_c\,(r)\,[T-T_i] \qquad \text{at } Z=0$$

$$k \frac{\partial T}{\partial Z} = 0 \qquad Z=t$$

$$k \frac{\partial T}{\partial Z} = 0 \qquad r=D_s/2$$

$$k \frac{\partial T}{\partial r} = q \qquad r=R\ .$$

The heat transfer coefficient in the contact region, h_c, is a function of local pressure P(r) and is given by

$$h_c = 1.45 \frac{k \tan\theta}{\sigma} \left(\frac{P(r)}{H} \right)^{0.985} , \tag{4}$$

which is similar to Eq. (2).

Here, H = lesser of H_1, H_2.

Roca and Mikic define an overall resistance from the perimeter to a constant temperature (T_i) region on the other side of the interface

$$R = \frac{T(r=R, Z=t/2) - T_i}{F} , \tag{5}$$

where F is the perimeter heat flux. Their results are shown in Figures 12b, c, and d. Overall thermal resistances are seen to vary with roughness (with the group $\sigma E/tP$) in a complex way. The greater σ, the longer the constricted conduction path from surface to surface, and hence the greater contact resistance. However, as roughness increases, the contact radius increases by virtue of roughness height, compensating for plate deformation (Figure 10). This tends to decrease the overall thermal resistance by allowing the radial heat inflow to turn downward toward the other plate at a greater radius, decreasing the average heat flux in the contact region. The group E tan θ/H is the inverse of the group previously encountered for surfaces under uniform pressure.

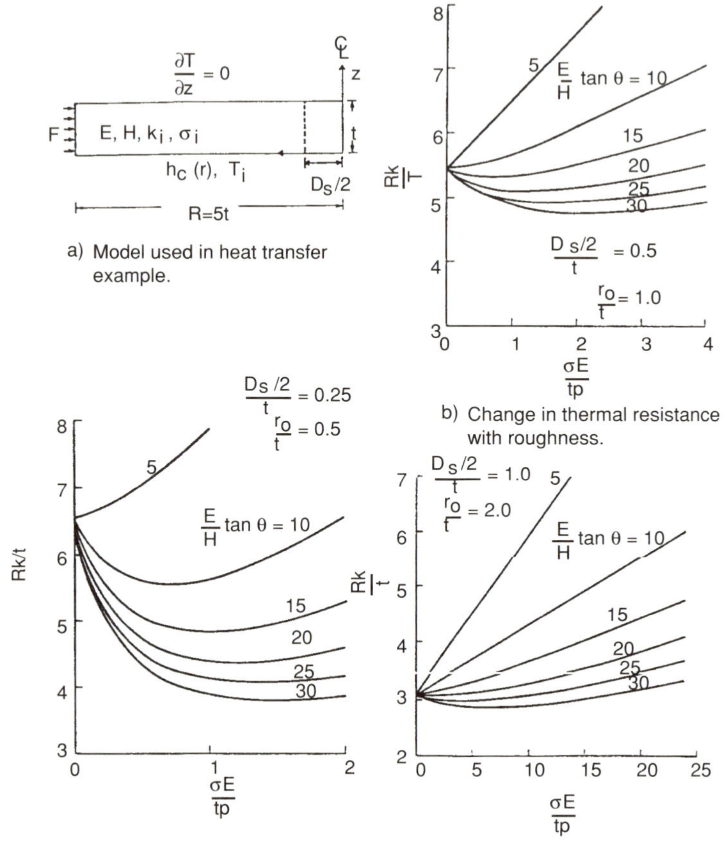

a) Model used in heat transfer example.

b) Change in thermal resistance with roughness.

c) Change in thermal resistance with roughness.

d) Change in thermal resistance with roughness.

Figure 12. Overall interface resistance

While the theoretical treatment of Roca and Mikic is the most complete that we know of, it is not particularly practical. Use of overall resistance comingles lateral plate conductance and contact conductance, providing many pages of graphical results, but has few design and analysis aids for the engineer. Bevans et al., (Ref. 20) use a simpler model (Figure 13). Two plates are bolted together with a contact region A_b with radius R_o. A uniform heat flux, F, is incident on the top plate and exits the bottom plate. Heat flows radially inward in the top plate (the lateral-plate flow) until the contact region is reached. Heat flows from the top plate to the bottom plate in this region. Heat flow in the bottom plate is the reverse of that in the top plate. Resistors are shown by jagged lines.

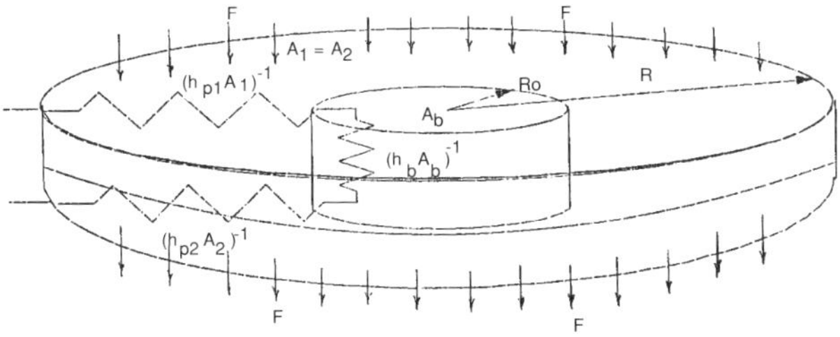

Figure 13. Bolted interface model

The steady-state heat-conduction equation for a differential element in the region between R and R_o for the top plate can be written

$$q_{out} - q_{in} = q_{absorbed} \tag{6}$$

or

$$k2\pi rt\frac{dT}{dr} - \left[k2\pi rt\frac{dT}{dr} + 2\pi kt\frac{d}{dr}\left(r\frac{dT}{dr}\right)dr\right]$$

$$= F\,2\pi rdr\ , \tag{7}$$

with the following boundary conditions

$$\text{At } r = R,\ dT/dr = 0$$

$$\text{At } r = R_o,\ T = T_o\ .$$

Eq. (7) can be integrated and solved for the temperature distribution across the plate in the region between the edge and the outer radius of bolt contact, yielding

$$T_o - T = \frac{FR^2}{kt}\left[\frac{1}{4}\left(\eta^2 - \eta_o{}^2\right) + \frac{1}{2}\ln\frac{\eta_o}{\eta}\right]\ , \tag{8}$$

where $\eta_o = R_o/R$. Following the work of Bevans, the conduction equation can be recast in integral form as

$$Q = h_p \int_{R_o}^{R} (T - T_o)dA \ , \tag{9}$$

where h_p can be considered a "heat-transfer coefficient" for the region between R and R_o.

Noting that $Q = F\pi(R^2 - R_o^2)$, there results

$$h_p = \frac{F\pi\left(R^2 - R_o^2\right)}{\int_{R_o}^{R} (T - T_o)2\pi r dr} \ . \tag{10}$$

Substituting Eq. (8) into Eq. (10) and integrating, the heat-transfer coefficient in the plate region from R to R_o becomes

$$h_p = \frac{2kt\left(R^2 - R_o^2\right)}{R^4\left[\eta_o^2 - \frac{\eta_o^4}{4}\ln\eta_o - \frac{3}{4}\right]} \ . \tag{11}$$

This heat-transfer coefficient is fictitious, as heat does not flow from the top to bottom plate in the region $R > R_o$. More properly, this is the coefficient that would exist if the uniform heat flux F flowed from the top to bottom plate by virtue of the temperature profile of Eq. (8).

Figure 13 shows a representation of two plates bolted together, containing annular regions and connected by a contact region of radius R_o. The overall resistance of this configuration is given by

$$\frac{1}{hA} = \frac{1}{h_{p1}A_1} + \frac{1}{h_b A_b} + \frac{1}{h_{p2}A_2} \ , \tag{12}$$

where the subscripts 1 and 2 refer to the top and the bottom plates, respectively. After replacement of h_{p1} and h_{p2} with Eq. (11), and noting that $A_2 = A_1$, the overall heat transfer coefficient of the approximated bolted joint becomes

$$h = \frac{1}{\dfrac{AR^4\left[\eta_o^2 - \eta_o^4/4 - \ln \eta_o - \dfrac{3}{4}\right]}{2A_1\left(R^2 - R_o^2\right)}\left[\dfrac{1}{k_1 t_1} + \dfrac{1}{k_2 t_2}\right] + \dfrac{A}{h_b A_b}} \quad . \tag{13}$$

If both plates are of the same material, $k_1 = k_2$, and using $A = \pi R^2$, $A_1 = \pi(R^2 - R_o^2)$, $A_b = \pi R_o^2$, and $I = \left[\eta_0^2 - \dfrac{\eta_0^4}{4} - \ln \eta_o - \dfrac{3}{4}\right]$,

$$h = \frac{1}{\dfrac{R^6 I(t_1 + t_2)}{2\left(R^2 - R_o^2\right)^2 k\, t_1 t_2} + \dfrac{1}{h_b\, h_o^2}} \quad . \tag{14}$$

The terms in this equation which are not known are R_o and h_b.

Practice, Contact Region

Bratkovich (Ref. 21) has recently developed a correlation for h_b. It is based on data from vacuum tests for aluminum alloy plates fastened by stainless-steel bolts or screws. The area on which h_b is based is $A_b = \frac{\pi}{4}(2D_s)^2$, where D_s is the bolt shaft diameter and equals R_o. Correlations were developed for both heat-transfer coefficients, h_b, and conductance, $C_b = h_b A_b$, and are given in Figures 14 and 15. The power law curves fits are, respectively,

$$\frac{h_b F_n}{k} = 44\left(\frac{T_q\, t_m\, \alpha T}{Y A_b^2}\right)^{0.437} \quad , \tag{15}$$

and

$$\frac{C_b}{k\, t_m} = 6.3\left(\frac{T_q\, \alpha T}{Y F_n^2\, t_m}\right)^{0.511} \quad . \tag{16}$$

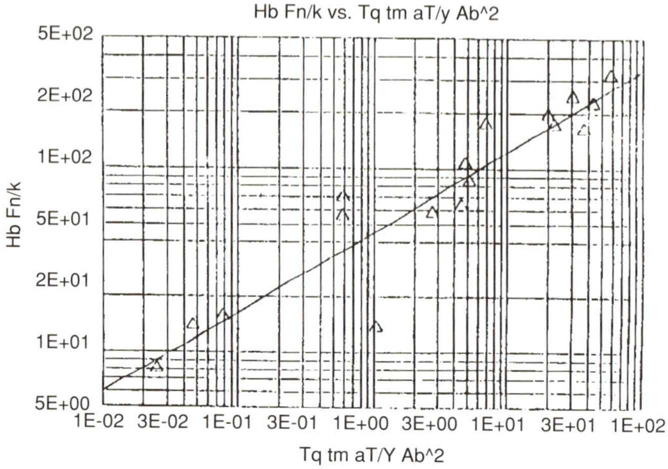

Figure 14. Final correlation: bolt region heat-transfer coefficients

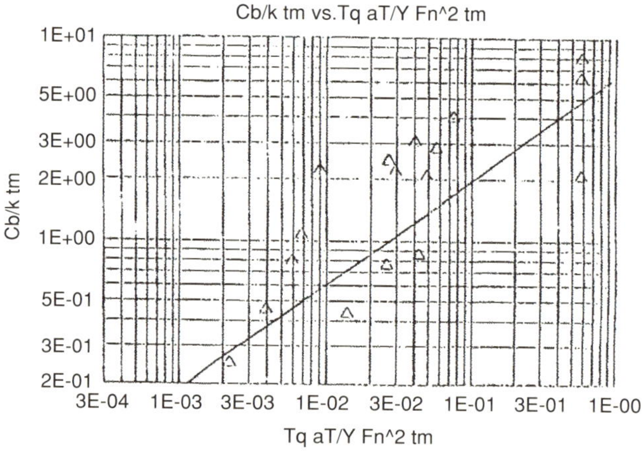

Figure 15. Final correlation: bolt contact-region conductance

Correlation coefficients for these two equations are, respectively, 0.936 and 0.859. Note that F_n is used instead of σ to represent root mean square (RMS) roughness. Also note that a and F_n lack the 10^{-6} multiplier. This is an oversight that will eventually be rectified, with equations and figures to be modified accordingly. However, these correlations can be used in their present form. A user's guide giving discrete values of the parameters in the correlations and sample problem to be used with Figure 15 are provided in Figure 16. Additional results obtained using the correlation are given in Table 2. It is recommended that the correlation be used within the range of the database provided in Figure 16.

Extrapolation beyond this range of parameters should be done cautiously. The strong dependence on roughness (-1.022 power) may need to be modified when a larger database becomes available.

Table 2. Bratkovich Bolted Joint Conductance

Bolted Joint Conductance from Bratkovich Correlation
T = 622R, Y = 33725 lb/in.^{2}, k = 97 Btu/hr ft°F

Torque in.-lb	Applicable Bolt	Finish microinch	Mean Thickness tm, inch	Conductance Btu/hr-°F	Resistance °C/W
19.8	8-32	25	0.125	1.57	1.2
		50	0.125	0.79	2.4
		25	0.5	3.14	0.6
		8	0.125	4.90	0.38
		50	0.5	1.57	1.2
31.7	10-32	25	0.125	2.5	0.75
		25	0.5	5.0	0.38
		8	0.5	15.6	0.12
		50	0.125	1.25	1.5
9.6	6-32	25	0.125	1.10	1.72
		25	0.5	2.20	0.86
		50	0.125	0.55	3.44
		50	0.5	1.10	1.72

Some discussion of the terms in Eqs. (15) and (16) is warranted. The term t_m is the arithmetic-mean plate thickness. The factor αT, where α is the coefficient of thermal expansion and T is the arithmetic-mean absolute temperature of the two plates, corrects the bolt torque, T_q, for differential thermal expansion. The differential thermal expansion is due to use of stainless-steel screws to fasten the aluminum alloy plates. If Eq. (15) is used, the area A_b should be equal to $\frac{\pi}{4}(2D_s)^2$. Use of Eq. (16) is recommended, as the conductances so calculated are independent of contact area.

It must be pointed out that these correlations apply only to the bolt or screw contact region and do not characterize the lateral conductances within the two plates. Overall conductances will be addressed in the next section.

TRW (Ref. 22) has provided nominal values of thermal resistance/conductance across screwed/bolted joints (Table 3). They allow that these values can be modified up or down depending on such parameters as screw torque, materials, surface finish, and flatness. The recommendations for small stiff surfaces can be thought to apply to the bolt contact region and can therefore be compared to the correlations of Eqs. (15) and (16). The recommendations for large thin surfaces apply to overall thermal conductance/resistance as defined in Eq. (12). The TRW recommendations are compared to the Bratkovich correlation Eq. (16) in Table 4. The first case is that of the sample calculation of Figure 16. The second case was derived from the work of Bevans et al. The Bratkovich correlation predicts higher conductances for the two cases and is seen to depend on a number of parameters not included in the TRW method.

User's Guide

Bolt (18-8 St. St)

Type	Shaft Diam, D_s, inch	Max. Torque, T_q, in.-lb.
0-80	0.060	—
2-56	0.086	2.5
4-40	0.112	5.2
6-32	0.138	9.6
8-32	0.164	19.8
10-32	0.19	31.7
1/4-20	0.25	75.2

Plate

Thickness, t_m, inch	Finish, F_n, microinch	Temp. Level, T, Rankine
0.0625	8	430
0.084	16	546
0.125	25	620
0.145	63	717
0.20		
0.3125		
0.675		

Properties (Al 6061-T6)

Thermal Conductivity k, Btu/Hr ft°F	Coeff. Linear Expansion $\alpha x 10^{-6}$, R^{-1}	Yield Stress, Y, lb/in.^{2}
97	11.8 (@430R)	36920 (@430R)
	12.6 (@558R)	35145 (@558R)
	13.25 (@622R)	33725 (@622R)
	13.9 (@717R)	31240 (@717R)

Sample Calculation

Plate 1 is 0.10 inch thick, plate 2 is 0.15 inch thick: $t_m = \frac{1}{2}(t_1 + t_2) = \frac{1}{2}(0.10 + 0.15) = 0.125$ inch

Bolt is #8-32, torqued to max. value
Each plate has same finish, 25 microinch RMS
Ave. plate temperature is 622R

$$\frac{Tq\,\alpha T}{Y F_n^2 t_m} = \frac{(19.8)\,(13.25)\,(622)}{(33725)\,(25)^2\,(0.125)} = 0.0619 \text{ (note that } 10^{-6} \text{ not used for } \alpha \text{ or } F_n)$$

From final conductance correlation: $c_b/k\,t_m = 1.55$

$$c_b = (1.55)\,(97)\,(0.125)/12 = 1.57\,\frac{\text{Btu}}{\text{hr°F}}$$

Figure 16. Bolted joint conductance data

For completeness, recommendations for bolt thermal-resistance obtained informally from LMSC are presented in Table 5. TRW and LMSC results are presented in consistent units in Table 6.

Table 3. Thermal Resistance/Conductance, Design Guideline from TRW

Screw Size	Small Stiff Surfaces				Large Thin Surfaces			
	Resistance		Conductance		Resistance		Conductance	
	(HR=F/ BTU)	(C/Watt)	(BTU/ HR-F)	(Watt/C)	(HR=F/ BTU)	(C/Watt)	(BTU/ HR-F)	(Watt/C)
2-56	2.5	4.74	0.4	0.21	5.0	9.48	0.20	0.105
4-40	2.0	3.79	0.5	0.26	4.0	7.59	0.25	0.132
6-32	1.25	2.37	0.8	0.42	3.0	5.69	0.33	0.176
8-32	0.66	1.25	1.52	0.80	2.0	3.79	0.50	0.264
10-32	0.4	0.76	2.5	1.32	1.0	1.90	1.00	0.527
1/4-28	0.15	0.28	6.67	3.51	0.5	0.95	2.00	1.054

Table 4. Comparison of TRW Recommendations with Bratkovich Correlation for Contact Region Conductance

Case	Parameters	Conductance, BTU/hr°F: TRW Recommendation for Small Stiff Surfaces	Conductance, BTU/hr°F: Bratkovich Correlation for Bolt Region
Sample Calculation Figure 3.2-3	t_m = 0.125 inch No. 8 32 bolt T_q = 19.8 inch-lb T = 622 Rankin RMS Roughness = 25 microinch Plates: 6061-T6 A ℓ Screws: 18-8 St. St.	1.52	1.57
Bevans et. al.	t_m = 0.09375 inch No. 10-32 bolt T_q = 24 in. lb T = 540R RMS Roughness = 11 microinch Plates: 6061-T6 A ℓ Screws: 18-8 St. St.	2.5	3.84

Practice, Overall Conductances

Most problems involve lateral heat flow within plates as well as heat flow from plate to plate in the contact region. So far we have dealt only with the latter. For axisymmetric heat flow to the bolt region, overall conductance is given by Eqs. (13) or (14).

Table 5. Bolt Thermal Resistance Estimate from G. D. Roads of LMSC

Maximum Resistance Versus Bolt Size and Plate Thickness (°F/Watt Bolt)

Steel Bolt Diameter (Threads/Inch)		0.062 Aluminum	0.125 Aluminum	0.250 Aluminum	0.375 Aluminum
NC 4-40	0.112	22.7	–	–	–
NC 6-32	0.138	11.9	3.96	–	–
NC 8-32	0.164	8.1	2.7	1.35	–
NF 10-32	0.190	5.4	1.8	0.9	0.6
NF 1/4-28	0.250	3.78	1.26	0.63	0.42
NF 5/16-24	0.312	2.7	0.9	0.45	0.3
NF 3/8-24	0.375	–	0.7	0.35	0.23
NF 7/16-20	0.437	–	–	0.29	0.19
NF 1/2-20	0.500	–	–	–	0.16

Assumptions:
- Typical Spacecraft Bolted Aluminum Interface in Vacuum
- Bare Clean Mill Rolled Surface Finish
- Standard Steel Bolts Torqued to Specification
- Primary Heat Transfer through Compressed Area Near Bolt

Note: Confirmation Measurements Suggested for Thermal Design Purposes

Reference: NASA CR119933 June 71 and Other Limited Measurements

Table 6. TRW and LMSC Bolted Joint Conductance Data

◆Resistance Values from Several Sources, °C/Watt

		*TRW-Large	±LMSC (Plate Thickness, In.)				*TRW–Small
Bolt	Diam., In.	Thin Surfaces	(0.062)	(0.125)	(0.250)	(0.375)	Stiff Surfaces
2-56		9.48					4.74
NC 4-40	0.112	7.59	12.6				3.79
NC 6-32	0.138	5.69	6.61	2.2			2.37
NC 8-32	0.164	3.79	4.5	1.5	0.75		1.25
NF 10-32	0.190	1.90	3.0	1.0	0.5	0.333	0.76
NF 1/4-28	0.250	0.95	2.1	0.7	0.35	0.233	0.28
NF 5/16-24	0.312		1.5	0.5	0.25	0.167	
NF 3/8-24	0.375			0.39	0.194	0.128	
NF 7/16-20	0.437				0.16	0.106	
NF 1/2-20	0.500					0.089	

∗ TRW, March 1984

± LMSC, George D. Rhoads, 20 July 1988

◆ Bolted aluminum interface in vacuum, bare clean mill rolled surface finish (LMSC), standard steel bolts torque to specification (LMSC) primary heat transfer through compressed area near bolt (LMSC)

Perimeter Bolt Pattern

For configurations where a perimeter bolt pattern is used, the method of Bevans is recommended. The plate is divided into pie-shaped sections (Figure 17) whose dimensions are used to calculate contact conductances. Eqs. (6) to (14) are applicable, where R_O (Figure 18) is now twice the bolt-shaft diameter (previously R_O equaled the bolt-shaft diameter). For more complex shapes or where thicknesses are not constant, the overall thermal network can be modeled using finite differences or finite element methods. Contact-region conductances from Figures 14 and 15 can be incorporated into such models.

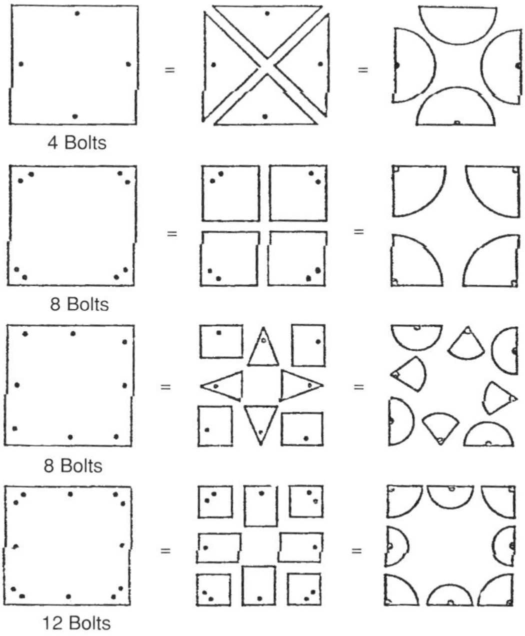

Figure 17. Division of plates with perimeter bolt mounting from Bevans et al.

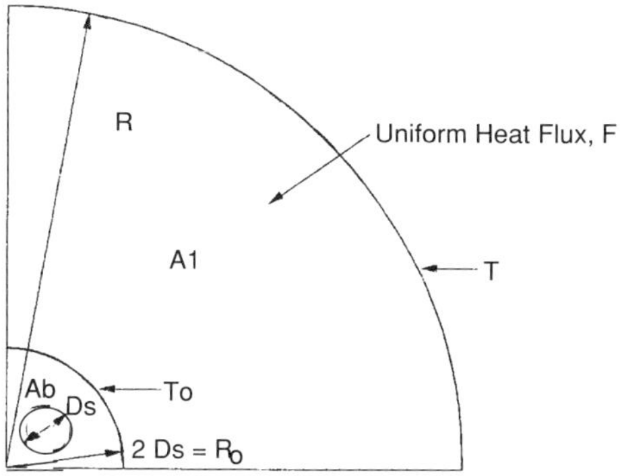

Figure 18. Elementary conduction element, four bolts, perimeter mounted

Where the perimeter-bolt pattern employs bolts on two opposed flanges, a rectangular version of Bevans' equation can be used. Here we have a strip between two bolts, subjected to a uniform flux, F (Figure 19). Following Bevans for a half slice

$$T - T_o = \frac{F}{2kt}\left(L^2 - x^2\right) \tag{17}$$

$$h_p = \frac{FW\,L}{\int_o^L (T - T_o) w\,dx} \tag{18}$$

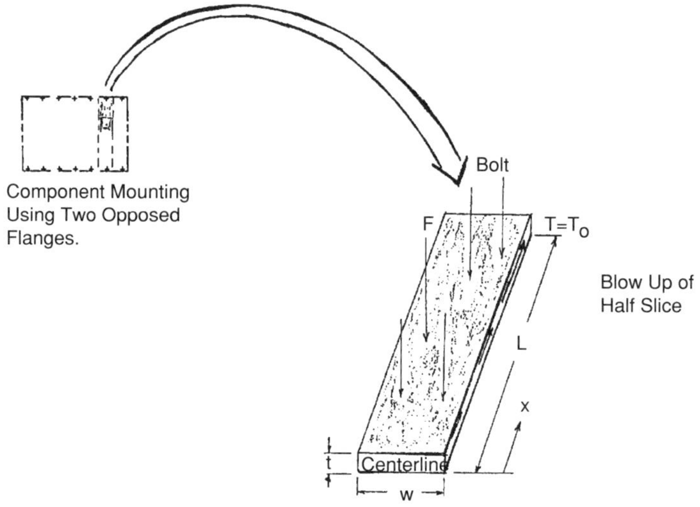

Figure 19. Model of conduction heat flowing in a slice

Substituting (17) into (18) and integrating

$$h_p = \frac{3kt}{L^2}\ , \tag{19}$$

plate conductance is

$$c_p = h_p A_p = \frac{3kt}{L^2} WL = \frac{3ktw}{L}\ . \tag{20}$$

It is interesting to note that if heat entered the half slice entirely at the centerline end, the conductance would be

$$c_p = \frac{3tw}{L} \quad . \tag{21}$$

This is one-third the conductance for the uniform-heat-flux case. Eqs. (19) or (20) can be used with bolt-region heat-transfer coefficient or conductance to obtain overall heat-transfer coefficients (see Eqs. 12 through 14) or overall conductances.

A design recommendation is available from TRW (Ref. 22) for average overall heat-transfer coefficients for perimeter-bolt patterns (Figure 20). This derives from the work of Bevans et al. for the configurations of Figure 21. Plates are relative thin, with t_m = (0.0625 + 0.125)/2=0.09375 inch. Results are characterized by inverse screw density (square inches per screw), with the caveat that screw spacings closer than 1 inch on center provides little further benefit. Heat-transfer coefficients for the bare interface are seen to be small, generally below 20 Btu/hr ft^2 deg F. This is a consequence of the lateral plate resistances for the long spans between bolts, i.e., bolt-contact-region conductances are relatively high compared to the lateral-plate conductances.

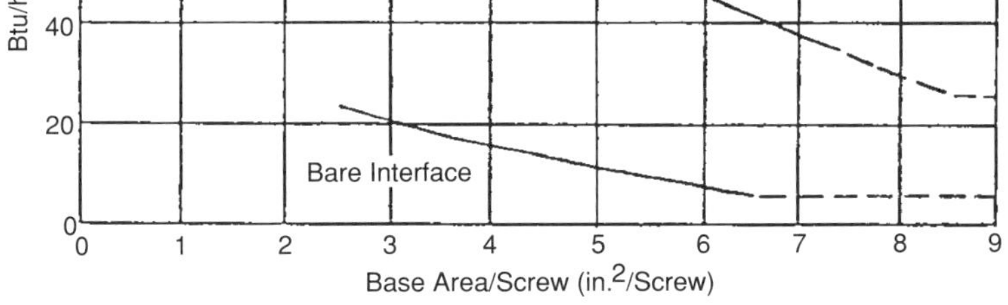

Figure 20. Recommended overall heat transfer coefficients for perimeter bolt pattern from TRW

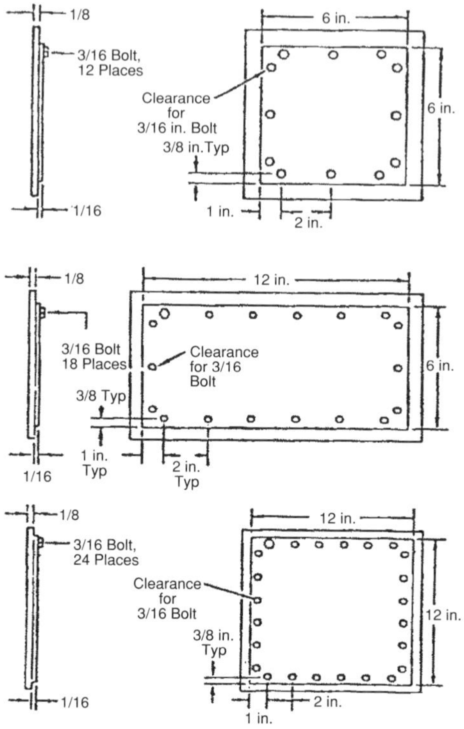

Figure 21. Bolted joint configurations tested by Bevans

This is verified and supplemented by the work of Welch and Ruttner (Ref. 23). They studied a configuration similar to that of Figure 22 with aluminum 6063-T6 plates that were each 5/16 inch thick. The 11-inch by 6-inch plates were fastened by sixteen no. 8-32 stainless-steel screws. Welch provided average heat-transfer coefficients for the entire plate. These are given in Table 7.

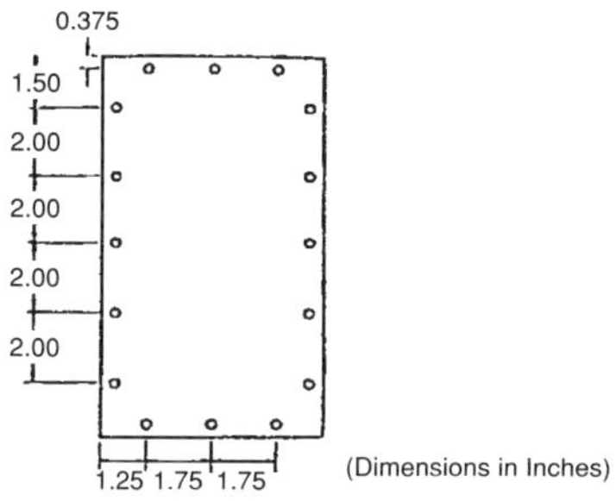

Figure 22. Bolted joint configuration tested by Welch

Table 7. Welch and Ruttner Bare Interface Results Summary

Torque in-lb.	Temperature °F	Ave. Heat Transfer Coefficients, BTU/hr ft^2 deg F
10	-30	50
10	160	65
20	-30	58
20	160	70

The Welch and Ruttner screw-spacing results in 4.125 in^2/screws, which for the Bevans configuration yields a heat-transfer coefficient of about 16 BTU/hr ft^2 deg F. The Welch and Ruttner coefficients are a factor of three to four greater than those of Bevans. This is explained to a large degree by the ratio of mean plate thicknesses for the two investigators:

$$\frac{(t_m)\text{Welch}}{(t_m)\text{Bevans}} = \frac{\frac{5}{16}}{\frac{1}{2}\left(\frac{1}{8}+\frac{1}{16}\right)} = \frac{0.3125}{0.09375} = 3.33 \ .$$

Therefore, for a perimeter-bolt pattern fastening an electronic unit to a mounting plate, the bare interface curve of Figure 20 should be used for mean plate thickness of about 0.1 inch. Table 7 should be used for thicker plates in the neighborhood of 0.3 inch thick. Results obtained this way can be cross-checked by computing overall heat-transfer coefficients (e.g., Eqs. 13 or 14) or by developing SINDA or NASTRAN thermal models. Contact region conductances are to be obtained for either approach from the Practice, Contact Region section.

Non-Perimeter Bolt Patterns

Figure 4 shows a bolt pattern that combines perimeter and inboard bolts as described in the Conduction Cooling Section. This configuration can be analyzed using a thermal math model and employing either the correlations of the Practice, Contact Region section or the TRW recommendations of the same section for short stiff surfaces. If a quick, rough estimate of overall conductances for a uniform bolt pattern is sought, Gluck's earlier correlation (Ref. 24) can be used (Figure 23). This earlier correlation predicts higher conductances than do the more detailed methods.

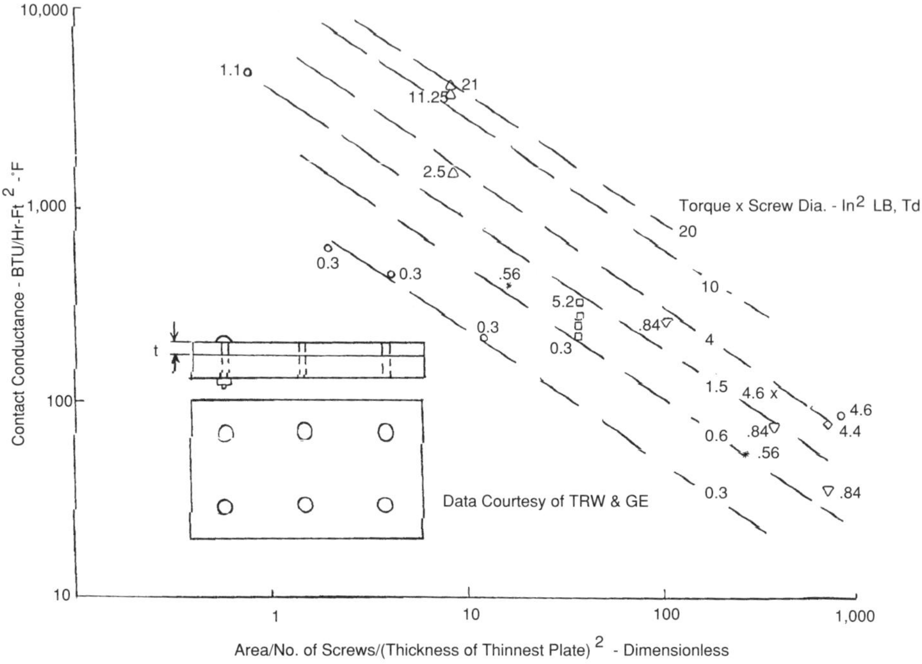

Figure 23. Rough correlation for overall conductance for a uniform bolt pattern

Complex Configurations for Bare Interfaces

Honeycomb Mounting Plates

Often the spacecraft side of the interface is of facesheet/honeycomb construction (Figure 24). Such construction provides high ratios of stiffness and strength to weight. However, facesheet thickness for practical applications can be well below 0.1 inch, typically 15 to 45 mils. Lateral-plate resistance is relatively high, and overall conductances can be of the order of those of the lower curve of Figure 20. Additionally, threaded inserts must be embedded in the honeycomb to receive the screws. Honeycomb face sheets tend to be less flat than metal plates. This can reduce contact-region conductance. However, the face-sheet honeycomb combination tends to be stiffer than thin mounting plates. This can result in large contact-region radius and increased contact-region conductance.

There is little experimental data on conductances where honeycomb-mounting-panel construction is used, either in the contact region or for the entire mounting region. Unless data is available, it is recommended that overall conductance values no higher than those for the bare interface in Figure 20 be used.

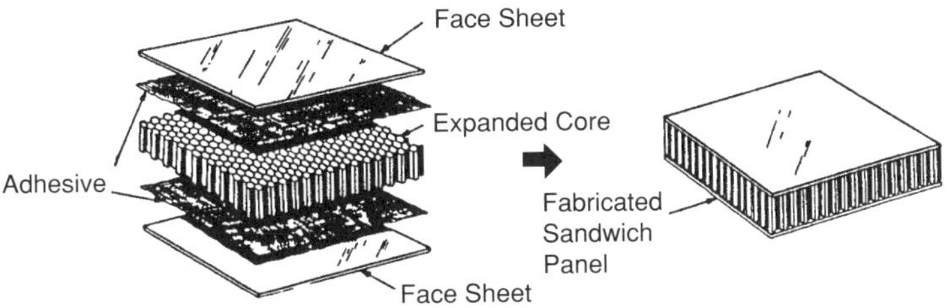

Figure 24. Honeycomb mounting panel

If heat pipes are embedded in the honeycomb below the component, panel conductance increases. General results are not yet available. However, the problem can be treated by developing thermal math models that account for contact conductance, face-sheet lateral conductance, and honeycomb and heat-pipe conductance.

Thermal Doublers

For electronic components where power dissipation per unit surface area is large, excessive temperatures can occur if heat is directly conducted from the baseplate to the spacecraft mounting plate. Temperatures can be reduced if a high-conductivity heat spreader is used between the component and the heat sink. Such spreaders, or thermal doublers, function by conducting heat laterally from high-power dissipation regions before final transport to the spacecraft mounting plate. Bobco and Starkovs (Ref. 25) analyzed a rectangular doubler of uniform thickness (Figure 26). Starkovs (Ref. 26) expanded the analysis to two heated footprints on a rectangular doubler (Figure 25), while Bobco (Ref. 27) analyzed various types of terraced doublers (Figure 27).

In Reference 25, Bobco and Starkovs develop and solve the equation

$$k\delta_o\left[\frac{\partial^2 T}{\partial x^2} + \frac{\partial^2 T}{\partial y^2}\right] - h(T - T\infty) = -q(x, y) \ , \qquad (22)$$

which goes with Figure 26. The term h is an overall heat-transfer coefficient from the doubler to the mounting plate. In this formulation, the doubler is assumed to be sufficiently thin so that there is no temperature gradient in the Z direction. Not included in the above three analyses is the additional contact interface associated with use of a doubler.

Considerable analytic results were obtained in these three investigations. Typical results are shown in Figure 28 for Ref. 25 and Figure 29 for Ref. 26.

Bobco (Ref. 25) points out that thermal doubler design is an intricate task involving constraints of component baseplate area, available mounting-plate area, and surrounding environment. The intent is to provide the lightest-weight practical doubler design consistent with these constraints that satisfies the maximum-allowable component-temperature requirement. Analytic solutions should be confirmed by finite-difference or finite-element thermal math-model results.

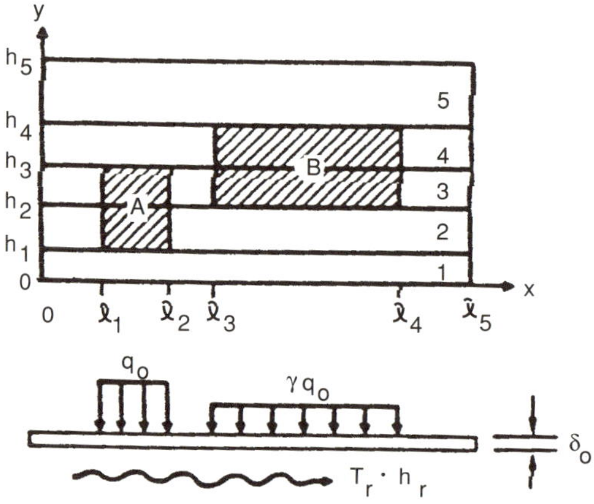

Figure 25. Schematic of a rectangular doubler with two heated footprints

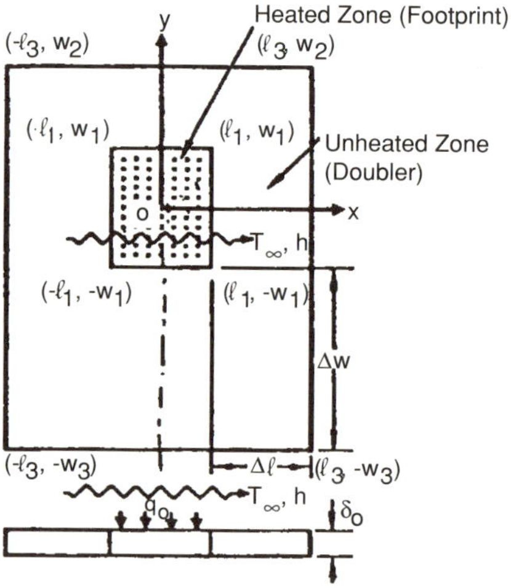

Figure 26. Schematic of thermal doubler with single symmetry

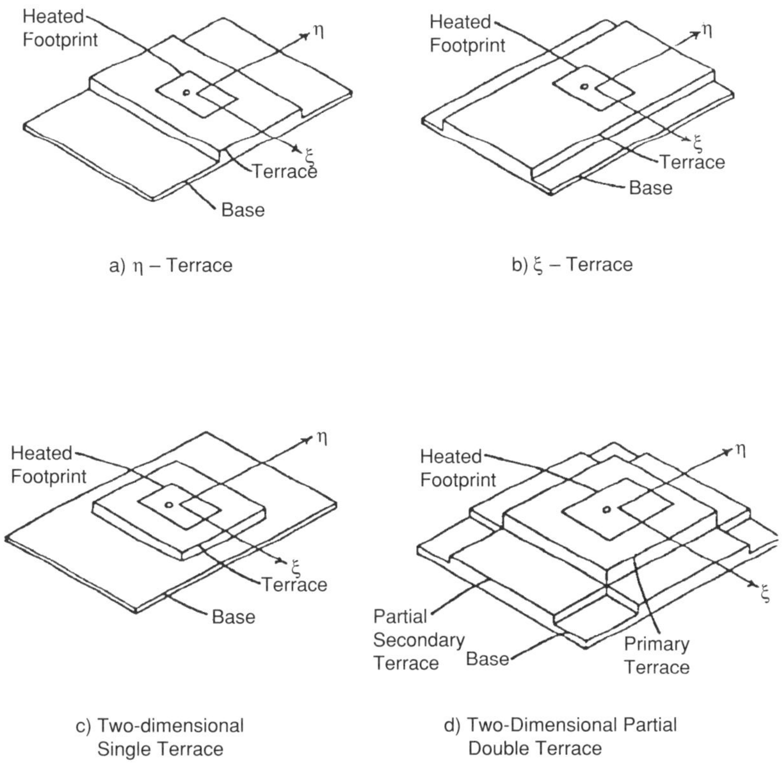

Figure 27. Four terraced double configurations that allow closed-form, separable solutions

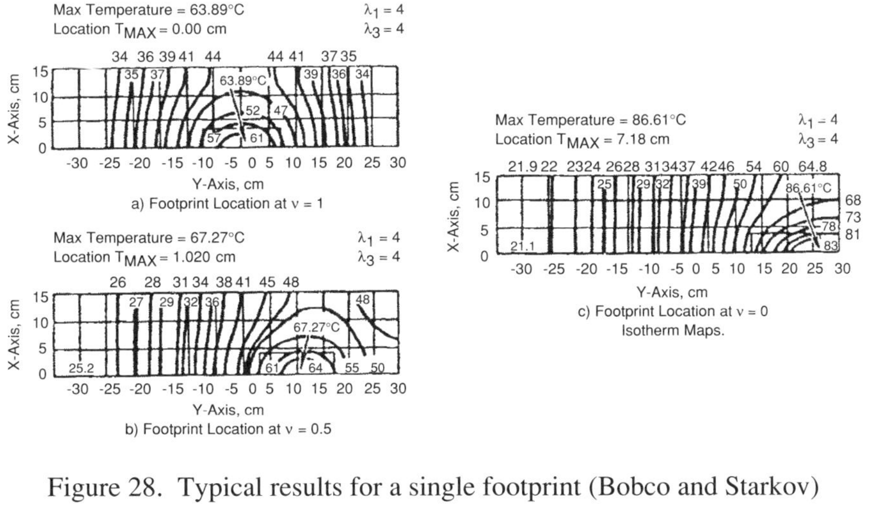

Figure 28. Typical results for a single footprint (Bobco and Starkov)

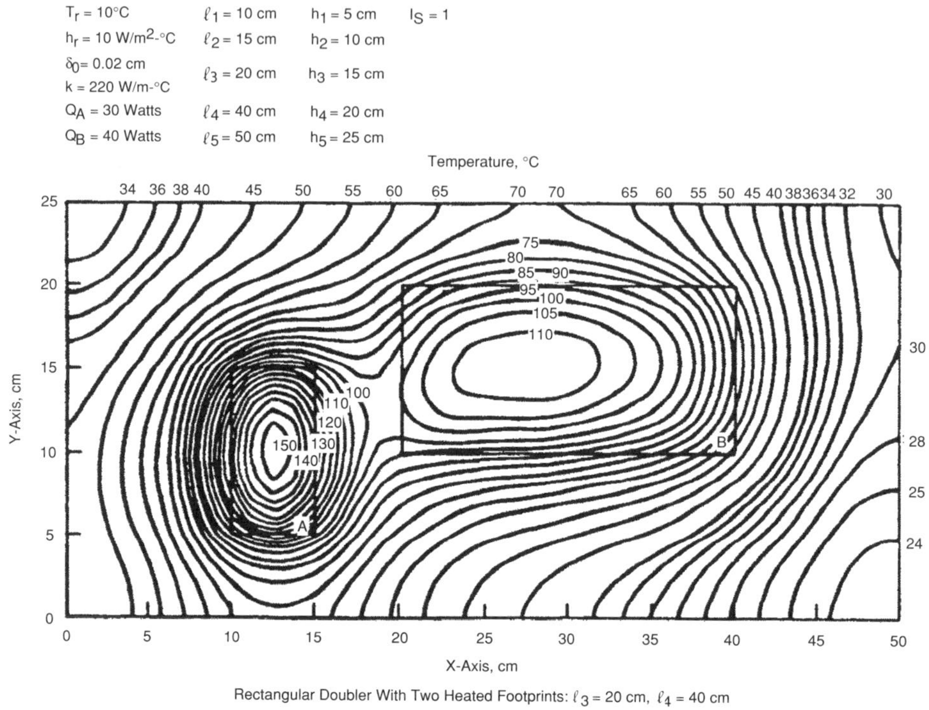

Figure 29. Typical results for two footprints on a rectangular doubler

Gluck and Young (Ref. 28) relieved the assumption of no vertical temperature gradient for an axisymmetric doubler (Figure 30). Here k is thermal conductivity, h is the contact heat-transfer coefficient, $\phi = T - T_o$, T is temperature within the doubler, and T_o is the mounting-plate (sink) temperature.

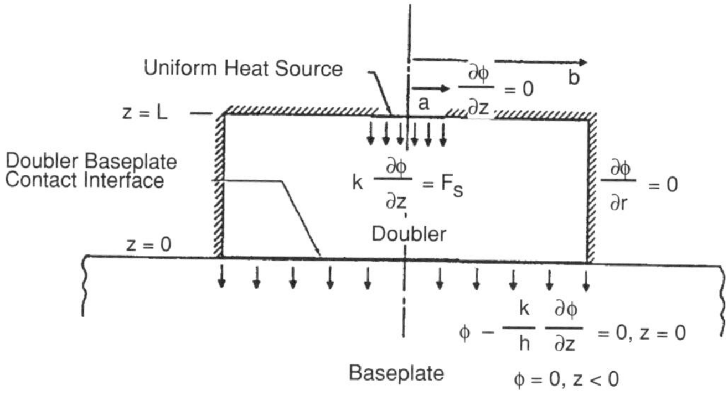

Figure 30. Axisymetric doubler model (Gluck and Young)

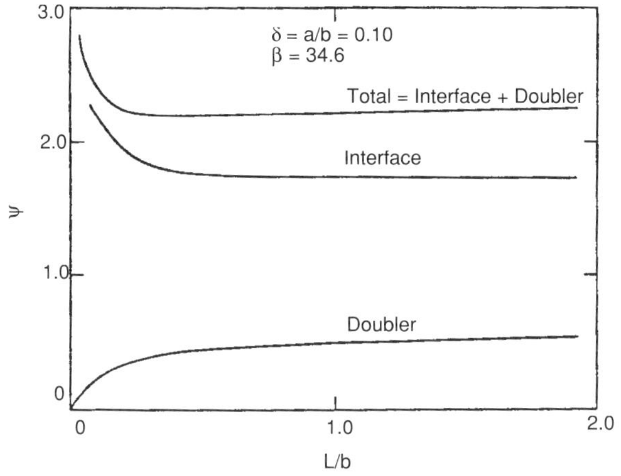

Figure 31. Dimensionless centerline temperature rise vs. doubler height

For this problem, formulation of an optimum doubler thickness results, beyond which component temperature increases. This is a consequence of the combined effects of increased spreading (which reduces the temperature rise across the contact interface) and increased doubler thermal resistance with doubler thickness (Figure 31). Results are governed by an inverse Biot-like group, β = k/hb, for small values of which heat flows tends to be columnar (no spreading) and for large values of which heat flow diffuses radially (perfect spreading), Figure 32a. The term F_o is the average heat flux across the crossection. Performance results are presented in Figure 32b, c, and d, where ϕ and ψ are the centerline values at Z=L. For L/B=0.01, collimation is noted for $\beta < 10^{-2}$, and nearly perfect diffusion is noted for $\beta > 10^{-2}$. As L/b increases, the collimation region diminishes and the perfect diffusion region begins at about $\beta = 10^{-2}$. Figure 32e and f present optimization results. The former presents a plot of the minimum value of $\psi_{r=o,z=h}$ versus β for eight values of δ. The latter presents the values of L/b that correspond to these minimum values.

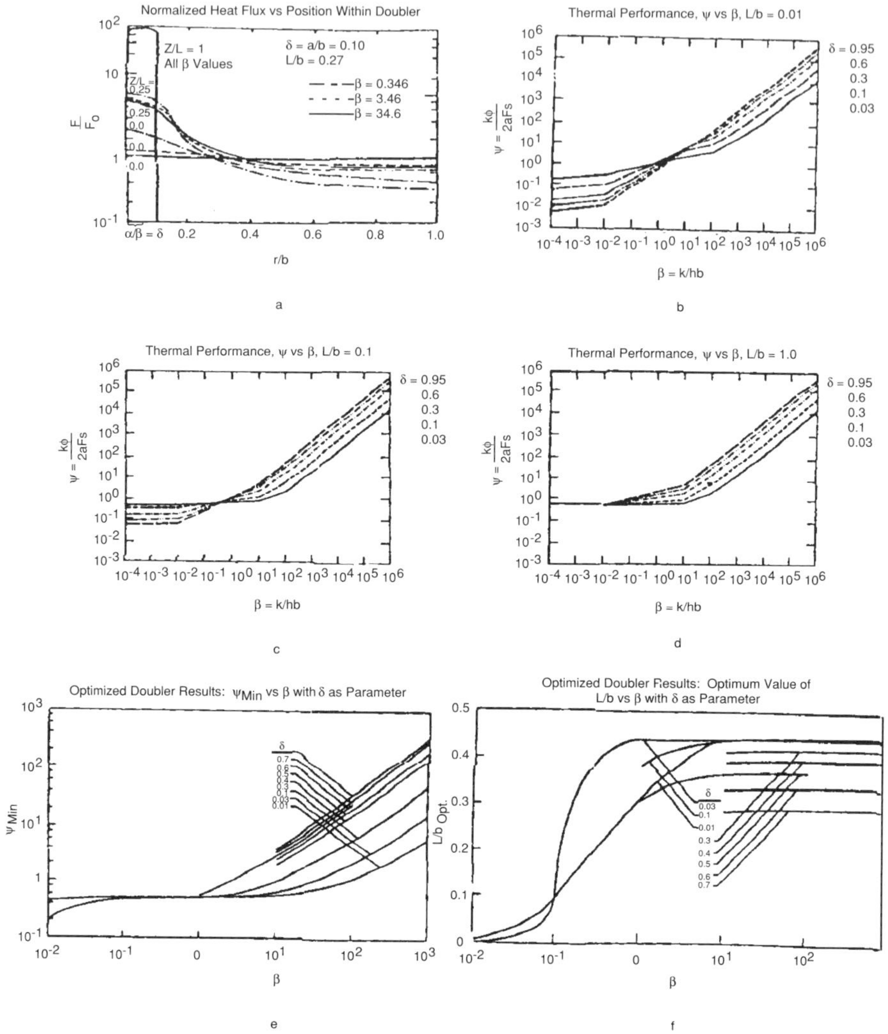

Figure 32. Doubler performance

Thermal/Structural Analysis

An example of a combined thermal and structural analysis was previously discussed with regard to the work of Roca and Mikic. More recently, finite-element codes have been used to conduct such analysis. Layton (Ref. 29) conducted a thermal/structural study of a traveling wave tube (TWT) using ABACUS and NASTRAN, with PATRAN used for graphical display. As part of that analysis, local heat-transfer coefficients were determined between the TWT baseplate and the cold plate to which it was mounted. Fastening was by two screws through a flange on one side of the TWT. It was assumed that both surfaces were perfectly flat and smooth. Computed pressure profiles and heat-transfer coefficients (Btu/hr-in^2-deg F) are shown in Figures 33a and b. The highest pressures are observed nearest each of the bolts, while pressure decreases to nearly zero at some distance away. Layton characterized local heat-transfer coefficient as a function of pressure from the work of Swartz (Figure 9) and others. He used the integration method of Goit (Ref. 30), and a multivariable interpolation routine to determine from the pressure profile average heat-transfer coefficients for each element.

a. Pressure profiles

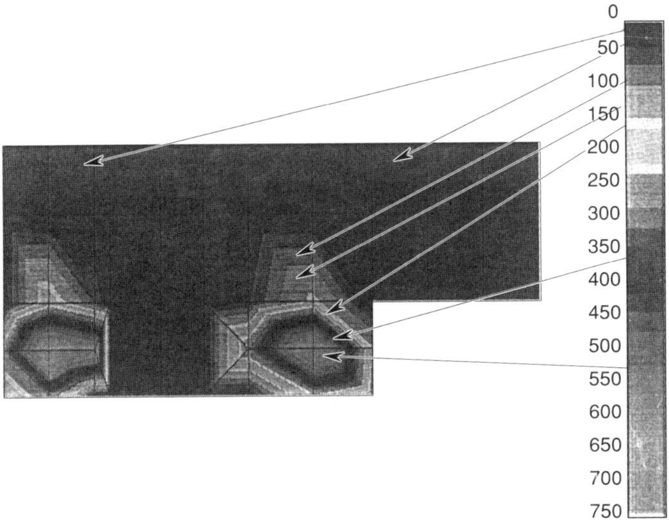

b. Local heat transfer coefficients (Btu/hr-in^2-°F)

<table>
<tr><td>0.89</td><td>0.86</td><td>0.90</td><td>0.85</td><td>0.82</td><td>0.86</td><td>0.91</td><td>0.86</td><td>0.83</td><td>0.80</td><td>0.80</td></tr>
<tr><td>1.82</td><td>1.95</td><td>0.93</td><td>0.95</td><td>0.93</td><td>0.86</td><td>1.75</td><td>1.68</td><td>0.82</td><td>0.82</td><td>0.80</td></tr>
<tr><td>10.25</td><td>13.69</td><td rowspan="2">5.70</td><td rowspan="2">1.00</td><td rowspan="2">1.01</td><td rowspan="2">2.64</td><td>9.68</td><td>8.12</td></tr>
<tr><td>12.29</td><td>14.26</td><td>9.82</td><td>7.77</td></tr>
</table>

Figure 33. Finite-element thermal/structural analysis result from Layton

BOLTED JOINT CONDUCTANCE, WITH INTERFACE FILLER

Contact conductance can be improved through the use of a filler material between the two plates. Such materials fill the microscopic voids present because of surface roughness (Figure 5). As the dimensions are small, even low-conductivity material can provide an improvement over the previous radiation mode of heat transfer. A variety of fillers have been studied and used over the years. They fall into three classes: greases, gaskets, and cured-in-place room-temperature-vulcanized (RTV) silicon compounds. Use of such fillers can create problems not present with bare interfaces. These problems include inadequate component grounding, inability or difficulty in removing component for rework, contamination, and structural loads. Fletcher et al. (Ref. 31), who studied a multitude of fillers, found that silicon grease provided the best thermal performance (Figure 34). It was subsequently found that such greases can become major sources of contamination. This has resulted in their essential elimination from most spacecraft applications.

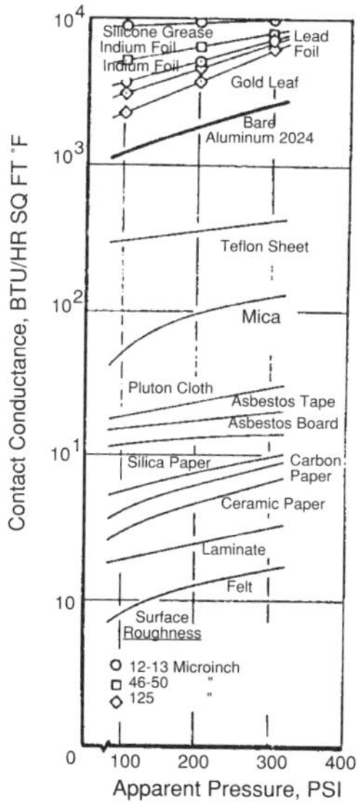

Figure 34. Thermal contact conductance of selected interstitial materials

Thermal Gaskets

A variety of thermal gaskets have been considered for interface filler applications. They include the gaskets shown in Figure 34 as well as Calgraph®, and tailored composites such as CHO-THERM® and COHR lastic®. Application of such gaskets is shown in Figure 35a, b, and c. To provide the desired thermal performance, many of these gaskets must be subjected to high pressures

(Figure 36). This impacts structural loads and can cause bowing of the mounting panel. Moreover, separation (zero pressure) may occur at some distance from the bolt (Figure 37). Because of this, use of such gaskets has been limited to small-scale applications, where the span between bolt centerlines is not large.

Chomerics, a division of Grace, Inc., provides a variety of thermal gaskets under the trade name CHO-THERM® (Table 8). They typically use thermally conductive but electrically isolating compounds (aluminum oxide, magnesium oxide, boron nitride) within an elastomeric binder (silicone, fluorosilicone, urethane). They are tailored to provide a variety of special capabilities: high- or low-dielectric strength, EMI shielding, highly conformal, solvent resistant, temperature resistant, and cut-through resistant. Table 9 provides properties for a number of CHO-THERMs®. Elastomeric properties that are important in the usage of thermal gaskets are: compressive deflection, stress relaxation, and compressive set.

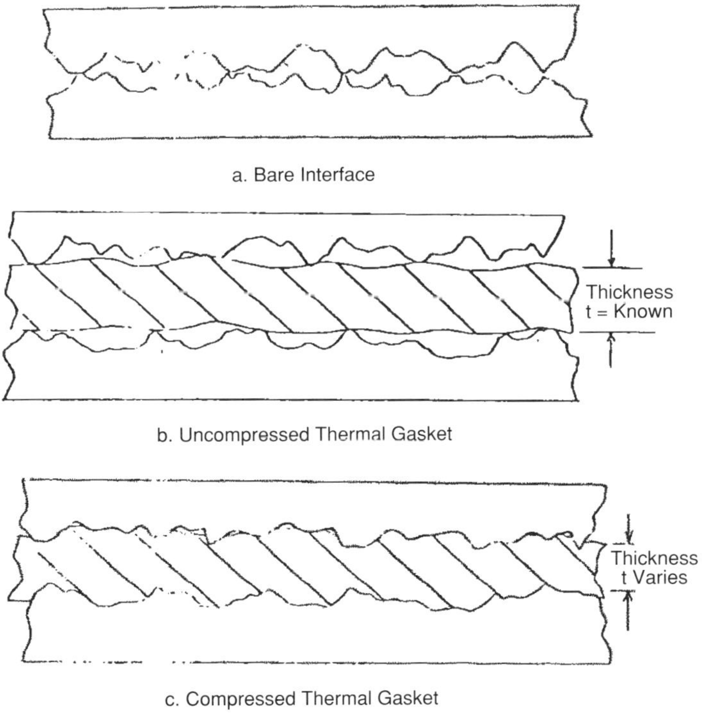

Figure 35. Use of thermal gaskets as an interface filler

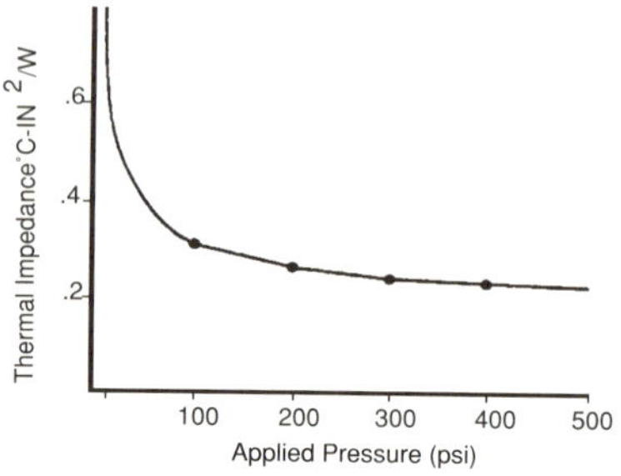

Figure 36. Thermal impedance vs. pressure for CHO -THERM 1671 material

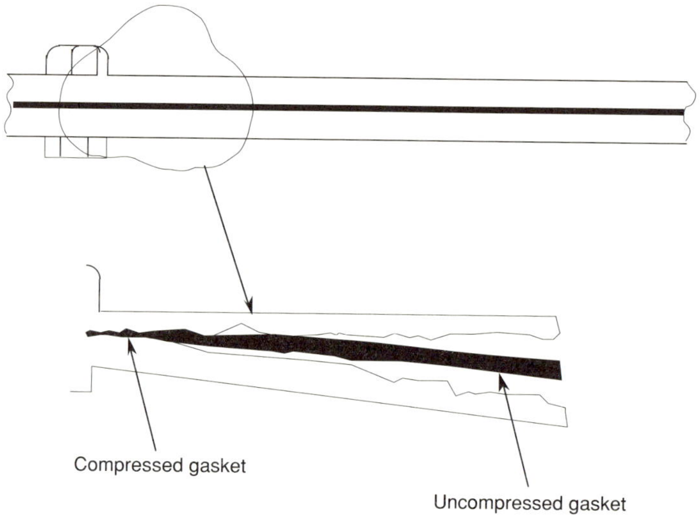

Figure 37. Bolted joint configuration with gasket

CHO-THERM® has a serious limitation when applied as a thermal gasket for component mounting in that it has an extremely high electrical resistivity, of the order of 10^{14} to 10^{15} ohm-cm. A continuous sheet may preclude meeting component electrical-grounding requirements. A typical requirement is that the electrical resistance from component mounting is to be less than 2.5 milliohms. Therefore, if CHO-THERM® is used, an auxiliary grounding method should be considered. This could be gasket cut-outs in the bolt region where compressible wire-grounding mesh is installed, or the use of grounding straps.

Table 8. CHO-THERM® Properties and Features

L Low M Moderate H High	Properties						Features						
	Thermal Impedance Low Moderate	Dielectric Strength High Moderate	Fiberglass Reinforced	Polyimide Reinforced	Polyester Reinforced	Aluminum Reinforced	No Mounting Pressure Required	No Silicone Contained	Continuous Rolls for Automation	Custom Molded Pads	Thermal Adhesive	Resists Petrochemicals and Hydrocarbons	NASA Outgassing Approved
1679 Boron Nitride, Silicone	L	M	•										•
1671 Boron Nitride, Silicone	L	M	•										•
1661 Boron Nitride, Silicone	L	M											•
1678 Boron Nitride, Silicone	L	M	•										
1674 Aluminum Oxide, Silicone	M	M	•						•				
1677 Boron Nitride, Fluorosilicone	M	M	•									•	•
1682 Magnesium Oxide, Urethane	M	M			•			•	•				
1684 Magnesium Oxide, Urethane	M	H						•	•				
1688 Boron Nitride, Urethane	L	H		•				•	•				
1694 Magnesium Oxide, Silicone	M	H		•					•				
1698 Boron Nitride, Silicone	L	H		•					•				
1680 Boron Nitride, Kapton*, Silicone	M	H		•			•		•				
1646 Boron Nitride, Silicone	L	None		•		•			•				
T274 Aluminum Oxide, Silicone	M	H	•							•			
1641 Aluminum Oxide, Silicone	M	M									•		
1642 Aluminum Oxide, Silicone	M	M									•		

*Tradename

Table 9. CHO-THERM® Typical Properties

Typical Properties	1679	1671	1677	1674	1678	1661	Test Method
Binder	Silicone	Silicone	Fluorosilicone	Silicone	Silicone	Silicone	—
Filter	Boron Nitride	Boron Nitride	Boron Nitride	Aluminum Oxide	Boron Nitride	Boron Nitride	—
Color	Yellow	White	White	Blue	Red	White	—
Thermal Conductivity $\left(\frac{\text{Btu-in.}}{\text{hr-ft}^2\text{-°F}}\right)$	18	16	14	10	13	26	Chomerics Test Method No. 28
$\left(\frac{\text{Cal-cm}}{\text{cm2-sec-°C}}\right) \times 10^3$	6.5	5.7	5.0	2.8	4.5	9.0	
Thermal Impedance $\left(\frac{\text{°C-in.}^2}{\text{watt}}\right)$	0.15-0.18	0.18-0.22	0.38-0.42	0.30-0.34	0.22-0.24	0.24-0.28	Typical Flat Plate Test Values
Voltage Breakdown Rating (VAC)	4000	4000	4000	2500	2500	4000	ASTM D149
Outgassing (% TML) (% CVCM*)	0.40 0.10	0.76 0.07	0.57 0.01	0.45 0.20	0.55 0.12	0.76 0.08	ASTM E 595-77
Thickness (mils)	10±2	15±2**	20±4	10±2	10±2	20±3***	—
Tensile Strength (psi)	1000	1000	400	1500	1000	200	ASTM D412
Tear Strength (lb/in.)	100	100	60	100	100	10	ASTM D624
Elongation, (%)	10	2	10	2	10	2	ASTM D412
Hardness (Shore A)	95	90	85	90	90	90	ASTM D2240
Specific Gravity	1.55	1.55	1.70	2.20	1.60	1.60	ASTM D792
Maximum Use Temperature (°C)	–60 to 200	–60 to 200	–60 to 200	–60 to 200	–60 to 200	–60 to 200	—
Volume Resistivity (ohm-cm)	10×10^{14}	10×10^{14}	10×10^{14}	2×10^{14}	10×10^{14}	10×10^{14}	ASTM D257

*Collected volatile condensable materials (0-10% acceptable)
**CHO-THERM 1671 is available up to 35 mils on custom orders
***CHO-THERM 1661 is available up to 100 mils on custom orders

Polycarbon, Inc.®, a member of the SIGRI Group, provides a flexible graphite gasket, Calgraph®, with promising properties for thermal gasket applications. Typical properties are given in Table 10. By comparing with Table 9, the thermal conductivity (normal to surface) is seen to be a factor of two or three greater than that for CHO-THERM®. Electrical resistivity is seen to be 15 to 16 orders-of-magnitude less. That is, Calgraph® is a sufficiently good electrical conductor that it can, perhaps, be used as a continuous gasket and still meet component-grounding requirements. Polycarbon, Inc.® claims that Calgraph® can achieve 40 percent compression, with 20 percent recovery, and less than 5 percent stress relaxation (creep).

Table 10. Calgraph Properties

Typical Properties of Calgraph

Property	Units	Value
Electrical Resistivity		
"a" Direction (parallel to surface)	Ohm-in.	0.0004
"c" Direction (normal to surface)	Ohm-in.	0.025
Bulk Density	lb/ft^3 (GM/cc)	70.0 (1.1)
Thermal Conductivity		
"a" Direction (parallel to surface)	Btu-in./hr-ft^2°F	1532
"c" Direction (normal to surface)	Btu-in./hr-ft^2°F	48
Thermal Expansion		
70–1800 °F		
(bulk density 1.7–1.9 GM/cc)	10^{-6}/°F	2.8–4.4
Hardness (Shore Scleroscope)		
At 1.0 GM/cc	—	30
At 1.3 GM/cc	—	40
Tensile Strength		
At 1.0 GM/cc and .015" Foil	psi	700
Permeability		
Air	cm^2/GM	<0.00001
Emissivity		
At 932 °F	—	0.4
Sublimation Temperature		
(does not melt)	°F	6600
Temperature Limit (in air)	°F	1000

Welch and Ruttner (Ref. 23) tested Calgraph® in the configuration shown in Figure 22. They divided the test plate into four regions as shown in Figure 38. Using a thermal math model of the test set-up and correlating test results to math-model predictions, they determined local heat fluxes and heat-transfer coefficients.

A comparison of local heat-transfer coefficients so calculated for bare and Calgraph®-filled interfaces is shown in Table 11. Results are shown for a -30-deg F and 160-deg F cold plate. It is seen that use of Calgraph® improved heat-transfer coefficients in all regions except the center region. In that region, it is clear that separation has occurred and pressure is zero with and without the Calgraph®. The largest improvement is seen in the screw region, where a factor of 2.8 to 3.1 enhancement was observed. Most investigators do not use thermal math models and present their results as average heat-transfer coefficients based on an assumed uniform heat flux from top to bottom plate (e.g., Bevans et al.). Therefore, Welch and Ruttner present results in this form (Table 12). On this basis, enhancement by a factor of 1.5 to 1.9 is observed.

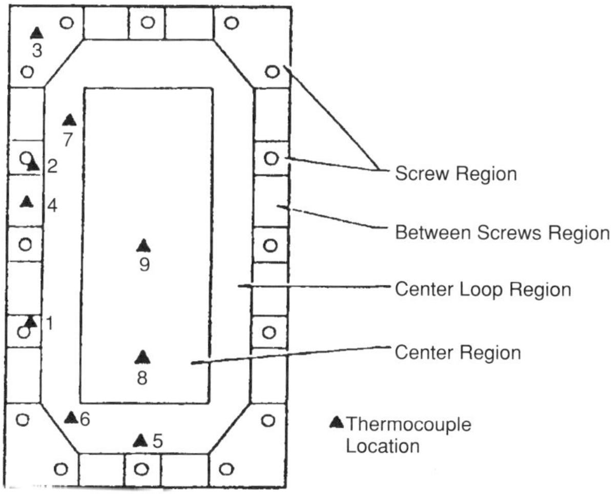

Figure 38. Test plate showing thermocouple and region location

Table 11. Unconstrained Heat Flux Heat Transfer Coefficients (Btu/hr-ft^2-deg F) That Match Experimental Data for the 20 in-lb Test

Region	Interface			
	Bare		Calgraph	
	Temperature (F)		Temperature (F)	
	–30	160	–30	160
Screw Region	250	450	700	1400
Between Screws	150	150	250	200
Center Loop	100	100	200	200
Center Region	0.1	1	0.5	6

Table 12. Area-Averaged Heat Transfer Coefficients (Btu/hr-ft^2-deg F) Based on Uniform Heat Flux Assumption

Region	Screw Torque (in-lb)			
	10		20	
	Temperature (F)		Temperature (F)	
	-30	160	-30	160
Bare	50	65	58	70
Calgraph	90	124	89	124

Calgraph® was also tested by Taylor (Ref. 32). An aluminum block, 2 inches by 2.75 inches by 1.5-inches high was mounted to a 0.5-inch thick aluminum plate by four no. 8-32 screws. Average heat-transfer coefficients reported were 2609 Btu/hr ft^2-deg F with Calgraph®, and 876 Btu/hr ft^2-deg F without, a factor-of-three improvement.

Cured-In-Place RTV Silicone Compounds

Thermal gaskets are seen to provide a factor of 1.7 improvement on an overall basis for 0.3125-inch-thick plates for a typical perimeter mounting configuration, with the largest improvement in the region of the bolts. No improvement is found in the center region, a consequence of bowing of the plates. To avoid this problem and provide near-continuous contact between the two plates, a cure-in-place method is widely used in the industry. A process specification was kindly provided by TRW. It calls for surface cleaning and drying, use of primer (both surfaces) or mold-release compound (at least one surface), installation of a stainless-steel mesh screen with gold-plate finish (for grounding) to be engaged by the mounting hardware, torqing of bolts, extrusion of filler material from at least 75 percent of the periphery of the mating surfaces (for coverage), and cure-in-place. Hughes Aircraft uses 10-mil Belleville washers at each screw instead of the wire mesh to achieve grounding. It should also be noted that some contractors forgo the use of a primer to allow easy removal of the box and room-temperature vulcanization (RTV).

This method creates an RTV mold that conforms to the profile of the cavity created between component baseplate and mounting plate by screw torque. Centerline gap due to bowing can be of the order of 10 to 20 mils. A variety of RTV compounds are used. Choice depends on cure time, viscosity (in order to extrude from between mating surfaces), low volatiles, etc. Hughes Aircraft often uses an RTV566 kit consisting of RTV566A and RTV566B, supplied by General Electric. Some contractors in their spacecraft applications use RTV filled with thermally conductive particles.

Taylor (Ref. 32), for a small stiff configuration, reported an average heat-transfer coefficient with RTV filler of 2685 Btu/hr ft^2-deg F, a factor of three greater than that for a bare interface. Bevans (Ref. 20) reported factors of 4 to 6 improvement in average heat-transfer coefficient using RTV-11 for the configurations of Figure 21. These results have been used by TRW in establishing the upper curve of Figure 20. Average coefficients are in the range of 35 to 85 Btu/hr ft^2-deg F.

Usage Recommendation for Filled Interfaces

For filled interfaces, a practice similar to that used for bare interfaces is recommended. Separate treatment should be used for the region near the bolt and for the entire plate. Near the bolt or for small stiff plates, as studied by Taylor, a factor of 2.5 to 3 improvement in heat-transfer coefficient over bare-interface values is justified for thermal-gasket (Refs. 23, 32) and RTV (Ref. 32) interfaces. Overall heat-transfer coefficients for large, thin plates with a perimeter-bolt pattern are considerably less for thermal-gasket filler than for RTV filler. This is a consequence of bowing and lack of gasket contact in the center region. Thermal gaskets are not recommended for such applications. For thin plates or for mountings where the spacecraft side of the interface is a honeycomb panel (without heat pipes), Figure 20 is recommended.

For thick plates with a perimeter-bolt pattern, the data of Welch and Ruttner can be used for heat-transfer coefficients for bare interfaces and those with a thermal gasket. On an overall basis the values in Table 12 can be used. In conjunction with a thermal math-model analysis, the local values from Table 11 can be used. For similar thick plates employing RTV filler, it is recommended that a thermal math-model analysis be conducted using the Calgraph® data of Table 11, except that center-region coefficients should be between 50 and 75 Btu/hr ft^2-deg F.

INTERFACE CONTROL DRAWINGS AND REQUIREMENTS

The mounting interface is not only a physical one, but is also a programmatic and subcontractor interface. The integrating-spacecraft contractor and the component-supplier subcontractor must jointly establish interface requirements and develop physical and functional Interface Control Drawing (ICD) details. The spacecraft contractor must provide the thermal environment extremes (mounting-plate temperature, radiation-sink temperatures) and physical mounting details, while the subcontractor must specify component dimensions, weight, bolt pattern, finishes, power dissipation, and modes of operation. If the component has a high average-power dissipation and/or local hot spots, the spacecraft contractor may find that extensive provisions (thermal doublers, heat pipes) may be required to control the spacecraft mounting panel to the specified temperatures. To preclude this, the spacecraft contractor may specify maximum-allowable average and local power (Figure 39). In the figure, type #1 and type #2 components are to be mounted to panels with and without heat pipes, respectively. Hence, type #2 components are constrained to a lower average-power dissipation. If one or two modules within the component have excessive power dissipation, a local hot spot may occur, causing the mounting panel locally to exceed the specified maximum temperature. To establish the adequacy of the interface the subcontractor may be required to supply a thermal math model of the component to the spacecraft contractor. The model need not be a full "box" model, but must have sufficient detail in the baseplate region that the spacecraft contractor can incorporate it into his detailed mounting-panel model and make credible temperature predictions. In this way, the spacecraft contractor can "guarantee" that he can meet the contractual interface requirements.

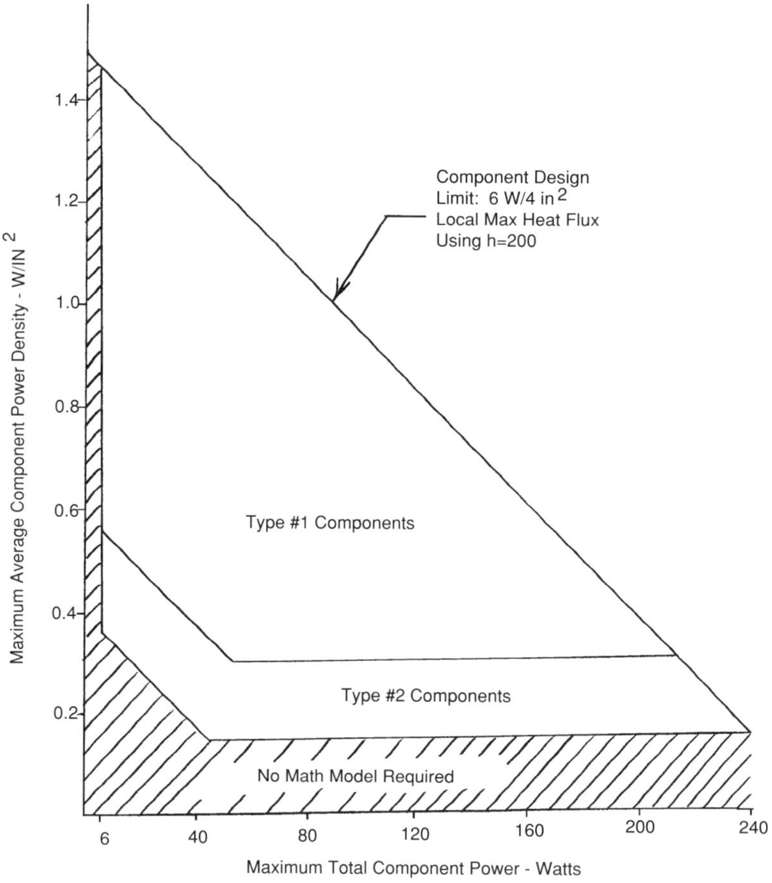

Figure 39. Component power density requirements, typical average and local values

COMPONENT TESTING

Most components are required to experience a thermal-vacuum test. This test provides environmental stress screening and demonstrates that component-performance specifications are met in vacuum at high- and low-temperature extremes. MIL-STD-1540B (Ref. 33) requires that the component be mounted "on a thermally controlled heat sink or in a manner similar to its actual installation in the space vehicle... The component heat transfer to the thermally controlled sink and the radiation heat transfer to the environment shall be controlled to the same proportions as calculated for the flight environment." For a component cooled primarily by conduction (the type of component addressed in this report), a schematic of the test set-up is shown in Figure 40. It is recommended that engineering development tests be conducted at this level for any component that analysis indicates is sensitive to contact conductance.

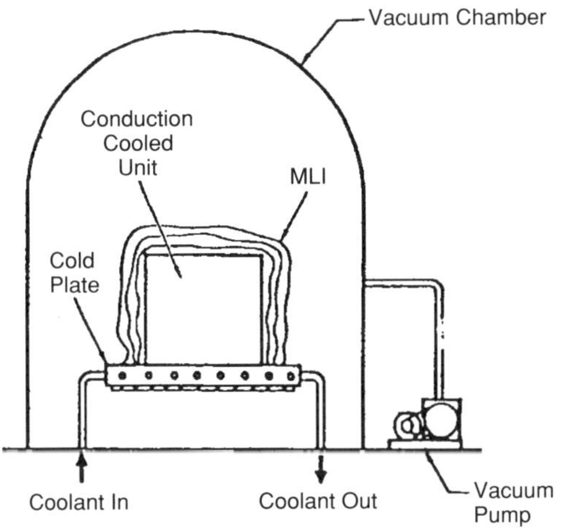

Figure 40. Test setup schematic, component thermal vacuum test

As a practical matter, replication of the flight mounting in the test set-up presents difficulties. Flight mounting panels tend to be lightweight, having lower lateral conductance than do test cold plates. The latter tend to be thick and stiff and usually employ forced liquid cooling. The size and construction of the test cold plates not only yield higher lateral conductance, but can result in higher contact conductances as well.

A simple thermal model was used to better understand the temperature profiles for the flight and the test environments (Figure 41). The three thermal models in the figure show the left-hand side of the bolted interface of a typical component slice for the flight mounting and two test mountings. A heat dissipation of 2.05 Btu/hr node (Q) is impressed on each node. Heat is ultimately rejected from the mounting plate to a sink (same for each case). Each model shows conductance in Btu/hr deg F and temperature results. The first test mounting has increased mounting-plate lateral conductance (6.15 Btu/hr deg F rather than 2.05 Btu/hr deg F). The second test mounting has both the increased lateral conductance and increased contact conductance (twice the local values used for the flight and for the first test mounting).

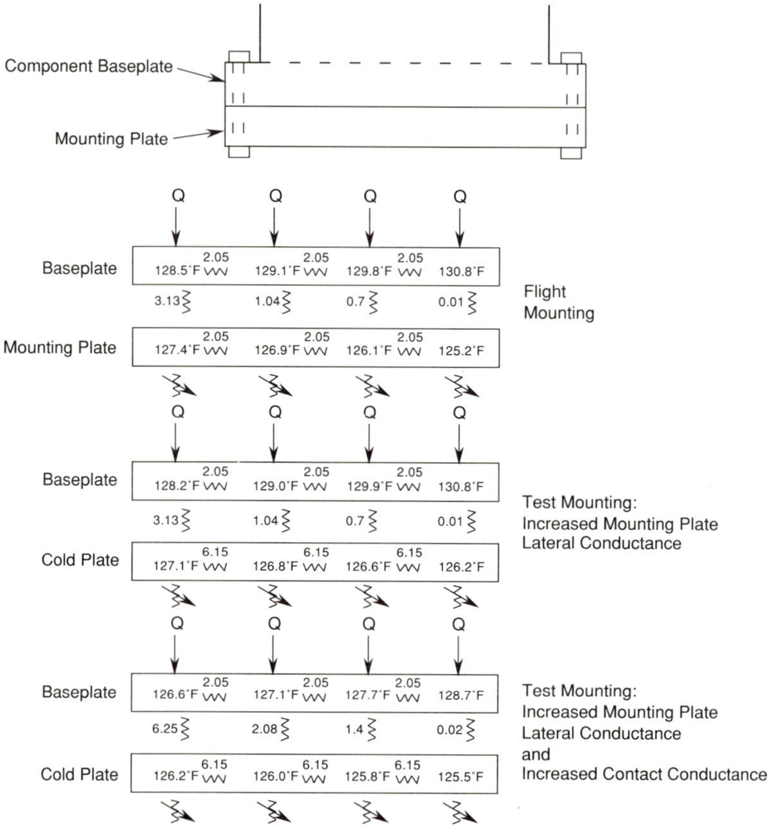

Figure 41. Thermal math results for flight and test mountings

The thermal model results show that increased mounting-plate lateral conductance by itself does not substantially change the baseplate temperature profile. However, the final figure shows that an increase in contact conductance does reduce baseplate temperatures, resulting in an undertest of the component.

It is noted then that high-contact heat-transfer coefficients in the test set-up impose lower temperatures on the component than would be experienced in flight. If such an effect is anticipated and is determined to be significant, it may be necessary to increase the cold-plate temperature, to provide compensation.

SUMMARY FOR CONDUCTION COOLING

Mounting of components to spacecraft panels and the associated heat removal has been treated for a variety of configurations and bolt patterns. Theory and prior art have been reviewed in order of increasing complexity, beginning with uniform contact pressure through bare bolted joints to thermal doubler applications and bolted joints with thermally enhancing filler. The interface-control process between prime contractor and subcontractor is addressed, as are interface issues during component testing.

Bolted or screwed joints (the standard method of component mounting) without interface filler result in two distinct thermal regions—a contact region in the vicinity of the bolt and a separated region at some distance for the bolt. For the former region, it is recommended that the Bratkovich correlations (Eqs. 15 and 16) or the TRW recommendations for short stiff surfaces (Table 2 or 3) be used to predict heat-transfer coefficients and conductances. The latter region is governed by lateral-plate conductance and can be treated by closed solutions, e.g., Eqs. 11 or 19, or by thermal math-model analysis. Alternatively, the combined problem can be treated by the use of empirical data for thin plates (Figure 20) and thick plates (Table 7).

There are two main types of materials used as interface fillers for bolted joints: thermal gaskets and cured-in-place RTV compounds. The former are well suited for stiff configurations without long distances between bolts. However, for most lightweight, perimeter-bolt pattern mountings, bowing is sufficient to cause zero gasket pressure (separation) to occur in the central region. Cured-in-place RTV compounds can fill these wide center spans, and therefore offer the capability of complete filling for perimeter-bolt patterns with long expanses between bolts. Use of such compounds, however, requires careful process control and can present problems if component removal is necessary.

CHO-THERM® has been used extensively as a thermal filler. However, its extremely high electrical resistivity may prevent component grounding from the baseplate to the mounting. Calgraph® flexible graphite gaskets have considerably lower electrical resistivities than does CHO-THERM®, and can potentially meet the grounding requirements. Thermal data is available from Welch and Ruttner for local heat-transfer coefficients (Table 11), and average heat-transfer coefficients (Table 12) for thick plates. For thin plates with a perimeter-bolt pattern, Figure 20 is recommended for RTV fillers.

If the mounting plate is of honeycomb/face sheet construction, the recommended heat-transfer coefficients are those for thin plates (Figure 20). If heat pipes are incorporated in the honeycomb below the component, the overall heat-transfer coefficient is expected to increase. However, such coefficients are not generally predicted at this time.

Average and local power density at the component's baseplate are generally controlled by interface-control documentation from the prime contractor, such as Figure 39. For components with higher power dissipations, thermal doublers can be used to spread heat laterally prior to transfer to the mounting plate. Design data are provided for rectangular (Refs. 25, 26, and 27) and axisymmetric (Ref. 28) thermal doublers.

Interface considerations associated with component thermal-vacuum testing are presented. It is seen that the high heat-transfer coefficients concomitant with stiff, thick test cold plates can result in component temperature reduced below those of flight. Such an occurrence is not conservative, can result in component under testing, and in some circumstances requires remediation.

THERMAL ISOLATION

Thermal isolators are often used to limit conduction-heat transfer through a mechanical connection. Typical applications include propellant-line supports, isolation under the mounting feet of instruments, and hydrazine-thruster catalyst-bed supports. These isolators can be made of a wide variety of low-conductivity materials, including fiberglass, stainless steel, titanium, or plastics. The choice of material is dictated by the conductivity, temperature range, and mechanical properties required for the particular application.

Design of a thermal isolator that supports a significant mass should be coordinated with mechanical/structural designers on the program. The idealized thermal requirements of minimum crossectional area and maximum height are generally the opposite of what is needed for structural stability. A typical isolator is shown in Figure 40 and includes isolation both between the components being bolted together and under the bolt head and nut to avoid a thermal "short" through the bolt. It is also recommended that the isolators have a "lip" to prevent the bolt from shifting under launch vibration and contacting the isolated component. It should be noted that this type of isolator requires careful control of tolerances on hole diameters and locations so that all the pieces come together without interference for all of the "feet" on the device.

The thermal conductance across an isolator must consider both the direct path between the foot and mounting surface and the path down the bolt. Contact resistances at the interfaces are generally ignored since they are small compared to the resistance through the isolator material itself. The resistance down the bolt can be increased by using titanium or, for very small devices, plastic bolts.

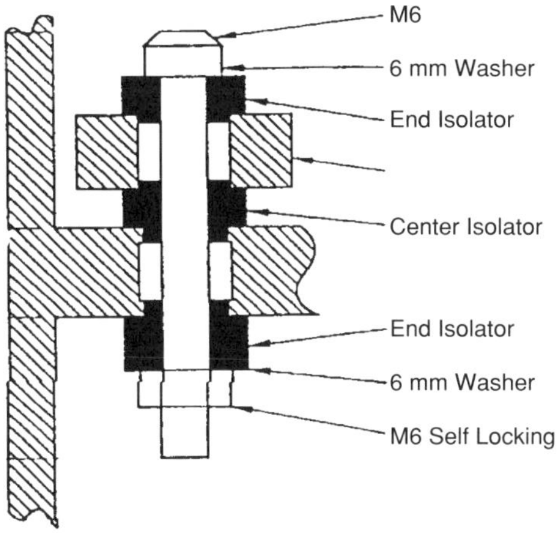

Figure 40. Thermal isolator at bolted interface

BEARING CONDUCTION

Conductance across bearings is one of the most uncertain parameters in spacecraft thermal analysis. The large dependence of the conductivity upon factors such as bearing design, speed, lubricant type and quantity, load, and temperature gradients from the inner to outer race make it impossible to identify "generic" conductivities for bearings.

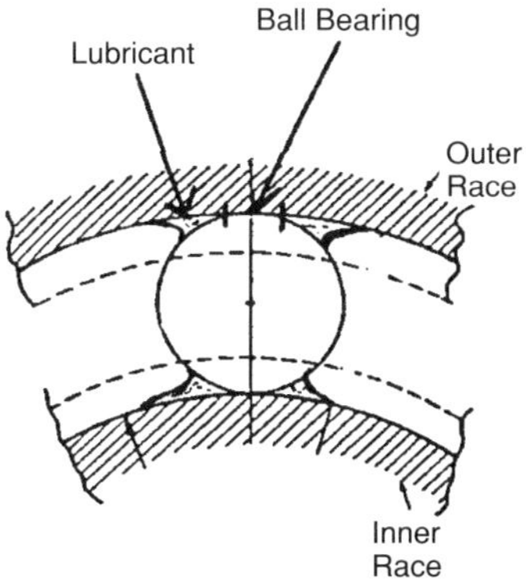

Figure 41. Bearing crosssection

The bearing crossection shown in Figure 41 illustrates the conduction mechanisms for a ball bearing in vacuum. There is a conduction path through the ball/race contact regions as well as through the lubricant. The contact conductance is affected by lubrication and the load, which is itself driven by preload, gravity effects, speeds, and temperature differences between the races. The conduction through the lubricant is complex and highly dependent upon the type and amount of lubricant and the rotational speed. Figures 42 through 44 contain measured data from Ref. 34 for a particular set of bearings, which illustrate the considerable effect of some of these factors on bearing conductance. The reader may wish to consult this reference for additional discussions on this subject. There have also been a number of other reports and papers, e.g., Refs. 35 and 36, which discuss the theoretical and experimental evaluation of the various factors effecting bearing conduction.

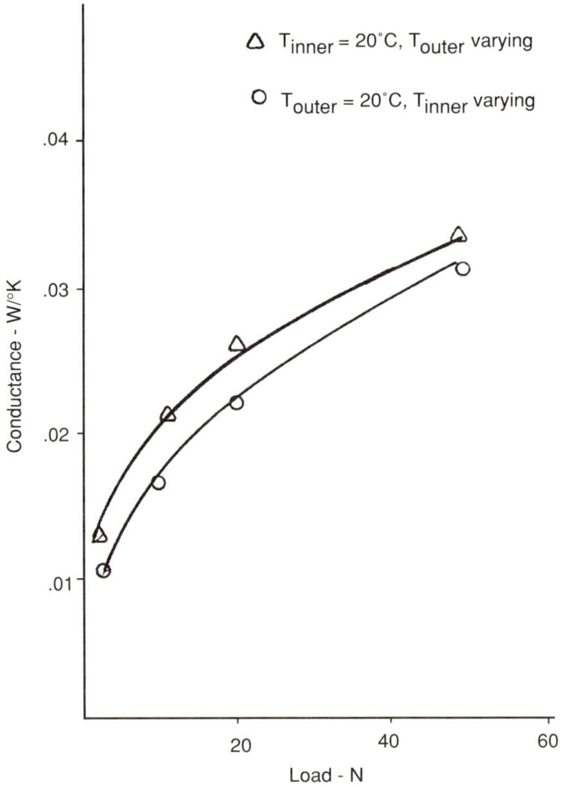

Figure 42. Mean conductance as a function of load for a lead-lubricated 42 nm O/D bearing. Speed: zero

Despite the considerable research done in this area, it is still not possible to provide a generalized set of conduction values to be used in thermal analysis involving bearings. The analyst is therefore left with the options of performing tests such as those discussed in Refs. 34, 35, and 36 to measure the conductivities of the bearings in question, or bounding the problem by looking at a wide range of conductances. If a test is performed, it must accurately simulate the lubrication, load, speed, vacuum temperature ranges, and gradients expected in flight, while ensuring that any gravity effects are accounted for. If a bounding analysis is conducted, a suitably wide range of conductances must be considered, e.g., from zero to a fairly high contact conductance across the entire area of the races. If the analysis shows a considerable sensitivity to bearing conductance, test measurements on the bearings early in the program are recommended. Do not rely on system-level thermal tests that may not provide the right conditions and that will usually occur too late in the program for design changes to be practical.

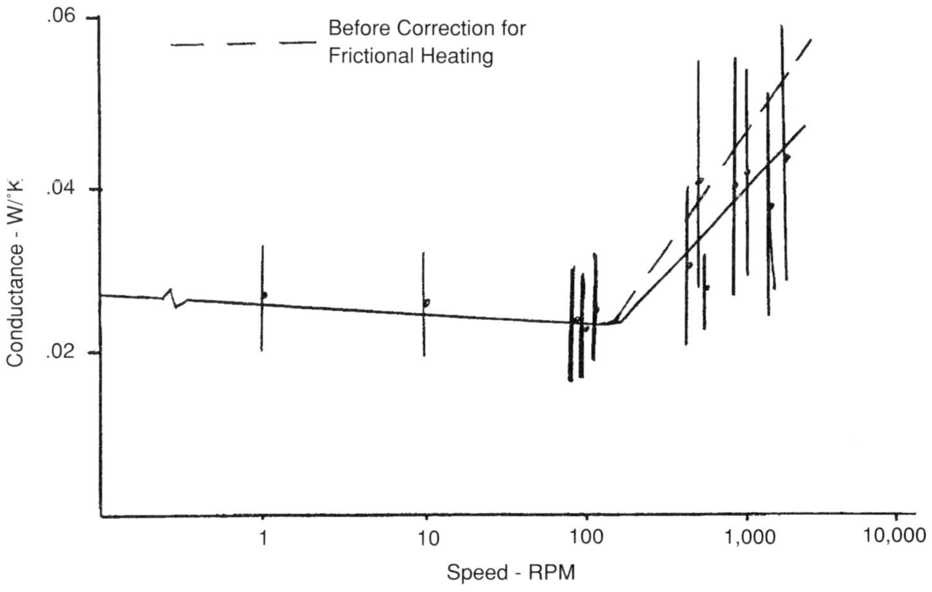

Figure 43. Conductance vs. speed for a 42 mm O/D lead-lubricated bearing. Load: 20 N, T_1: 40-deg C, T_c: 20-deg C

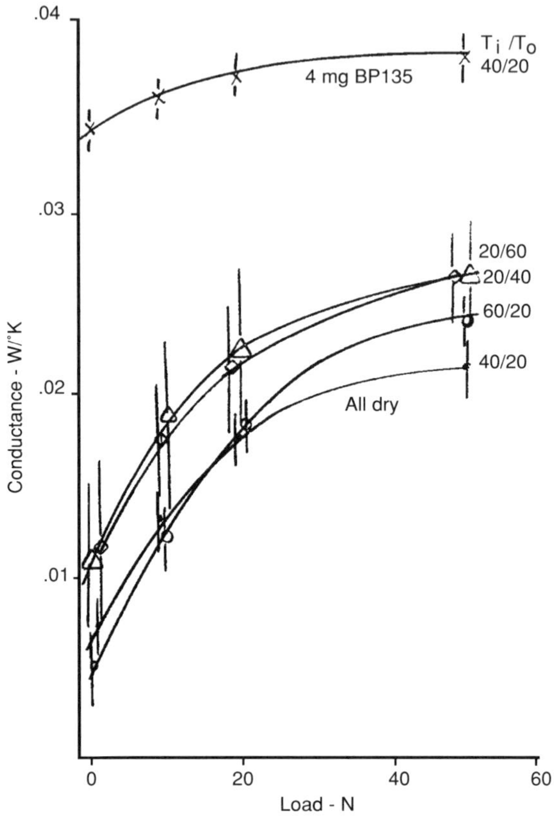

Figure 44. Conductance as a function of load for a 16 mm O/D bearing
Speed: zero

Nomenclature

a	coefficient of thermal expansion (Figures)
a	heat source radius at top of doubler
A	area of heat transfer
A_b	area of contact region near bolt
A_1	area of heat conduction element in no contact region for top plate
A_2	area of heat conduction element in no contact region for bottom plate
b	doubler radius
C	conductance
D_s	bolt shaft diameter
E	elastic modulus
F	heat flux

F_n	RMS surface roughness (finish)
F_o	uniform heat flux at base of doubler
F_s	uniform heat flux source at top of doubler
h	heat transfer coefficient, overall heat transfer coefficient
h_b, h_c	heat transfer coefficient in bolt contact region
h_e	heat transfer coefficient for elastic deformation
h_p	heat transfer coefficient for plastic deformation
h_p	heat transfer coefficient in plate conduction region
H	hardness
k	thermal conductivity
L	doubler height
N_b	number of bolts
r	radial coordinate
r	correlation coefficient
p	applied pressure
p, P	power input
q, Q	rate of heat flow
r_o	radius of applied pressure
R_o	radius of contact region near bolt
R	radius of thermal network model element
R	overall resistance
t	thickness of plate
t_m	mean thickness of plates
t_1	thickness of top plate
t_2	thickness of bottom plate
$\tan\theta$	mean of absolute slope of profile
T	temperature
T_i	reference temperature
T_o	temperature at boundary of resistance network element
T_q	bolt torque
x, y, z	coordinates
Y	yield strength
α	coefficient of thermal linear expansion
β	k/hb
δ	a/b
η	r/R
η_o	Ro/R, ratio of contact radius to element radius
ν	Poisson's ratio
σ	RMS surface roughness (same as F_n)
ϕ	$T-T_o$ (for doubler)
ψ	$k\phi/2aF_s$ (for doubler)

REFERENCES

1. M. G. Cooper, B. B. Mikic, M. M. Yovanovich, "Thermal Contact Conductance," *Int. J. Heat Mass Transfer*, Vol. 12, pp. 279-300, 1969.

2. B. B. Mikic, "Analytic Studies of Contact of Nominally Flat Surfaces; Effect of Previous Loading," ASME Paper 71-Lub-M, 1971.

3. B. B. Mikic, "Thermal Contact Conductance; Theoretical Considerations," *Int. J. Heat Mass Transfer*, Vol 17, pp. 205-214, 1974.

4. M. Sridhar and M. Yovanovich, "Critical Review of Elastic and Plastic Thermal Contact Conductance Models an Comparison with Experiments," AIAA-93-2776, 1993.

5. S. Song, M. M. Yovanovich, K. Nho, "Thermal Gap Conductance: Effect of Gas Pressure and Mechanical Load," AIAA-89-0429, 1989.

6. P. J. Schneider, "Conduction," Section 3 of *Handbook of Heat Transfer*, eds. by W. M. Rohsenow and J. P. Hartnett, McGraw-Hill, 1973.

7. E. Fried and F. A. Costello, "Interface Thermal Contact Resistance Problem in Space Vehicles," ARS Journal, Vol. 32, pp. 237-243, 1962.

8. M. E. Barzelay, et al., "Effect of Pressure on Thermal Conductance of Contact Joints," NACA TN 3295 (Syracuse Univ.), Washington, D. C. , May 1955.

9. E. Fried, "Study of Interface Thermal Contact Conductance," General Electric Co. 64SD652, May 1964 (also ARS Paper No. 1690-61, April 1961).

10. H. Fenech and W. H. Rohsenow, "Prediction of Thermal Conductance of Metallic Surfaces in Contact," *J. Heat Transfer, Trans. ASME*, Vol. 85, pp. 15-24, 1963.

11. B. B. Mikic and W. M. Rohsenow, "Thermal Contact Resistance," MIT Rep. No. 4542-41, 1966.

12. I. J. Henry, "Thermal Contact Resistance," AEC Rep. No. 2079-2, MIT, 1964.

13. M. M. Yovanovich and H. Fenech, "Thermal Contact Conductance of Nominally Flat Rough Surfaces in a Vacuum Environment," *Prog. Astronautics & Aeronautics,* Vol 18, 1966.

14. C. Swartz, "Thermal Conductance in Aluminum Bolted Joints," Aerospace Corporation ATM 82(9975)-29, 19 January 1982.

15. E. Fried and M. J. Kelly, "Thermal Conductance of Metallic Contacts in a Vacuum," in *Thermophysics and Temperature Control*, edited by G. Heller, Academic Press, 1966.

16. E. Fried and H. L. Atkins, "Thermal Joint Conductance in a Vacuum," ASME paper 63-AHGT-18, 1963.

17. R. T. Roca and R. E. Mikic, "Thermal Contact Resistance in a Non-Ideal Joint," Massachusetts Institute of Technology, Report No. DSR 71821-77, Nov. 1971.

18. R. T. Roca and R. E. Mikic, "Thermal Conductance in a Bolted Joint," AIAA Paper No. 72-282, April 1972.

19. Ingemar Fernlund, "A Method to Calculate the Pressure Between Bolted or Riveted Plates," Report No. 17, Chalmers University of Technology, Gothenberg, Sweden, 1961.

20. J. T. Bevans, T. Ishimoto, B. R. Loya, and E. E. Luedke, "Prediction of Space Vehicle Thermal Characteristics," Air Force Flight Dynamic Laboratory Technical Report AFFDL-TR-65-139, August 1965.

21. T. F. Bratkovich, "Correlations for Contact Region Bolted Joint Conductance in Vacuum," The Aerospace Corporation, ATM No. 91(9975)-15, 10 April 1991.

22. TRW, Project MILSTAR Engineering Design Standards, Section 2.2, "Thermal Environments," March 1984.

23. J. W. Welch and L. E. Ruttner, "An Experimental and Computational Analysis of the Thermal Interface Filler Material Calgraph," AIAA-89-1658, June 1989.

24. D. F. Gluck, "Tentative Bolted Joint Contact Conductance Correlation: Aluminum to Aluminum in Vacuum," IOC 86.5462.9-13, 11 July 1986.

25. R. P. Bobco and R. P. Starkovs, "Rectangular Thermal Doublers of Uniform Thickness," AIAA Journal, Vol 23, No. 12, pp. 1970-1977, Dec. 1985.

26. R. P. Starkovs, "Rectangular Thermal Doublers with Two Heated Footprints," AIAA-87-1612, June 1987.

27. R. P. Bobco, "Thermal Doublers: Closed Form Solution for Three Single Terrace Categories and a Partial Double Terrace," Hughes Aircraft, Space and Communications Report 4132.15/2594, Jan. 25, 1984.

28. D. F. Gluck and P. H. Young, "Performance and Optimization of an Axisymmetric Thermal Doubler," Proceedings, 1988 Summer Computer Simulation Conference, Seattle, Washington, 25-28 July, 1988.

29. W. C. Layton, "Traveling Wave Tube Thermal/Structural Study," Aerospace Report No. TOR-0088(9975)-1, 1 August 1988.

30. E. H. Goit, "Third Degree Integration Formulas with Four Points in Two Dimensions," Master of Science Thesis, State University of New York at Buffalo, Buffalo, N. Y., February 1969.

31. L. S. Fletcher, P. A. Smuda, and D. A. Gyorog, "Thermal Contact Resistance of Selected Low-Conductance Interstitial Materials," AIAA Journal, Vol. 7, No. 7, pp. 1302-1309, July 1969.

32. P. F. Taylor, "TWT Collector Thermal Interface Test," Hughes Aircraft, Interdepartmental Correspondence to W. A. Hagemeyer, 4 December 1985.

33. MIL-STD-1540B, "Military Standard, Test Requirements for Space Vehicle," 10 October 1982.

34. M. Yovanovich, "Analytical and Experimental Investigation on the Thermal Resistance of Angular Contact Instrument Bearings," Report No. E-2215, MIT Instrumentation Laboratory, December 1967.

35. "Thermal Conductance of ASAT Bearing," H. J. Deacon, The Aerospace Corporation, ATM 87(2426-02)-3, 20 November 1986.

36. K. T. Stevens and M. J. Todd, "Thermal Conductance Across Ball Bearings in Vacuum," European Space Tribology Laboratory, Report No. ESA (ESTL) 25, February 1977.

Section 3

Multilayer Insulation and Barriers

Martin Donabedian
David G. Gilmore
The Aerospace Corporation

Multilayer insulation (MLI) and single-layer radiation shields are among the most common thermal control elements on spacecraft. MLI blankets are used either to prevent excessive heat loss from a component or excessive heating from environmental fluxes, rocket plumes, etc. Most spacecraft flown today are covered with MLI blankets, with cut-outs provided for radiator areas to reject internally generated waste heat. MLI blankets are also typically used to protect internal propellant tanks, propellant lines, solid rocket motors and cryogenic dewars. Single-layer radiation barriers are sometimes used in place of MLI where a lesser degree of thermal isolation is required, since they are lighter and cheaper to manufacture. The reader is referred to Chapter III for specific examples of how these blankets and barriers are used in typical thermal designs.

Multilayer insulation is composed of multiple layers of low-emittance films. The simplest construction is a layered blanket assembled from embossed, thin Mylar sheets (1/4 mil thick) with a vacuum-deposited aluminum finish on one side of each sheet. The embossing results in the sheets touching only at a few points, thereby minimizing conductive heat paths between the layers. The layers are aluminized on one side only, so that the Mylar can act somewhat as a low-conductivity spacer. More complex, high-performance construction uses Mylar film metallized on both surfaces (aluminum or gold) with silk or Dacron net, Tissuglas paper, or "Super-Flock" whiskers as the low-conductance spacers.

Heat transfer through multilayer insulation is a combination of radiation, solid conduction, and, under atmospheric conditions, gaseous conduction. The gaseous-conduction heat transfer is minimized by allowing the insulation to vent to space after the vehicle is launched, or by using the insulation in an evacuated wall (such as in the space between a cryogenic pressure vessel and the external vacuum jacket shell). Solid-conduction heat transfer is minimized by keeping the density of the low-conductance spacers between the reflective surfaces as low as possible and making the blanket "fluffy" to minimize contact between layers. Radiation heat transfer is minimized by interposing as many enclosing reflective surfaces (metallized sheets) as is practical.

Because these heat-transfer mechanisms operate simultaneously and interact with each other, the thermal conductivity of an insulation is not strictly definable, analytically, in terms of variables such as temperature, density, or physical properties of the component materials. It is therefore useful to refer to either an apparent thermal-conductivity, K_{eff}, or an effective emittance, ε^* (referred to as E-STAR through the blanket). Both of these values can be derived experimentally during steady-state heat transfer.

The low thermal conductivity of evacuated-insulation systems can largely be attributed to removal of gas from the void spaces within the insulation. The degree of vacuum necessary to achieve the desired effectiveness can be established by considering the mechanism by which the heat flows. The gas conduction can be divided into two regions: the region ranging from atmospheric pressure down to a few Torrs (1 Torr = 1 mm of mercury) in which gas conduction is independent of pressure, and the region at pressures below a few Torrs in which gas-conduction depends on pressure. The transition from one type of gas conduction region to the other depends upon the dimensions of the system with respect to the mean-free path of the gas molecules. The effect of gas pressure on conductivity can be characterized by the curves in Figure 1. The effective conductivity begins to decrease sharply between 1 and 10 Torr until about 10^{-4} to 10^{-5} Torr, where the heat conducted by the gas is only a small portion of the residual heat transfer. A finite value of effective thermal conductivity remains at lower pressures due to heat transfer by solid conduction and radiation between the elements of the insulation.

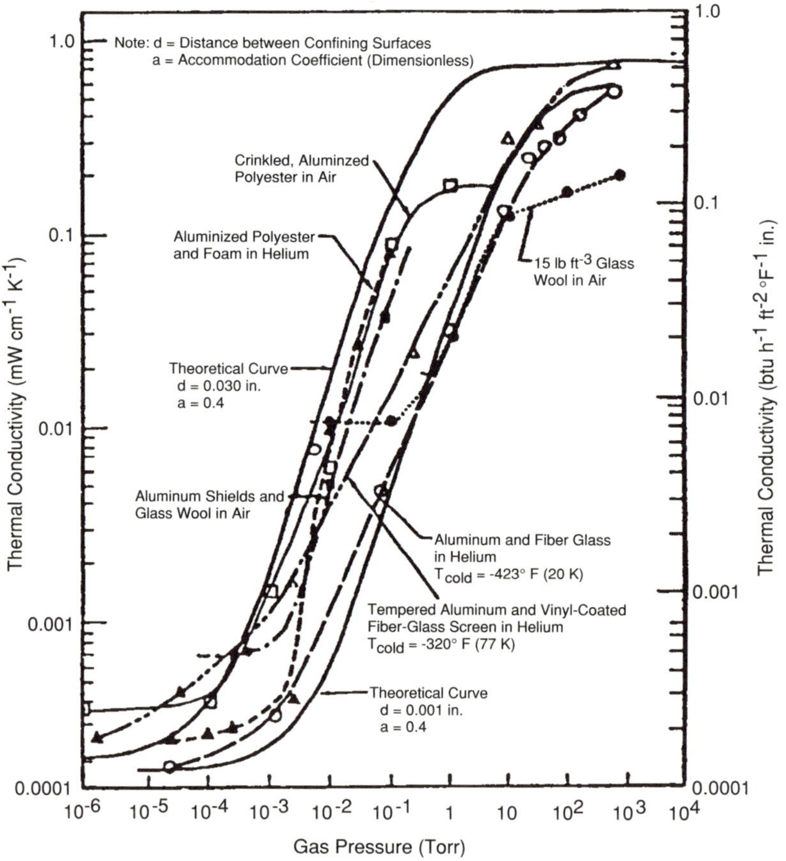

Figure 1. Effect of has pressure on thermal conductivity

In theory, for highly evacuated MLI systems (i.e., with gas pressures of 10^{-5} Torr or less), the theoretical emittance, ε, for a blanket of N non-contacting layers of emissivity of ε_1 and ε_2 on opposite sides, is computed as

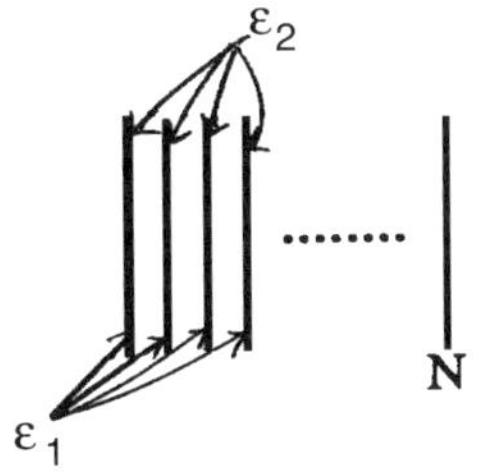

$$\varepsilon = \frac{1}{\frac{1}{\varepsilon_1} + \frac{1}{\varepsilon_2} - 1} \left(\frac{1}{N+1} \right) \tag{1}$$

In practice, the effective emittance of a MLI blanket is generally derived from experimental tests at gas pressure of 10^{-5} Torr or less calculated from

$$\varepsilon = \frac{Q}{A\sigma\left(T_H{}^4 - T_C{}^4\right)} , \tag{2}$$

where T_H and T_C are the hot and cold boundary temperatures in °R, A is the surface area of the blanket in square feet, Q is the net heat transferred in Btu/hr, and σ is the Stefan-Boltzmann constant in units of Btu/hr-ft^2 - R^4.

Alternately, the use of an effective thermal conductivity, K_{eff}, is also used. When K_{eff} is stated in units of Btu-ft-/hr-ft-^{2}-°F, ε^* is related to K_{eff} by

$$\varepsilon^* = \frac{(K_{eff})(12)(T_H - T_C)}{(\ell)(\sigma)\left(T_H{}^4 - T_C{}^4\right)} , \tag{3}$$

where ℓ is the thickness of the MLI between the hot and cold boundaries, stated in inches.

Figure 2 illustrates theoretical and experimental data for embossed aluminized (one surface) Mylar insulation versus number of insulation-blanket layers. As indicated by Eq. (1), the emittance for a multilayer blanket theoretically varies inversely with one over one plus the number of layers. However, in practice, simply increasing the number of layers past a certain value will not improve performance. As the number of layers increases, radiative heat transfer becomes small compared to conductive "shorts" between layers and other losses. Thus, a point of diminishing returns is reached. Considering these trends, about 25 layers are usually sufficient to obtain a minimum overall conductance value.

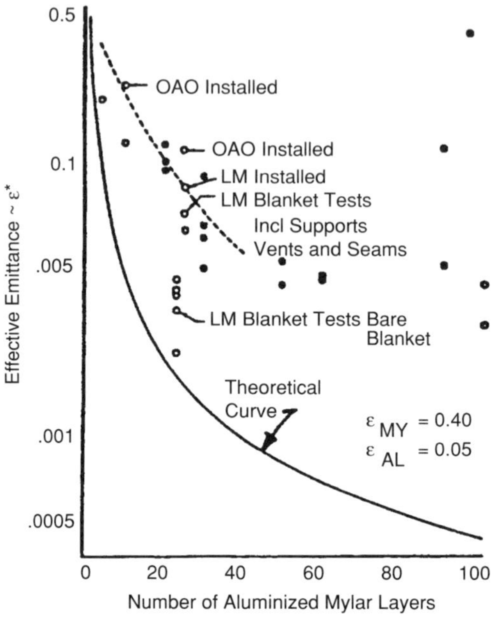

Figure 2. Effective emittance vs. number of single aluminized layers

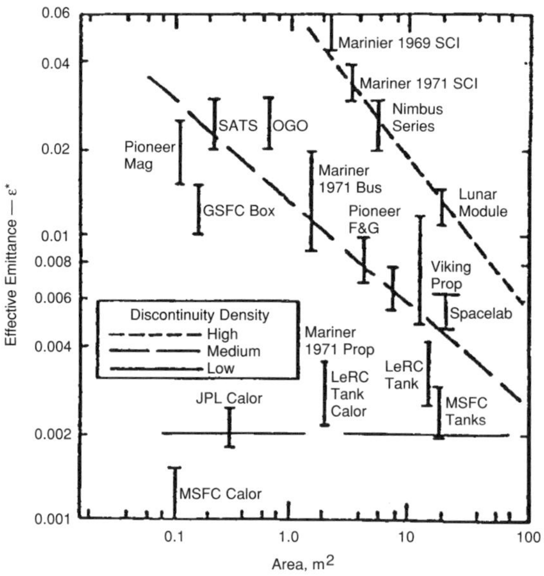

Figure 3. Effective emittance vs. installed area
(Stimpson and Jaworski)

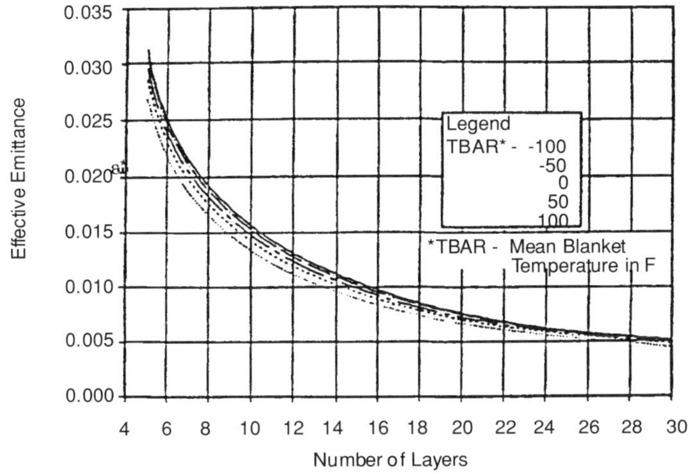

Figure 4. MLI blanket effective emittance derived from the spacelab thermal test data

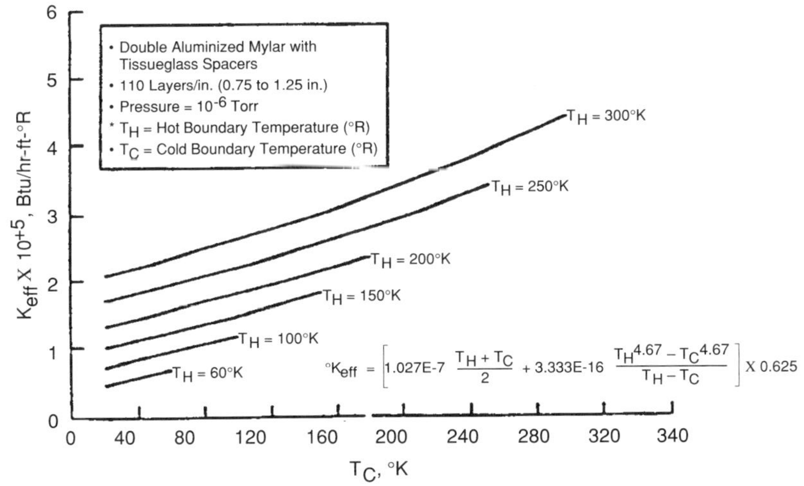

*Note: T_H and T_C are in units of °R in the equation but plotted in °K on the graph

Figure 5. Effective thermal conductivity of MLI blankets

In well-controlled laboratory tests, it is possible to achieve values of 0.005 or lower for ε*. However, when a blanket is configured for spacecraft application, experience has shown that an ε* more like 0.015 to 0.030 is representative of current design, manufacturing, and installation methods for relatively small-area applications. As the size of the application increases, the relative performance generally increases, as indicated by the data on Figure 3. This results from the smaller relative influences of heat leaks due to edge effects, seams, cable penetrations, etc. For very-large-area applications with minimal penetrations, such as on the Spacelab, the laboratory performance approaches 0.005 at 30 layers, as shown by the test data on Figure 4. Performance data from cryogenic tankage and controlled calorimeter tests typically also show better performance of ε* down to 0.002, and K_{eff} down to 1×10^{-5}, as shown in Figures 3 and 5. The data on Figure 5, although given in terms of K_{eff}, can be found to be quite comparable to Figure 4 when the average temperatures are similar and Eq. (3) is used. The data of Figure 3 shows that the control of discontinuities through the design and fabrication of insulation joints and penetration is crucial to the problem of reducing the effective emittance of multilayer blankets. Small area blankets show high effective emittance along with considerable manufacturing variation.

The performance of an MLI system can be severely degraded by the pressure of even very modest amounts of gas. Data from General Dynamics (Figure 6) shows that a blanket gas-pressure increase from 1×10^{-5} Torr to 1×10^{-4} Torr increased the system heat-leak by 33 percent. This demonstrates further the importance of reducing outgassing and the prevention of contamination of the blankets that may account for some cases of thermal performance degradation of MLI systems.

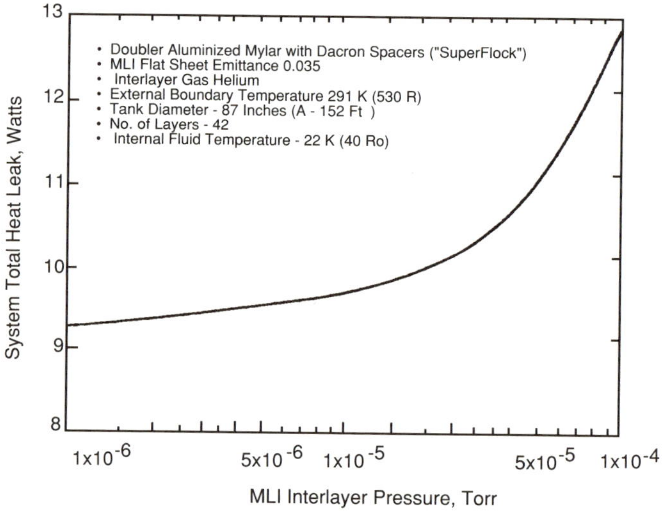

Figure 6. Effect of gas pressure on MLI blanket performance of a cryogenic tank (from General Dynamics data)

BLANKET CONSTRUCTION

MLI blankets typically consist of three to thirty layers of metalized plastic sheets. The inner layers are typically as thin as practical (1/4 mil) to minimize weight and are usually made of aluminized Mylar, which is mass produced for commercial applications and is inexpensive. The innermost and outermost layers of the blanket, however, are usually made of 1- or 2-mil aluminized Kapton which is much more rugged than the Mylar and provides protection during installation and handling. Mylar cannot be used as an outer layer material since it disintegrates under prolonged UV exposure. The aluminized Kapton also provides a desirable α/ε ratio for the exposed outside layer of the blanket. Aluminized Kapton should also be used for the inner layers in applications where blanket temperatures are between 250-deg F and 550-deg F for long periods, or up to 750°F for short periods.

Sometimes glass-fiber cloths such as Beta Cloth or Astroquartz are used for the outer layer of blankets. Astroquartz is a fabric constructed of woven quartz fiber that is white in appearance and can withstand very high temperatures, which makes it ideal for protection against rocket-motor plume heating. Beta Cloth, which lines the Space Shuttle payload bay, is similar to Astroquartz except that the glass fibers are coated with Teflon, which increases their emittance. Both Beta Cloth and Astroquartz, however, are difficult to clean and work with, and can be a significant source of contaminants on orbit.

As was mentioned earlier, the layers of an MLI blanket can be separated either by "crinkling" each layer (by hand or an embossing process) or by the use of separator nets made of Nomex, Dacron, silk, or other materials. The layers of three typical blankets are shown in Figure 7.

In order to evacuate the air trapped between blanket layers before launch, vent paths have to be provided. Sometimes this is done by making all blanket layers from a material with small perforations (e.g., 0.030 inch holes on 0.25 inch centers). In other blankets this may be done by leaving one or more edges of the blanket unsealed or by cutting small X shaped slits through the blanket at some regular interval. If a blanket is not adequately vented, it will billow out during the depressurization of launch ascent and may be severely damaged or torn loose from the vehicle. The vent paths must also be sufficient to insure that residual pressure between blanket layers drops to below 10^{-4} Torr within a few hours of launch so that the blankets will be fully effective.

The many layers of an MLI blanket are typically held together by either stitching around the edges or the use of small tabs or buttons spaced at some regular interval. In addition, the blanket edges are typically finished off by taping

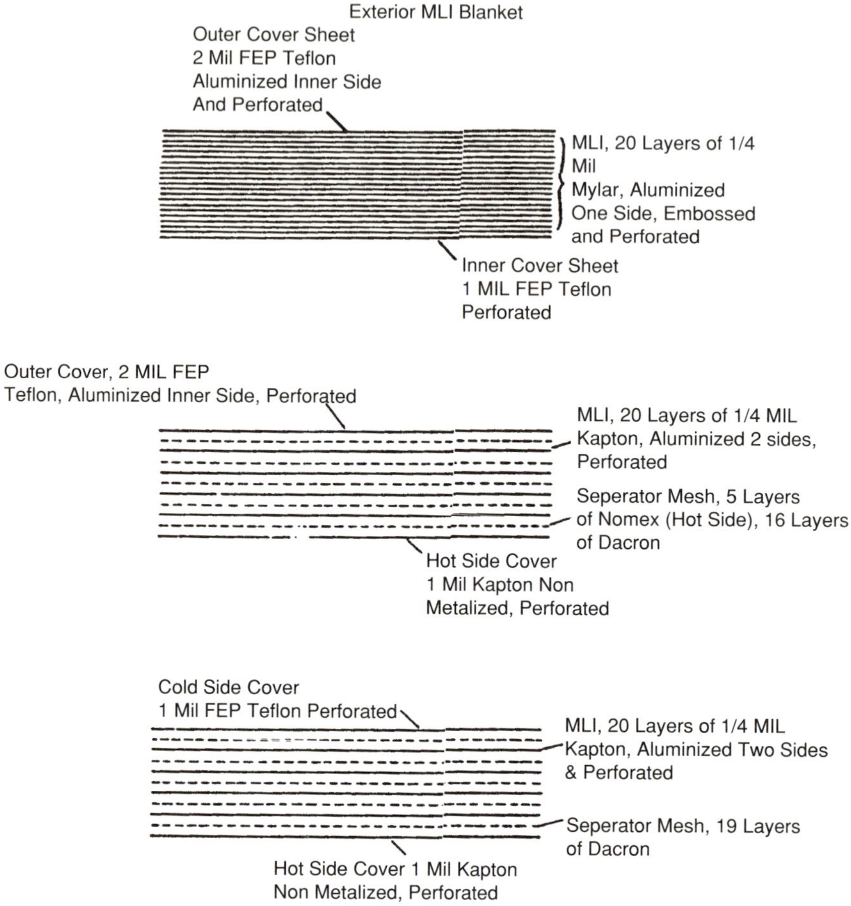

Figure 7. OIS Motor Cavity Internal MLI Blanket

them closed (Kapton tape with acrylic, not silicone, adhesive), unless they are being used as a vent path. The blanket is then usually attached to the spacecraft using velcro strips. A sample blanket drawing showing stitching and velcro installation is shown in Figure 8. The overall blanket dimensions must be sized to provide a loose fit over the hardware to be covered and must account for shrinkage when the blanket cools off in the worst cold-case environment. A blanket that is too tight can be severely degraded due to compression and conductive "shorting" of its inner layers. MLI effective emittance as high as 0.3 has been observed in some poorly designed blankets.

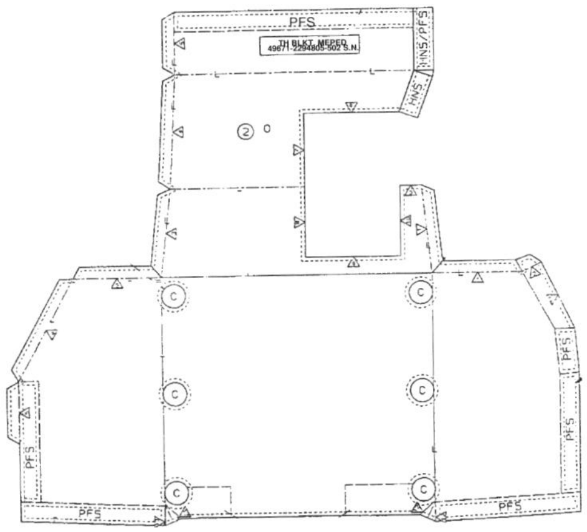

Figure 8. MLI blanket example

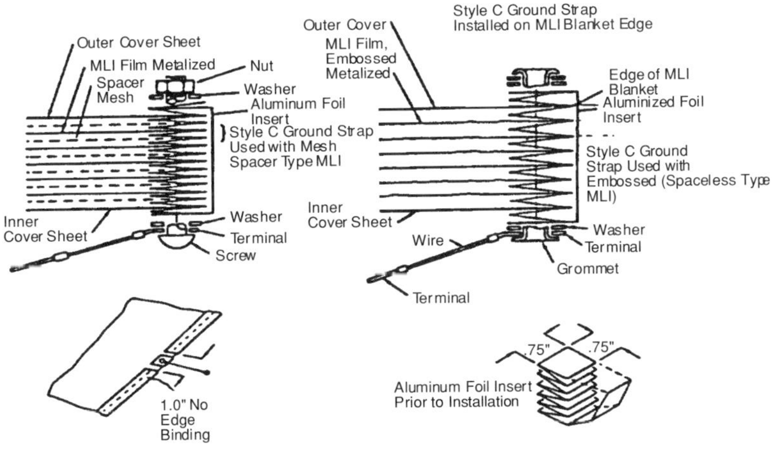

Figure 9. MLI electrical ground strap

Most spacecraft programs require that all metalized layers in an insulation blanket be grounded to prevent electrostatic charge build-up and subsequent arcing. There are different means of achieving this, one of which is shown in Figure 9. In this design a metal-foil strip is accordion-pleated between the layers at the edge of the blanket and held in place with a bolt-like fastener. A short length of wire is then used to ground the blanket to the spacecraft structure. On larger blankets several of these ground straps may be required to minimize the conduction distance through the very thin vapor-deposited metal coatings on the blanket layers. If the outside surface of the blanket must also be grounded, as is sometimes required, aluminized Kapton with a thin layer of electrically conductive indium-tin oxide (ITO) may be used as the outer layer.

Section 4

Heaters, Thermostats, Solid State Controllers

David G. Gilmore
The Aerospace Corporation

INTRODUCTION

Ideally, thermal control of a satellite or component would be achieved using only passive techniques, such as surface finishes. Unfortunately, variations in environment and component heat-generation rates, along with degradations of surface finishes over time, can drive temperature variations in a passive design to ranges larger than some components can withstand. Because of this, heaters are sometimes required to protect components under cold-case environmental conditions or to make up for heat that is not dissipated when an electronic box is turned off. Heaters may also be used with thermostats or solid-state controllers to provide precise temperature control of a particular component. A third common use for heaters is to warm-up components to their minimum operating temperatures before they are turned on. Each of the above applications will be illustrated later.

HEATER TYPES

The most common type of heater used on spacecraft is the patch heater, several of which are shown in Figure 1. It consists of an electrical resistance element sandwiched between two sheets of flexible electrically insulating material, such as Kapton. The patch may have one circuit, or more than one, depending on whether redundancy is required within the patch. Redundancy is generally required on spacecraft systems since heater circuits can fail. Sometimes the redundancy is provided within the patch and sometimes it is provided by using two separate patches. The patch heaters shown in the figure illustrate the custom shapes to which these heaters may be made. In most instances, however, a simple rectangular patch of some standard dimension is used.

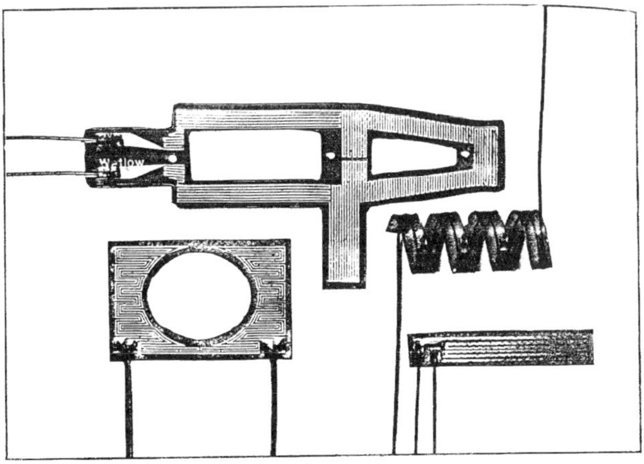

Figure 1. Patch heaters

A cartridge heater is another type that is often used to heat blocks of material or high-temperature components such as hydrazine-thruster catalyst beds. Such a heater is shown in Figure 2 and consists of a wound resistor enclosed in a cylindrical metallic case. A hole is typically drilled in the component to be heated and the cartridge potted into the hole. Another attachment technique involves the use of a clamp or small bracket to hold the heater. These heaters are typically a quarter inch diameter or less and up to a few inches long.

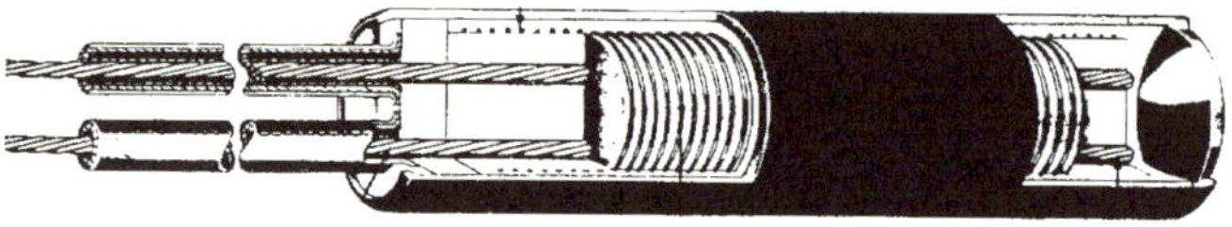

Figure 2. Cartridge heater

CONTROL

Almost all heaters have some sort of control over their operation. This typically involves a relay that is commandable from the ground to enable or disable power to the heater, a fuse to protect the spacecraft from a short circuit, and, usually, a thermostat or solid-state controller to turn the heater on and off at predetermined temperatures. More sophisticated satellites sometimes use their on-board computer to monitor temperatures and turn heaters on and off as appropriate using relays.

The simplest arrangement involves only the heater, a fuse, and a ground commandable relay to turn the heater on and off. This arrangement is typically used only for heaters, that are activated only for special events or for heaters that can be left on all the time. A typical application is heating up the catalyst beds on hydrazine thrusters to around 100-deg C before the thruster is fired. (Firing the thruster with a low initial catalyst-bed temperature decreases the catalyst life.) The heater is commanded on, the catalyst-bed is heated, the thruster is fired, and the heater is turned off until the next maneuver, all under ground control. Such a heater is illustrated in Figure 3.

Most applications of heaters on spacecraft require some automatic control of the heater to keep a component at a desired temperature and to minimize the amount of time the heater is on so as to reduce power consumption. Historically, the most common control device is a mechanical thermostat, such as the one shown in Figure 4. These typically consist of a small hermetically sealed can containing a switch driven by a snap-action bimetallic actuator. The temperature at which the thermostat clicks on, known as its set point, is fixed for any given thermostat. The engineer can select from an array of standard thermostats available from the manufacturer to get a set point close to what is desired, or a custom device can be ordered. In addition to the set point, the dead band, or the difference between the temperatures at which the thermostat turns on and turns off, is important. A smaller dead band reduces the temperature swing of the device being heated and reduces power consumption a little (since the average temperature is lower). On the other hand, the smaller dead band also increases the number of cycles on the thermostat itself and decreases its reliability. In any event, dead bands less than 4-deg C are not recommended due to problems that have occurred in the past. Small dead bands have been known to increase the chance of "dithering," in which the thermostat rapidly cycles on and off. This is a failure condition that can cause the set point to drift lower, resulting in an undertemperature of the component being controlled.

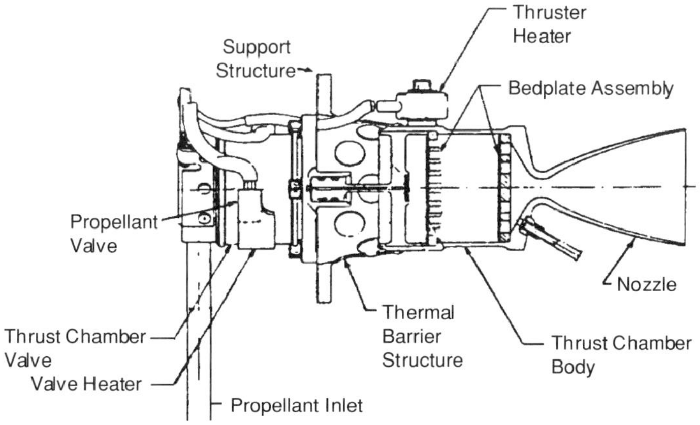

Figure 3. Hydrazine thruster

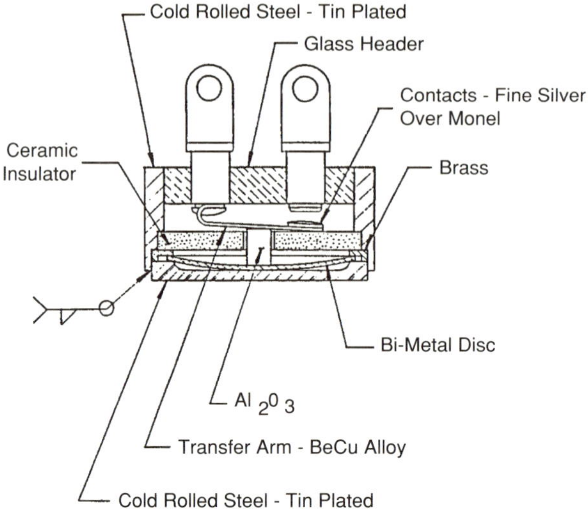

Figure 4. Elmwood thermostat

Even though thermostats are fairly reliable, the large number of them that may be present on a typical satellite (up to several hundred) results in occasional on-orbit failures. Because of this, and the increasing life requirements of satellites, solid-state controllers are becoming more common. Such a controller, an example of which is shown in Figure 5, replaces the mechanical switch with an electronic device that has a higher reliability and life expectancy. Such controllers are used extensively on the Defense Meteorological Satellite Program, the Hubble Space Telescope, and the Space Station. They employ a temperature sensor that can be located either at the controller or at a remote location, as desired. Another advantage of solid-state controllers is that extremely tight dead bands (< 0.1-deg C) are possible for very precise temperature control, such as is required by the Hubble Space Telescope. Optical systems, some sensors, and electronic frequency standards often require precise temperature control, which cannot be achieved with a thermostat.

Some of the most advanced satellites, including MILSTAR and other national-security-related programs, use on-board computers to control heaters. Such systems read the temperatures from telemetry sensors placed throughout the vehicle and send signals to turn relay-controlled heaters on and off as required. This allows enormous flexibility since the control set points and dead bands can be adjusted on orbit by uplinking new tables and/or logic to the spacecraft computer. In one instance, the loss of an entire satellite was averted because of the flexibility of its computer-controlled heaters.

Temperature Controller Module

Specifications	
Package:	Hermetically sealed can, .65 x .85 x .95 inches
Control Power (Heater):	0 to 100 watts, higher power available.
Quiescent Power Consumption (Standby Power):	30 mW
Input Power:	28 VDC nominal, 15 VDC to 45 VDC range
Efficiency:	98% minimum
Set Point Accuracy:	=.25-deg C, closer tolerances available
Weight:	Less than 30 grams
Loop Gain:	Provisions for external adjustment of control loop gain
Compensation:	Provisions for addition of loop compensation.
MTBF:	4.7 million hours minimum @ 25-deg C controller ambient
Electronic Components:	Meet requirements of JAN TXV, MIL-8838, MIL-R-55182, and MIL-C39014 (Commercial model also available)
Module Ambient (Heat Sink Temp.):	-55-deg C to +75-deg C

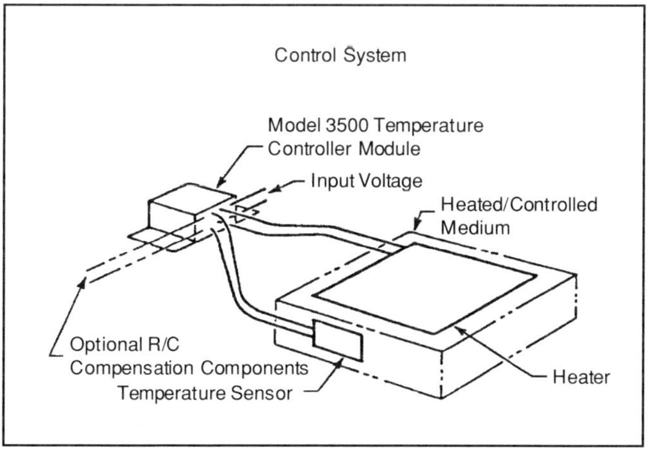

Figure 5. TAYCO solid-state controller

CIRCUITS

A typical satellite will have dozens of heaters controlled by different thermostats, relays, solid-state controllers, or computers. Many different types of redundancy schemes may be employed, even on the same satellite, depending on the criticality of the heater in question.

A representative heater is shown in Figure 6. It consists of redundant resistance elements in a single-patch heater. Each element is powered by a separate spacecraft power bus (satellite power systems are normally redundant), and each element has its own enable/disable relay, which is commandable from the ground. Series-redundant thermostats provide single-fault tolerance on each element for a thermostat failed closed. If one of these thermostats fails open, however, the circuit is dead. A number of these heaters are used on the satellite. A typical panel of equipment with heater and thermostat locations is shown in Figure 7. The heaters are the dark rectangular patches and the thermostats are black dots. It should be noted that the branch of the bus that supplies power to these heaters is fused, although this is not shown in Figure 6.

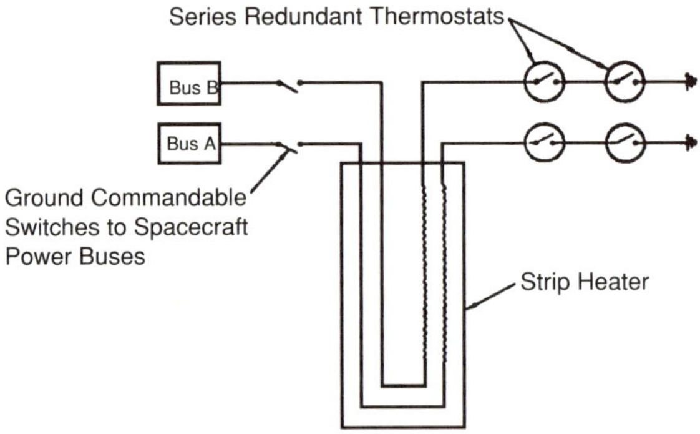

Figure 6. Heater circuit example

There is a large number of ways to lay out heaters and thermostats, depending on the level of reliability required. Figure 8 shows four different schemes used on one satellite. The most reliable (Type I in the figure) consists of redundant resistance elements working off of different power buses, each element employing "quad-redundant" thermostats. A quad-redundant arrangement requires at least two failures to disable thermostatic control. The other arrangements represent lower reliability designs, but require fewer thermostats.

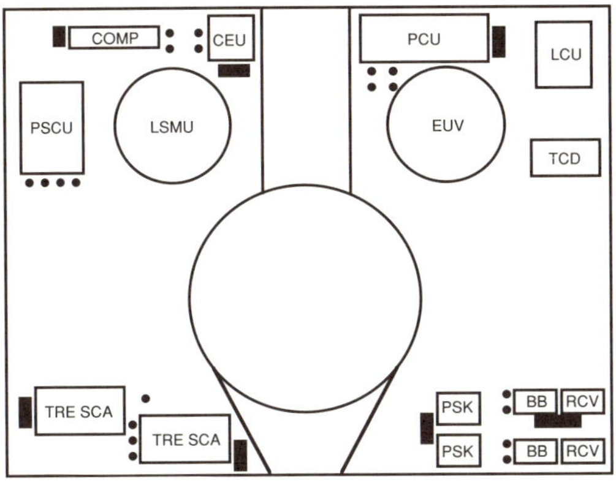

Figure 7. Heater and thermostat layout

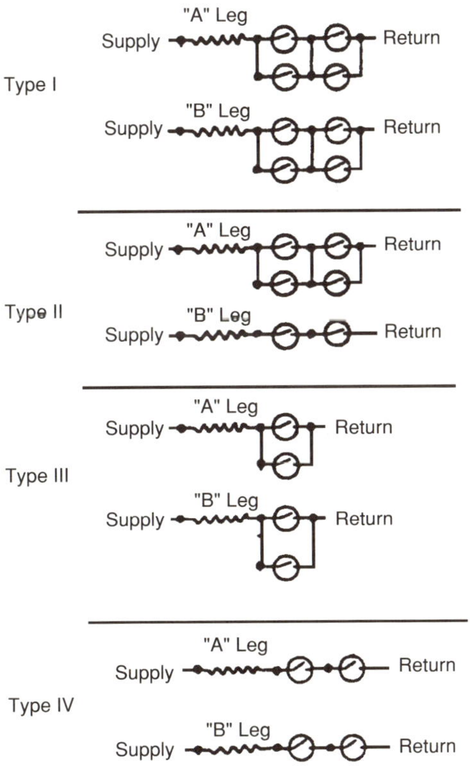

Figure 8. Some heater wiring schemes

A schematic for the heater system used on one panel of equipment on the Defense Satellite Communication System (DSCS) spacecraft is shown in Figure 9 to illustrate a typical application. Two sets of heaters are used: one set of survival heaters with a set point of -18-deg C, which are used during launch before the spacecraft is fully powered-up in its operational orbit; and a second set of heaters with a set point of 13-deg C, which are used during normal on-orbit operations. The survival heaters have a lower set point to reduce their power draw (less heat is radiated away from the spacecraft at the lower temperature). The operational heaters, however, need a higher set point, since the electronics boxes will not function properly at the survival temperature. It should be noted here that two sets of heaters would not be required if the satellite used a computer-controlled heater system where a different set of set points could be accessed by the computer software during survival-mode operation.

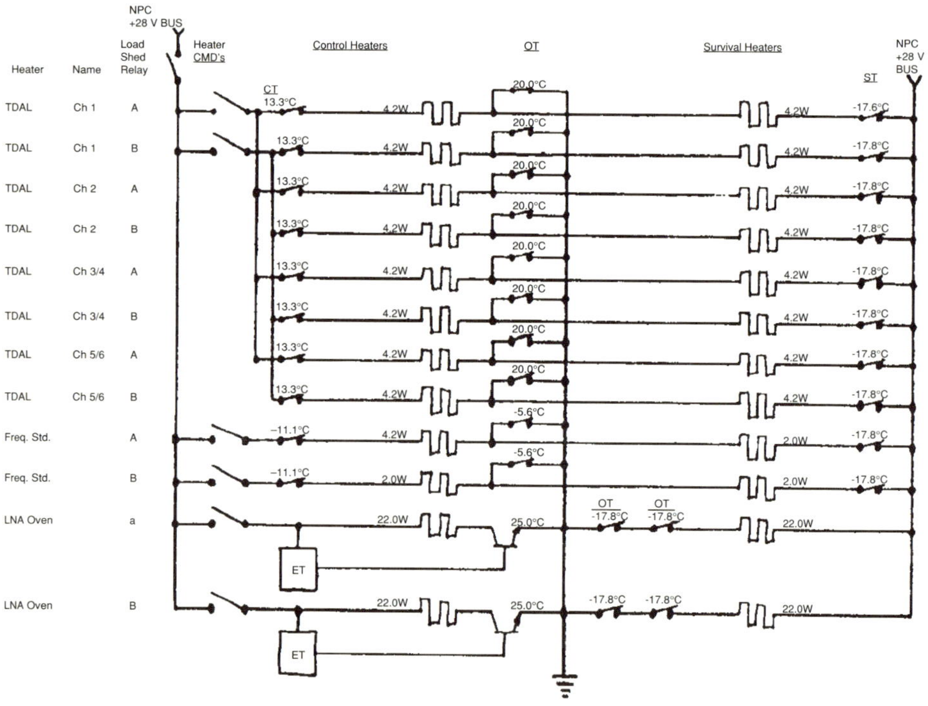

Note: CT = Control Thermostat
OT = Overtemperature Thermostat
ST = Survival Thermostat
ET = Electronic Thermostat
Temperature Listed is Thermostat Set Point

Figure 9. DSCS satellite north panel heater schematic

The survived heaters are not redundant because they are not normally used and because the failure of a single heater would not result in a loss of the mission. They are, however, always connected to the power bus, without relays, to protect the spacecraft at all times. The control heaters, on the other hand, are completely redundant in circuitry, with one control thermostat on each circuit. Each circuit also has an over-temperature thermostat that switches off the heater at 20-deg C if the primary thermostat fails closed. Some of the "A-side" heaters are grouped together on a single commandable enable/disable relay, as are some of the "B-side" heaters. In addition, there are two heaters controlled by electronic thermostats that are used on the low noise amplifier oven to precisely control the temperature of an oscillator crystal. The DSCS heater schematic is offered only as an example and there are wide variations in heater circuit layouts found on different satellites.

Section 5

Louvers

Brian E. Hardt
The Aerospace Corporation

R. D. Karam
R. J. Eby
Fairchild Space Co.

INTRODUCTION

Louvers are active elements that have been used in different forms on numerous spacecraft. In general, louvers can provide about a six-to-one variation in heat rejection from fully closed to fully open without any power consumption. Thus, they find applications where the internal power dissipation varies rather widely due to duty-cyclic considerations. The most widely used louver assembly is the bimetallic spring-actuated rectangular-blade (venetian blind) type. Hydraulically activated and pinwheel louvers are being used less and less a factor. The reliability question can be "solved" by the design technique of making each louver blade independently actuated with a bimetal clock spring. Thus, a single-point failure exists for only one blade, not the entire assembly. The spring can be integrated with a heater/controller to decrease the closed-to-open temperature range from 10-deg C to 17-deg C, to as little as 1-deg C.

Louver assemblies consist of five main components: baseplate, blades, actuators, sensing elements, and structural elements, as illustrated in Figure 1. The baseplate is a surface of low-absorptance-to-emittance ratio that covers the critical set of components whose temperature is being controlled. The blades, that are driven by the actuators, are the elements of the louvers that give variable-radiation characteristics at the baseplate. While closed, the louvers shield and decouple the baseplate from the surroundings, but when open allow for a radiative coupling between the baseplate and the surroundings. The radiation characteristics of the baseplate can be varied over the range defined by these two extreme positions. The actuators are the elements of the louvers that drive the blades according to the temperature seen by the sensors placed in the baseplate. Actuators of the louvers flown on satellites have been bimetal spirals or bellows, although other types could be used, such as Bourdon spiral and electrical devices. In a single actuation system all the blades are driven by a single actuator. In the multiple-blade actuator system several actuators are required to operate the system. Generally, bimetals are used in multiple-actuation systems, and bellows in single-blade systems.

The actuator is intended to drive the blade angle as determined by the temperature of the baseplate. A strong conductive path between the actuator and baseplate is therefore desired in order to minimize the temperature gradient between the actuator and baseplate. The thermal coupling between a bimetal actuator and baseplate is composed of both radiative and conductive paths. Bellows or Bourdon actuators use a tank or tube of liquid or liquid/vapor to actuate the blades. The tank or tube is typically soldered to the baseplate to ensure a strong conductive coupling.

Louver assemblies have been designed for both shadow or sunlight operation. The two design approaches for operation in sunlight are to use a sun shield or to modify the louver assembly for high-temperature operation.

VENETIAN (VANE)

The most widely used louver assembly is the bimetallic spring-actuated rectangular-blade (venetian blind) type. Hydraulically activated and pinwheel louvers are becoming less and less common. The arrangement of activators, housing, blades, and structure for a venetian-blind type louver assembly is shown schematically in Figure 1. In addition to supplier-particular differences, design features vary depending on whether or not the assembly is exposed to solar illumination and whether actuation is a bimetallic spring alone or a bimetallic spring in conjunction with a heater/controller. Principal suppliers are Fairchild Space & Electronics Company and Northrop Corporation, whose designs are quite similar, and RCA. Characteristics for flight-qualified rectangular blade louver assemblies are given in Table 1.

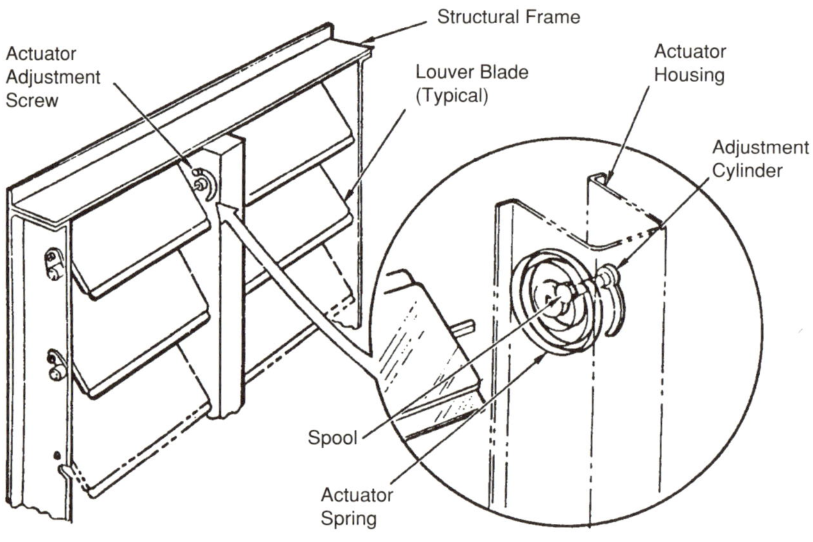

Figure 1. Fairchild and Northrop louver assembly schematic

Table 1. Characteristics of flight qualified rectagular blade louver assemblies

Supplier	Northrop	Fairchild	Fairchild	Fairchild	RCA	RCA
Type	Rectangular Blade	Rectangular Blade	—	High Temp., Rectangular Blade	Rectangular Blade	Rectangular Blade
Control Technique	Passive, Radiative and Conduction Coupling with Mounting Plate	Passive, Radiative and Conduction Coupling with Mounting Plate	—	Passive, Coupling with Mounting Plate, MLI Between Actuator Cover & Housing	Active Using Solid State Thermal Switch & Heater	—
Temperature Change, Open to Closed, °C	34	17	—	10	1	1
Effective Emittance						
Closed	0.1	0.1	—	0.1	0.15	0.15
Open	0.7	0.7	—	0.7	0.85	0.85
Size, inch	8.0x16.0x2.5	8.6x16.5x2.5	21.9x42.6x2.5	23.8x24.5x2.5	10x10x5	10x10x5
Weight, lb	0.65	1.00	5.50	5.33	0.25	1.8
Area, ft^2	0.89	0.98	6.50	6.50	0.70	4.20
Specific Weight, lb/ft^2	0.73	1.02	0.84	0.82	0.36	0.43
Number of Blades per Assembly	8	8	42	24	2	8
Number of Actu-ators per Assembly	4	4	21	22	2	8
Blade Rotation	Unidirectional	Unidirectional	—	Alternating Direction	Unidirectional	—
Materials & Finishes						
Frame	Aluminum	Aluminum	—	Aluminum	No Frame, Aluminum Support	Aluminum
Actuator Cover	-	-	-	Fiberglass	-	-
Actuator Housing	Aluminum	Aluminum	—	Aluminum	Aluminum	—
Actuator	Bimetallic Spring Chase 6650	Bimetallic Spring Chase 6650	—	Bimetallic Spring Chase 6650	Bimetallic Spring Chase 6650	—
Blades	Polished Aluminum	Polished Aluminum	—	White Striped Polished Aluminum	Foam Sandwiched Fiberglass Quill; Al/Kapton, First Surface Tape, Both Sides	—
Shaft	Anodized Aluminum	Anodized Aluminum	—	Tespel	Fiberglass	—
Bearings & Supports						
Inboard	Teflon Spool/ Aluminum Bearing	Teflon Spool/ Aluminum Bearing	—	Teflon Spool/ Aluminum Bearing	Supported & Contacted by Bimetallic Spring, No Bearing	—
Outboard	Teflon Bearing Between Al. Shaft & Housing	Teflon Bearing between Al. Shaft & Housing	—	Teflon Bearing Between Al. Shaft & Housing	Delvin AF Bushing, Presumable No Contact On-Orbit	—
Base	Aluminum	Aluminum	—	Ag/Teflon Second Surface Tape on Aluminum	No Base Louver Supports Attach to Equipment Mounting Plate	Aluminum

In the Fairchild and Northrop designs blade rotation is effected by the expansion or contraction of a spiral bimetallic actuator, by virtue of heat gained or lost in exchange with the equipment-mounting plate (Figure 2). One end of the actuator is attached to the frame structure and the other to the Teflon spool. A square cutout in the spool supports the inboard louver blade end. The actuator is coated black to enhance radiative interchange. The conduction path is through the aluminum housing. The actuator is adjusted relative to the frame to obtain the desired fully-closed to fully-opened temperature range. Each blade is supported inboard and outboard by a bearing assembly (see Figure 2). Inboard, the Teflon spool bears against and rotates with respect to the aluminum support structure. The outboard end of the louver blade shaft rotates within and is supported by a Teflon bearing, with end play established by the distance between the Teflon thrust pad and the set screw. Each louver blade consists of a central torque tube bonded to flanges. The louver blade crossection forms a hollow, thin-walled rectangle of high-aspect ratio. The blades are highly polished to reduce emittance.

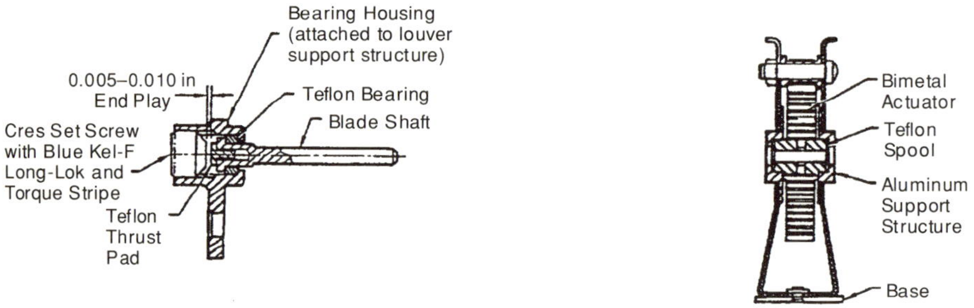

Louver Outboard Bearing Assembly

Louver Inboard Bearing Assembly

Figure 2. Louver bearing assemblies

Louver assemblies of the type described above have been used in applications where direct solar illumination is avoided. Such satellite applications include Pegasus, OAO, NIMBUS, ERTS, V075, Voyager, Seasat, and GPS. As each louver assembly contains several independently actuated blades, a degree of redundancy is inherent in this design approach.

RCA's design approach employs active control of blade position though a bimetal/heater assembly (Figure 3). Frame structures are used for the larger louver assemblies, while the smaller assemblies are frameless. In this latter case, the actuator and the end support bracket are aligned and then attached to the equipment mounting plate with a foamed closeout used at the edges. The blades are supported and centered inboard by the bimetal/heater assembly. The fiberglass shaft with bonded-on, ball-end pivot is supported outboard by a Delrin AF bushing in the end-support bracket. The blades, comprised of a foam sandwich about the fiberglass quill, have a 1-mil, first-surface-aluminized Kapton film on each side.

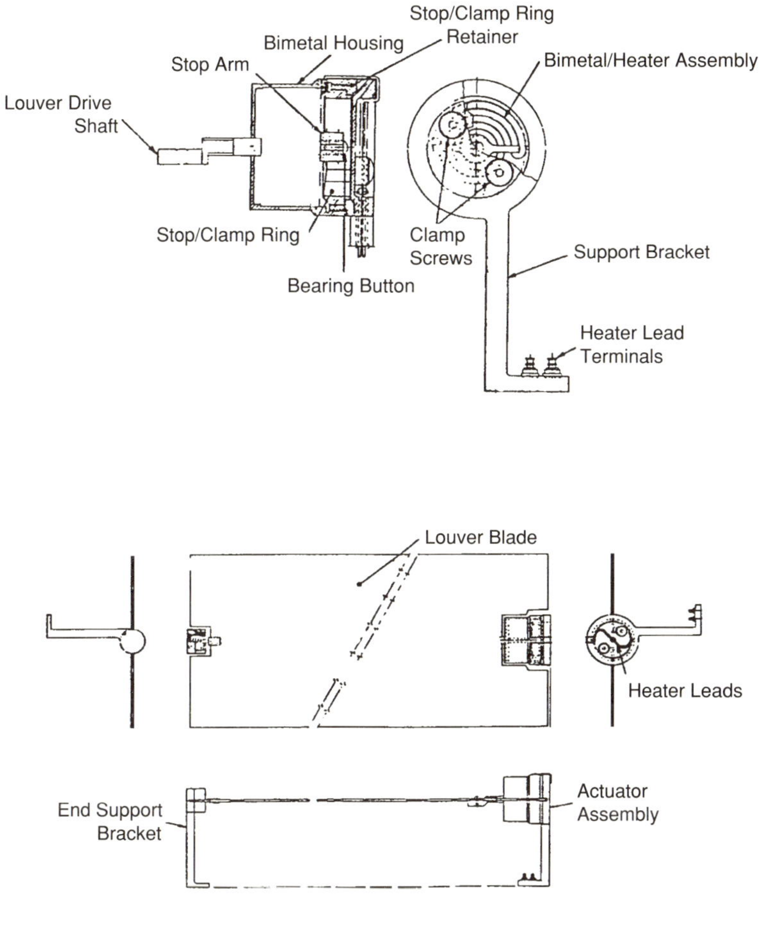

Figure 3. RCA louver assembly

The bimetal/heater assembly drives the blade from fully closed to fully open over only a 1-deg C temperature change. The louver begins to open passively (by conduction from the mounting plate to the bimetallic spring) at 10-deg C range. This provides backup if the active controller fails off. The failed-on case can be corrected by ground disabling of the heater circuit. The bearing/support system provides a load-carrying capability during ground testing and, if alignment is true, the absence of friction on-orbit.

ANALYSIS OF VANE-TYPE LOUVERS

Heat Transfer Characteristics

Radiation through louvers is characterized by an effective emissivity, ε_{eff}, and an effective absorptivity α_{eff}, which satisfy the steady-state energy equation for an isothermal body in a solar-space environment:

$$\frac{Q}{A} = \varepsilon_{eff}\sigma T^4 - \alpha_{eff}S \quad , \tag{1}$$

where Q (watts) is net heat transfer from louvered area A(m^2), T(K) is absolute temperature, S (W/m^2) is solar constant, and σ is Stefan-Boltzmann constant (5.668×10^{-8} W/m^2K^4).

Effective Emissivity

Effective emissivity may be viewed as the ratio of net-heat transfer from a louvered surface to the radiation from an equivalent black area ($\varepsilon = 1.0$) at the same temperature but in the absence of louvers. When no external heat sources are present, the definition reduces to

$$\varepsilon_{eff} = \frac{Q}{A\sigma T^4} \quad . \tag{2}$$

The motivation for writing Eq. (2) in this form is the simplified equivalence to gray-body radiation. Actual heat transfer in louvered systems involves conduction along the frame and actuator housing cover, heat loss through actuator insulation and blade shafts, and variable feedback from reflections off the specular blades. In addition, frictional effects are inconsistent and generally result in nonuniformity in the blade's angular positions. When the definition given by Eq. (2) is used to obtain test values of effective emissivity, they will inherently contain these distortions and other deviations from a purely radiative system.

Effective emissivity tests are conducted in cryogenic vacuum chambers (Refs. 4 and 7). The louvered panels are instrumented with heaters and thermocouples, and the unlouvered backside is covered by multilayered insulation and guard heaters. This practically eliminates radiation from the back and provides better estimates of the energy escaping through the louvers. The assembly is suspended in the chamber by low-conducting wires, and a series of steady-state temperatures are recorded corresponding to various heater power levels. Effective emissivity is calculated from the data by using Eq. (2).

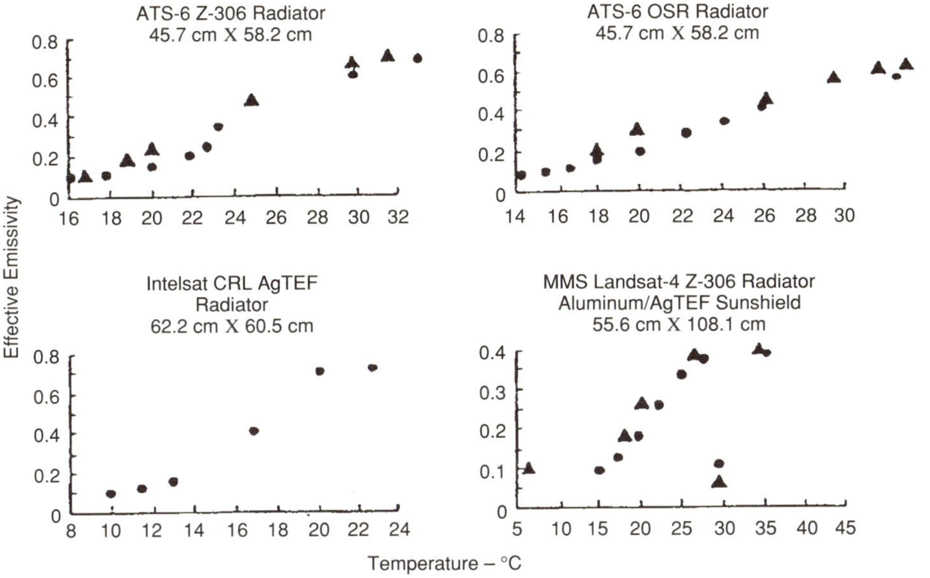

Figure 4. Louvers effective emissivity variation with temperature (test data)

Some test results are shown in Figure 4, with comments in Table 2. All of the panels were uniformly heated and maintained nearly isothermal. The data were adjusted to account for a vacuum-chamber wall temperature higher than zero absolute (usually - 190-deg C).

The discrepancy in the values of effective emissivity as equilibrium is approached from above versus from below is attributed mainly to frictional effects. But the test points can be generally contained within two straight lines that bound the louver's performance. A linear variation between closed and open positions is commonly assumed in thermal analysis, thus

$$\varepsilon_{eff} = \varepsilon_c \text{ (constant)}, T \leq T_c$$

$$\varepsilon_{eff} = \varepsilon_o - \frac{\varepsilon_o - \varepsilon_c}{1 - T_c/T_o} \left(1 - T/T_o\right), T_c \leq T \leq T_o \qquad (3)$$

$$\varepsilon_{eff} = \varepsilon_o \text{ (constant)}, T \geq T_o \quad ,$$

where the subscripts o and c refer to fully open and fully closed, respectively. A quadratic form is sometimes found to represent more accurately temperature dependence in the active region:

$$\varepsilon_{eff} = \varepsilon_o - \frac{\varepsilon_o - \varepsilon_c}{(1 - T_c/T_o)^2} \quad (1 - T/T_o)^2, T_c \le T \le T_o \tag{4}$$

Table 2. Louver Effective Emissivity (Test Data)

Program Louver Size	Radiator Hemispherical Emittance	ε_{eff}		Radiator* ΔT
		Open	Closed	
ATS-6** 45.7 cm x 58.2 cm	OSR $\varepsilon = 0.77$	0.62	0.114	18 K
ATS-6 45.7 cm x 58.2 cm	Z-306 $\varepsilon = 0.88$	0.71	0.115	18.6 K
GPS 40.6 cm x 40.5 cm	Z-306 $\varepsilon = 0.88$	0.70	0.09	18 K
Intelsat CRL↑ 62.2 cm = 60.5 cm	AgTEF $\varepsilon = 0.76$	0.67	0.08	10 K
MMS Landsat-4↑↑ 55.6 cm x 108.1 cm	Z-306 $\varepsilon = 0.88$	0.39	0.10	17 K

* ΔT = T(OPEN) - T(CLOSED) in vacuum there is usually about 2-deg C lag between radiator temperature and bimetallic temperature.

** ATS-6 has white stripes on the blades and a fiberglass actuator housing treated with aluminized Kapton tape (Kapton out).

↑ Counter-rotating blades ATS-6 housing and blades and enlarged springs.

↑↑ Aluminum shield with AgTEF exterior and Z-306 interior.

Analysis (Refs. 3 and 8) shows variations that can be represented by segments of sine or cosine curves. This is an expected result from the idealized mathematical models that incorporate assumptions of infinite blade length and heat transfer only by radiation. In this case, effective emissivity varies almost linearly with projected open area, which is a trigonometric function of blade angle.

Equation (2) is also used to calculate from test data the effective emissivity, F_ε, of shielded louvers. Theoretically, F_ε is related to e_{eff} without a shield by the equation

$$F_\varepsilon = \frac{\varepsilon_i \, \varepsilon_{eff}}{1 + \varepsilon_{eff} \, \varepsilon_i / \varepsilon_e} , \tag{5}$$

where ε_i and ε_e are internal (facing the louver) and external emissivities of the shield. Since test values include conduction and other heat loss, they are found to be slightly higher than those obtained by using Eq. (5).

Effective Absorptivity

Effective solar absorptivity of a louvered panel may be defined as the fraction of incident solar energy absorbed per unit area of a louvered surface. The definition appears in Eq. (1) as

$$\alpha_{eff} = \frac{\varepsilon_{eff} \sigma T^4 - Q/A}{S} , \tag{6}$$

where, again, Q may be regarded as actual net heat loss by a louvered surface having area A. The impinging solar flux S (nominally 1350 W/m^2) is not modified by the direction cosines, so that the values of α_{eff} inherently contain the effects of sun multireflection off specular components.

Equation (6) is used to obtain the effective absorptivity by test. The test configuration used for finding ε_{eff} is modified to include a sun simulator and a means for varying the angle of incidence. A detailed description of a typical facility is given in Ref. 4.

Few solar-simulation tests have been performed on louvers to date and published data remain scanty. The results of tests on two early versions of the type eventually used on ATS-6 are reported in Ref. 4. These units, however, did not have many of the design features later added to enhance the thermal performance, and the results appear to be somewhat different from those obtained with later louvers. Data from two sets that represent current designs are given in Figure 5. The effective absorptivity was calculated using Eq. (6), with effective emissivity (as function of fixed-blade angle) already known from tests without sun input. The tests were conducted in solar-simulation chambers using xenon compact arc type lamps with a beam half-angle collimation of less than 1.5 deg. The mounting radiators were uniformly heated and were sufficiently conductive (1.27-cm-thick aluminum) to limit lateral gradients to less than 2-deg C. All data were obtained shortly after application of radiator coating and hence are undegraded values. For long-term performance evaluation, and at least for fully open louvers, the practice has been to increase the values of ε_{eff} by the ratio of the radiator's estimated degraded absorptivity to the as-applied value.

Experimental data on the trend of variation of effective absorptivity with blade angle at zero azimuth agree reasonably well with analysis (Refs. 3 and 4). Correlation with the predicted values is also good when the mathematical models incorporate the gap that exists between the blade louver edge and the mounting panel (Ref. 9). However, for azimuths other than zero, correlation becomes erratic, particularly at blade angles less than fully open. The major reasons for this probably lie in the false assumption made in the analysis that the blades are infinitely long (length-to-width ratio of a blade is generally less than 5.0) and in ignoring the presence of frame and the actuator assembly.

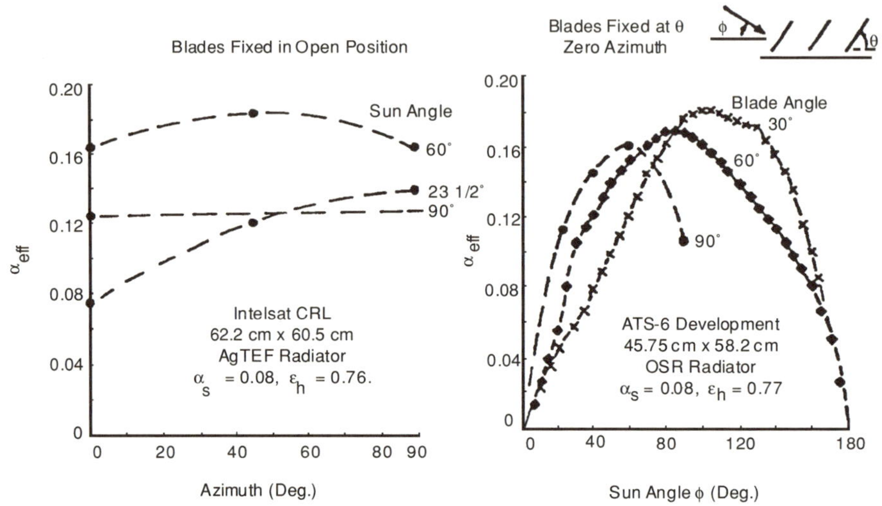

Figure 5. Louver effective solar sbsorptivity variation with azimuth and blade angle (test data)

Effective absorptivity is not defined for shielded, louvered radiators. An efficient sunshield design requires a low-solar-absorptance, high-emittance coating on the exterior. The interior surface should have high emittance to enhance heat exchange with the radiator when the blades are open. When spacecraft envelopes permit, an oversized shield of potentially very low temperature should be considered.

Performance Curves

Performance curves of louvered radiators relate the heat rejection rate to the radiator temperature. The curves are usually generated for steady-state, isothermal conditions in order to reflect maximum and minimum heat-rejection capabilities. This information used in the initial phase of a thermal design to determine the louver size that will accommodate the required heat-rejection rates at specified temperatures and environment.

Eq. (1) can be modified to include heat inputs from infrared sources (such as Earth) and reflected solar energy (albedo):

$$\frac{Q}{A} = \varepsilon_{eff}\left(\sigma T^4 - IR\right) - \alpha_r fS - \alpha_{eff} S \tag{7}$$

Here IR is infrared and $\alpha_r fS$, f being the albedo factor, is reflected solar energy (usually diffuse), which filters through the louver and is absorbed by the radiator with solar absorptance α_r. In practice, there is uncertainty in including this effect due to lack of experimental and flight data. One approach is to modify incident albedo by the effective emissivity before multiplying by α_r. This, in a sense, adjusts the diffuse input by a view factor prior to impingement on the radiator. A more conventional but conservative approach is to replace α_r by α_{eff}.

The effective emissivity, ε_{eff}, in Eq. (7) implicitly contains the radiator emissivity and is a predetermined function of radiator temperature. Effective absorptivity is assumed to be a known function of blade angle (which is linearly related to temperature) and sun angle. Infrared radiation, as it appears in Eq. (7), is the value adjusted by view factors to an imaginary louver plane parallel to the radiator. Incident solar flux, S, is not modified by the angle of incidence, in conformity with the definition of α_{eff}.

Net heat rejected from a shielded louvered surface is found from the equation

$$\frac{Q}{A} = F\varepsilon\left[\sigma T^4 - \left\{\frac{\alpha_e}{\varepsilon_e} S(\sin\theta + f) + IR\right\}\right] , \tag{8}$$

where θ is sun incident angle and the subscript e refers to the exterior of the shield. Infrared flux is adjusted by script F from sources to shield.

For many satellite systems, the performance of louvered radiators may be evaluated by considering net-heat rejection averaged over an orbital period. In this case Eqs. (7) and (8) become very useful. This averaging technique has been found to be valid for many thermal designs in which massive electronic components are mounted to a louvered honeycomb tray (Refs. 2 and 10).

Shielded vs. Unshielded Louvers - Special Cases

It is essential in a louvered design that the utmost in heat rejection be achieved in the open-blades position, while only negligible heat be transferred when the blades are closed. While it is inevitable that the mere presence of louvers will lead to some blockage to radiation with open blades, the situation can be improved by making the thicknesses of blades and frame and the width of the actuator housing as small as is practical, and by providing a highly reflective finish on the surfaces viewed by the radiator. But, as noted earlier, the specularity of louver parts lead to sun-ray entrapment and a reduction in heat rejection. Shielding will eliminate this effect but at the expense of introducing an additional resistance to radiation to space.

In the following, a comparison is made between the hot-case operations (open louvers) of shielded and unshielded louvers based on orbital average performance under various environments. The shielded system is assumed to be an MMS flight-type configuration having the following characteristics:

Sunshield $\alpha_e = 0.14$
Sunshield $\varepsilon_e = 0.76$
Sunshield $\varepsilon_i = 0.88$
Radiator $\varepsilon_r = 0.88$
Effective emissivity without shield $\varepsilon_{eff} = 0.71$
Effective emissivity with shield (Eq. 5) $F_\varepsilon = 0.34$

The properties of the unshielded system are assumed to be those of the INTELSAT counter-rotating louver (CRL), with effective emissivity of 0.67 and effective absorptivity as described in Figure 5. In the calculations, the solar constant was taken as 1350 W/m^2, albedo factor 0.35, and Earth radiation 200 W/m^2.

Solar and Albedo Equal Zero

For satellites where louvers are mounted on shadowed or anti-sun sides having negligible albedo input, heat rejection is

$$\left(\frac{Q}{A}\right)_{unshielded} = \varepsilon_{eff}\left(\sigma T^4 - IR\right)$$

$$\left(\frac{Q}{A}\right)_{shielded} = F_\varepsilon\left(\sigma T^4 - IR\right) .$$

Since for the given data ε_{eff} is nearly twice F_ε in the open position ($\varepsilon_{eff} \sim F_\varepsilon$ in the closed position), unshielded louvers are much preferred in a purely infrared environment. The orbital average-heat rejection as a function of radiator temperature is given in Figure 6. An IR value of 200 W/m^2 corresponds to Earth flux input in a low-altitude orbit. The case IR = 0 corresponds to a geosynchronous orbit with perfectly aligned north/south faces.

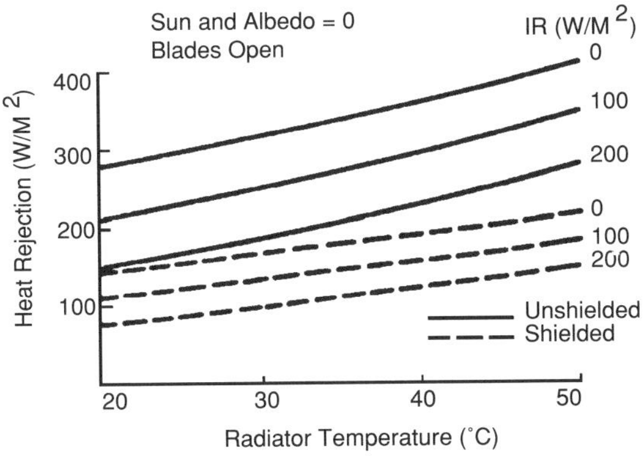

Figure 6. Louver heat rejection in IR environment

Sun-Oriented Low Earth Orbit

A similar comparison can be made for a spacecraft that is sun-oriented in a 370-km, 30-deg inclined orbit. Two orientations are considered: the first, shown in Figure 7, is such that the sun vector is parallel to the open louver blades during the sunlight portion of the orbit; the second, shown in Figure 8, has the sun vector inclined 30 deg off the normal to the plane of the louver. This second orientation represents the maximum solar input to an open louver array as indicated in Figure 5. It is noted in Figures 7 and 8 that the orbital average heat rejection is greater for unshielded louvers in both cases.

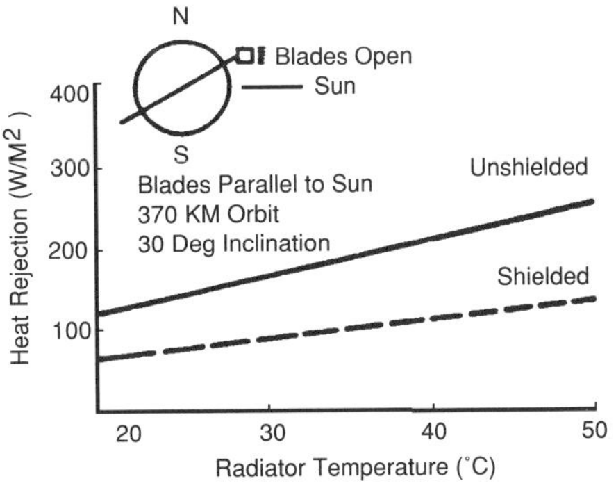

Figure 7. Shielded vs. unshielded louvers in sun-oriented near Earth orbit, sun parallel to blades

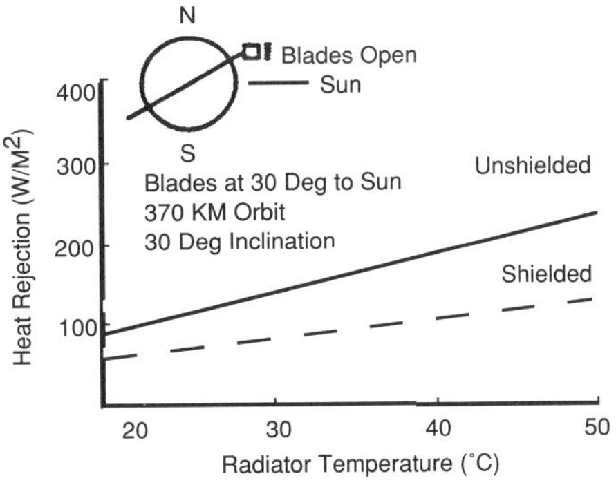

Figure 8. Shielded vs. unshielded louvers in sun-oriented near Earth orbit, sun at 30 deg to blades

Earth-Oriented Low Earth Orbit

A condition in which the combined absorbed solar and Earth flux is maximum during a near-Earth polar orbit is given in Figure 9. In this case, the sum of the Earth and solar flux inputs to exposed louvered radiators is relatively large and remains nearly constant when the angle between the sun and the plane of the louvers varies between 30 deg and 60 deg. At the smaller angles, close to 30 deg, heating from the Earth represents about 50 percent of the total absorbed input. At larger angles, although Earth flux diminishes, solar flux increases. In this particular orbit, the exposed louver system can exhibit less heat rejection capability than a shielded configuration. It may be noted, however, that a change in the orbit hour, equivalent to a rotation of the plane of the orbit, introduces a shadow period that reduces the net solar input so that, even for short shadow time, exposed louvers are more efficient on an orbital average basis. The cross-over point depends on radiator temperature. For 30-deg C it is shortly after 8:00 hours.

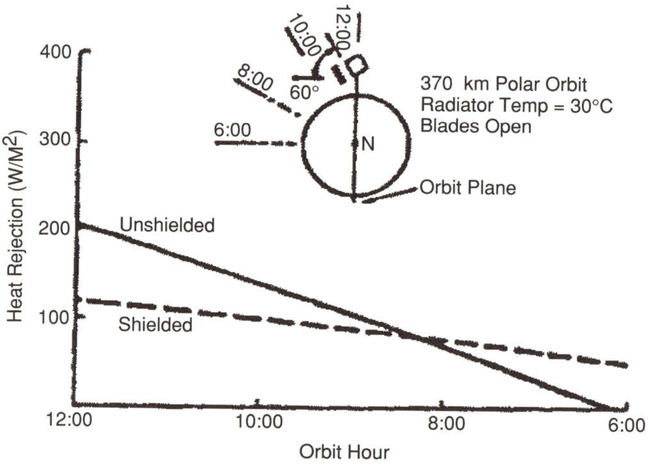

Figure. 9. Shielded vs. unshielded louvers in near-Earth polar orbit, louver blades at 60 deg to orbit plane

Earth-Oriented Geosynchronous Orbit (No Shadow)

Consider an Earth-oriented geosynchronous orbit during which the solar vector moves at various angles inclined to the plane of the louver. The effective absorptivity for exposed louvers varies continuously with azimuth, as shown in Figure 5. The orbital average value of α_{eff} for open blades at a given sun angle can be calculated from CRL test data.

Sun Angle (deg)	Orbit Average α_{eff}
0	≈0
23.5	0.109
60	0.178
90	0.124

Figure 10 shows heat rejection as a function of radiator temperature for no shadow periods. It can be seen that even under worst heating angles, unshielded louvers are more efficient than shielded at the radiator temperatures usually encountered in spacecraft temperature control.

It is evident from this study that an exposed louver system offers greater heat rejection in most practical cases. An exception is a near-Earth orbit in which the louver continuously views the sun 30-deg to 60-deg off normal. Of course, other orbits and trajectories are possible, and comparison studies must include orbital transient variations. But it appears that with the introduction of even small shadow periods, most systems exhibit greater heat-rejection rates when the louvers are exposed.

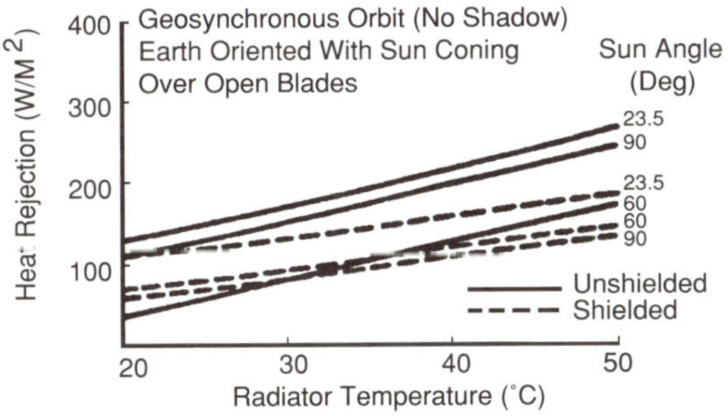

Figure 10. Shielded vs. unshielded louvers in Earth-oriented geosynchronous orbit

Louvers in Sunlight - Blade Temperature

Available test data on louvers in the sun confirm the prediction that high temperatures can be reached on the blades (Ref. 4). This presents problems of outgassing and delamination in designs where adhesives are used to bond plate sections. In addition, conduction-heat transfer from heated blades can reduce the long-term reliability of the bearings and actuator spool.

The temperature may be lowered by introducing strips of white paint (low absorptivity and high emissivity) on the blade surface. The equivalent absorptance-to-emittance ratio is thus reduced, leading to cooler temperatures. This effect appears in Figure 11, which reproduces the results obtained at NASA GSFC. Figure 12 shows how the blade temperature can be related to the percentage of blade surface area (one side) that is painted. Standard applications limit the amount to about 15 percent, and the stripes are usually located near the blade edge that is farthest from the radiator with the white exposed when the blades are closed. The effect of increased blade emissivity on effective emissivity is shown in Figure 13.

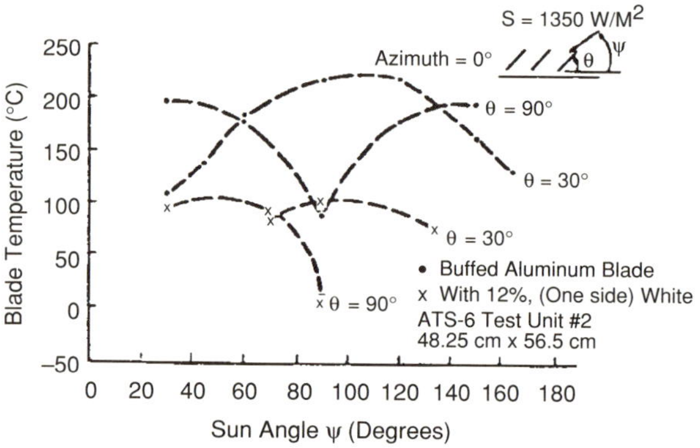

Figure 11. Louver blade temperature in the sun (test data)

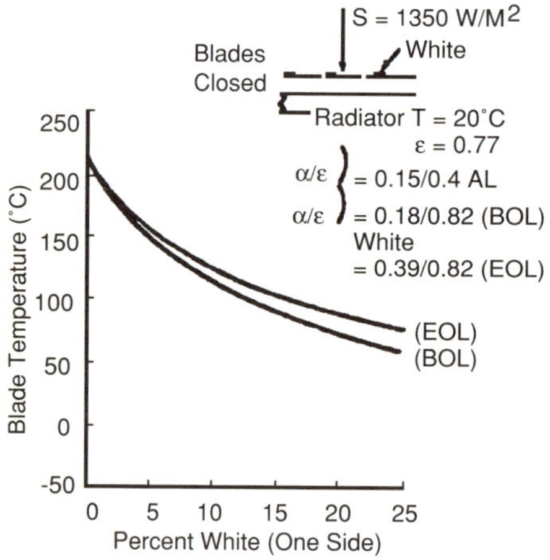

Figure 12. Effect of white paint on blade temperature

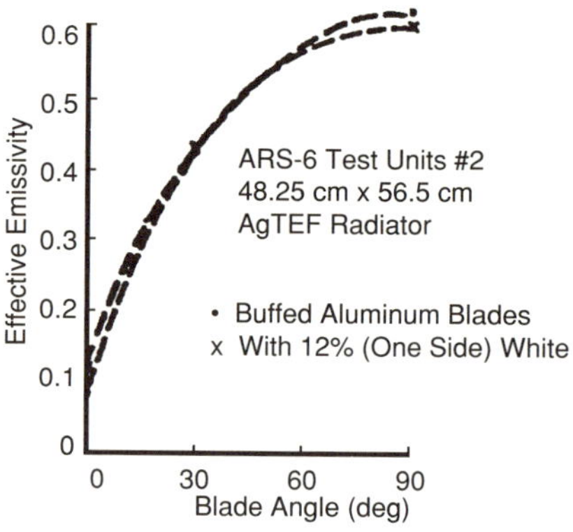

Figure 13. Effect of painted blades on effective emissivity (test data)

Louvers Transient Response

In most applications, louvers are mounted to equipment baseplates with large thermal masses. Hence the orbital temperature change of radiators is generally not radical and bimetallic response follows closely.

Actuator response time may be quantified by considering hypothetical cases in which there is an instantaneous step change in radiator temperature. Response time can then be characterized by the time required for a louver to complete a half cycle, from fully closed to fully open, or vice versa.

The transient problem is usually treated by constructing a small nodal model with conduction coupling between the radiator and the actuator housing and radiation couplings between the radiator, the actuator, and the external environment. There is significant sensitivity of response time to the values of conductances and it may become necessary to conduct simulation tests if response time is a critical factor in predicting performance.

Typical profiles are given in Figures 14 and 15. These represent the results of analysis of shielded louvers in a near-Earth high noon orbit. The mounting flange of the actuator housing was assumed to contact the radiator with an interface conductance of 140 W/m^2K. The actuators were radiatively coupled to the inner structure of the housing, which is coupled to the exterior structure through the surrounding multilayer insulation. The effective emissivity of the louver was assumed to vary linearly between 0.115 (at 10-deg C) and 0.70 (at 28-deg C). It is seen that the actuator temperature reaches the value for which the louver opens or closes in about 24 minutes.

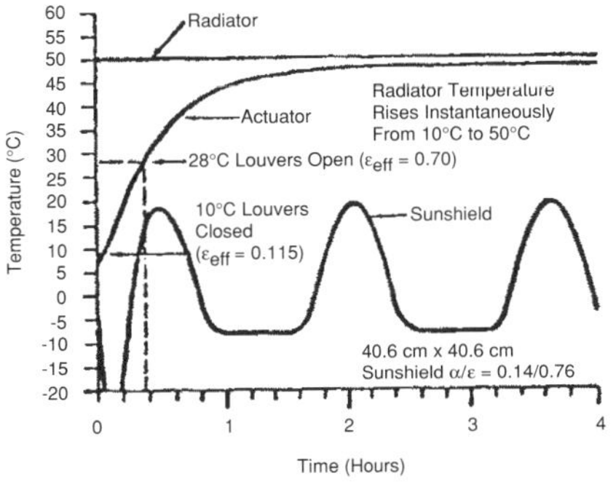

Figure 14. Louver heat-up response (450-km high noon orbit)

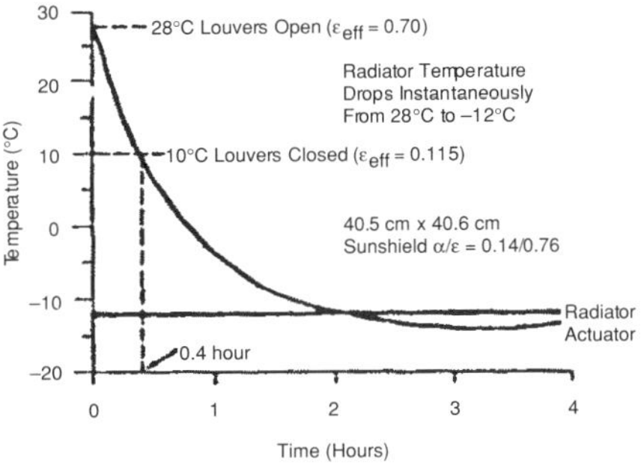

Figure 15. Louver cool-down response (450-km high noon orbit)

Non-Isothermal Transient Analysis

Thermal analysis of non-isothermal and transient radiators is performed by resorting to nodal computer models(10). A typical arrangement is shown in Figure 16. The nodes labeled 1 through 6 are located immediately facing the actuators and are used to set the emissivity of the nodes with respect to their location within blade pairs or within a blade and an end frame. As an example, the temperature of nodal point 1 sets the emissivity value of 7 and 8 in accordance with a given temperature-emissivity matrix. The input data file may also contain the effective absorptivity as function of sun angle (in plane ψ and azimuth ϕ) and blade angle θ, or temperature.

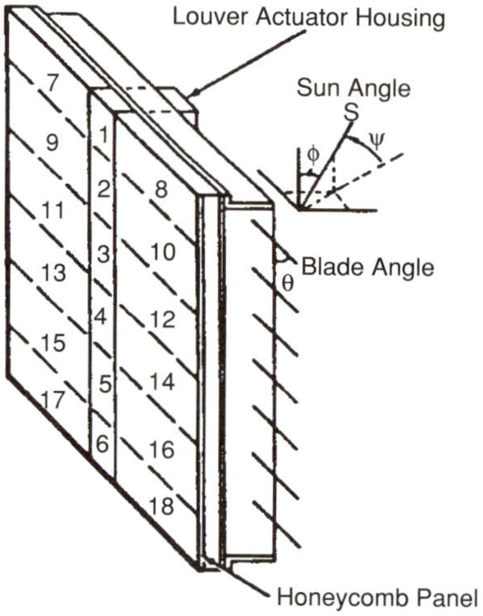

Figure 16. Nodal model of a louvered panel

For exposed louvers, the quasi-steady state equation

$$\frac{Q}{A} = \varepsilon_{eff}(\theta)\left[\sigma T^4 - IR\right] - \alpha_r(\theta)\, fS - \alpha_{eff}(\psi, \phi, \theta)S$$

is used with applicable orbital parameters to generate values of Q as function T (or θ) and (ψ,ϕ). The values are tabulated as bivariate arrays and input to the transient thermal model (usually SINDA format). The computer program performs a first interpolation using orbit position (time) to select for each θ the appropriate T and Q array. A second interpolation using the value of sun angles produces heat rejected at a particular nodal point. The double interpolation for each node can require lengthy computer time, and consideration may be given to generating orbital average (steady state) data for approximate representation of actual behavior.

Shielded louvered radiators are similarly modeled with nodes on the shield having a one-to-one radiation coupling to opposing nodes on the radiator. Cross-viewing among nodes does not generally exist due to the close proximity of louver blades and shield.

An analytical treatment of non-isothermal louvered radiators is presented in Ref. 11. An important conclusion from that study is that the usual profiles of effective emissivity vs. temperature, generated for isothermal panels, apply in non-isothermal systems, provided that the distance between centers of adjacent blades is not too large. It is also concluded that the use of louvers leads to significant reduction in potential lateral gradients.

SUN SHIELDED/HIGH-TEMPERATURE LOUVERS

Three designs capable of operation in direct sunlight have been developed: the RCA pinwheel, NASA's shielded louver assembly for the Multimission Modular Spacecraft, and Fairchild's high-temperature louver assembly. Successful operation in a solar environment requires isolation of the activator from solar heating, avoidance of material or bond degradation due to overheating, avoiding of solar trapping by louver blades, and increased surface area due to decreased effective emissivity in the fully open position. The RCA four-lobed pinwheel, discussed in the following sections, employs a single bimetal/heater activator assembly in the recessed configuration, with stop elements at 45 deg. When fully open the aluminized Teflon radiator surface constitutes only 50 percent of the total circular area.

NASA's MMS rectangular-blade louver assembly employs a silvered Teflon second-surface mirror shield bonded to an aluminum sheet that covers the louver assembly. While solar energy blockage is certain, the effective emissivity of the open configuration is only 0.39. Fairchild's high-temperature louver assembly, discussed in the preceeding section, employs an insulated housing, blade white-striping to reduce blade temperature, a Vespel shaft to isolate actuator from blades at a maximum predicted temperature of 220-deg C, and silvered Teflon second-surface tape over the base. While requiring somewhat less frontal area than the two other designs, this design requires extensive solar ray-tracing analysis, and testing over wide combinations of blade angle and solar elevation and azimuthal angle, and can therefore be more difficult to implement.

PINWHEEL LOUVERS

The pinwheel louver consists of a lobed louver blade, an actuator assembly, a guard ring, and a special radiator pattern, as shown in Figure 17. This type of louver may be selected due to its low mechanical profile (less than 0.5 inch) or its tolerance of solar loads. The louver will open passively through the action of a bimetallic spring or may be driven open using an electronic controller and a small heater on the spring. The RCA pinwheel louver blade is shown in detail in Figure 18 and consists of a fiberglass hub, foam sandwich blades, a fiberglass support framework, and a single aluminized-Kapton-film outer shield. The outer shield shields the hub and blades from most of the external environment. This is necessary to prevent wide variations in hub and blade temperatures, which in turn would affect the bimetal temperature and thus its response.

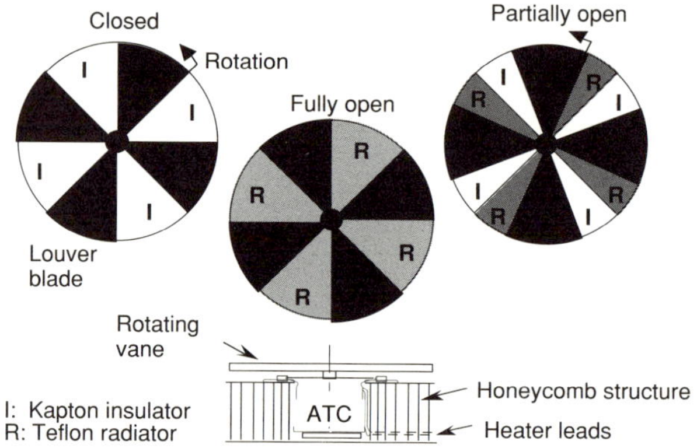

Figure 17. Pinwheel thermal louver

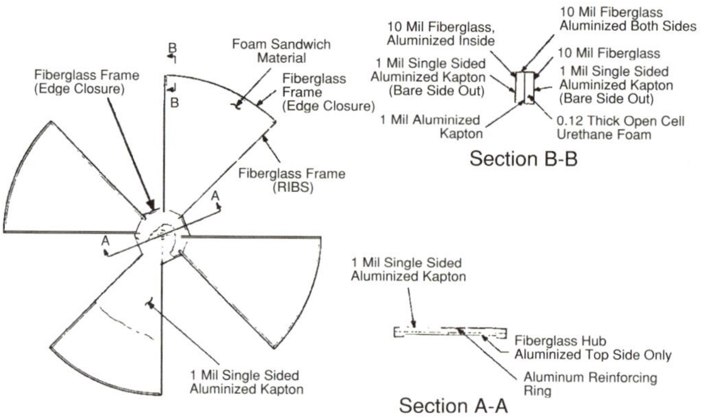

Figure 18. Pinwheel louver blade

The pinwheel actuator assembly is shown in detail in Figure 19. The actuator assembly consists of a bimetal element, bimetal heater strip, drive shaft assembly, bimetal housing, outer housing, clamp ring, stop element, and two Delrin AF bushings. The bimetal heater strip is bonded directly to the bimetal element, which in turn is bonded into the bimetal housing.

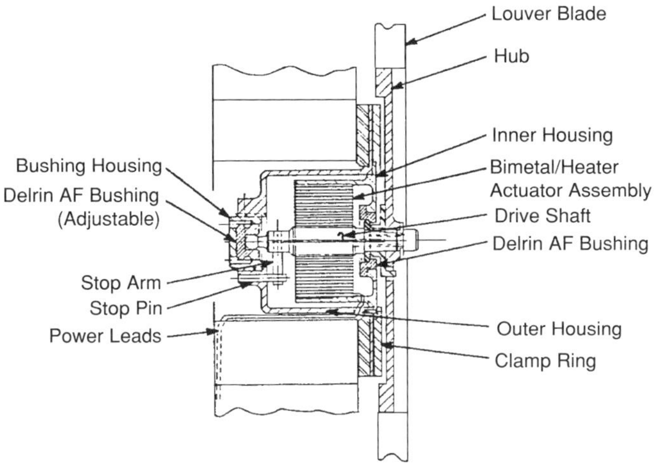

Figure 19. Pinwheel louver actuator

The drive shaft assembly is attached to the inner coil of the bimetal element and carries the stop arm and two bearing surfaces that ride in the Delrin AF bushings, one of which is mounted in the bimetal housing. The bimetal housing mounts inside the outer housing, which contains the stop element and an adjustable Delrin AF bushing. The whole assembly mounts in a hole in the spacecraft honeycomb panel external wall and is held in place with a clamp ring. The actuator passive set point is adjusted by loosening the clamp ring, rotating the bimetal housing, then retightening the clamp ring.

The stop elements limit the blade rotation at the fully closed and fully open positions (45 deg of angular rotation). The actuator operation is the same as for a vane louver actuator. A temperature change of 15-deg C is required to drive the louver from the fully closed to fully open position.

The Delrin AF bushings are adjusted at assembly to limit the drive shaft axial movement to 10 mils. The bushings also provide low-torque louver-blade support during 1-g testing; this minimizes the 1-g testing error.

The radiator/guard ring assembly is shown in detail in Figure 20. The radiator consists of a guard ring for louver blade protection and alternating radiator segments of second-surface aluminized Teflon and aluminized Kapton. The Teflon areas are the radiating areas and have a low-solar-absorptance value ($\alpha < 0.2$). The aluminized Kapton areas are the low emittance areas, which act as insulation when the louvers are closed. The louver blade covers the Teflon areas when in the closed position, and covers the Kapton areas when in the open position.

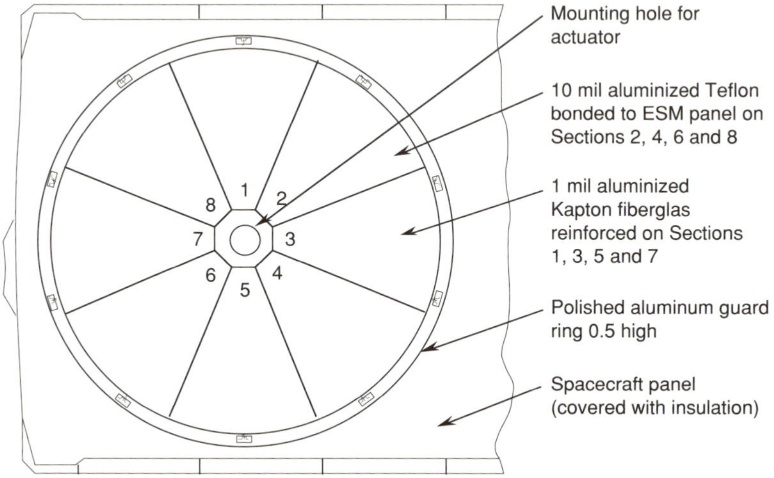

Figure 20. Pinwheel louver radiator/guard ring assembly

Each RCA pinwheel louver has a heat-rejection capacity of approximately 25 to 30 watts when open and a heat leakage of approximately 5 to 7 watts when closed. The heat-rejection rate is linearly proportional to the louver blade position.

REFERENCES

1. H. Frankel and R. Eby, "SMM Hardware Evaluation," Proceedings of the SMRM Degradation Study Workshop, NASA Document 408-SMRM-79-0001, May 9-10, 1985, pp. 105-123.

2. R. O. Wales (ed.), "ATS-6 Final Engineering Performance Report," Vol. I, NASA Reference Publication 1080, Nov. 1981.

3. M. Furukawa, "Analytical Studies on Design Optimization of Movable Louvers for Space Use," *Journal of Spacecraft and Rockets*, Vol. 16, Nov.-Dec. 1979, pp. 412-425.

4. T. J. Michalek, E. A. Stipandic, and M. J. Coyle, "Analytical and Experimental Studies of an All Specular Thermal Control Louver System in a Solar Thermal Vacuum Environment," AIAA Paper 72-268, 1972.

5. A. J. Boscia, "A Laboratory Method for the Determination of Effective Emittance of Spacecraft Thermal Control Louvers," Joint National Meeting of American Astronautical Society (15th Annual) and Operations Research Society (35th National), Paper No. VIC. 3, June 17-20, 1969, Denver, Co.

6. Design Catalog, Chase Thermostatic Metals, GTE Metal Laminates Division, 1704 Barnes Street, Reidsville, NC 27320

7. W. H. Kelly, J. H. Reisenweber, and H. W. Flieger, "High Performance Thermal Louver Development," AIAA Paper 76-460, 1976.

8. J. A. Plamondon, "Analysis of Movable Louvers for Temperature Control," *Journal of Spacecraft and Rockets*, Vol. I, Sept.-Oct. 1964, pp. 492-497.

9. COMSAT Louver Thermal Analysis for INTELSAT, Fairchild Space Company Report no. 310-SR-2001 (with appendix), May 1975.

10. H. Hwangbo, J. H. Hunter, and W. H. Kelly, "Analytical Modeling of Spacecraft with Heat Pipes and Louvers," AIAA Paper 73-773, 1973

11. R. D. Karam, "Temperature Distribution in Louvered Panels," *Journal of Spacecraft and Rockets*, Vol. 16, March-April 1979, pp. 92-97

Section 6

Radiators

David G. Gilmore
The Aerospace Corporation

INTRODUCTION

Satellite waste heat is ultimately rejected to space by radiator surfaces. Radiators occur in several different forms, such as satellite structural panels, flat-plate radiators mounted to the side of the satellite, or panels that are deployed after the spacecraft is on orbit. Whatever the configuration, all radiators reject heat by IR radiation from their surfaces. The radiating power is dependent on the emittance of the surface and its temperature. The radiator must reject both the satellite waste heat plus any radiant-heat loads from the environment or other spacecraft surfaces that are absorbed by the radiator, as shown in Figure 1. Most radiators are therefore given surface finishes with high IR emittance (> 0.8) to maximize heat rejection and low solar absorptance ($\alpha < 0.2$) to limit heat loads from the sun. Typical finishes, which are discussed in more detail in the Thermal Surface Finishes section of this chapter, include quarts mirrors, silvered or aluminized Teflon, and white paint.

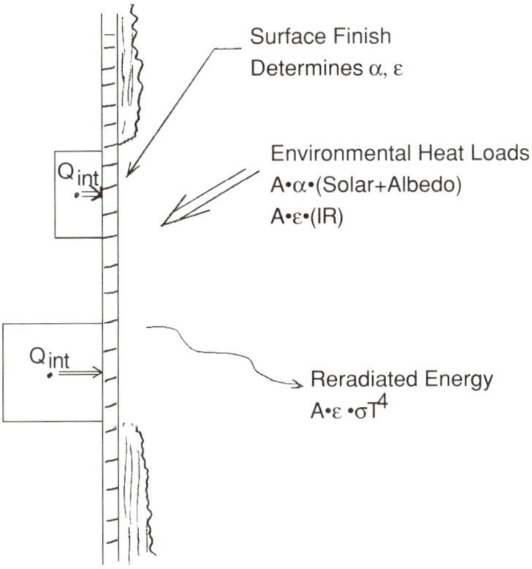

Figure 1. Environmental loads $+\Sigma Q_{int}$ = reradiated energy (steady state, no external blockage)

The radiating power of a radiator is a strong function of temperature. The total heat leaving a radiator surface is given by the simple expression

$$Q = A \varepsilon \sigma T^4 \tag{1}$$

where A = surface area

ε = emittance

σ = Stefan-Boltzman constant = 5.669×10^{-8} w/m^2 °K^4

= $.1714 \times 10^{-8}$ Btu/hr ft^2 °R^4.

The T^4 term results in a large increase in radiating capability with temperature, as shown in Figure 2. The radiating power at 50-deg C is about twice that at 0-deg C. At cryogenic temperatures the effect is even more pronounced, with a 70-deg K radiator having only 1/300th the heat rejection capability of a room-temperature radiator. This makes cryogenic radiators extremely sensitive to environmental heating and heat leaks through insulation and supports, and leads to special design considerations, which are discussed in Chapter VIII.

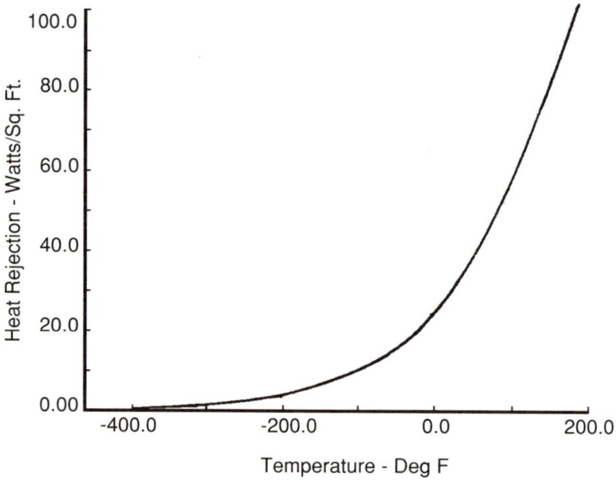

Figure 2. Blackbody radiator heat rejection

Most spacecraft radiators reject between 10 and 30 watts of spacecraft internally generated electronics waste heat per square foot. The upper end of this range is typical of a radiator that runs at a fairly high temperature (say 40-deg C) and experiences relatively modest heat backload from the environment or other spacecraft surfaces. The lower end of the range might represent a radiator running below room temperature in low Earth orbit, where environmental backloads can be substantial. The actual sizing is determined by a thermal analysis that considers the desired operating temperature, worst-case satellite waste heat, environmental heating, and radiative and conductive interactions with other spacecraft surfaces. Weights for radiators typically vary from almost nothing, if an existing structural panel is used as a radiator, to around 2 1/2 pounds per square foot for a deployable radiator and its support structure.

PASSIVE STRUCTURE RADIATORS

The most common and simplest radiator is illustrated in Figure 1. An existing aluminum honeycomb panel wall of the satellite is used. The panel serves both as part of the structure and as a radiator, with its weight normally charged to the structures subsystem. The panel face-sheets spread heat out from the electronics boxes with an area on the outside face acting as the radiating surface. Heat is conducted fairly well from the inner to outer face sheets through the aluminum honeycomb core. Lateral heat conduction, however, occurs mainly in the face sheets. Appendix C includes equations that can be used to calculate the conductance of honeycomb cores in different directions.

Sometimes the face sheets are made thicker than required for structural reasons to help spread the heat out from the boxes and give a greater "fin efficiency." Separate plates of aluminum or other material may also be placed under high-power boxes to help spread the heat out on the panel. These plates are called "doublers" and are discussed in the Mountings and Interfaces section of this chapter. Weights due to increased face-sheet thickness or the use of doublers are generally charged to the thermal control subsystem.

STRUCTURAL PANELS WITH HEAT PIPES

If a honeycomb panel radiator has some electronics boxes mounted to it that have very high heat dissipation, then the lateral conduction in the face sheets may not be sufficient to spread the heat out over an area large enough to radiate it to space. This would result in very large temperature gradients in the panel and cause the high-power boxes to exceed their upper temperature limits. Doublers or increased face-sheet thickness may be used to overcome this problem; however, there comes a point at which these techniques would result in an unacceptably large weight increase. To avoid this weight penalty, heat pipes are often used to spread the heat. The results of one trade study comparing heat pipes to doublers on a communications satellite are shown in Figure 3.

For an application that has a fairly constant heat loads, such as a panel of TWT amplifiers on a communications satellite, constant-conductance heat pipes may be used, as shown in Figure 4. Variable-conductance heat pipes (VCHP) may be used in a situation where there is a wide variation in equipment or environmental heat loads, or a requirement to minimize cold-case heater power, or to tightly control the temperature range of a component. A variable-conductance heat-pipe radiator panel is shown in Figure 5. Refs. 1, 2, and 3 discuss applications of fixed- and variable-conductance heat-pipe radiator panels on satellites, and a more detailed discussion of heat pipe technology may be found in Chapter VII.

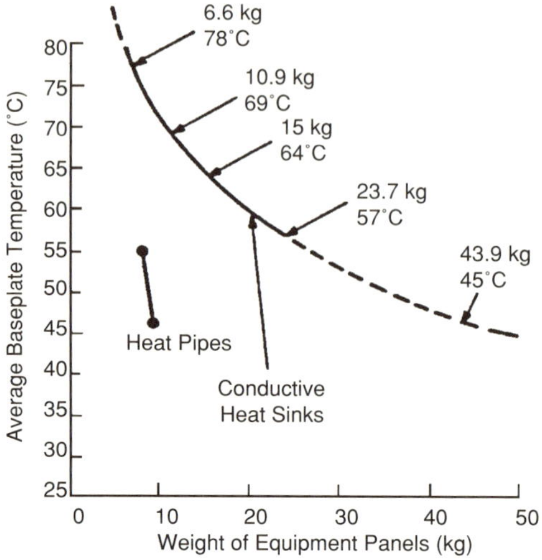

Figure 3. Weight of conductive doublers versus heat pipes

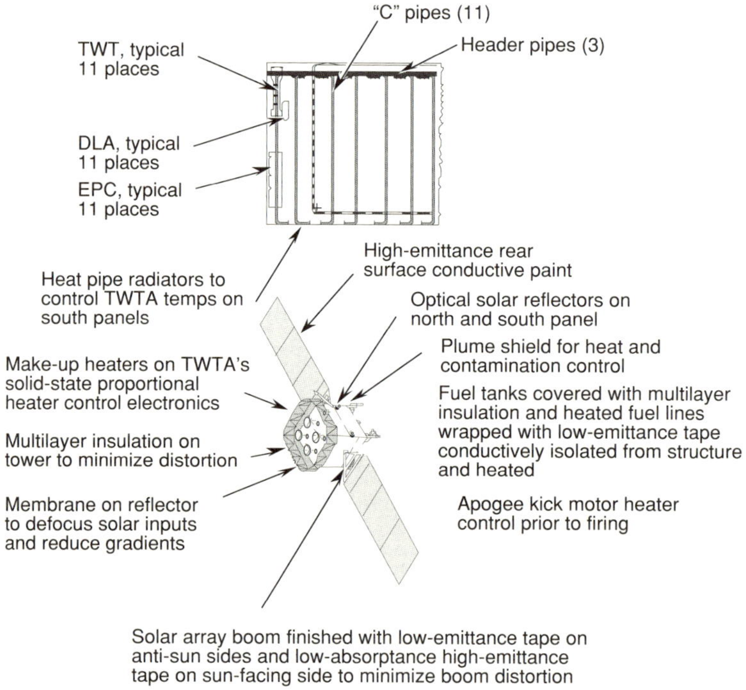

Figure 4. Heat pipe radiator panel (Martin Marietta)

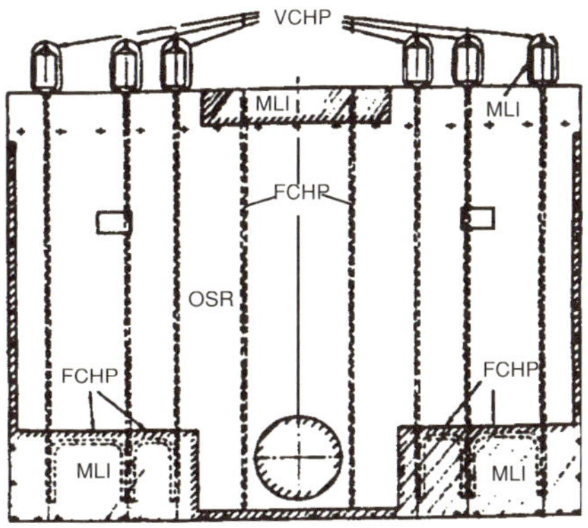

Figure 5. Variable-conductance heat pipe radiator (Mitsubishi Electric)

BODY MOUNTED RADIATORS

Some applications require a radiator that is not part of the vehicle structure. This may occur if the radiator must run at a temperature different from that of the rest of the spacecraft or if there simply are no vehicle structural panels that can be conveniently used as a radiator. In such a situation a "body-mounted" radiator, such as that shown in Figure 6, may be used. The radiator itself may be a honeycomb panel or a stiffened aluminum plate. Heat is transported from the heat dissipating components to the radiator using fixed- or variable-conductance heat pipes, and additional heatpipes may be used to spread the heat out in the radiator panel itself. Low-conductance mountings and multilayer insulation may be used to thermally isolate the radiator panel from the spacecraft.

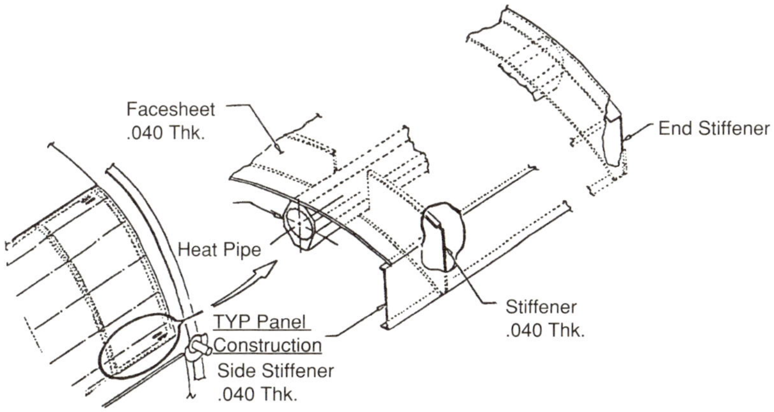

Figure 6. Body-mounted radiator (Loral)

DEPLOYABLE RADIATORS

At the time of this writing, most satellites are capable of rejecting their internal waste heat using the structural panel-or body-mounted radiators discussed above. As satellite power levels (and therefore waste-heat levels) increase in the future, however, or as satellite size is reduced through the use of high-density electronics packaging, there will come a point where there is simply not enough

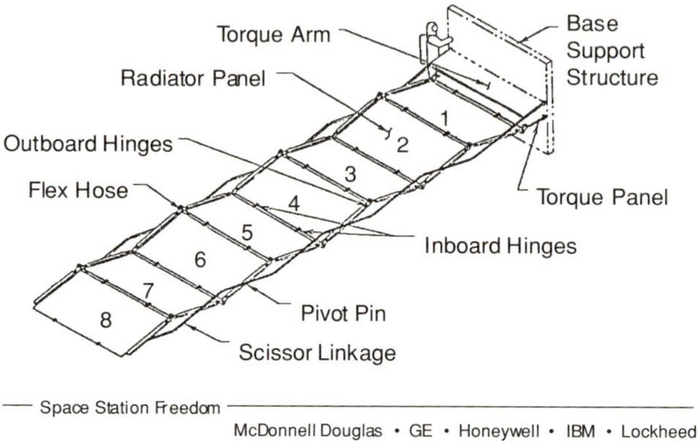

Heat Rejection System (HRS) Fact Sheet

Purpose
- Cools Space Station Crew, Subsystems, and Experiments Heat Loads

Programmatics
- Customer: McDonnell Douglas Space Systems Company/NASA-JSC
- Contract Duration: 5/91 - 1/98

Deliverables
- 3 Dev. Panels
- 1 Qual. Burst Panel
- 1 Full Scale Engr. Proto.
- 1 Qual Unit
- 6 Flight Units
- 9 Shipping Containers

Characteristics
- Each Unit Consists of 8 Panels, 9' x 11'
- Deployed by Scissors Mechanism
- 75' Deployed Length
- 2200 Lbs Per Unit
- Two Cooling Temperatures:
  - 29-deg F Units - 11 kW Cooling Each
  - 52-deg F Units - 16 kW Cooling Each
- Condensing Ammonia Two-Phase Cooling Fluid
- Bonded Honeycomb Panel Construction
- White Ceramic Thermal Paint

Figure 7. Space station deployable radiator (Loral)

area on the satellite bus to reject the internally generated waste heat. In such a situation, deployable radiators will be required to increase the available radiating area.

Loral/Vought Systems has developed several deployable radiator systems. Very large deployable radiators are currently being developed for the NASA Space Station, as shown in Figure 7. This system uses a pumped, two-phase ammonia fluid loop to transport up to 16 kw out to each radiator assembly. A smaller, one-phase ammonia version of this radiator is used for cooling the Space Station electrical-power subsystem. The Space Shuttle uses deployable radiators with a pumped freon loop, as shown in Figure 8. A deployable radiator system for use by payloads sitting in the Shuttle bay or by free flying satellites is shown in Figure 9. Further information on pumped-fluid-cooling systems can be found in the Pumped Fluid Loops section of this chapter.

All of the pumped-fluid-loop radiators discussed above were developed for manned systems that either have short mission durations (Space Shuttle) or are massively redundant and can be serviced by astronauts (the Space Station has six main radiator assemblies). Satellites, however, are designed for long-duration missions with no servicing. Pumped fluid loops are generally a concern for such systems because of the potential for failures of mechanical pumps.

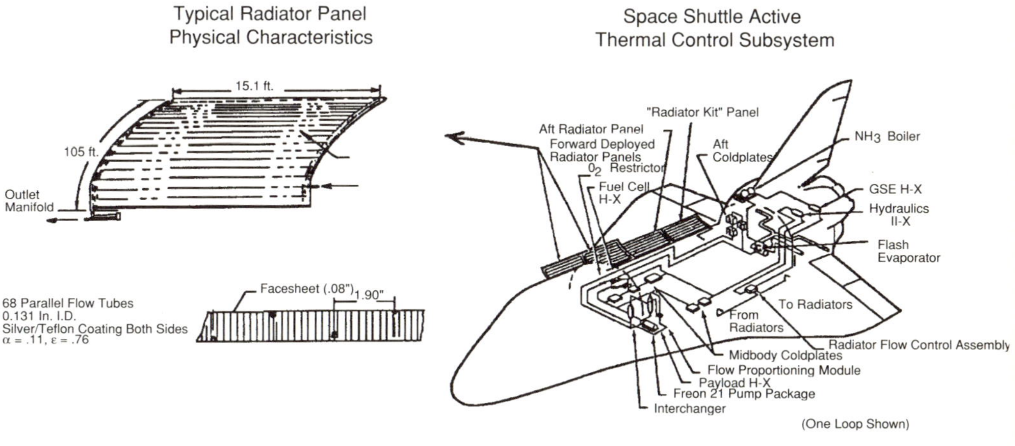

Figure 8. Space shuttle active cooling system (Loral/Rockwell)

The self-contained heat rejection module (SHRM) has been conceived for use on future spacecraft that may be carried into orbit by the shuttle. Either as sortie missions or deployable. The SHRM contains deployable radiators and flow control equipment. Heat transfer to the payload is through a contact heat exchanger. This concept also includes incorporation of a refrigeration system that will increase heat rejection capacity. Two wings of the SHRM can be deployed from the Orbiter using an ATM deployment mechanism to provide a radiating area of nearly 1000 FT^2 ($93M^2$) and heat rejection capability over 30 KW (100,000 BTU/HR)."

Extensive thermal vacuum testing has been completed demonstrating the SHRM concept with deployable radiator panels use of fluid swivels to transfer fluid between panels and dual mode operation (i.e., as a pumped fluid radiation system or a direct condensing panel refrigerator system).

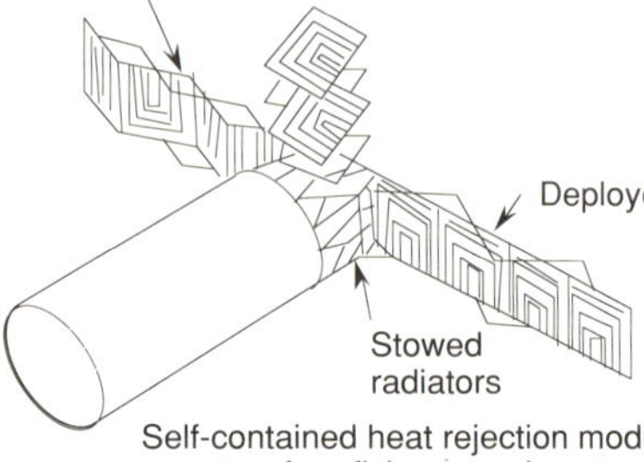

Self-contained heat rejection module applied to a free-flying experiment module

Figure 9. Self-contained heat rejection module-development program (Loral)

Figure 10. Deployable CPL radiator (Martin Marietta)

Because of these reliability concerns, several organizations, including Boeing, Martin Marietta, and TRW, have developed or studied deployable radiators using heat pipes or capillary pumped loops (CPL) instead of pumped fluid loops. Heat pipes and CPLs are essentially passive devices with no moving parts and, potentially, are more reliable than pumps. An example of a small deployable radiator built by Martin Marietta using a CPL is shown in Figure 10. Further information on heat pipes and CPLs can be found in Chapter VII. At the time of this writing, no deployable radiator using heat pipes or CPLs has been flown in an operational satellite.

RADIATOR EFFECTIVENESS

In the design of a radiator employing parallel heat pipes or coolant tubes, the spacing of the pipes/tubes and the fin thicknesses must be determined. The smallest radiator area will be achieved if one were to use very thick fins and close pipe spacing for maximum fin efficiency. Despite its small size, however, such a radiator would be very heavy due to the large number of pipes and the thick fins. Since weight is usually the critical driver for satellites, a somewhat less-efficient, but lighter, radiator may be preferred.

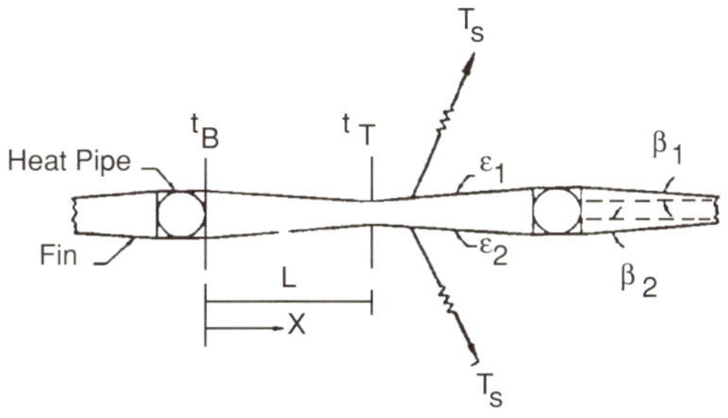

Figure 11. Radiator analysis terminology

For any radiator, an optimum combination of heat pipe spacing and fin thickness exists, which gives the minimum total radiator weight. The generalized heat-balance equation for a fin radiating to an effective sink temperature T_S is:

$$\frac{d^2T}{dx^2} - \frac{(t_B - t_T)}{Lt_T + (t_B - t_T)(L - x)} \frac{d_T}{d_x} - \frac{\sigma\left(\frac{\varepsilon_1}{\cos\beta_1} + \frac{\varepsilon_2}{\cos\beta_2}\right)L}{2K\left[Lt_T + (t_B - t_T)(L - x)\right]}\left(T^4 - T_s^4\right) = 0, \quad (1)$$

and the boundary conditions to be satisfied are:

$$T|_{x=0} = T_B$$

$$\left.\frac{dT}{d_x}\right|_{x=L} = 0$$

(see Figure 11 for definitions of the parameters). This equation was solved numerically by Chang (Reference 4) to derive the following expression for fin effectiveness for a fin of uniform thickness:

$$\begin{aligned} \eta_e &= \left(1 - 1.125\zeta + 1.60\zeta^2\right)\left(1-\theta^{*4}\right) \quad 0.01 \le \zeta \le 0.2 \\ &= \left(-.405 \log \zeta + .532\right)\left(1-\theta^{*4}\right) \quad 0.2 \le \zeta \le 2.0 \end{aligned} \tag{2}$$

$$\text{where } \zeta = \frac{\sigma L^2 T^3 B(\varepsilon_1 + \varepsilon_2)}{kt}$$

$$\theta^* = \frac{T_s}{T_B}$$

$\varepsilon_1, \varepsilon_2$	=	Radiator emittance, side 1 and Side 2
L	=	Heat pipe spacing ÷ 2
T_B	=	Temperature at fin base
T_S	=	Radiative sink temperature
σ	=	Stefan Boltzmann constant
k	=	Fin conductivity
t	=	Fin thickness

It is important to note that this expression for fin effectiveness is not the same as the usually definition of fin efficiency. Here it is the ratio of the net heat rejected by the fin to the heat that would be rejected by an isothermal fin to a 0°R sink. This definition of effectiveness therefore accounts for the thermal backload to the fin from the sink as well as the efficiency of the fin itself. The heat rejected from the radiator is therefore calculated as $Q = A\varepsilon\eta_e\sigma T_B^4$ instead of the usual $Q = A\varepsilon\eta_e\sigma\left(T_B^4 - T_s^4\right)$.

Equation 2 can be used to calculate the effectiveness of the radiator for various combinations of heat pipe spacing (L) and fix thickness (t). Once the effectiveness is known, the total area required to radiate the satellite waste heat, and the resultant weights of heat pipe, fin material, and radiator optical coating, can be easily calculated. Figure 12 shows the results of such an analysis for a two-sided flat-aluminum-heat-pipe radiator rejecting 1000 watts at 70°F to an effective sink temperature of -125°F. (The effective sink temperature accounts for the backloads on the radiator caused by environmental heating and radiative interchange with other spacecraft surfaces.) The heat pipes were assumed to weigh .075 pounds per linear foot.

It can be seen that the minimum radiator weight occurs for a fin thickness of approximately 7 mils, a heat-pipe spacing of approximately 8 inches, and an overall fin effectiveness of only 0.5. If thicker fins and closer pipe spacing is used, the fin effectiveness goes up and the size goes down, but the total weight is much greater. If 20-mil fins and 4-inch pipe spacing is used, the fin effectiveness goes up to .78 and the size is reduced by 36 percent, but the weight increased by 50 percent. The actual total weight of the radiator must, of course, include any support structure and, for a deployable radiator, deployment-mechanism weight. The minimum fin thickness may also, in some designs, be driven by structural considerations. The above calculation does, however, illustrate the fact that maximum fin efficiency does not give a minimum radiator weight.

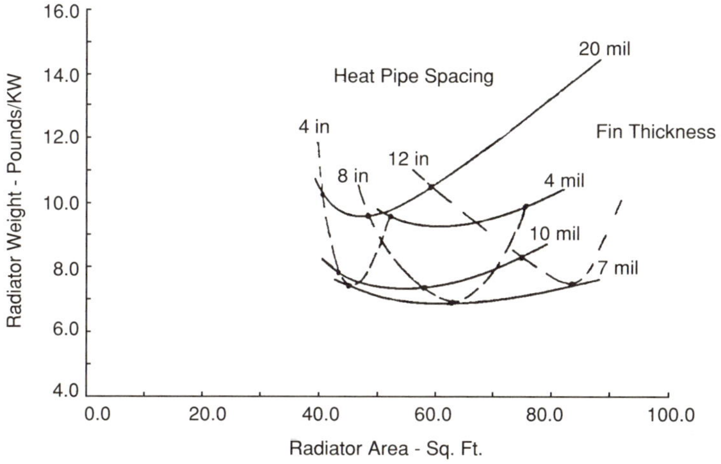

Figure 12. Radiator analysis results

EXPERIMENTAL RADIATORS

Before leaving this subject, it should be noted that a number of more exotic radiator technologies have been studied including flexible, moving belt, and liquid droplet radiators. The flexible radiator shown in Figure 13 can be conveniently stowed in a small volume. The moving-belt radiator shown in Figure 14 transports heat by moving the radiator surface and thereby eliminates the need for heat pipes or fluid loops in the radiator itself. Liquid-droplet or liquid-sheet radiators, shown in Figure 15, eliminate the radiator fin and tubes entirely, and radiate heat directly from a low-vapor-pressure fluid that is sprayed out into space and then collected and recirculated. The use of heat pumps to boost the radiating temperature of any radiator and thereby reduce its size has also been studied. More information on each of these experimental technologies can be found in references 5 through 12.

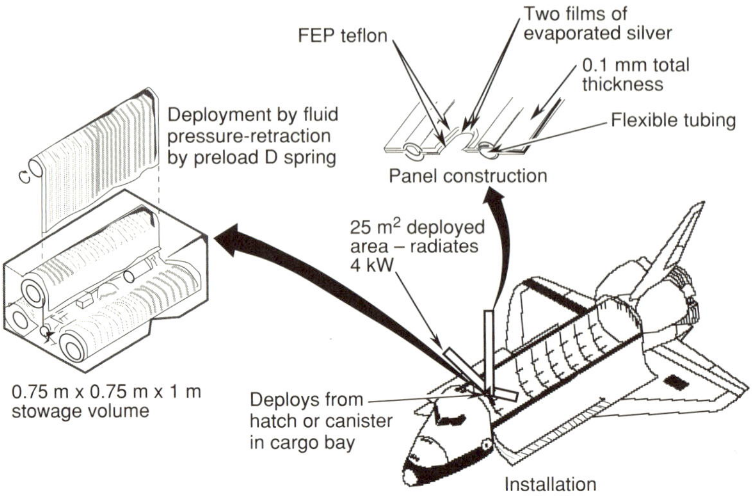

The Flexible Radiator System is a heat rejection module containing a pumped fluid loop and deployment system and interfaces with the payload through a compact heat exchanger. Flexible composite materials, that can be conveniently stored until deployed in orbit are being developed.

The soft tube, two-sided radiator design is a laminate of 2 layers of silver backed Teflon adhesively bonded to each other which provides a 0.85 radiating fin effectiveness. Deployment is by gas pressurization of manifolds in either side of the panel while retraction is by a preloaded spring. The design is capable of rejecting 160 watts/m^2 (51 BTU/HR-FT^2) of heat with a 40-deg F return temperature.

Figure 13. Flexible radiator system (Loral)

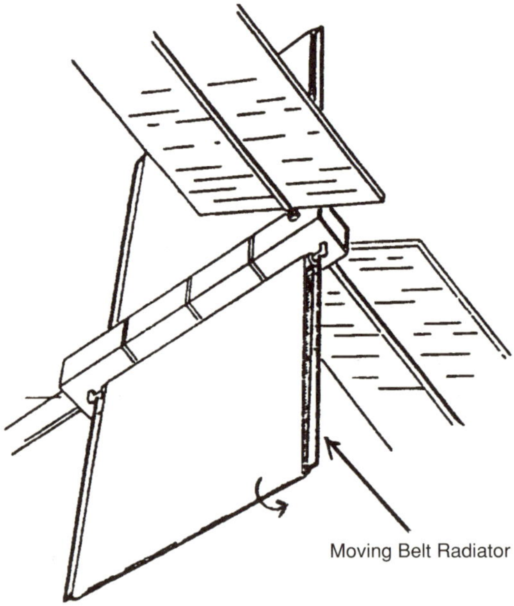

Figure 14. Moving-belt radiator

Radiative "Fins and "Heat Pipes" of Conventional Radiators
Replaced by Multiple Streams of Liquid Droplets

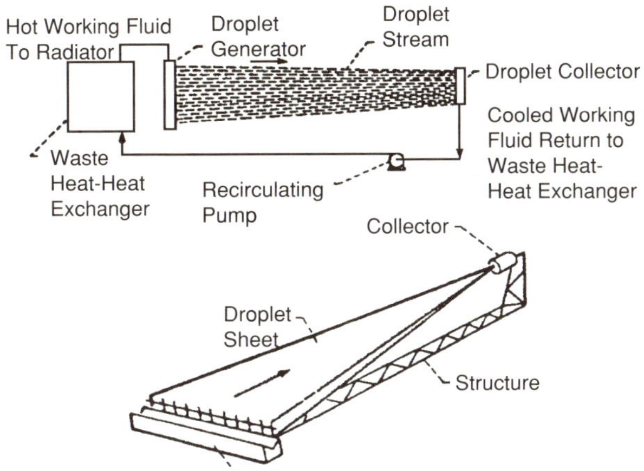

LDR Working Fluid Requirements

- Space compatibility
  - Vapor pressure - Less that 10^{-9} torr to minimize evaporation losses.
  - Optical properties - Low absorptivity to solar radiation (near IR) to minimize solar heating.
    - High emissivity in far IR for optimum radiator cooling.
  - Chemical properties - Resistant to chemical reaction by energetic (5 eV) atomic oxygen resulting in molecular changes to fluid.
    - Resistant to ultraviolet absorption which can enhance oxygen reaction.

- Generator/collector compatibility
  - Viscosity - Low to minimize fluid pressure losses in generator, collector, and pumps.
  - Surface tension - High for rapid droplet formation.
    - Low to inhibit wetting of orifice and collector surfaces.
  - Chemical properties - Negligible corrosion of collector and generator.

Figure 15. Liquid-droplet radiator

REFERENCES

1. P. C. Wise, J. Raisch, W. Kelly, and S. P. Sharma, "Thermal Design Verification of a High Power Direct Broadcast Satellite," AIAA-86-1339, June 1986.

2. P. Joy and L. Goliaszewski, "Advanced Thermal and Power Systems for the Satcom-Ku Satellites," AIAA-86-0618, June 1986.

3. H. Masumoto et al., "Development of a VCHP/FCHP Radiator System for 3-Axis Stabilized Geostationary Satellite Application," AIAA-85-1012, June 1985.

4. H. V. Chang, "Optimization of a Heat Pipe Radiator Design," AIAA-84-1718, June 1984.

5. R. L. Cox and J. W. Leach, "Flexible Deployable - Retractable Space Radiators," AIAA-77-764 June 1977.

6. P. F. Dexter and W. L. Haskin, "Analysis of Heat Pump Augmented Systems for Spacecraft Thermal Control," AIAA-84-1757, June 1984.

7. B. Drolen, "Heat-Pump Augmented Radiator for High Power Spacecraft Thermal Control," AIAA-89, January 1989.

8. K. A. White, R. Ponnappan, J. E. Beam, and E. T. Mahefkey, "Conceptual Design of an 1 m Long 'Roll out Fin' Type Expandable Space Radiator," AIAA-86-1323, June 1986.

9. K. A. White, "Liquid Droplet Radiator Development Status," AIAA-87-1537, June 1987.

10. D. L. Chubb and K. A. White, "Liquid Sheet Radiator," AIAA-87-1525, June 1987.

11. A. F. Presler et al., "Liquid Droplet Radiator Program at the NASA Lewis Research Center," ASME 86-HT-15, June 1986.

12. Lt. J. Feig, "Radiator Concepts for High Power Systems in Space," AIAA-84-0055, January 1984.

Section 7

Thermoelectric Coolers

INTRODUCTION

Thermoelectric coolers have been used for modest, 10-deg C to 20-deg C, local cooling applications. Single-stage devices operating at room temperature to -100-deg C are considered state-of-the-art devices. More widespread use of multistage devices is limited by the very low efficiency when a large temperature difference is required between the hot and cold stage. The problem of providing redundancy can be solved by connecting multiple devices in parellel; however, this results in more elements or reduced performance. There are numerous devices that have been used in various commercial applications. Adaptation to a spacecraft application has not presented any unsolvable problems.

Thermoelectric coolers have the advantages of simplicity, reliability, compactness, and noiseless operation. Major problems are structural integrity of bismuth telluride, thermal stresses due to different thermal coefficients of expansion of joined materials, and low thermodynamic coefficient of performance (COP). These coolers use the Peltier effect, i.e., the passage of an electric current through a junction formed by dissimilar metals, resulting in cooling of the junction. A typical single-stage thermoelectric cooler consists of a p-type and n-type semiconductor connected by a metallic conductor, as depicted schematically in Figure 1. The heat removed from the cold junction is the difference between the Peltier cooling effect and the Joulian heat generated by the current, plus the heat conducted from the hot to the cold junction. The result is the net cooling capacity of the couple. In practice, bismuth telluride, Bi_2Te_3, is the most commonly used semiconductor for both the p-type and n-type pellets of the circuit.

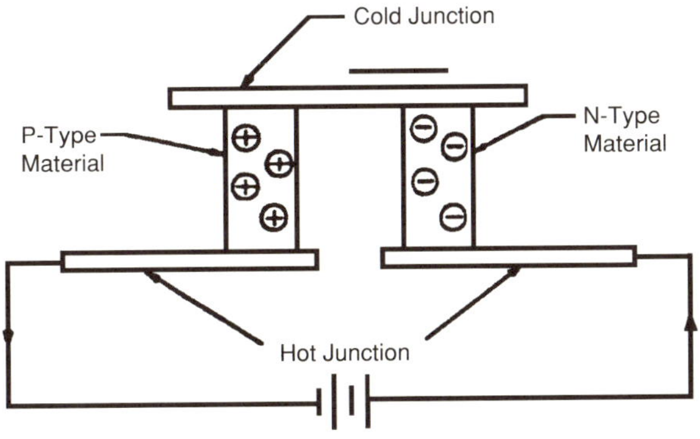

Figure 1. Peltier thermoelectric couple

Thermoelectric couples are often fabricated in parallel to increase heat-pumping capacity by multiples of the number of couples used (Figure 2). This is usually done by soldering alternate p-type and n-type pellets of bismuth telluride to copper straps so that pellets are electrically in series and thermally in parallel. Delrin spacers are sometimes used to provide structural support for the pellets. A beryllium-oxide ceramic wafer is then soldered across each of the junctions, hot and cold, forming a module. Beryllium oxide is used because of its high thermal conductivity and electrical resistivity, facilitating circuitry internal to the cooler. Modules so formed can be fabricated into a cascaded assembly, such as the four-stage Borg-Warner cooler. Multistaging can provide a greater temperature difference or a higher COP, for a given heat-pumping rate and overall temperature difference. Specific power (Figure 3) is seen to increase very rapidly as cold-junction temperature decreases for a given hot-junction temperature. A rapid but less marked increase in specific power is observed as hot-junction temperature is increased.

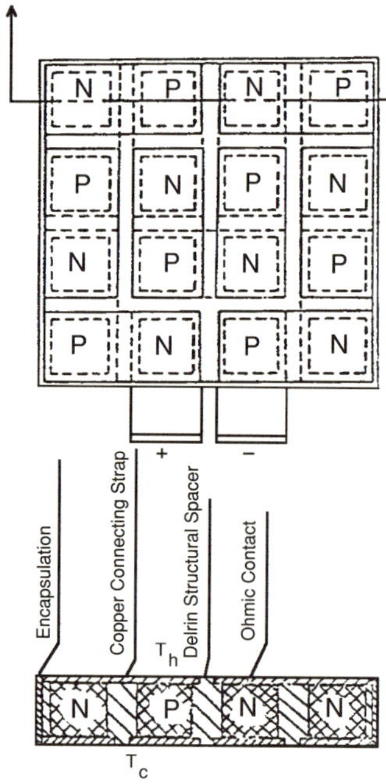

Figure 2. Construction features of eight-couple, single-stage module

The data points on Figure 3 represent results for a variety of real coolers. The Cambridge, SBRC, Marlow, and the majority of the Borg-Warner coolers are commercially available units. A wide variation is seen in the specific-power requirements of these units, but they all fall above the quasi-theoretical 300-deg C hot-junction curve. Among other things, this is due to variation in bismuth-telluride properties from unit to unit, unit design differences, and uncertainty in cooling load. Three Borg-Warner coolers, indicated by the square symbol within a plus sign, are high-performance laboratory units employing improved bismuth-telluride materials. The Comsat six-stage unit, while a ground-based design with ultimate heat rejection to air via a pins-on-plate heat exchanger, illustrates the thermal problems associated with integrating a payload, a preamplifier in this case, with a thermoelectric cooler (Figure 4). For low-power usage, parasitic heat leaks to the cooler and the preamplifier must be maintained as small as possible. Hence, two thermal shields are used to reduce both conduction heat transfer along the wave guides and radiant heat transfer.

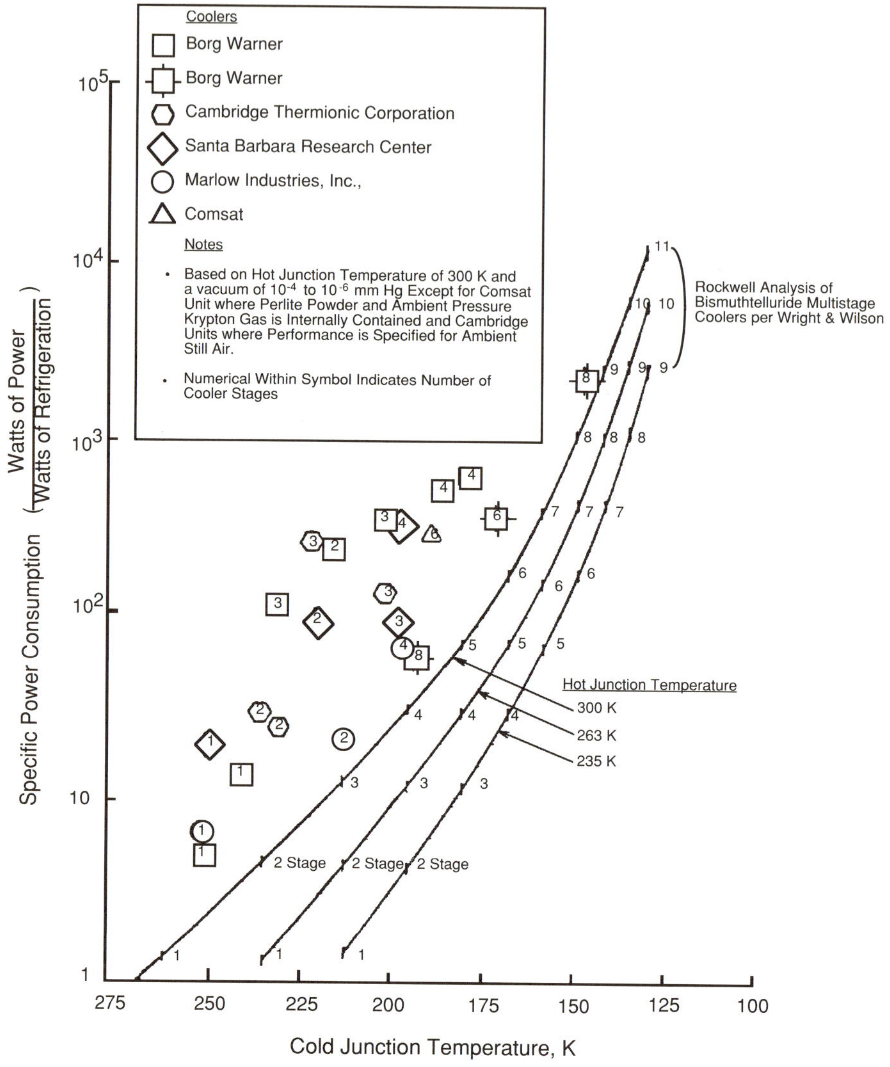

Figure 4. Performance of thermoelectric coolers

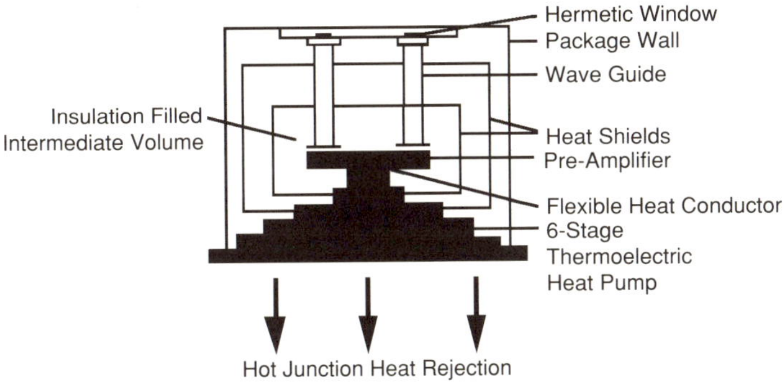

Figure 5. Schematic of preamplifier/TE cooler package

Section 8

PCMs and Heat Sinks

Leon Bledjian
The Aerospace Corporation

D. V. Hale
M. J. Hoover
M. J. O'Neill
Lockheed Missiles and Space Co.

PHASE-CHANGE MATERIAL APPLICATIONS

The use of phase-change materials (PCM) for thermal control is not a new concept; consider the ice box used to store perishable foods before efficient, inexpensive active refrigeration systems were developed. In recent years, however, research and development effort in PCM technology has resulted in novel PCM applications, for both terrestrial and space environments.

The numerous electrical components used on a spacecraft present challenging thermal control problems. High-power-dissipating components must be prevented from overheating, without the Earthbound benefits of free convection to the air or conduction to a cold plate. Conversely, components that are only activated occasionally must be prevented from cooling below an operational temperature, and the lack of atmosphere precludes normal convective methods. Such components present an ideal application for PCM thermal control.

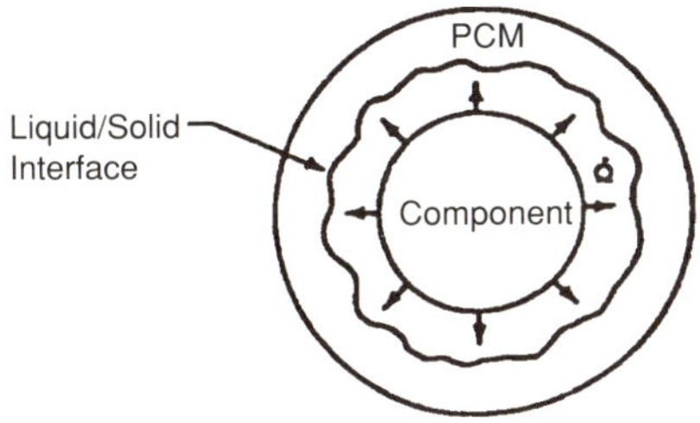

Figure 1. PCM thermal control system for one duty cycle electronic component on launch or reentry vehicle

The simplest form of PCM thermal control for electronic components is for short-duty-cycle components in launch or reentry vehicles. Although such components are used only once, they generate large quantities of heat that must be removed from the component to prevent overheating and subsequent failure. By utilizing PCM, such a component can be thermally protected as shown in Figure 1. The generated heat is absorbed via latent heat of fusion by the PCM without an appreciable temperature rise of the component. Such a system is totally passive and very reliable.

A more general application of PCM-TCS for electronic components is for cyclically operating components, i.e., components that operate in repeating on-off cycles. Such a system is shown in Figure 2. In this arrangement, when the component generates heat during the "on" portion of the cycle, the energy is stored via phase change in the PCM. During the "off" portion of the cycle, the heat of fusion energy is removed via radiator, heat pipe, thermal strap, etc., to refreeze the PCM in preparation for the next "on" portion of the cycle. The alternate melting and freezing of the PCM enables the component to operate very nearly isothermally at all times.

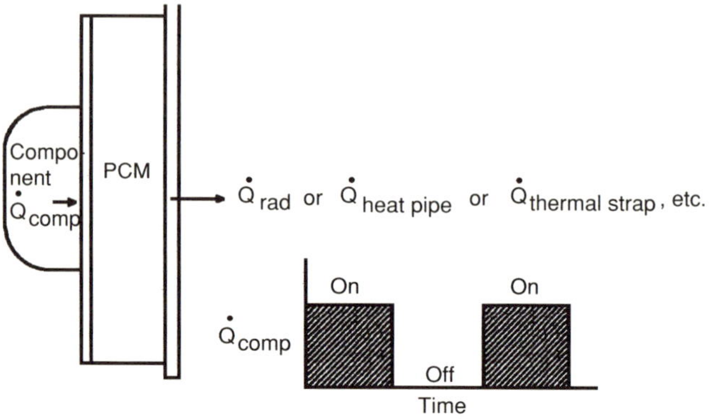

Figure 2. PCM thermal control system for cyclically operating components

Three PCM thermal control systems were used on the Lunar Roving Vehicle (LRV) during the mission of Apollo 15. The first PCM system was attached to the signal processing unit (SPU) and to batteries via thermal straps. During each LRV sortie, the heat produced by the SPU was absorbed through phase change in the PCM package. The thermal mass of the batteries offered additional heat-storage capability. After the sorties, louvers were opened on a radiator that was also connected to the PCM package via thermal straps. The PCM was thus refrozen by dumping the latent heat of fusion to space via radiation, thereby preparing the PCM package for the next sortie. The second PCM system was integrally bonded to the drive control electronics (DCE). During a sortie, the heat dissipated within the DCE was absorbed by the PCM. After the sortie, the PCM was refrozen via a thermal strap-louvered radiator system. The third PCM was integrally bonded to the lunar communications relay unit (LCRU). During a sortie, heat generated by the LCRU was stored within the PCM. After the sortie, insulation pads covering radiator surfaces on the LCRU were manually removed to allow heat radiation to space to refreeze the PCM. This concept of storing and releasing energy via phase change can be extended to a larger scale for space missions that will present varying spacecraft thermal environments. For example, a satellite orbiting the Earth encounters drastically different thermal environments cyclically as it passes in and out of the Earth's shadow. During such a mission, solar energy can be stored and released by a PCM package to dampen the otherwise large temperature changes that would be experienced during the orbit cycle. A specific example might be a crew compartment or refrigeration compartment that is required to remain nearly isothermal throughout the orbit.

The compartment could be enveloped by a layer of PCM to absorb and release solar energy during the orbit to provide isothermal conditions at the melt temperature of the PCM.

Another example of variable spacecraft thermal environment is encountered by landing vehicles on atmosphereless planets or moons. The day/night cycle on such a cosmic body presents a thermal environment that changes radically during the rotation of the body. If the landing craft is enshrouded in PCM, the solar energy can be conserved from the day for use during the night, thus ensuring an operational inner environment for crew and/or components.

Another example of PCM thermal-energy storage devices is shown in Figure 3, in schematic form. The dissipative losses of energy in electronic components, which generally are treated as wasted energy, can be heat-piped to a central PCM-TES for later use in thermal control or energy production. This recovery of usually wasted energy for conversion into reusable energy should be beneficial for long space voyages.

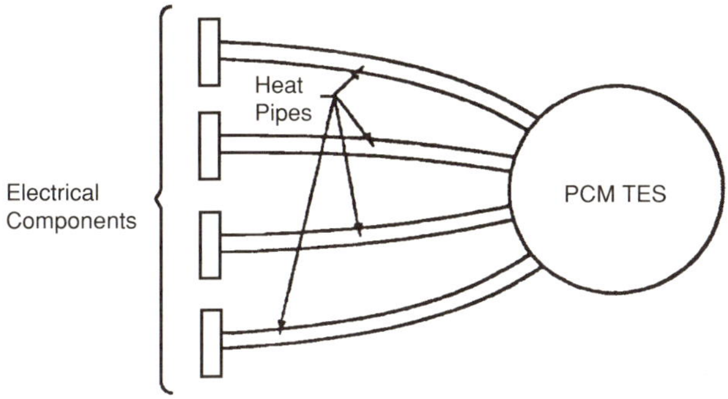

Figure 3. Central PCM thermal energy storage system

PCMs with high melt temperatures can be used in conjunction with electronic power-producing systems. Radiators used to collect solar energy can be packed with PCM to store the energy via phase change at the melt temperature. This stored energy can then be converted into electrical power by using the large temperature difference between the radiator and deep space to drive either thermionic or thermoelectric devices. If the power-producing devices are sized for the mean solar power received during the planetary orbit, production of electrical power can continue during the shadow portion of the orbit because of the capability of the PCM to store heat and maintain the radiator at a constant temperature. With conventional photoelectric radiators, the power production ceases during the shadow portion of the orbit, and energy is stored in cells or batteries for use during the "off" portion of the cycle. The thermionic or thermoelectric concept, when fully developed, may offer significant increases in efficiency over the conventional concepts. Preliminary analytical and experimental studies reported in Ref. 1 indicate the feasibility of this PCM application, and materials have been found with suitable properties for such PCM systems.

Phase-change material systems can be used in conjunction with space-flight experiments. Many delicate experiments rely upon precisely calibrated instruments. PCM packages can be integrated into these instruments (Figure 4) to maintain thermal stability and/or to ensure isothermal conditions during the experiment.

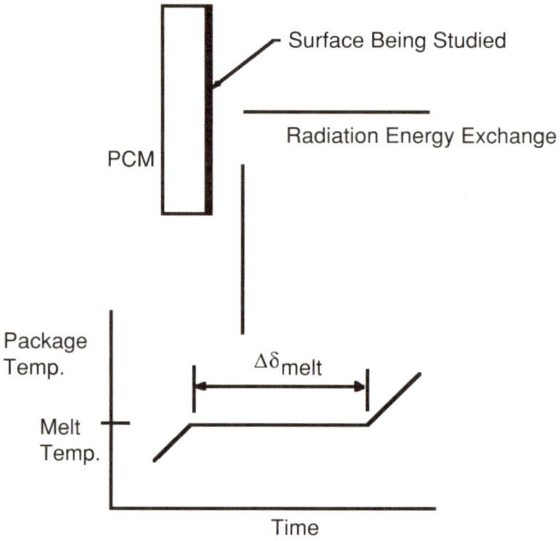

Figure 4. PCM storage for radiometric property device

PCM can be used to great advantage in fluid-loop/radiator systems because of its unique isothermal heat-storage capacity. One such application was used on Skylab. The coolant fluid returning from the external radiators experienced sizable temperature variations during the course of an orbit cycle. For efficient operation of the heat exchangers, through which the fluid passed, these temperature variations were too large. A thermal capacitor utilizing PCM damped these temperature variations by alternate melting and freezing, as illustrated in Figure 5. Thus, the thermal capacitor was used to maintain the fluid entering the heat exchanger within an allowable temperature range.

Another thermal capacitance application for PCM is in the radiator itself. If the radiator is subject to cyclical heat-removal requirements during the course of a mission, it must ordinarily be sized for the peak load for successful operation. If

PCM is integrated into the radiator, the radiator can be sized for the mean heat-removal requirement, since it can store the energy at peak load via phase change for later dumping to space by radiation. Significant savings in area and weight can be obtained with such a PCM radiator.

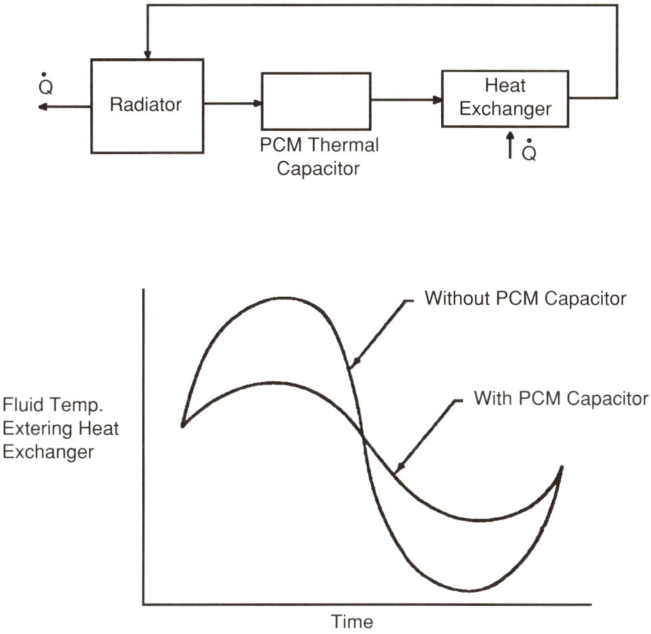

Figure 5. PCM Thermal capacitor in fluid loop

PHASE CHANGE MATERIALS

The most common phase change transformations are solid-liquid (melting/freezing), liquid-gas (vaporization), solid-gas (sublimation), and anhydrous salt transformations. Because of the very large volumetric changes involved in vaporization or sublimation, it is impractical to consider these two phase-change transformations for reversible heat storage, although it is possible to design a device, such as a heat pipe, that will act as a reversible heat-storage unit. Usually vaporization and sublimation are used in an open-loop fashion, where the vaporized or sublimed vapor is vented overboard (expendable cooler). Water is a very effective expendable coolant and is used in several space applications such as Gemini, Apollo, Space Shuttle, etc. Water melts at 32-deg F, absorbing 143 Btu/lbm. The amount of heat required to raise the temperature of water from 32-deg to 212-deg (sensible heat) is 180 Btu/lbm. The amount of heat required to vaporize water at one atmosphere of pressure is 970 Btu/lbm. The total energy required to vaporize ice is the sum of the sensible and phase-change heats involved which is 1293 Btu/lbm. Most known expendable coolants absorb considerably less heat, ammonia (NH_3) being the second-best expendable coolant that is used extensively. The design and fabrication of expendable cooling devices for aerospace and military uses is a specialized field. Some of the organizations that have this experience are Hamilton Standard, Allied Signal Corp., Fairchild, McDonnell Douglas, and others.

The design and fabrication of thermal-storage devices for solid-liquid transformation is considerably simpler because of the absence of a large volumetric change in melting or freezing. The selection of a suitable phase-change material, however, is often a project of considerable size and complexity. A good PCM should possess the following characteristics: high heat of fusion per unit mass; proper melting point temperature (or temperature range); reversible solid-liquid transition; noncorrosive; non-toxic; high flash-point; low coefficient of expansion; stable; high thermal conductivity in both phases; little or no supercooling; and low vapor pressure at room temperature. A realistic figure of merit (FOM) should take all these desirable characteristics into consideration. It is quite erroneous to assume that once a good match on melting temperature is found, high heat of fusion per unit mass is the controlling characteristic. However, an effort should be made to select a PCM with a heat of fusion of greater that 60 Btu/lb, for room-temperature applications, and greater than 20 Btu/lb, for low-temperature applications.

There are a number of classes of materials that have been investigated for use in phase-change devices. Some of the more important are:

(1) Inorganic salt hydrates, such as $Na_2SO_4.10H_20$ (glauber's salt) and $C_aCl_2.6H_20$
(2) Organic compounds, such as paraffins (C_nH_{2n+2}), alcohols, phenols, aldehydes, and organic acids
(3) Eutectics of organic materials, such as 88-mole percent acetic acid + 12-mole percent benzoic acid
(4) Natural inorganic elements, such as sulphur (S).

Table 1 gives a representative list of candidate PCMs in the temperature range of -14-deg F to 142-deg F, with most of the candidate materials having a melting-point temperature near room temperature. This temperature range is of interest to temperature control of electronic equipment and to environmental control of manned spacecraft. Materials such as polyethylene glycol (Carbowax 600), technical eicosane, and various other materials were marketed by Cryo-Therm Inc., in the mid to late 60s. Lately, Energy Science Laboratories, Inc., has been active in PCM research and development. Ref. 1 provides a discussion of thermal-energy-storage technologies having applications in renewable energy systems.

Table 1. Typical PCMs in the Range of -14-deg F to 142-deg F

Material	Melting Point (F)	Heat of Fusion (Btu/lb)
Technical aicosane ($C_{20}H_{42}$)	92-98	66-78
Polyethylene glycol (Carbowax 600)	68-77	63
Transit heat 60	60	100
Transit heat 86	86	130
Nitrogen pentoxide (N_2O_5)	86	138
Phosphonium chloride (PH_4C1)	82	324
Dibasic sodium phosphate ($Na_2HPO_612H_20$)	98.98	120.2
Sodium sulfate ($NA_20_4.10H_20$)	87.8	92.3
Glycerol [$C_3H_5(OH)_2$]	64.4	85.5
Calcium chloride ($CaC1_2.6H_20$)	84.2	73.2
p-Xylene [$C_6H_4(CH_3)_2$]	60.8	70.7
Sodium chromate ($Na_2CrO_4.H_20$	73.4	70.5
n-Undecane ($C_{11}H_{24}$)	-14.1	60.8
n-Dodecane ($C_{12}H_{26}$)	14.7	90.6
n-Tridecane ($C_{13}H_{28}$)	22.3	66.6
Tetradecane ($C_{14}H_{30}$)	42	98
n-Hexadecane ($C_{16}H_{34}$)	64.4	101.3
n-Heptadecane ($C_{17}H_{36}$)	71.16	72.5
n-Octadecane ($C_{18}H_{38}$)	82.4	104.9
n-Nonadecane ($C_{19}H_{40}$)	89.78	80.6
Octacosane ($C_{28}H_{58}$)	142	109
1-Tetradecanol [$CH_3(CH_2)_{12}.CH_2)OH$]	100.40	98.8
Acetic acid (CH_3COOH)	62.06	77.7
Water	32	143

Table 2 gives a representative list of candidate PCMs for lower-temperature applications, such as infrared detectors and other solid-state electronic devices. Modern material databases can quickly provide lists of several candidate organic or inorganic PCMs for any given melting temperature range. Refs. 3 to 5 provide useful information in preparing and conducting a PCM heat-storage development program.

Refs. 6 and 7 provide a "good start" for the analytical treatment of phase-change heat-conduction problems in one dimension. Multidimensional phase-change heat conduction can be solved by the use of numerical techniques, Refs. 9 and 10.

PCM DESIGN CONSIDERATION

In the design of PCM thermal control systems, a number of engineering considerations must be examined, including temperature range, heat of fusion, specific heat, density, vapor pressure, thermal conductivity, toxicity, packaging, the use of filler materials, etc. A large number of specifics can be found in Reference 11.

Table 2. Typical PCMs for Low Temperature

Material	Melting Point (F)	Heat of Fusion (Btu/lb)
Methyl propyl ketone (C_5H_{10})O	-108.0	44.8
Amyl alcohol ($C_5H_{12}O$)	-110.0	47.9
1-Methyl 1-2 ethylbenzene (C_9H_{12})	-113.4	38.0
Ethyl acetate ($C_4H_8O_2$)	-116.3	50.8
Methyl ethyl ketone (C_4H_8O)	-122.6	47.5
N-Butylbenzene ($C_{10}H_{14}$)	-127.3	35.1
Isopropyl alcohol (C_3H_8O)	-127.3	37.9
Butyl alcohol ($C_4H_{10}O$)	-128.6	53.9
N-Heptane (C_7H_{16})	-131.1	60.3
Tolueme (C_7H_8)	-137.2	30.9
Ethyl Benzene (C_8H_{10})	-138.8	37.1
N-Hexane (C_6H_{14})	-139.5	65
Isopropylbenzene (C_9H_{12})	-140.1	34.6
N-propylcyclopentane (C_8H_{16})	-178.6	37.8
1-Neptune (C_7H_{14})	-182.2	54
2.4 Dimethyl pentane (C_7H_{16})	-182.2	28.8
Chloropropane (C_3H_7C1)	-189.4	36.0
Butane (C_4H_{10})	-211	32.6
Ethane (C_2H_6)	-278	40.1
Methane (CH_4)	-297	25.2

HEAT SINKS

Thermal-energy storage may also be accomplished with solid materials. For sensible heat storage the figure of merit (FOM) is the specific heat of the material, c_p (Btu/lbm-deg F) when weight is critical, or the product of density and specific heat, ρc_p (Btu/in^3-deg F) when volume is critical. In order to avoid large and objectionable temperature gradients, it is also desirable for a heat sink to have a relatively high thermal conductivity, k (Btu/hr ft-deg F). One of the best known heat-sink materials for spacecraft applications is beryllium and its alloys, with c_p = 0.45 (Btu/lbm-deg F) and a ρc_p = 0.03 (Btu/in^3-deg F). Typical aluminum and magnesium alloys, used often in spacecraft designs, have a c_p = 0.2 (Btu/lbm-deg F) and ρc_p = .016 to .02 (Btu/in^3-deg F) approximately. Beryllia and alumina ceramics have been used as sensible heat-storage materials for electronic parts. Sensible heat-storage is a fairly simple and well-understood engineering concept, and not much can be stated about it. There is a large selection of handbooks and other documents from which values of density, specific heat, and thermal conductivity can be looked up. At very low temperatures (less than 30-deg K), the specific heat of most solid materials becomes very low, thus lead and its alloys have been used for heat storage at low temperatures.

REFERENCES

1. W. R. Humphries and E. I. Griggs, "A Design Handbook for Phase Change Thermal Control and Energy Storage Devices," NASA Technical Paper 1074 November 1977.

2. F. Baylin, "Low Temperature Thermal Storage: A State of the Art Survey," Solar Energy Research Institute (SERI)/RR-54-164 July 1979.

3. J. F. Keville, "Development of Phase-Change Systems and Flight Experience on an Operational Satellite," *Progress in Astronautics and Aeronautics*, Volume 56, 1977.

4. L. Bledjian, J. R. Burden, and W. H. Hanna, "Development of a Low-Temperature Phase Change Thermal Capacitor," *Progress in Astronautics and Aeronautics,* Volume 65, 1979.

5. P. G. Grodzka, E. Picklesimer, and L. E. Conner, "Cryogenic Temperature Control by Means of Energy Storage Materials," AIAA-77-763, 1977.

6. T. R. Knowles, G. W. Webb, "M/PCM Composite Thermal Storage Materials," AIAA-87-1489, 1987.

7. F. Kreith, *Principles of Heat Transfer*, Second Edition International Textbook Co., 1966.

8. L. Bledjian, "Phase-Change Heat Conduction in General Orthogonal Curvilinear Coordinates," Progress in Astronautics and Aeronautics, Volume 70, 1980.

9. N. Shamsundar and E. M. Sparrow, "Analysis of Multidimensional Conduction Phase Change Via the Enthalpy Model," *ASME Journal of Heat Transfer*, Volume 97, No. 4, August 1975, pp. 333-340.

10. G. E. Schneider, "Computation of Heat Transfer with Solid/Liquid Phase Change Including Free Convection," AIAA-85-0404, 1985.

11. D. V. Hale, M. J. Hoover, and M. J. O'Neill, "Phase Change Materials Handbook," NASA Contractor Report CR-61363, September 1971.

12. T. R. Knowles and G. W. Webb, "Metal/Phase-Change Material Composite Heat Sinks," Air Force document AFWAL-TR-88-3069, October 1988.

Section 9

Pumped Fluid Loops

Tung T. Lam
The Aerospace Corporation

INTRODUCTION

Spacecraft thermal control techniques can be divided into two categories: passive thermal control (PTC) and active thermal control (ATC). The passive thermal control systems can be achieved by control of the conductive and the radiative heat paths through the selection of the proper geometrical configurations, insulation blankets, sun shields, radiating fins, surface thermo-optical properties, thermal coatings, heat sinks, and phase-change materials. Such a system does not involve any moving parts or fluids. The spacecraft component temperatures are maintained within the desired temperature range by properly controlling the dissipated energy between all spacecraft elements through the conduction and radiation heat paths. However, when the passive thermal control technique is unable to deal with the environmental extremes, or to accommodate equipment dissipating high power, it may be more efficient to employ active thermal control techniques in the design. The design can be achieved by the use of heaters, louvers, heat pipes, thermoelectric coolers, pumped fluid loops, cryogenic cooling and venting.

The pumped fluid loops (PFLs) are devices that provide efficient transfer of a large amount of thermal energy between two points by means of forced liquid convective cooling. A simplified pumped fluid loop, as shown in Fig. 1, consists of a pumping device, a heat exchanger, and a space radiator. The cooling technique can be accomplished by the use of a coolant as the thermal-energy transport agent. The coolant absorbs the dissipated thermal energy from a component and transfers it to a heat sink. The final heat-rejection process depends on whether the coolant is expendable or non-expendable. As for the expendable coolant, the working fluid is rejected from the space vehicle once it has accomplished its mission. In the case of the non-expendable coolant, the working fluid is recirculated within the system once its thermal energy has been radiated to space via a radiator. The PFLs for space applications are the subject of this chapter.

Recent space vehicles, such as the DSP and Orbiter, have used pumped fluid loops in their ATC subsystems. The schematic Orbiter ATC subsystem is depicted in Figure 2. The ATC collects excess heat from the cabin interchanger, the fuel-cell heat exchanger, the hydraulics heat exchanger, the GSE heat exchanger, and the payload heat exchanger, and ejects heat from the Orbiter to space. The ATC payload cooling elements consist of the radiator panels, the flash evaporator, the AFT cold plates, and the ammonia boiler. The radiators provide cooling for the payload while the payload bay doors are open on orbit. As it can be envisioned from the flow diagram as shown, the thermal control design of an ATC subsystem with PFL is normally more difficult than that for a PTC subsystem. Subsequent sections explain the pumped-fluid-loop design in detail.

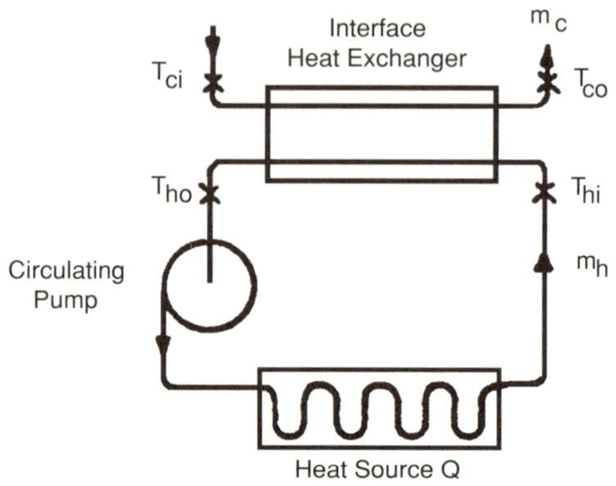

Figure 1. Schematic of a simplified pumped fluid loop system

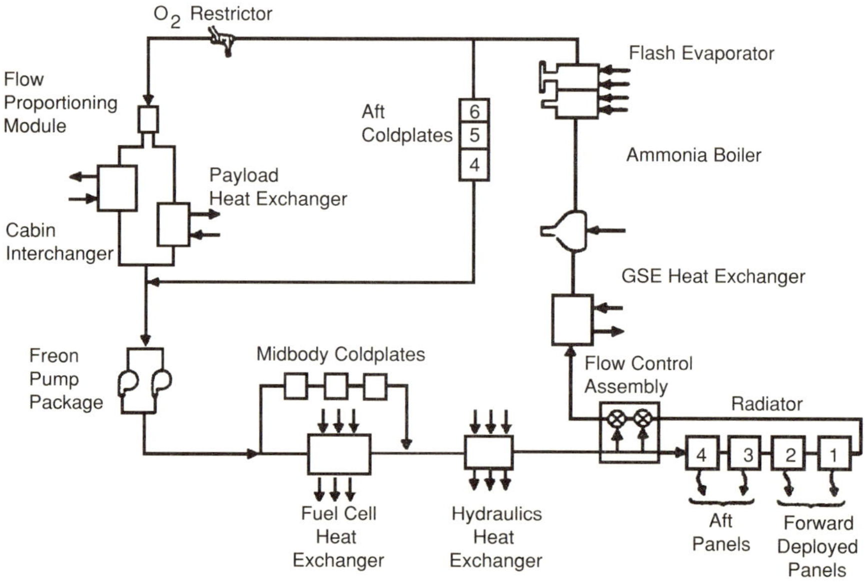

Figure 2. Orbiter active thermal control system

In what follows, some basic fluid-flow equations and friction analysis on pipe and tubes are presented first, followed by discussions of the forced-convection heat-transfer process, PFL hardware, working fluids, engineering design guidlines, and computer analysis tools. It is impossible to provide all the applicable details and analytical equations for the design of a pumped-fluid-loop;

therefore, numerous references have been listed at the end of this section. Interested readers should refer to the references for more detail.

FLUID-FLOW CONCEPTS AND BASIC EQUATIONS

Introduction

As discussed previously, the various components (pump, heat exchanger, radiator, etc.) within a PFL system are connected mainly by conduits. The flows through pipes or tubes may be laminar or turbulent. The pipe flow regime is primarily a function of the Reynolds number, $Re=\rho VD/\mu$, where ρ is the density of the fluid, V the average flow velocity, D the pipe diameter, and μ the dynamic viscosity of the fluid. A critical Reynolds number distinguishes the flow regimes between laminar or turbulent flow in pipes. The flow becomes turbulent when $Re>2300$. However, there is a range of Reynolds numbers for transition flow that has been observed experimentally. Depending on the pipe roughness and smoothness, the flow changes from laminar to turbulent in the range of the Reynolds numbers from 2000 to 4000. In the study of fluid flow in a piping system, the conservation principles are used to set up the governing equations. These equations are the conservation of mass, the conservation of momentum, and the conservation of energy. In the following sections some basic equations used in pipe-flow analysis are briefly presented.

Fundamentals of Pipe Flow

Continuity Equation

Consider steady flow through a portion of the stream pipe; the conservation of mass states that the net mass outflow from sections 1 to 2 of the control volume must be zero. Since there is no flow through the wall of the stream pipe, the continuity equation for one-dimensional flow in a pipe can be written as

$$\dot{m} = \rho_1 V_1 A_1 = \rho_2 V_2 A_2 \quad , \qquad (1)$$

where the subscripts 1 and 2 refer to the flow parameters at the inlet and the outlet, respectively, $\dot{m}$ is the mass flow rate, V the mean velocity, and A the cros-sectional area.

Bernoulli Equation

Steady-state one-dimensional internal flow is generally known as internal pipe flow, and the governing equation is called Bernoulli's equation. A flow network is simulated by specifying flow-passage connections, which include pipes, pumps, and/or valves. Associated with a flow passage are the upstream and downstream pressures, and the characteristics that govern the pressure drop.

The pressure drop across a pipe is given by Bernoulli's equation. The general form of Bernoulli's equation, including head loss due to irreversibilities for pipe flow along a streamline, is

$$\left(\frac{P_1}{\rho} + \frac{V_1^2}{2} + gZ_1\right) - \left(\frac{P_2}{\rho} + \frac{V_2^2}{2} + gZ_2\right) = h_{l_t} , \qquad (2)$$

where the subscripts 1 and 2 refer to the flow parameters at the inlet and the outlet, respectively, P is pressure, V is average fluid velocity, ρ is density, g is the gravitational constant, Z is elevation, and h_{l_t} is total head loss. This equation shows that the total head loss is the difference in potential energy, flow energy, and kinetic energy. The importance of the total head loss will be discussed in detail in the next section.

Head Loss

The main purpose in the analysis of flow in pipes and tubes is to evaluate the pressure changes that result from incompressible flow in the system. The pressure changes in a flow system result from friction and changes in elevation and flow velocity. In a frictionless flow, the Bernoulli equation could be used to account for the effects of changes in elevation and flow velocity. In the analysis of real flows, the effect of friction must also be included. This acts to decrease the pressure, causing a pressure "loss" compared to the frictionless flow case. The pressure "loss" h_{l_t} (total head loss) in Eq. (2) contains two constituents: the major head loss h_l (due to friction in fully developed flow in constant-area portions of the system) and the minor head loss h_{l_m} (due to flow-through valves, tees, elbows, and frictional effects in other nonconstant-area parts of the system). Thus, the total head loss in a piping system can be written as

$$h_{l_t} = h_l + h_{l_m} . \qquad (3)$$

The details of the major and minor losses in fluid flow are the subjects of discussion in the following section.

Major Losses

Flow through a piping system causes a reduction in static head, which may be expressed in terms of velocity head $V^2/2$. The major head loss is given by

$$h_l = f\frac{L}{D}\frac{V^2}{2} , \qquad (4)$$

where L and D are the length and diameter of the pipe, respectively. The friction factors are

$$f = \frac{64}{Re} \qquad (5)$$

for laminar flow, and for turbulent flow in smooth pipes

$$f = \frac{0.079}{Re^{0.25}} \qquad 4 \times 10^3 < Re < 2 \times 10^4 \quad \text{[Ref. 5]} \qquad (6)$$

$$f = \frac{0.184}{Re^{0.2}} \qquad 2 \times 10^4 < Re < 3 \times 10^5 \quad \text{[Ref. 7]} \qquad (7)$$

$$f = \frac{0.046}{Re^{0.2}} \qquad 2 \times 10^4 < Re < 10^6 \quad \text{[Ref. 5]} \qquad (8)$$

$$f = \frac{1}{[1.581 \ln(Re) - 3.28]^2} \qquad 10^4 < Re < 10^7 \quad \text{[Ref. 6]} \qquad (9)$$

In laminar flow, the friction factor is a function of Reynolds number only; it is independent of roughness. The widely used frictional factor f is determined from empirical correlation of the Moody diagram (Ref. 2), shown in Figure 3 as

$$f = \frac{0.25}{\left[\log\left(\frac{\frac{\varepsilon}{D}}{3.7} + \frac{5.74}{Re^{0.9}}\right)\right]^2} , \qquad (10)$$

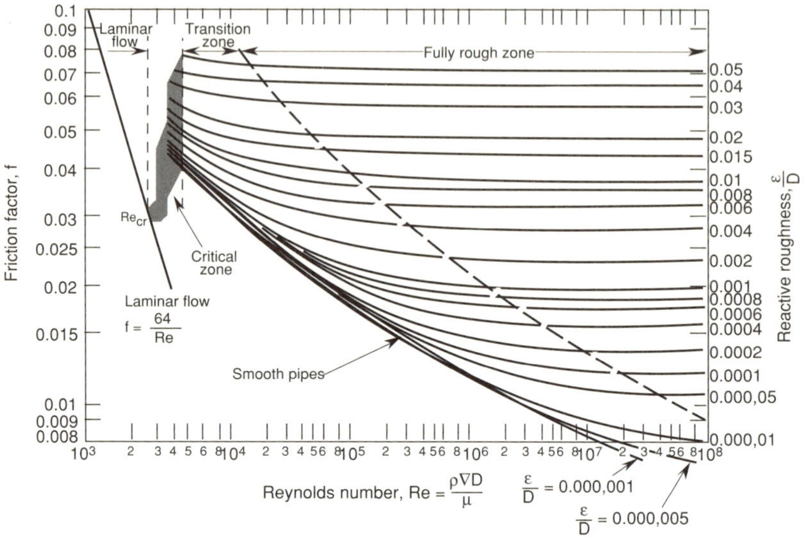

Figure 3. Friction factor for fully-developed flow in circular pipes

where Re (=ρVD/μ) is the Reynolds number, μ is the dynamic viscosity, and ε/D is the relative roughness. The choice of the cut-off Reynolds number between the two regimes (laminar and turbulent) is somewhat arbitrary.

Minor Losses

Flow in a piping system may be required to pass through a variety of fittings, bends, or abrupt changes in area, which usually results in flow separation. Energy in the fluid is dissipated by the mixing of fluid in the separated zones. This results in additional head losses, which are primarily due to flow separation. These losses are small compared to the frictional losses and are, therefore, called minor losses. The minor head loss may be expressed by either

$$h_{l_m} = K\frac{V^2}{2} \tag{11}$$

or

$$h_{l_m} = f\frac{L_e}{D}\frac{V^2}{2}\ , \tag{12}$$

Table 1. Loss Coefficients for Pipe Exits

Exit Type	Diagram	Loss Coefficient, K*
Projecting Pipe		1.0
Square-Edged		1.0
Rounded	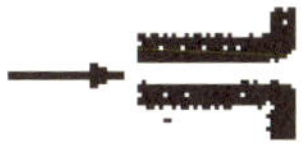	1.0

*Based on $h_{lm} = K(\overline{V}^2/2)$. Calculation of Head Loss

where L_e is the equivalent length and K, the loss coefficient, must be determined experimentally for each situation. Loss coefficients for various types of entrances and exists are shown in Tables 1 and 2. Minor loss coefficients for sudden area changes (enlargements and contractions) are given in Table 3, and Figures 4 and 5.

Table 2. Loss Coefficients for Pipe Entrances

Extrance Type	Diagram	Loss Coefficient, K*
Re-entrant		0.78
Square-edged		0.34+
Slightly rounded		0.2-0.25
Well Rounded†	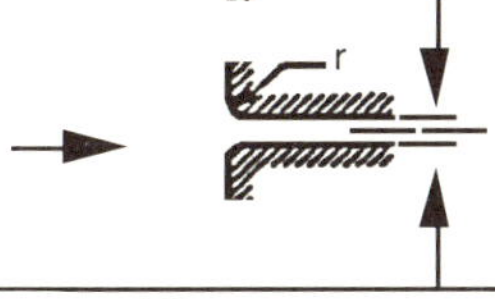	0.04

*Based on $h_{l_m} = K(\overline{V}^2/2)$, where $\overline{V}$ is the mean velocity in the pipe.
† $r/R \cong 0.35$

Table 3. Loss Coefficients for Gradual Contractions

Diagram	Included Angle, θ, Degrees	Loss Coefficient, K*
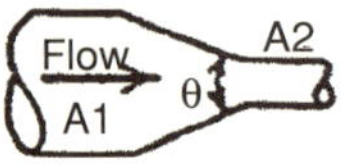	30	0.02
	45	0.04
	60	0.07

*Based on $h_{lm} = K\,(\overline{V}_2^2/2)$.

The minor loss of a pipe bend is nomally expressed by an equivalent length of straight pipe. The equivalent lengths of a 90-deg pipe bend and miter bend are shown in Figures 6 and 7. The representative equivalent lengths for valves and fittings are also given in Table 4.

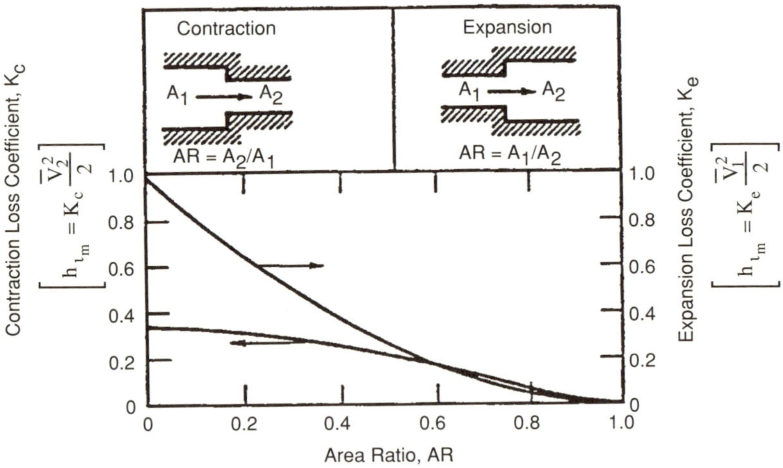

Figure 4. Loss coefficients for flow-through sudden area changes

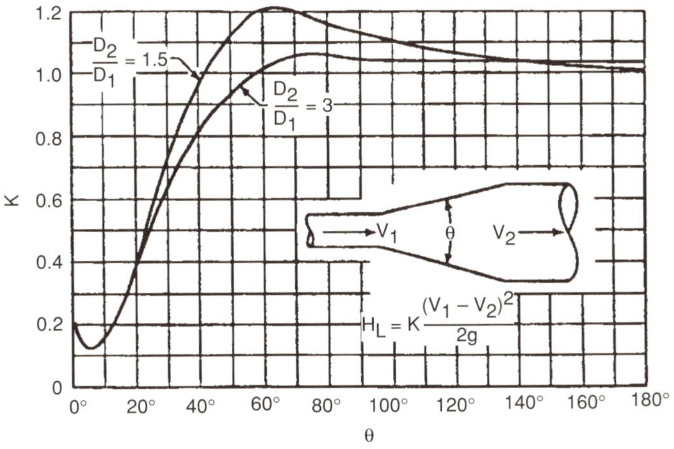

Figure 5. Loss coefficients for conical expansions

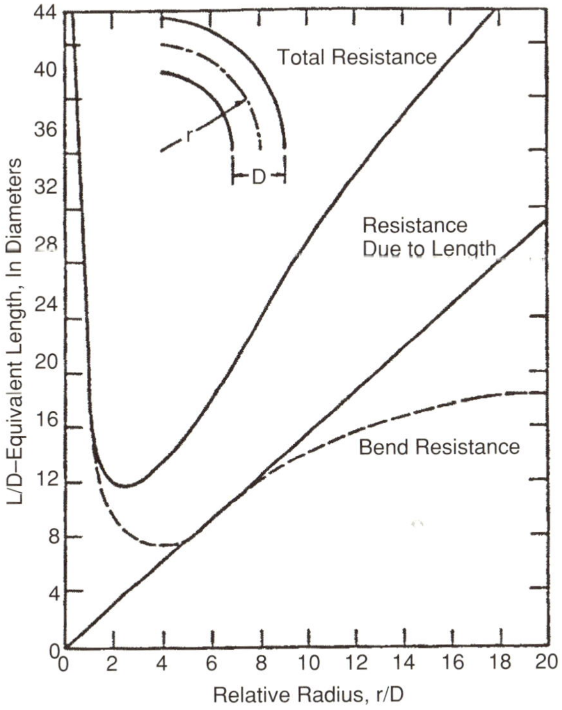

Figure 6. Design chart resistance of 90-deg bends in circular pipe with fully developed turbulent flow

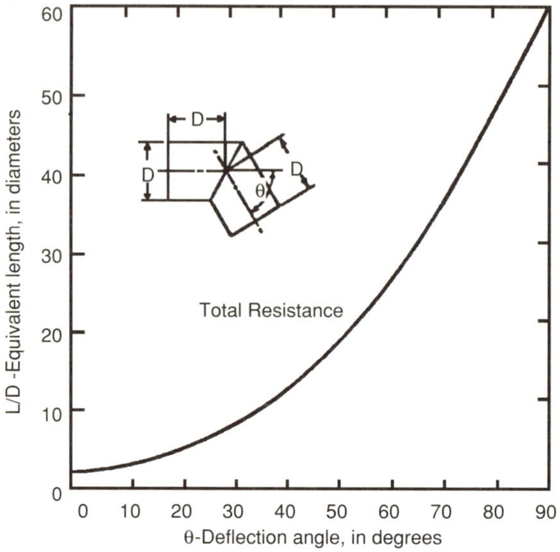

Figure 7. Design chart resistance of miter bends in circular pipe with fully developed turbulent flow

Table 4. Representative equivalent lengths in pipe diameters (L_e/D) for valves and fittings

Fitting Type	Description	Equivalent Length, L_e/D*
Globe Valve	Fully Open	350
Gate Valve	Fully Open	13
	3/4 Open	35
	1/2 Open	160
	1/4 Open	900
Check Valve		50-100
90 deg Std. Elbow		30
45 deg Std. Elbow		16
90 deg Elbow	Long Radius	20
90 deg Elbow		50
45 deg Street Elbow		26
Tee	Flow through run	20
	Flow through branch	60
Return bend	Close pattern	50

*Based on $h_{l_m} = f \dfrac{L_e}{D} \dfrac{\overline{V}^2}{2}$

Friction Factor and Pressure Drop

The Bernoulli equation can be interpreted as the difference of energy per unit weight between two points on a streamline. The change of kinetic energy is usually small compared to the total head loss. Hence, it will be neglected in the evaluation of the pressure drop. By using Eq. (2), the pressure drop across a pipe can be stated as

$$P_2 - P_1 = \frac{\rho h_{l_t}}{g_c} - \frac{\rho g}{g_c}(Z_2 - Z_1) \ . \tag{13}$$

By using Eqs. (3), (4), and (11) in Eq. (13), the pressure drop across a pipe becomes

$$P_2 - P_1 = -\frac{\rho V^2}{2g_c}\left(f\frac{L}{D} + K\right) - \frac{\rho g}{g_c}(Z_2 - Z_1) \ , \tag{14}$$

where g_c is a conversion factor. The value of g_c is 32.2 ft-lbm/lbf-sec^2 in the British Gravitational System, 1 kg-m/newton-sec^2 in the Systeme Internationale d'Unites (SI) System, and 1 gm-cm/dyne-sec^2 in the Absolute Metric System.

By incorporating the continuity equation (1), Eq. (14) can be rewritten as

$$P_2 - P_1 = -\frac{\dot{m}^2}{2\rho A^2 g_c}\left(f\frac{L}{D} + K\right) - \frac{\rho g}{g_c}(Z_2 - Z_1) \ . \tag{15}$$

The equation shows that the pressure drop across a piping system is the sum of the elevation difference of the two points under consideration and the total head loss. In case of neglecting the elevation change and the minor loss, Eq. (14) can be rewitten as

$$\Delta P = P_2 - P_1 = -\frac{\rho V^2}{2g_c} f\frac{L}{D} \ . \tag{16}$$

FORCED CONVECTION IN PIPES AND TUBES

Introduction

In convection heat transfer in internal flow in pipes or tubes it is customary to define an axially local heat-transfer coefficient h_x as

$$q_x = h_x \left(T_w - T_b\right) , \tag{17}$$

where T_W and T_b are the mean pipe-wall temperature and the fluid-bulk mean temperature, respectively. The flow-length average heat-transfer coefficient $\bar{h}$ is the integrated average of h_X for the total of the pipe from x=0 to x=L,

$$\bar{h} = \frac{1}{L}\int_o^L h_x dx . \tag{18}$$

For convenience, the heat-transfer coefficient is commonly related to the dimensionless Nusselt number in convective heat transfer. The Nusselt number, by its definition, is the ratio of the convective conductance h to the molecular thermal conductance k/D. The local Nusselt number Nu_x is then expressed by

$$Nu_x = \frac{h_x D}{k} . \tag{19}$$

The mean flow-length-average Nusselt number based on $\bar{h}$ is defined as

$$Nu_D = \frac{\bar{h} D}{k} = \frac{1}{L}\int_o^L Nu_x \, d x . \tag{20}$$

When the effects of axial heat conduction, viscous dissipation, and flow work are neglected within the fluid, the heat transfer within the system can be evaluated by

$$q = \bar{h}\left(T_w - T_b\right) = \left(\frac{Nu_D \, k}{D}\right)\left(T_w - T_b\right) . \tag{21}$$

Heat Transfer in Laminar Tube Flow

Let T(r,z) be the temperature distribution in the fluid, where r and z are the radial and axial coordinates, respectively. The heat flux from the fluid to the tube wall is governed by Fourier 's law of heat conduction

$$q(z) = -k\left[\frac{\partial T(r,z)}{\partial r}\right]_{wall} , \tag{22}$$

where k is the thermal conductivity of fluid. Combining Eq. (22) with Newton's law of cooling, Eq. (17), the heat-transfer coefficient can be written in terms of the dimensionless temperature gradient as

$$h = -k \left[\frac{\partial q(r)}{\partial r} \right]_{wall} . \quad (23)$$

The temperature profile for flow inside a circular tube can be obtained by performing an energy analysis on a fluid element (Refs. 8, 9, 10). From knowledge of the temperature profile, the heat-transfer coefficient may be shown to be of the form (Ref. 19)

$$Nu_D = \frac{h\,D}{k} = 4.364 \quad \text{for uniform heat flux at the tube wall,} \quad (24)$$

and

$$Nu_D = \frac{h\,D}{k} = 3.66 \quad \text{for constant tube wall temperature.} \quad (25)$$

The Nusselt number for laminar flow inside a circular tube was given above for two different boundary conditions, namely, the constant wall temperature and the constant wall heat flux. Shah and London (Ref. 18) have compiled a list of Nusselt numbers and the quantities f·Re (i.e., the product of the friction factor and the Reynolds number) for geometries other than the circular tube and parallel plates for the above boundary conditions. The results are listed in Table 5.

The Nusselt number and the friction factor for laminar flow in ducts of various crossections have been determined in the region where velocity and temperature profiles are fully developed. If the duct crossection for flow is not circular, then the heat transfer and friction factor, for many cases of practical interest, can be based on the hydraulic diameter D_h, defined as

$$D_h = \frac{4A_c}{P} , \quad (26)$$

where A_c and P are the crossectional flow area and the wetted perimeter, respectively. It is this diameter that should be used in the calculation of the Nusselt and Reynolds numbers.

Heat Transfer in Turbulent Tube Flow

A classical expression for computing the local Nusselt number for fully developed turbulent flow in a smooth circular tube can be obtained from the Chilton-Colburn analogy. The analogy relates the local drag coefficient C_f to the local Stanton number St (=Nu/Re·Pr) in the form

$$StPr^{2/3} = \frac{C_f}{2} = \frac{f}{8} , \quad (27)$$

Table 5. Nusselt Number and Friction Factor for Fully Developed Laminar Flow In Ducts of Various Cross Sections (Ref. 9)

Geometry ($L/D_h > 100$)	Nu_T [1]	Nu_H [2]	Nu_H [3]	f Re
	3.657	4.364	4.364	64.00
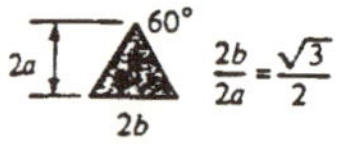	2.47	3.111	1.892	53.33
	3.34	4.002	3.862	60.22
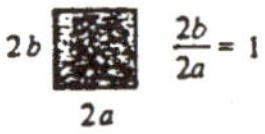	2.976	3.608	3.091	56.91
	3.391	4.123	3.017	62.20
	3.66	5.099	4.35	74.80
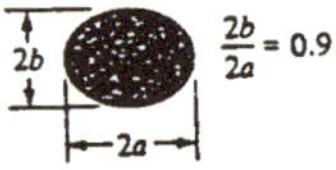	4.439	5.331	2.930	72.93
	5.597	6.490	2.904	82.34
	7.541	8.235	8.235	96.00
	4.861	5.385	-----	96.00

1 Nusselt number for uniform wall temperature.

2 Nusselt number for uniform wall heat flux in the flow direction while the wall temperature remains uniform around the periphery.

3 Nusselt number for uniform wall heat flux both in the flow direction and around the periphery.

where Pr ($=\upsilon/\alpha$) is the Pradtl number. It is the ratio of kinematic viscosity and thermal diffusivity of a fluid, which represents the relative magnitudes of diffusion and heat conduction in the fluid medium. Substituting the friction factor from Eq. (7) into Eq. (27) yields the Colburn equation for turbulent flow inside a smooth tube:

$$\mathrm{Nu} = 0.023\,\mathrm{Re}^{0.8}\,\mathrm{Pr}^{1/3} \quad . \tag{24}$$

Eq. (24) is applicable for $0.7 < \mathrm{Pr} < 160$, $\mathrm{Re} > 10000$ and $\mathrm{Le/D} > 60$ for smooth tubes. A large number of empirical correlations have been developed by many investigators in the past to determine the heat-transfer coefficient. Some of these correlations are presented in Table 6.

SYSTEM HARDWARE

Pumps

General

A pump is a machine that adds energy to liquid. It converts kinetic energy into pressure potential. A pump consumes more power than it gives off due to internal friction losses. Some major losses include hydraulic losses (flow friction and turbulence) and mechanical losses (friction in bearings and other internal mechanical parts). Depending on their design and mechanical action, most of the pumps used in space applications can be classified into two categories (Ref. 38):

(1) Rotodynamic pump. Adds energy to a liquid medium through the work done by a rapidly rotating vaned impeller. The radial-flow centrifugal pumps, mixed-flow pumps, axial-flow pumps, and propellers can be classified as rotodynamic pumps. Some typical rotodynamic pumps are shown in Figure 8.

(2) Displacement pump. This category includes the reciprocating (Figure 9) and rotary (Figure 10) pumps. It imparts energy by a positive displacement action.

Table 6 Summary of Correlations for Forced Convection Turbulent Flow Inside Duct

Correlation	Remarks
$f = (1.82 \log Re - 1.64)^{-2}$	Smooth tubes, $Re > 10^4$
$f = 0.316\ Re^{-0.25}$	Smooth tubes, $Re < 2 \times 10^4$
$f = 0.184\ Re^{-0.2}$	Smooth tubes, $2 \times 10^4 < Re < 3 \times 10^5$
$Nu = 0.023\ Re^{0.8}\ Pr^{1/3}$	$0.7 < Pr < 160$; $Re > 10{,}000$; $L/D > 60$; smooth pipes
$Nu = 0.023\ Re^{0.8}\ Pr^{n}$ $n = 0.4$ for heating $n = 0.3$ for cooling	$0.7 < Pr < 160$; $Re > 10{,}000$; $L/D > 60$; smooth pipes
$Nu = 0.027\ Re^{0.8}\ Pr^{1/3}\left(\frac{\mu_b}{\mu_w}\right)^{0.14}$	$0.7 < Pr < 16{,}700$; $Re > 10{,}000$; $L/D > 60$; smooth pipes
$Nu = \frac{Re\ Pr}{X}\left(\frac{f}{8}\right)\left(\frac{\mu_b}{\mu_w}\right)^{n}$ where $X = 1.07 + 12.7(Pr^{2/3} - 1)\left(\frac{f}{8}\right)^{1/2}$ $n = 0.11$ heating with uniform T_w $n = 0.2$ cooling with uniform T_w $n = 0$ uniform wall heat flux or gases	Smooth or rough pipes $10^4 < Re < 5 \times 10^6$ $0.5 < Pr < 200$ with 5 to 6 percent error $0.5 < Pr < 2000$ with 10 percent error Properties, except m_w,, are evaluated at bulk mean temperature
$Nu = 0.036\ Re^{0.8}\ Pr^{1/3}\left(\frac{D}{L}\right)^{0.055}$	$10 < \frac{L}{D} < 400$
$Nu = 5 + 0.016\ Re^{c}\ Pr^{d}$ where $c = 0.88 - \frac{0.24}{4 + Pr}$ $d = 0.33 + 0.5e^{-0.6Pr}$	$0.1 < Pr < 10^4$ $10^4 < Re < 10^6$ $\frac{L}{D} > 25$

a, *b*: fluid properties evaluated at the bulk mean temperature.
w: fluid properties evaluated at the wall temperature.

Class	Sketch of a Typical Model
Radial Flow	
Mixed Flow	
Axial Flow or Propeller	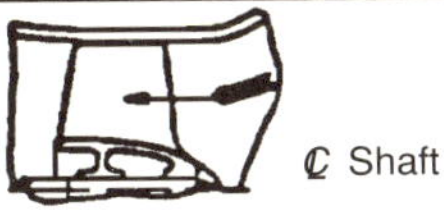

From Hydraulic Institute (1975).

Figure 8. Rotodynamic pumps

Class	Sketch of a Typical Model
Piston	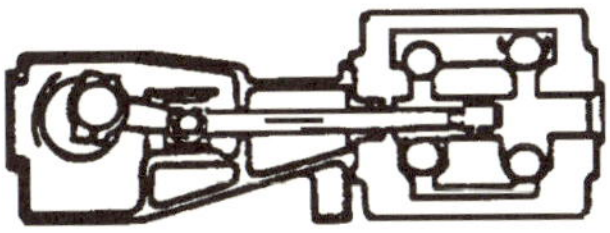
Diaphragm	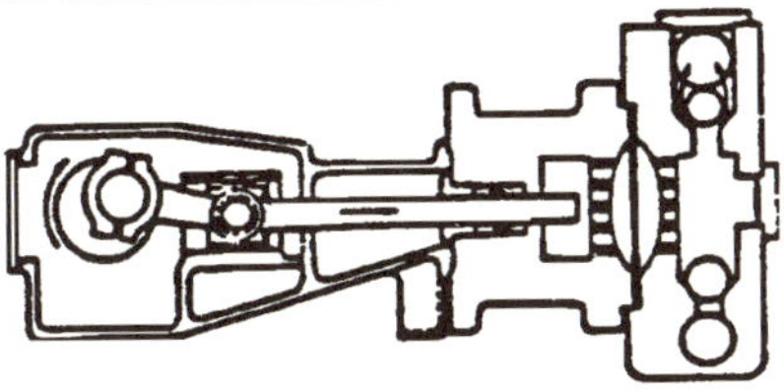

From Hydraulic Institute (1975).

Figure 9. Reciprocating displacement pumps

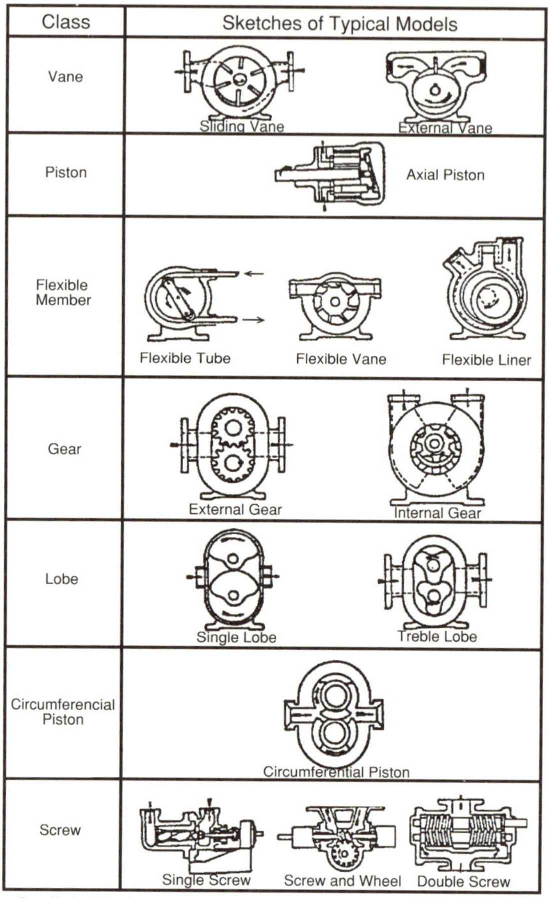

From Hydraulic Institute (1975)

Figure 10. Rotary displacement pumps

The European Space Agency (Ref. 38) has compiled a list of some important features of the rotodynamic and displacement pumps. They are listed in Table 7 for easy reference.

The following requirements are normally imposed on pumps for space-application fluid loops:

(1) High efficiency
(2) Low mass
(3) Relative low mass to output-power ratio
(4) Hermetically sealed
(5) Minimum operational noise level

(6) Able to withstand mission vibration and shock loads
(7) Compatible with onboard electrical system
(8) Applicable for aerospace-environment usage
(9) Able to handle typical liquid coolants as working fluids
(10) High operational reliability

Table 7. Main Features of Typical Pumps

Rotodynamic Pumps	Positive Displacement Pumps
• Very efficient when operating at speeds from 1200 to 3600 rpm, within the range of usual alternating current electric motors.	• They operate at very low rotating speed (220 to 500 rpm).
• Cannot be run efficiently at low speeds to pump small quantities.	• Efficiencies, although can be high, are below those for rotodynamic pumps.
• Overall efficiency usually ranges from .7 to .85.	• Efficiency decreases when wear increases leakage.
• Not easily regulated. Regulation by throttling is simple but wasteful. Regulation by running speed adjustment maintains more or less the efficiency but requires auxiliary equipment which is expensive, complicated, and unreliable. The use of multiple winding motors and invertors to control motor speed considerations.	• The discharge characteristic is a pulsating one. A smoother discharge is obtained in double or treble-acting units. Rotary pumps exhibit greater uniformity of deliver than reciprocating pumps.
• Cannot deliver at high pressure unless a large and heavy type is used. The pressure rise increases with increasing the peripheral speed.	• The delivery is substantially constant, regardless of the pressure rise developed.
• Cannot handle viscous liquids.	• Can be used with very viscous liquids. This is particularly true in the case of rotary pumps. The pressure rise drops with increasing viscosity.
• Not self-priming. This can be overcome in various ways.	• Self-priming and capable of coping with high suction lifts.
• No relief valves are to be used. Even complete throttling does not present any danger to the pump or loop as no further pressure rise develops.	• Relief or bypass valves are to be used. Unable to operate against a closed discharge. Even a slight decrease in delivery may cause a substantial pressure rise.

Table 7. Main Features of Typical Pumps (Continued)

Rotodynamic Pumps	Positive Displacement Pumps
• Limited by cavitation and power.	• Able to handle large proportions of vapor. Enough liquid must be present to provide a liquid sealing file for the clearances. Suitable for pumping hot liquid.
	• Limited by pressure and power.
• Leak through the shaft seal. Submerged pumps can be used to prevent this drawback.	• Problems of leakage are minimized particularly with diaphragm pumps.
• Smaller in size than other types for equal capacity.	• Much bigger than rotodynamic because of low rotating speed.
• Low cost. Rugged, reliable in operation.	• Complicated construction. Inlet and outlet valves must be used (not in the case of rotary pumps). Very sensitive to wear because comparatively large surface areas move in close contact. Diaphragm pumps do not present friction, but diaphragm materials are of limited use at elevated temperatures.

Arranged by the compiler (Ref. 38) after: Nekrasov (1969), London (1974), Pollak & Cruger (1974), Scoble (12974), Settles, et al. (1977).

Characteristic Curves

The performance parameters for a typical pump consist of the pressure head (H), the input power (P), and the machine efficiency (η) under some specific operating conditions. These parameters are the pump dependent variables. The volumetric flow rate (Q), angular speed (ω), impeller diameter (D), and fluid properties (i.e., density, ρ) are the independent variables. Pump characteristic curves are generated by plotting a dependent variable as a function of one of the independent variables. Pump performance is difficult to predict analytically except at the design point of the specific machine. Hence, it is measured experimentally. Some typical characteristic curves showing head, efficiency, and horsepower as a function of the discharge for a typical centrifugal pump are shown in Figure 11. These curves are shown for a centrifugal pump tested at constant speed. When a pump with performance curves such as shown in Figure 11 is installed in a pumping system, its design operating point is controlled by the so-called system-components (e.g., piping, valves, and fittings) resistance. The system resistance, as defined in Eq. (16), is proportional to the square of the velocity. The friction factor and equivalent length vary somewhat with flow rate, accounting for the deviation from a parabolic velocity distribution. The head-capacity curve of a typical pump with the system-resistance curve superimposed is shown in Figure 12 as a function of the volume flow rate. The only possible system operating point is the intersection of these two curves where the head developed by the pump just balances the head loss due to friction in the system.

Pump Laws

The basic pump laws are derived using the principles of dynamic similarity and dimensional analysis. The word "similarity" in pump design refers to two machines operating under identical flow conditions. The results from the basic pump laws (Ref. 39) are presented as follows:

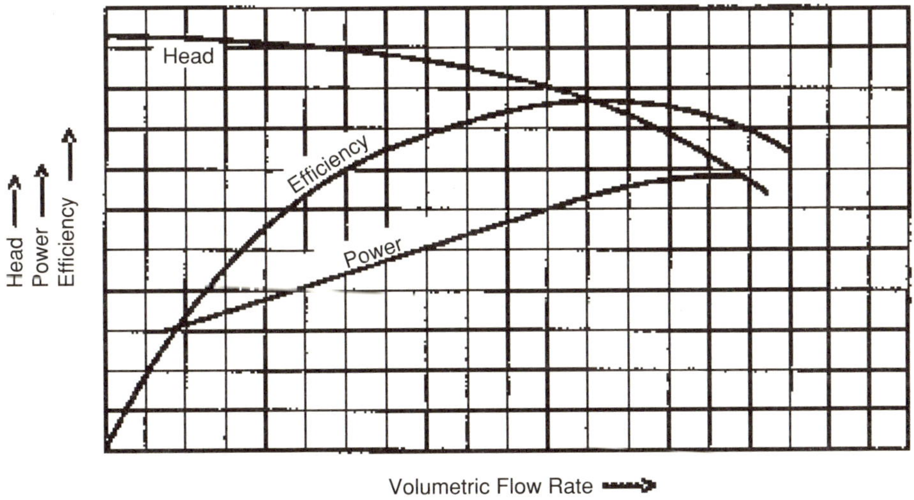

Figure 11. Characteristic curves for typical centrifugal pump

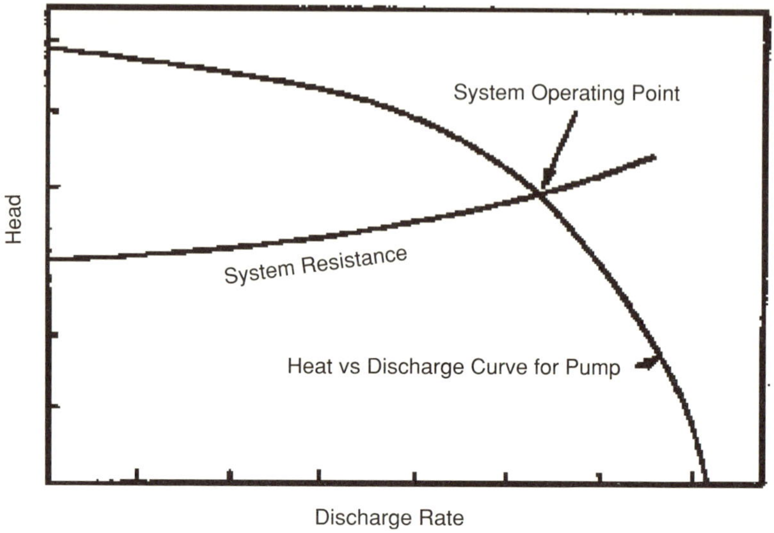

Figure 12. Pump and system curves

$$\frac{Q_1}{\omega_1 D_1^3} = \frac{Q_2}{\omega_2 D_2^3} , \tag{25}$$

$$\frac{H_1}{\omega_1^2 D_1^2} = \frac{H_2}{\omega_2^2 D_2^2} , \tag{26}$$

and

$$\frac{P_1}{\rho_1 \omega_1^3 D_1^5} = \frac{P_2}{\rho_2 \omega_2^3 D_2^5} , \tag{27}$$

and the specific speed is defined by

$$N_s = \frac{\omega Q^{1/2}}{H^{3/4}} . \tag{28}$$

These laws only hold true at different operating conditions as long as the pump efficiency is constant.

Heat Exchangers

Types of Heat Exchangers

The most common heat exchangers can be classified into three categories. They are:

(1) Flat-plate.
(2) Shell-and-tube. The simplest form is the double-pipe exchanger shown in Figure 13a. If the hot and cold fluids both flow in the same direction, it is referred to as a parallel-flow type; otherwise, it is a counterflow type. Some common type of shell-and-tube heat exchangers are shown in Figure 13b.
(3) Crossflow. The two fluids flow at right angles to each other, as shown in Figure 13c. The flow may be called mixed or unmixed within the crossflow arrangement.

Heat Transfer Calculations

The main objective in the thermal design of heat exchangers is to determine the necessary surface area required to transfer heat at a given rate for given fluid temperatures and flow rates. One important factor is the knowledge of the overall heat-transfer coefficient, U. The overall heat-transfer coefficient U can be related in the following fundamental heat-transfer relation

$$Q = UA(\Delta T_m) \ , \tag{29}$$

where

U = overall heat transfer coefficient
A = surface area for heat transfer consistent with definition of U
ΔT_m = mean temperature difference across heat exchanger

Overall Heat-Transfer Coefficient. The overall heat-transfer coefficient U is proportional to the reciprocal of the sum of the thermal resistances. Consider a wall exposed to a hot fluid A on one side and cooler fluid B on the other side. Some common configurations are:

plane wall;
$$U = \frac{1}{\frac{1}{h_o} + \frac{L}{K} + \frac{1}{h_i}} \ ,$$

and cylindrical wall;

$$U_o = \frac{1}{\dfrac{r_o}{r_i h_i} + \dfrac{r_o}{k} \ln\left(\dfrac{r_o}{r_i}\right) + \dfrac{1}{h_o}} \qquad U_i = \frac{1}{\dfrac{1}{h_i} + \dfrac{r_i}{k} \ln\left(\dfrac{r_o}{r_i}\right) + \dfrac{r_i}{r_o h_o}},$$

where subscripts i and o represent the inside and outside surfaces of the wall, respectively.

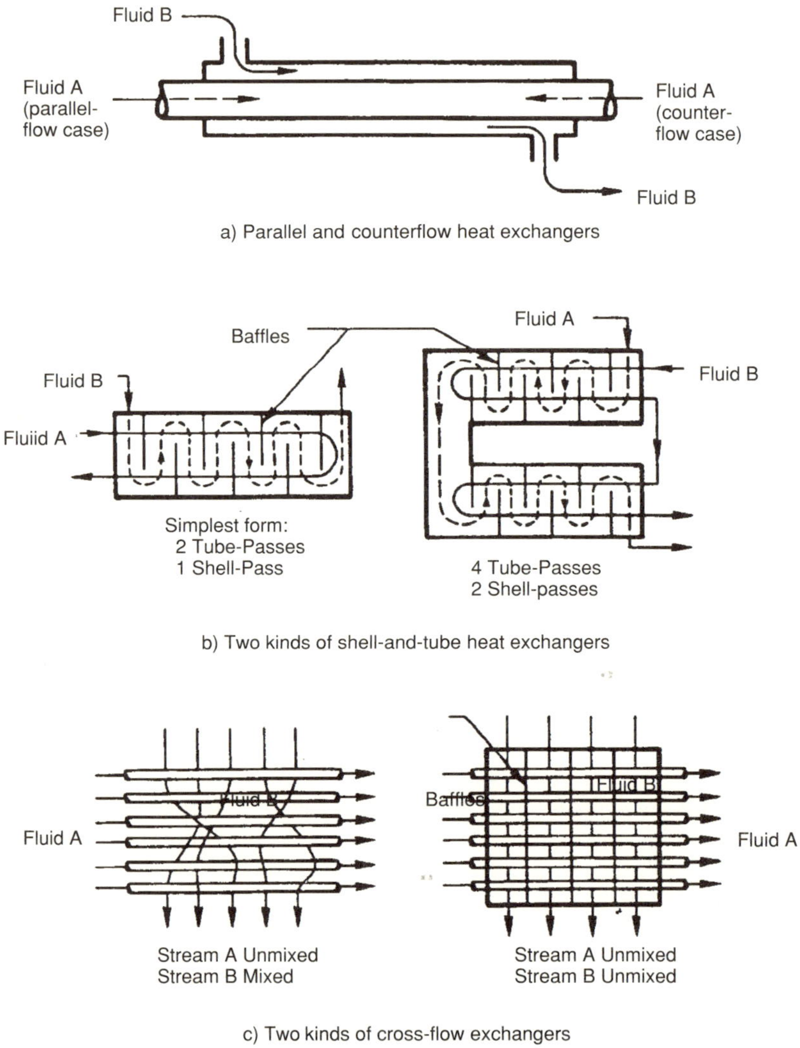

Figure 13. Basic types of heat exchanger

Log-Mean Temperature Difference (LMTD). The mathematical expression for the LMTD can be derived by considering a parallel-flow flat plate exchanger, the temperature profiles of which are shown in Figure (14). From an energy balance on a differential fluid element with length dx for each fluid, the mean temperature ΔT_m for either parallel or counterflow can be determined

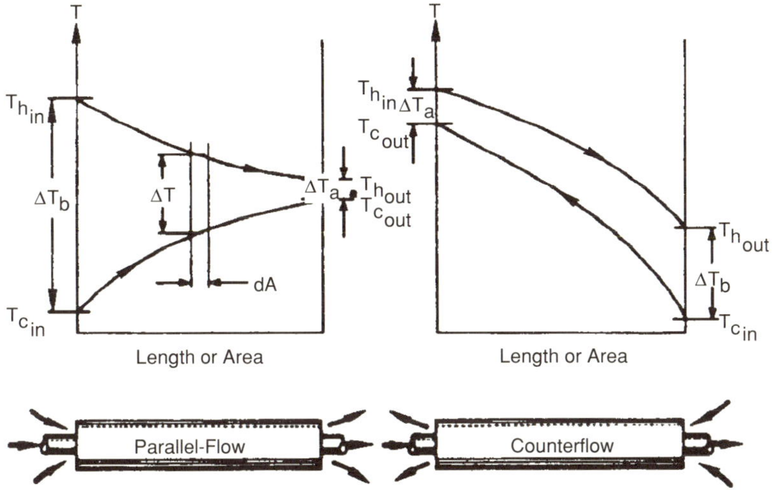

Figure 14. The temperature variation through single-pass heat exchangers

from the following expression

$$Q = UA\,\Delta T_m = UA\left[\frac{\Delta T_a - \Delta T_b}{\ln\left(\frac{\Delta T_a}{\Delta T_b}\right)}\right] . \tag{30}$$

Thus, the average effective temperature difference ΔT_m in Eq. (29) can be written as

$$\Delta T_m = \text{LMTD} = \frac{\Delta T_a - \Delta T_b}{\ln\left(\frac{\Delta T_a}{\Delta T_b}\right)} . \tag{31}$$

The expression defined by Eq. (31) is called the LMTD.

Correction Factors for Complex Heat Exchangers. For more complex heat exchangers, such as those involving multiple tubes, several shell passes, or crossflow, determination of the average effective-temperature difference is so difficult that the usual practice is to modify Eq. (29) by a correction factor F. Correction factors for several common configurations are given in Figure 15. In these figures the notation (T,t) to denote the temperatures of the two fluid streams has been introduced, since it is immaterial whether the hot fluid flows through the shell or the tubes.

Heat Exchanger Effectiveness (NTU Method)

The performance of an exchanger can be determined once its configuration and the imposed temperature difference are known. However, either the inlet or outlet temperature of the heat exchanger may not be a known until the design is complete. An iterative process requiring a trial-and-error approach of the heat transfer rate and the exit temperature is necessary. The so-called effectiveness method developed in full detail by Kays and London in a book titled *Compact Heat Exchangers* is useful in heat-exchanger design. The heat exchanger effectiveness is defined as

$$\varepsilon = \frac{\text{actual heat transfer}}{\text{maximum possible heat transfer}} = \frac{Q_{actual}}{Q_{max}} \quad . \tag{32}$$

The maximum possible heat transfer occurs if one fluid underwent a temperature change equal to the maximum temperature difference available within the system. This is equal to the temperature of the entering hot fluid minus the temperature of the entering cold fluid. The procedure uses the effectiveness ε to eliminate the unknown discharge temperature. As a result, the solution for the heat exchanger effectiveness becomes a function of the other known system parameters. These include the mass flow rate of the fluid ($\dot{m}$), heat capacity (c_p), heat-transfer area (A), and the overall heat-transfer coefficient (U). Letting the heat capacitance $C = \dot{m}c_p$,

$$Q_{actual} = C_h\left(T_{hi} - T_{ho}\right) = C_c\left(T_{co} - T_{ci}\right) \quad . \tag{33}$$

Equation (33) is an energy balance on both the hot and cold fluids. The maximum possible heat transfer occurs when the fluid of smaller heat capacitance undergoes the maximum temperature variation. This can be stated as

$$Q_{max} = C_{min}\left(T_{hi} - T_{ci}\right) \; . \tag{34}$$

By combining Eqs. (32) and (34), in terms of the heat exchanger effectiveness, the actual heat transfer is governed by

$$Q_{actual} = \varepsilon C_{min}\left(T_{hi} - T_{ci}\right) \; . \tag{35}$$

The heat-transfer rate can be determined once the specific value of the heat-exchanger effectiveness is known. The effectiveness (ε) for the parallel single-pass heat exchanger is

$$\varepsilon = \frac{1 - \exp\left[-\left(1 + C_{min}/C_{max}\right) NTU\right]}{1 + C_{min}/C_{max}} \; , \tag{36}$$

and the corresponding expression for the counterflow case is

$$\varepsilon = \frac{1 - \exp\left[-\left(1 - C_{min}/C_{max}\right) NTU\right]}{1 - C_{min}/C_{max} \exp\left[-\left(1 - C_{min}/C_{max}\right) NTU\right]} \; , \tag{37}$$

where C_{max} and C_{min} are the maximum and minimum values of the C ($=\dot{m}c_p$) for the hot or the cold fluid. Expressions for the effectiveness of other configurations are given in Table 8 and Figure 16, where $C = C_{min}/C_{max}$. Note that for an evaporator or condenser C=0, because one fluid remains at a constant temperature, making its effective specific heat infinite. The NTU appearing in the last two expressions is the so-called number of transfer units, which is defined as

$$NTU = \frac{UA}{C_{min}} \; . \tag{38}$$

The NTU may be considered as a heat exchanger size-factor.

Heat Exchanger Design

The preceeding sections have provided means for predicting heat-exchanger performance. There are other considerations in designing heat exchangers in addition to the prediction of heat transfer. The primary ones are the minimization of pumping power and the minimization of weight. The weight and size of heat exchangers used in space or aeronautical applications are very important parameters.

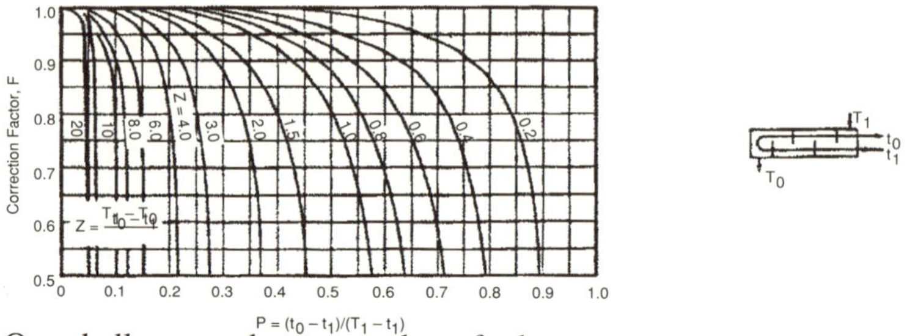

a. One shell pass and even number of tube passes

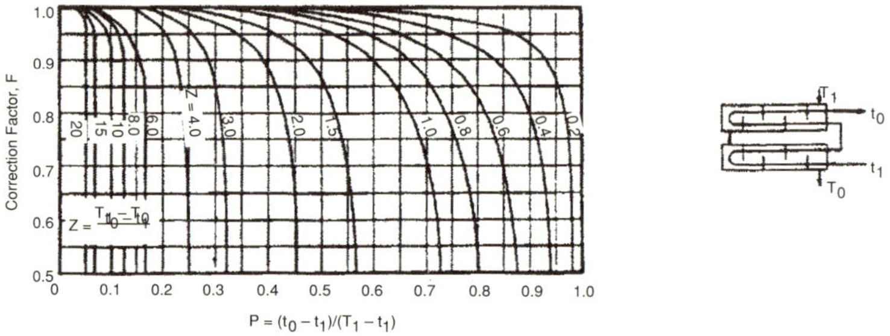

b. Two shell passes and twice an even number of tube passes

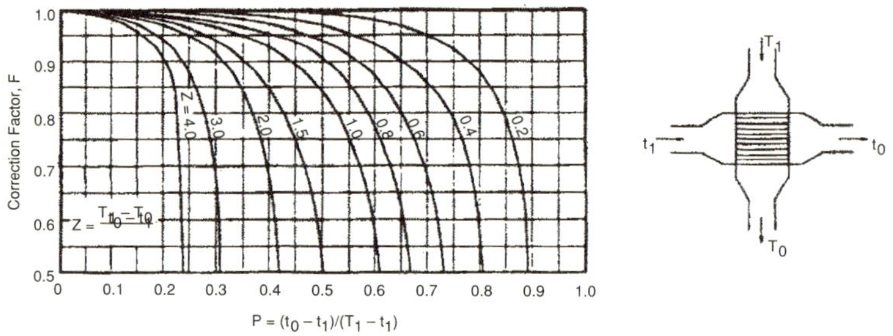

c. Crossflow with one fluid mixed

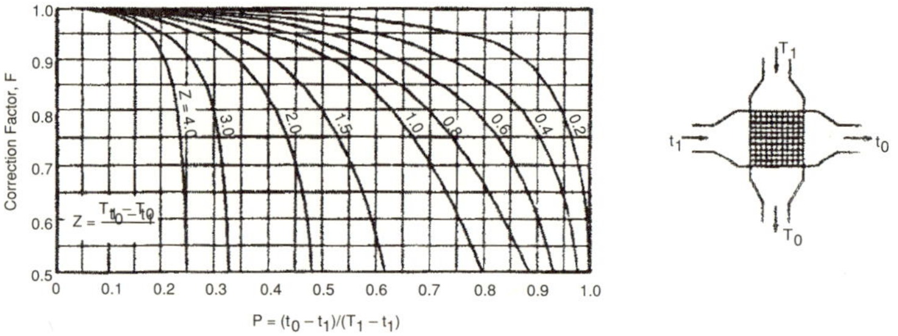

d. Crossflow with both fluids unmixed

Figure 15. Correction factors for some common heat exchanger configurations

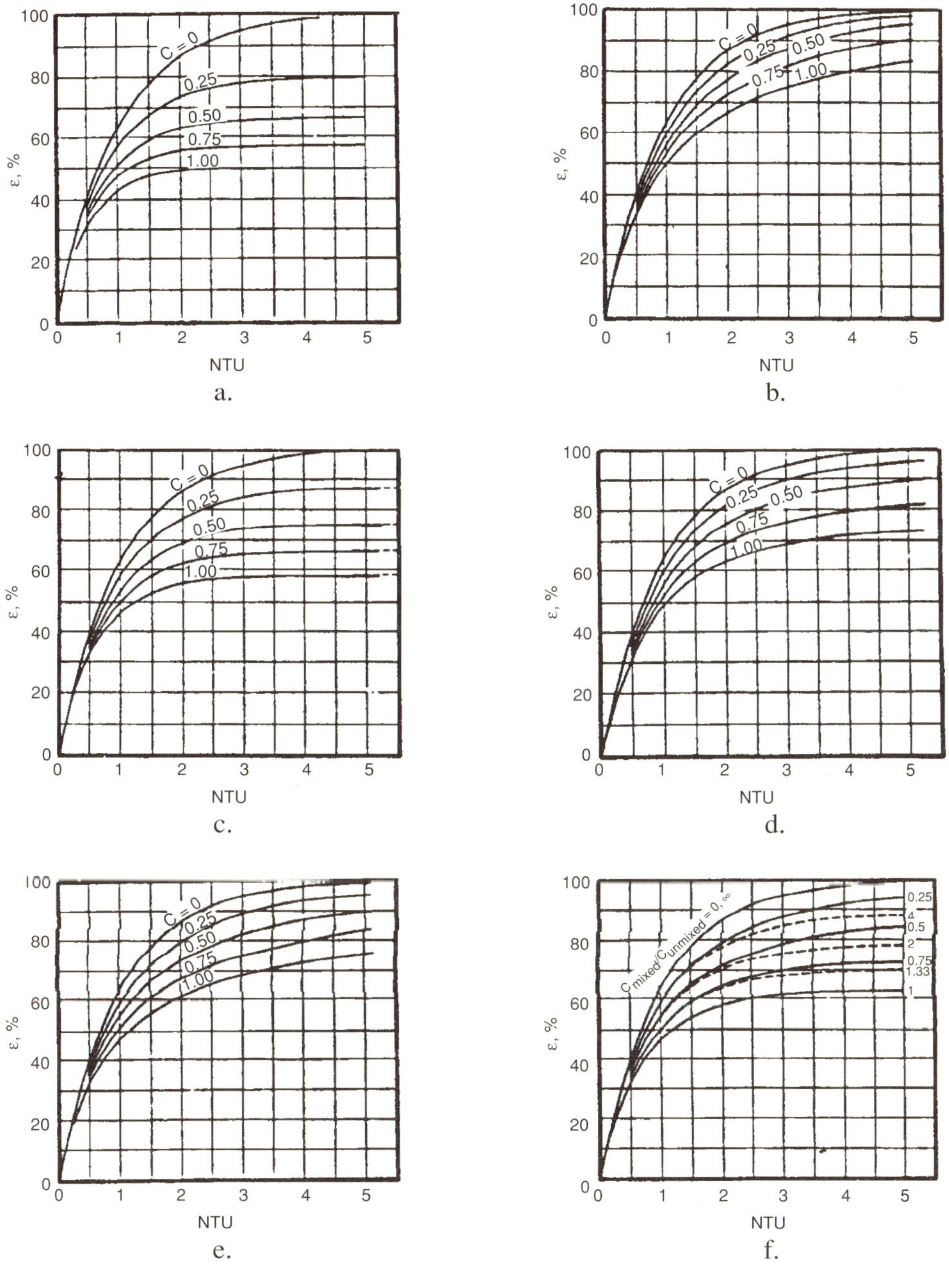

Figure 16. Heat-exchanger effectiveness

Table 8. Effectiveness for Various Heat Exchanger Configurations

Exchanger Type	Effectivness	See graph In
Parallel-flow: single-pass	$e = \dfrac{1 - \exp[-NTU(1 + C)]}{1 + C}$	Fig. 16a.
Counterflow: single-pass	$e = \dfrac{1 - \exp[-NTU(1 - C)]}{1 - C \exp[-NTU(1 - C)]}$	Fig. 16b.
Shell-and-tube (one shell pass; 2, 4, 6, etc., tube passes)	$e_1 = 2\left[1 + C + \dfrac{1 + \exp[-NTU(1 + C^2)^{1/2}]}{1 - \exp[-NTU(1 + C^2)^{1/2}]}(1 + C^2)^{1/2}\right]^{-1}$	Fig. 16c.
Shell-and-tube (n shell passes; 2n, 4n, 6n, etc., tube passes)	$e_n = \left[\left(\dfrac{1 - e_1 C}{1 - e_1}\right)^3 - 1\right]\left[\left(\dfrac{1 - e_1 C}{j - e_1}\right)^3 - C\right]^{-1}$	Fig. 16d. for n = 2
Crossflow (both streams unmixed)	$1 - \exp\{C\,(NTU)^{0.22}\,[\exp[-C\,(NTU)^{0.78}] - 1]\}$	Fig. 16e.
Crossflow (both streams mixed)	$e = NTU\left[\dfrac{NTU}{1 - \exp(-NTU)} + \dfrac{(NTU)(C)}{1 - \exp[-(NTU)(C)]} - 1\right]^{-1}$	
Crossflow (stream C_{min} unmixed)	$e = C\{1 - \exp[-C[1 - \exp(-NTU)]]\}$	Fig. 16f. (dashed curves)
Crossflow (stream C_{max} unmixed)	$e = 1 - \exp\{-C[1 - \exp[-(NTU)(C)]]\}$	Fig. 16f. (solid curves)

WORKING FLUIDS

Some of the most commonly used coolants in heat exchangers are discussed in this section. Their physical properties, such as vapor pressure (P_{sat}), density (ρ), specific heat (C_p), dynamic viscosity (μ), and thermal conductivity (k) are given in Figs. 17-27. These include:

Figure 17	Monsanto OS 59
Figure 18	FC 75
Figure 19	Freon E1, E2, E3, E4, E5
Figure 20	Freon 11, 12, 13, 21, 22, 113, 114, 142
Figure 21	Flutec PP-2, PP-9, PP-50
Figure 22	Water/Glycol solutions
Figure 23	Coolanol 15, 25, 35, 45
Figure 24	Carbon Tetrachloride
Figure 25	Water
Figure 26	Methanol/Water solution, DC-200
Figure 27	Air

Interested readers should refer to References 49 and 59 for other coolants.

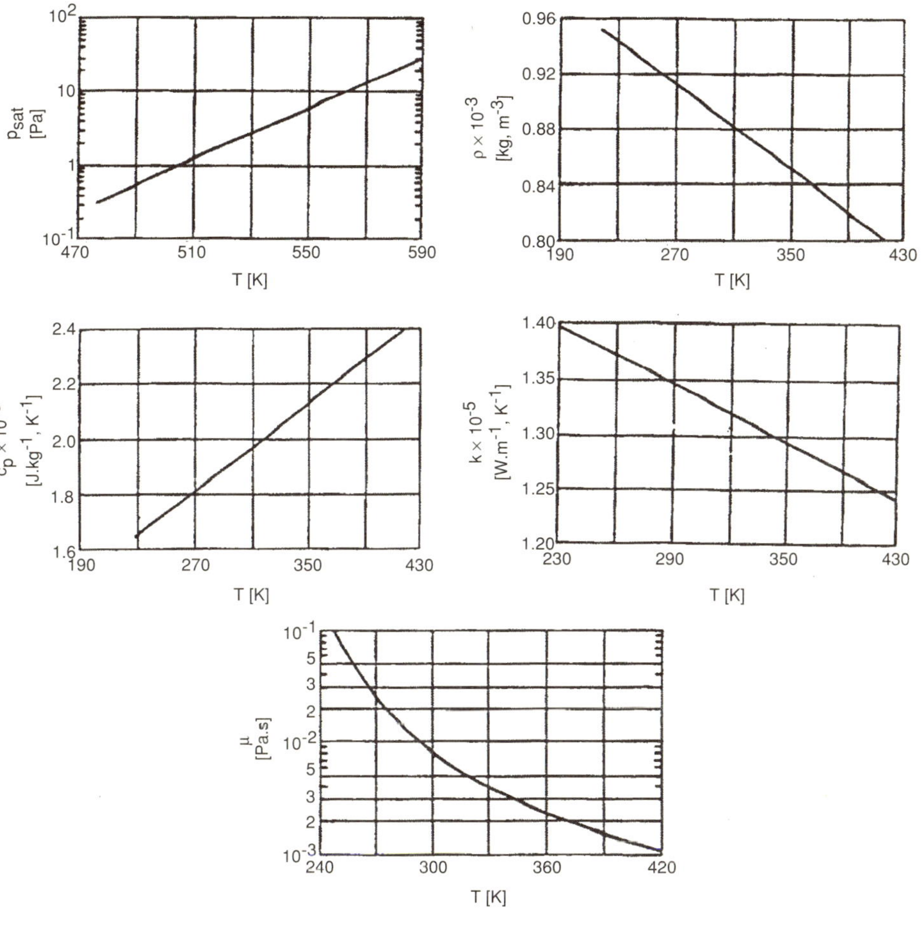

Figure 17. Physical properties of Monsanto OS 59

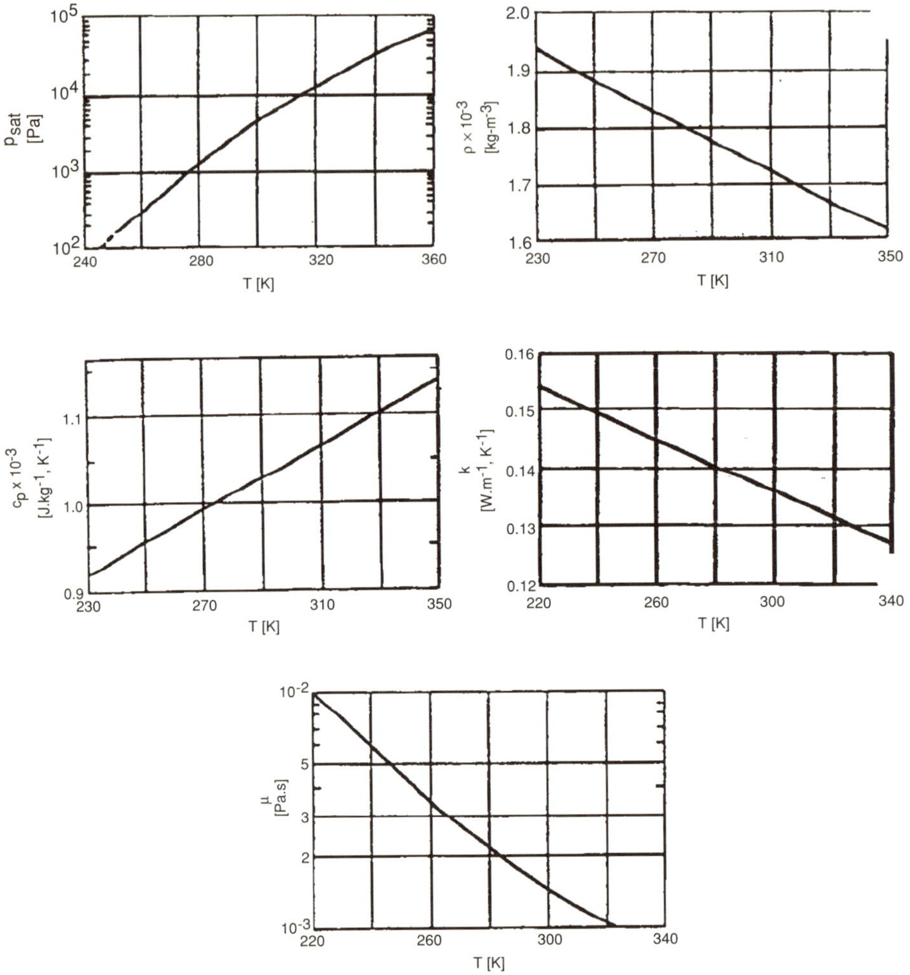

Figure 18. Physical properties of FC 75

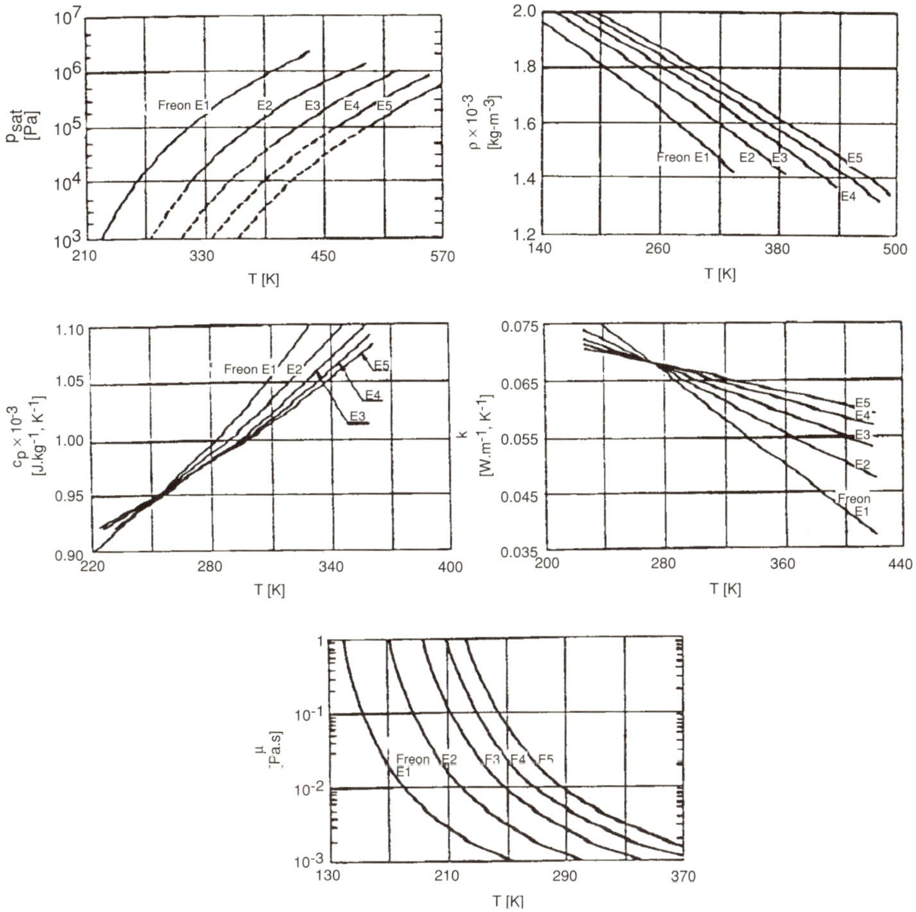

Figure 19. Physical properties of Freon E1, E2, E3, E4, E5

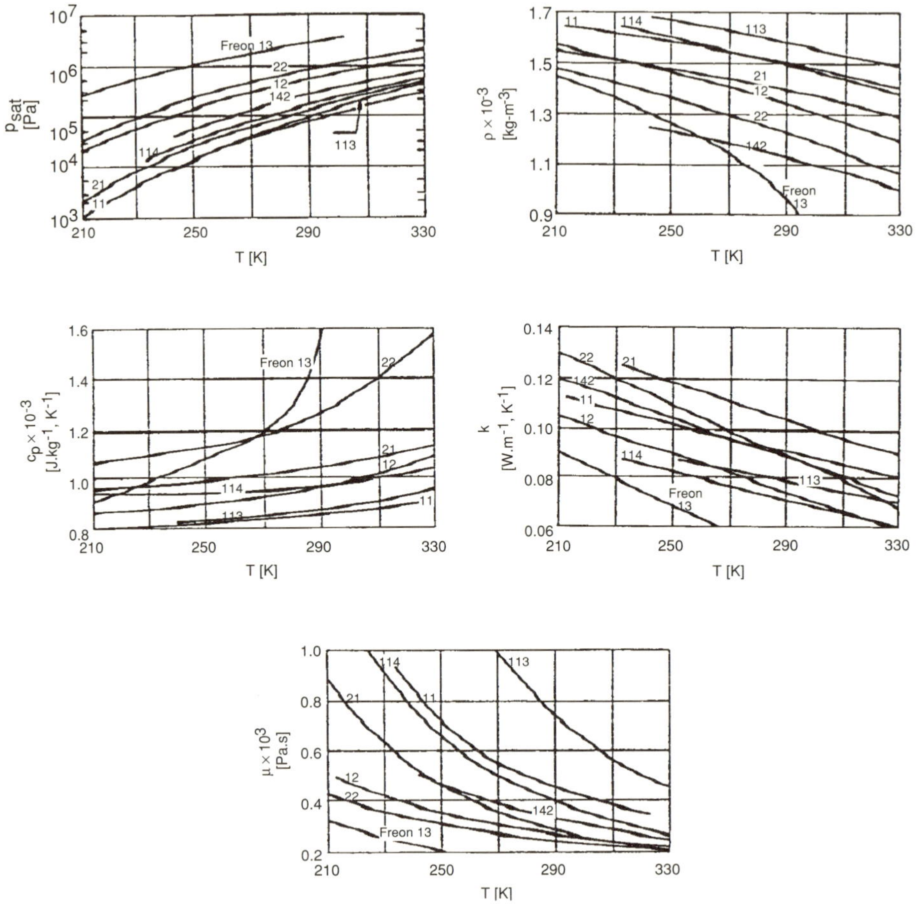

Figure 20. Physical properties of Freon 11, 12, 13, 21, 22, 113, 114, 142

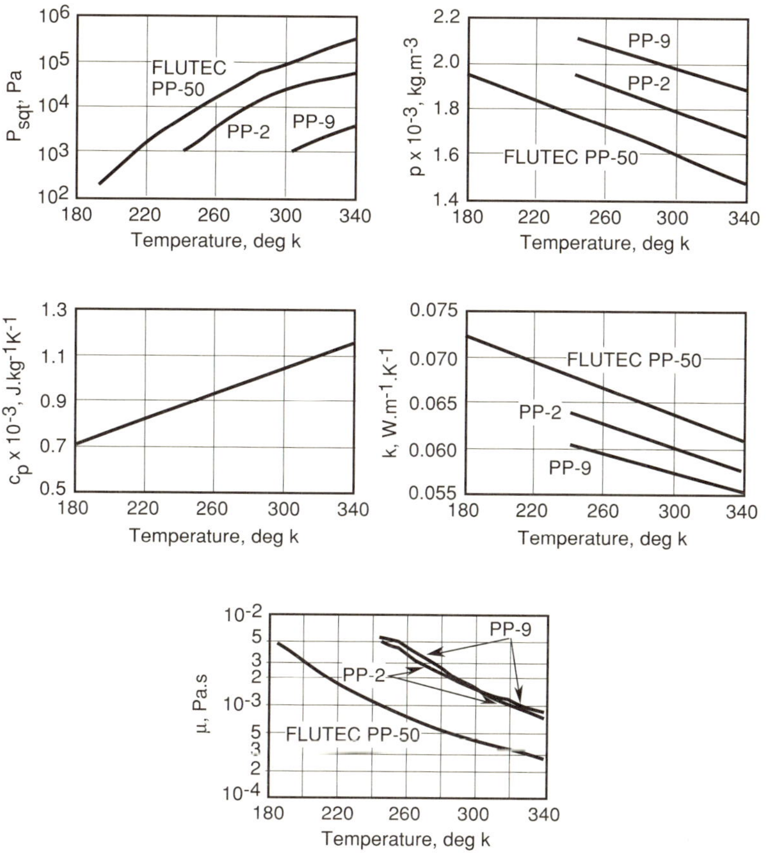

Figure 21. Physical properties of Flutec PP-2, PP-9, PP-50

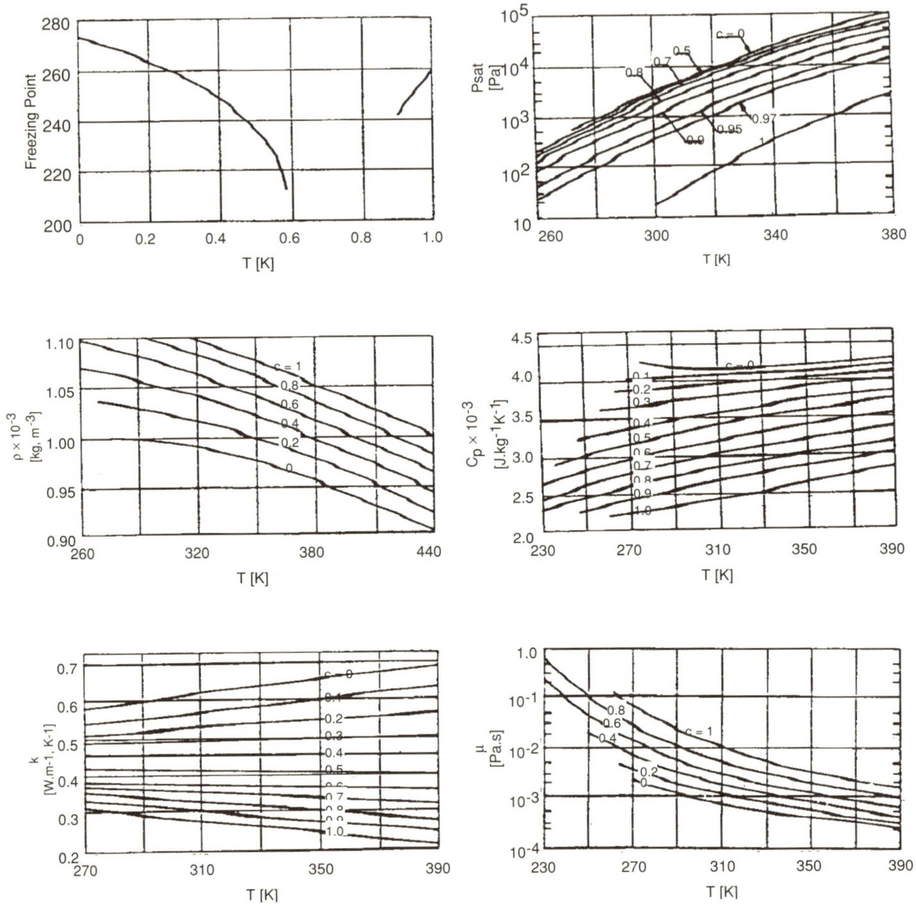

Figure 22. Physical properties of water/glycol solutions

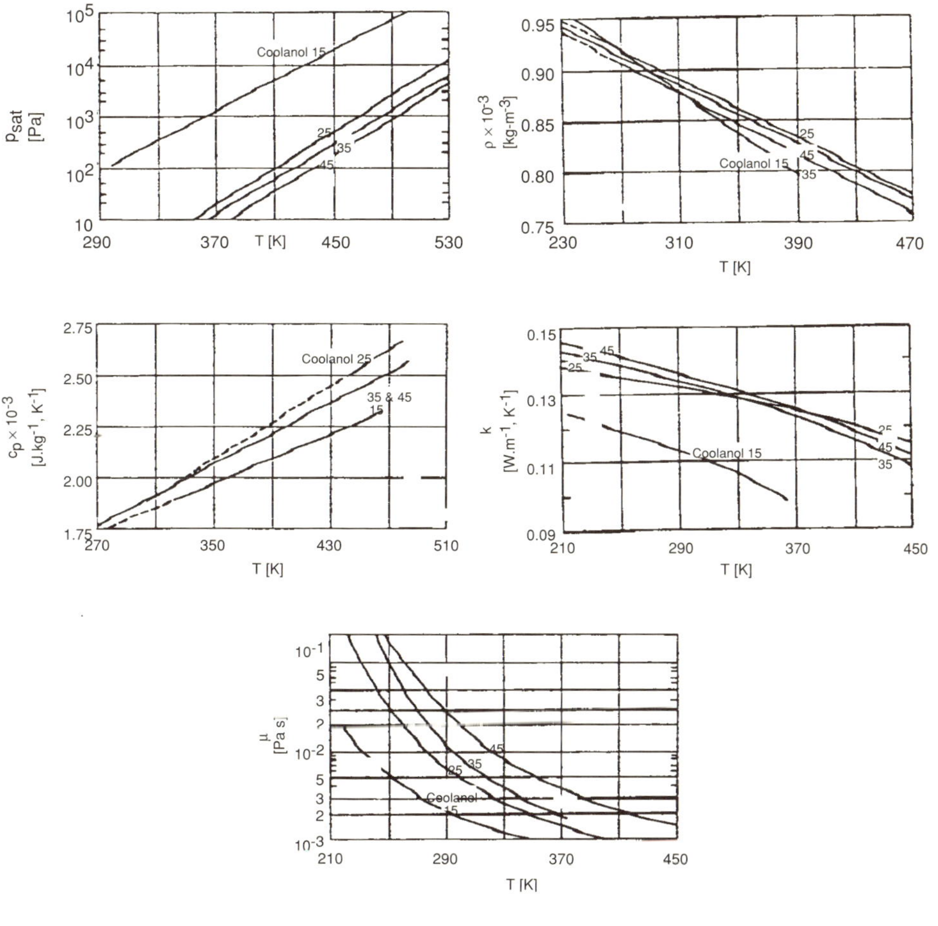

Figure 23. Physical properties of Coolanol 15, 25, 35, 45

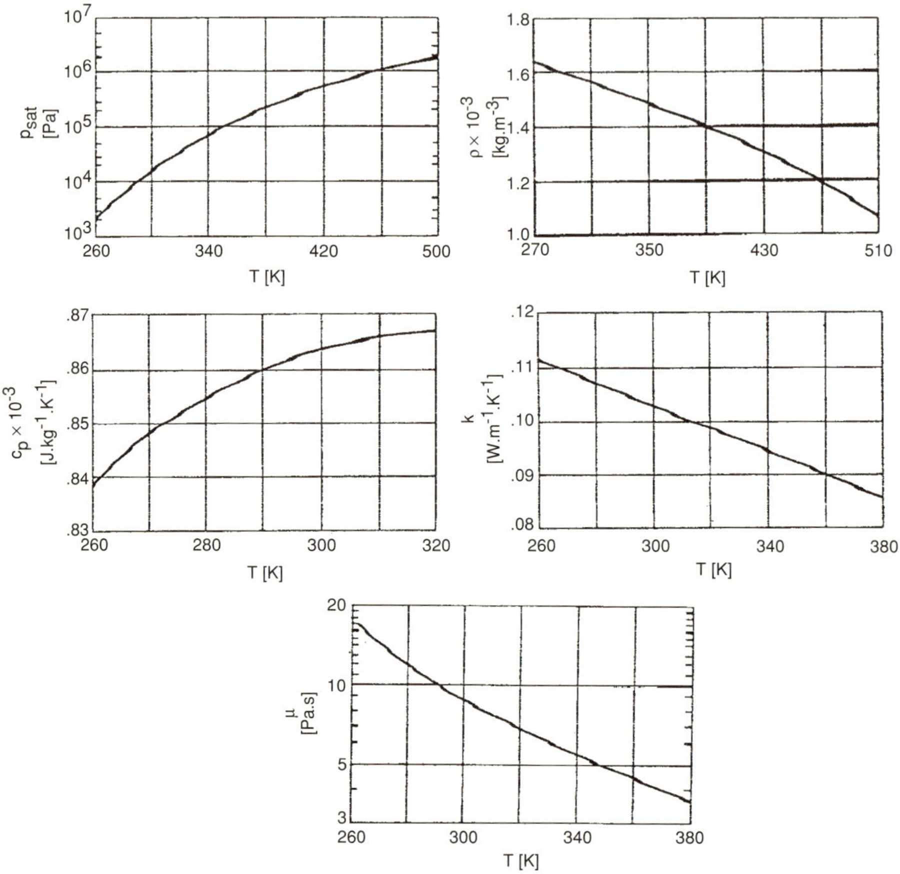

Figure 24. Physical properties of carbon tetrachloride

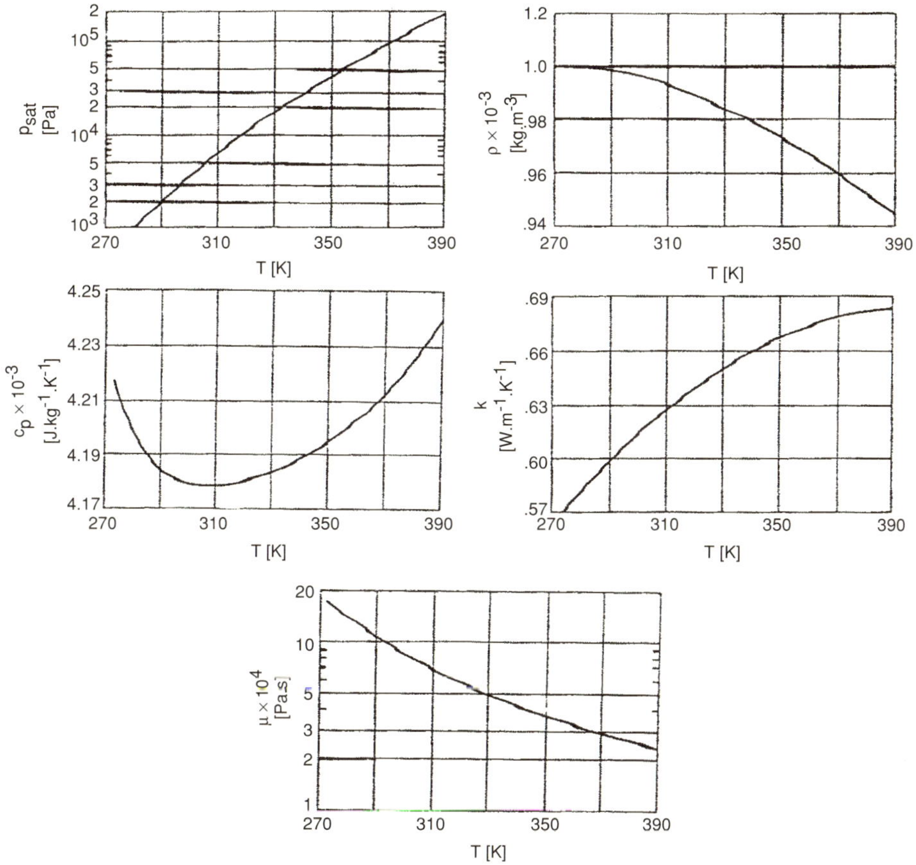

Figure 25. Physical properties of water

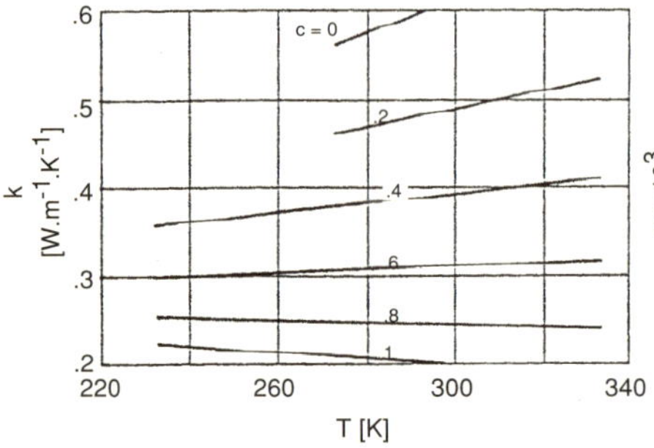

Methanol/Water Solution

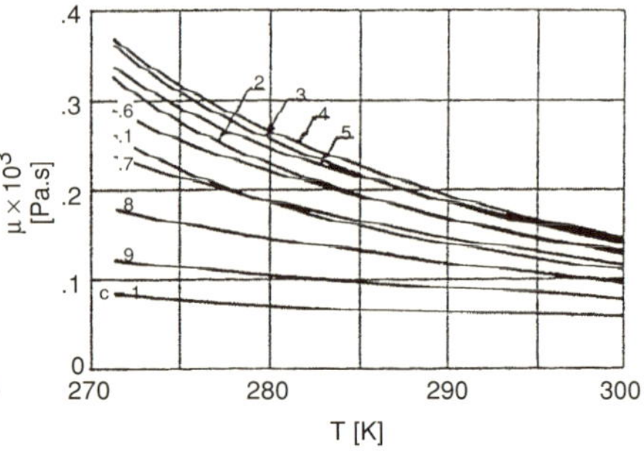

DC 200

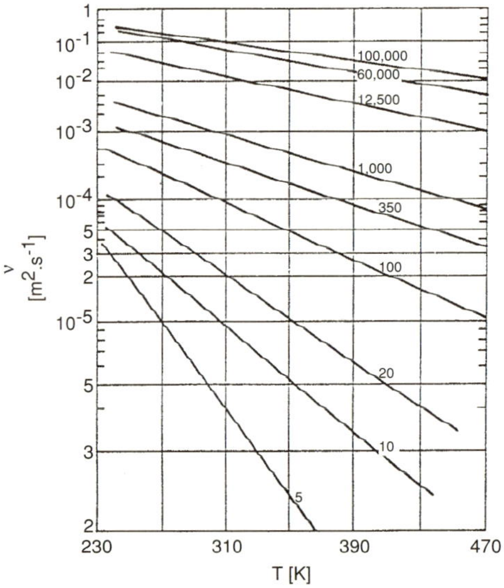

DC 200

Figure 26. Physical properties of methanol/water solution, DC-200

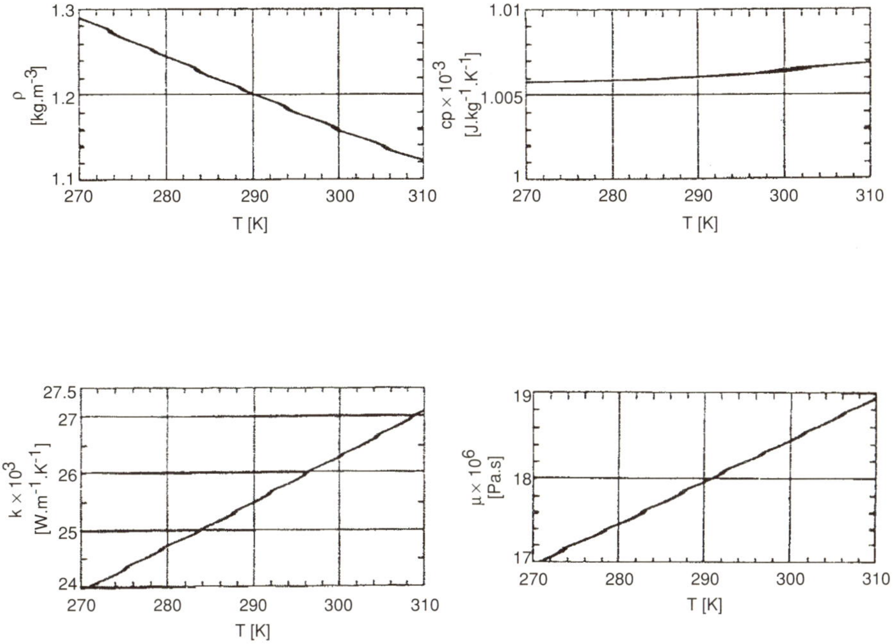

Figure 27. Physical properties of air

ANALYSIS OF A FLUID LOOP

The engineering background presented in the previous sections is adequate for analyzing the fluid loop in Fig. 1. The design procedure in this section follows closely as provided in Reference 60. The following engineering data are assumed to be known for the system:

(1) General layout of the fluid loop, including system geometry and dimensions.
(2) Thermal properties of the coolant, k, c_p, ρ, μ.
(3) Total heat-flow rate to be removed, Q.
(4) Mass flow rates in each loop and heat exchanger; m_h, m_c.
(5) Inlet temperature of the cold fluid in the heat exchanger, T_{ci}.

The temperatures throughout the loop and the pumping-power requirements can be determined by performing an energy balance on the system. To compute the temperatures in the loop, the heat-exchanger effectiveness must be calculated. The following steps are needed:

(1) Compute the required heat-exchanger heat-transfer surface areas.
(2) Compute fluid properties such as density, specific heat, thermal conductivity, dynamic viscosity, and Pradtl number.
(3) Compute Reynolds number (Re) for each section of the loop.
(4) Compute the Nusselt number (Nu) and the convective heat-transfer coefficient (h).
(5) Compute the temperature effectiveness (η) of all the extended surfaces.
(6) Compute the overall heat-exchanger thermal conductance (U).
(7) Compute the number of heat transfer units, NTU.
(8) Compute the heat-exchanger effectiveness, ε.

Once all of the above engineering data becomes available, the temperatures throughout the loop can be calculated from

$$T_{co} = T_{ci} + \frac{Q}{C_c} \tag{39}$$

$$T_{hi} = T_{ci} + \frac{Q}{\varepsilon C_1} \tag{40}$$

$$T_{ho} = T_{ci} + Q\left(\frac{1}{\varepsilon C_1} - \frac{1}{C_h}\right) , \tag{41}$$

where C_1 is the smallest of C_h and C_c.

The pumping power, P_p, required to operate the fluid system against the pressure drop ΔP can be calculated from the relation

$$P_p = \Delta P \frac{\dot{m}}{\rho} \frac{1}{\eta_p} , \tag{42}$$

where η_p is the overall pump efficiency, ΔP the pressure loss through the system, $\dot{m}$ is the fluid-mass flow rate, and ρ is the fluid density at the location of the pump. Thus, the term $\dot{m}/\rho$ represents the fluid-volume flow rate through the pump. The following steps can be used as a general guideline to compute the pressure losses within the system:

(1) Compute the Reynolds number in all flow conduits.
(2) Compute the friction factor f for the straight parts of the tubes.
(3) Compute pressure loss due to friction along the tube walls.
(4) Compute pressure loss for all the pipe bends.
(5) Compute pressure loss in all the fittings (e.g., valves, manifolds, and entrances, etc).
(6) Compute pressure loss in heat source and heat sink.

SYSTEM ANALYSIS - COMPUTER SOFTWARE

Three thermal-fluid network analyzers are presented in this section. The intent is to provide a brief overview of these codes. The interested reader should refer to the references at the end of this article for more detail. The general capabilities of each thermal/fluid analyzer are described in the following paragraphs.

ESATAN - FHTS

The Fluid Heat Transfer System (FHTS) was developed by GEC Engineering Research Centre in the United Kingdom as an extension of the European Space Agency's principal thermal-analysis package, ESATAN. The FHTS computer code can solve both steady-state and transient fluid-flow problems. It can be used to obtain thermal-hydraulic solutions to single or two-phase fluid-flow systems. Users are able to construct pumped fluid loops from basic node and conductance data to simulate hardware such as pumps and heat exchangers. By defining fluid nodes, fluid conductances, and mass flow links within the framework of ESATAN, engineering simulations can be performed for all-fluid, all-thermal, or combined fluid and thermal systems simultaneously. A variety of predefined models commonly used in fluid systems, known as fluid elements (e.g., pumps, heat exchangers, tee fittings, and valves etc.), have been included within the software to reduce the users' input effort. The FHTS has an internal library of fluid-property correlations that can be used to simulate various types of coolant. These include water, ammonia, R11, R12, R22, R114, R502, and air. The user can specify any of these fluids by assigning the appropriate one to the nodal entity. The final system solution gives pressure and temperature (or enthalpy) at each fluid node, and mass-flow rate on each fluid link. For more detail on the FHTS, the interested reader should refer to Ref. 63.

The Aerospace Corporation's SINDA

Recently, a flow network solution scheme has been implemented in The Aerospace Corporation's version of the SINDA thermal analyzer Ref. 64. The computer code can be used for standalone fluid flow and coupled heat-transfer/fluid-flow networks. For standalone flow problems, the flow network solution capability can be used as a design tool to size the various flow elements such as the pipes, valves, and pump. In coupled thermal/fluid problems the coupling arises from the temperature dependence of the fluid properties. The fluid is assumed to be single phase, viscous, and incompressible. In addition, the flow is one dimensional and completely bounded by solid boundaries. Another major assumption in the flow solution is that the flow is always at quasi-steady state. Hence, the transient pressure fluctuation is assumed to be negligible. However, the validity of this assumption breaks down for high speed flows when shock waves are formed or when the flow becomes choked. The solution to a flow network includes the pressure distribution and the mass flow rate across each flow passage.

SINDA '85/FLUINT

Under a NASA contract, Martin Marietta Corporation undertook the task of development of an advanced SINDA thermal analyzer computer program in 1983. The final product of the contract was the SINDA '85. This latest version of SINDA has been improved by a series of enhancements that include the fluid-flow network capability known as the fluid integrator (FLUINT). The combined new computer code SINDA '85/FLUINT has both thermal and fluid network capabilities. It can be used to perform pressure/flow analysis of a system containing an arbitrary tube network simultaneously with the thermal analysis with transient or steady-state boundary conditions. This permits the mutual influences of thermal and fluid problems to be included in the analysis.

FLUINT is intended to provide a general analysis framework for internal one-dimensional fluid systems. The computer code can be applied to any arbitrary fluid system; therefore it is not restricted to any specific geometries or configurations. There are twenty refrigerants immediately available as working fluids, or the user can specify their own fluid properties for any specific applications. The code can handle both single- and two-phase flow and the transitions between these states. FLUINT also includes some common fluid system components. FLUINT may be used for a wide variety of analyses; however, there are a few limitations in the code and the user should be aware of them. For example, no mixing of working fluids are allowed within the fluid systems. The program is specifically intended only for internal low-speed, incompressible viscous fluid flows in tubes. Fluid-flow phenomena such as compressibility, shock wave, and choked flow cannot be analyzed in SINDA '85/FLUINT.

REFERENCES

Fluid-Flow Concepts and Basic Equations

1. R. W. Fox and A. T. McDonald, *Introduction to Fluid Mechanics*, 3rd ed., John Wiley & Sons, Inc., New York (1985).

2. L. F. Moody, "Friction Factors for Pipe Flow," *Transaction of the ASME*, 66, pp. 671-684 (1944).

3. P. K. Swamee and A. K. Jain, "Explicit Equations for Pipe-Flow Problems," *Proceedings of the ASCE, Journal of the Hydraulics Division*, 102, HY 5, pp. 657-664 (1976).

4. "Flow of Fluids through Valves, Fittings, and Pipe," Crane Company, New York, N.Y., Technical Paper No. 410 (1982).

5. W. M. Kays and M. E. Crawford, *Convective Heat and Mass Transfer*, 2nd ed., McGraw-Hill, New York (1972).

6. B. S. Petukhov, "Heat Transfer and Friction in Turbulent Pipe flow with Variable Physical Properties," in *Advances in Heat Transfer*, Vol. 6, T. F. Irvine and J. P. Hartnett (eds.), Academic Press, New York, p. 504-564 (1970).

7. M. N. Ozisik, *Heat Transfer - A Basic Approach*, McGraw-Hill Book Company, New York (1985).

Forced Convection in Pipes and Tubes

8. J. P. Holman, *Heat Transfer*, 6th ed., McGraw-Hill Book Company, New York (1986).

9. M. N. Ozisik, *Heat Transfer - A Basic Approach*, McGraw-Hill Book Company, New York (1985).

10. F. P. Incropera and D. P. DeWitt, *Fundamentals of Heat and Mass Transfer*, 2nd ed., John Wiley & Son, Inc., New York (1985).

11. F. W. Dittus and L. M. K. Boelter, *University of California (Berkeley) Publications in Engineering*, 2, pp 443 (1930).

12. W. Nusselt, "Der Warmeaustausch zwischen Wand und Wasser im Rohr," *Forsch. Geb. Ingenieurwes.*, 2, pp. 309 (1931).

13. J. P. Hartnett, "Experimental Determination of the Thermal Entrance Length for the Flow of Water and of Oil in Circular Pipes," *Transaction of the ASME*, 77, pp. 1211 (1955).

14. R. W. Allen and E. R. G. Eckert, "Friction and Heat-Transfer Measurements to Turbulent Pipe Flow of Water (Pr=7 and 8) at Uniform Wall Heat Flux," *Journal of Heat Transfer*, 86, pp. 301- 310 (1964).

15. B. S. Petukhov, "Heat Transfer and Friction in Turbulent Pipe Flow with Variable Physical Properties," in J. P. Hartnett and T. F. Irvine, (eds.), *Advances in Heat Transfer*, Vol. 6, Academic Press, Inc., New York, pp. 504-564 (1970).

16. E. N. Sieder and C. E. Tate, "Heat Transfer and Pressure Drop of Liquids in Tubes," *Industrial and Engineering Chem*istry, 28, pp. 1429 (1936).

17. A. P. Colburn, "A Method of Correlating Forced Convection Heat Transfer Data and a Comparison with Fluid Friction," *AIChE Journal*, 29, pp. 174 (1933).

18. R. K. Shah and A. L. London, *Laminar Flow: Forced Convection in Ducts*, Academic Press, New York (1978).

19. S. R. Sellars, M. Tribus, and J. S. Klein, "Heat Transfer to Laminar Flow in a Round Tube or Flat Plate - The Graetz Problem Extended," *Transaction of the ASME*, 78, pp. 441-448 (1956).

20. R. H. Norris and D. D. Streid, "Laminar-Flow Heat-Transfer Coefficient for Ducts," *Transaction of the ASME*, 62, pp. 525-533 (1940).

System Hardware

21. R. A. Bowman, A. C. Mueller, and W. M. Nagle, "Mean Temperature Difference in Design," *Transaction of the ASME*, 62, pp. 283-294 (1940).

22. D. Q. Kern, *Process Heat Transfer*, McGraw-Hill Book Company, New York (1950).

23. K. Gardner and J. Taborek, "Mean Temperature Difference: A Reappraisal," *AIChE J.*, 23, pp. 770-786 (1977).

24. J. Taborek, "Evolution of Heat Exchanger Design Techniques," *Heat Transfer Engineering*, 1, pp. 15-29 (1979).

25. A. P. Fraas and M. N. Ozisik, *Heat Exchanger Design*, John Wiley & Sons Inc., New York (1965).

26. W. M. Kays and A. L. London, *Compact Heat Exchangers*, 2nd ed., McGraw-Hill Book Company, New York (1964).

27. E. U. Schlunder (ed.), *Heat Exchanger Design Handbook*, Hemisphere Publishing Corporation, Washington (1983).

28. G. F. Hewitt (ed.), *Hemisphere Handbook of Heat Exchanger Design*, Hemisphere Publishing Corporation, Washington (1989).

29. R. K. Shah, A. D. Kraus, and D. Metzger (eds.), *Compact Heat Exchangers - A Festschrift for A. L. London*, Hemisphere Publishing Corporation, Washington (1990).

30. J. P. Gupta, *Working with Heat Exchangers*, Hemisphere Publishing Corporation, Washington (1990).

31. G. Walker, *Industrial Heat Exchangers - A Basic Guide*, 2nd ed., Hemisphere Publishing Corporation, Washington (1990).

32. N. H. Afgan and E. U. Schlunder, *Heat Exchangers: Design and Theory*, McGraw-Hill, New York (1974).

33. S. Kakac, R. K. Shah, and A. E. Bergles (eds.) *Heat Exchangers: Thermal-Hydraulic Fundamentals and Design*, Hemisphere Publishing Corporation, Washington (1982).

34. *Standards of the Tubular Exchanger Manufacturers Association*, Tubular Exchanger Manufacturers Association, New York (latest edition).

35. W. Hyrnisak, *Heat Exchangers*, Academic, New York (1958).

36. D. H. Fax and R. R. Mills, Jr., "General Optimal Heat Exchanger Design," *Transaction of the ASME*, 79, pp. 653-661 (1957).

37. K. P. Singh and A. I. Soler, *Mechanical Design of Heat Exchangers and Pressure Vessel Components*, Arcturus Publishing, Inc., Cherry Hill, N. J. (1984).

38. "Spacecraft Thermal Control Design Data," Vol. 2, Section Q, European Space Agency, (1981).

39. R. W. Fox and A. T. McDonald, *Introduction to Fluid Mechanics*, 3rd ed., John Wiley & Sons, Inc., New York (1985).

40. D. R. Pitts and L. E. Sissom, *Heat Transfer*, Schaum's Outline Series, McGraw-Hill Book Company, New York, 1977.

41. J. H. Lienhard, *A Heat Transfer Textbook*, Prentice-Hall Inc., New Jersey, 1981.

42. Hydraulic Institute, *Hydraulic Institute Standards for Centrifugal, Rotary and Reciprocating Pumps*, latest edition, Cleveland, Ohio.

43. B. Nekrasov, *Hydraulic for Aeronautical Engineers*, translated from the Russian by V. Talmy, 1st ed., MIR Publishers, Moscow, Chapter VI, pp. 80-87, Chapter XII, pp. 182-221, 1969.

44. A.V. London, "Factors in the Selection of Pumps for Process and Chemical Duties," *Pumps-Pompes-Pumpen*, No. 88, pp. 25-31, 1974.

45. F. Pollak and C. O. Cruger, "Comparison of Applications and Characteristic of Positive Displacement and Centrifugal Pumps," *Pumps-Pompes-Pumpen*, No. 96, pp. 400-407, 1974.

46. G. Scobie, "Select the Pump that Meets Your Needs," *Chart. Mech. Eng.*, Vol. 21, No. 51, pp. 59-63, 1974.

47. G. S. Settles, J. T. Hamick, W. J. Barr, M. Summerfield, and M. Gunn, "Energy-Efficient Pumps Utilization," *J. Energy*, Vol. 1, No. 1, pp. 65-72, 1977.

Working Fluids

48. L. Haar and J. S. Gallagher, "Thermodynamic Properties of Ammonia," *Journal of Physical and Chemical Reference Data*, 7, pp. 635-792 (1978).

49. N. B. Vargaftik, "Tables of the Thermophysical Properties of Liquids and Gases," 2nd ed., Hemisphere Publishing Co., Washington (1975).

50. *Thermodynamic Properties of Refrigerants*, American Society of Heating, Refrigerating and Air-Conditioning Engineers (1969).

51. *ASHRAE Handbook - Fundamentals*, American Society of Heating, Refrigerating and Air-Conditioning Engineers, Inc., Atlanta, GA (latest edition).

52. R. C. Downing, "RefrigerantEquations," ASHRAE Paper 2313, *Transaction of the American Society of Heating, Refrigerating and Air-Conditioning Engineers*, 80, Part II, pp. 158 (1974).

53. E.I . DuPont de Nemours & Co., Bulletin T-11 (1972).

54. E. I. DuPont de Nemours & Co., Bulletin T-22 (1972).

55. J. J. Martin, "Thermodynamic Properties of Dichlortetrafluoromethane," *Journal of Chemical and Engineering Data*, 5, pp. 334-336 (1960).

56. J. J. Martin, "Thermodynamic Properties of Perfluorocyclobutane," *Journal of Chemical and Engineering Data*, 7, pp. 68 72 (1962).

57. E. I. DuPont de Nemours & Co., Bulletin T-502 (1969).

58. E. I. DuPont de Nemours & Co., Bulletin T-503 (1968).

59. "Spacecraft Thermal Control Design Data," Vol. 2, Section Q, European Space Agency, (1981).

Analysis of a Fluid Loop

60. "Spacecraft Thermal Control Design Data," Vol. 2, Section Q, European Space Agency, (1981).

System Analysis - Computer Software

61. "RETRAN 02 - A Program for Transient Thermal Hydraulic Analysis of Complex Fluid Flow Systems," Vol. 1: Theory and Numerics (Ref. 2), EPRI NP 1850 CCMA (1984).

62. T. M. Porsching, J. H. Murphy, and J. Redfield, "Stable Numerical Integration of Conservation Equations for Hydraulic Networks," *Nuclear Science and Engineering*, 43, pp. 218-225 (1971).

63. J. R. Turner, T. J. Swift, T. M. Andrews, and A. Lebru, "ESATAN FHTS - A Piped Fluid Network Capacity," Proceedings of the 3rd European Symposium on Space Thermal Control and Life Support Systems, Noordwijk, The Netherlands, 3-6 Oct. 1988 (ESA SP-283, December 1988).

64. "SINDA/1987/ANSI Code," User's Manual, The Aerospace Corporation, El Segundo, CA (1990).

65. "SINDA '85/FLUINT - Systems Improved Numerical Differencing Analyzer and Fluid Integrator," User's Manual, Version 2.3, Martin Marietta Denver Aerospace, Denver, CO (1990).

66. B. A. Cullimore, "FLUINT: Generalized Fluid System Analysis with SINDA '85," *AIAA Paper No. AIAA-87-1466* (1987).

Chapter V

Thermal Design Analysis

David G. Gilmore
Robert L. Collins
The Aerospace Corporation

NOMENCLATURE

A	=	area
C	=	thermal capacitance
C_p	=	specific heat
D	=	diameter
D_H	=	hydraulic diameter
E	=	voltage
F	=	radiation configuration (form) factor
G	=	thermal conductance
h	=	convective heat transfer coefficient
L	=	length or running length
I	=	current
k	=	thermal conductivity
$\dot{m}$	=	mass flow rate
n	=	arbitrary exponent
N	=	number of iterations
Q	=	heat rate
r	=	radius
R	=	resistance
t	=	time
T	=	temperature
T_∞	=	surrounding media or free stream temperature
U	=	velocity
U_∞	=	free stream velocity
V	=	volume
$\dot{w}$	=	flow rate
ws	=	sampling frequency
wc	=	maximum frequency
x	=	arbitrary distance
dT	=	temperature difference
$\mathcal{F}$	=	script F (grey body form factor)
α	=	thermal diffusivity
β	=	coefficient of volumetric expansion
ρ	=	density
σ	=	Stephen-Boltzmann constant
τ	=	stability factor
ε	=	emittance
θ	=	angle
λ	=	radiation linearization factor
δ	=	convergence criterion (relaxation criterion)
ζ	=	damping factor

Symbols, subscripts and units not specifically mentioned in the nomenclature are explained at the point of usage within the text.

INTRODUCTION

This chapter of the handbook is intended to provide an introduction to the thermal analysis/design process as applied to satellite projects. It begins with a summary overview of the overall process and then returns for a detailed discussion of how the analysis is performed and what computer codes are used and why. A descriptive narrative of the thermal analysis performed for a real program is included as an analysis example.

SATELLITE PROJECT PHASES

Before discussing the details of how a thermal analysis is performed, it may be useful to outline the phases that a typical satellite program will go through and the type of support required of the thermal engineer for each phase. We have broken the effort into Concept Definition, Validation, Full Scale Development, and Operational phases. The actual activities for each of these phases may vary from program to program, but the following discussion should give the reader a general idea of what is expected of the thermal engineer as a program matures.

The Concept Definition Phase

The concept definition phase is the first phase of a program for which engineering support is normally called upon. This phase usually consists of a customer-sponsored trade study examining the feasibility of various approaches to meet the intent of a concept. At this point, the "intent" may be little more than a qualitative description of a mission or capability that planners would like to see filled. It is normal to have the concept change during these studies in response to technical and fiscal realities, or competition from alternative systems. Usually the only constraint given to the specialists in various disciplines is the cut-off year for state-of-the-art technology projections. (This technology cut-off specifies that all required technologies are projected to be at a flight-ready level of maturity at the specified date.) The basic end products that result from these concept-definition phase studies are a parametric analysis of viable approaches, a subsystem state-of-the-art assessment, and the definition of a baseline spacecraft configuration concept.

The bulk of the thermal engineer's effort in this phase goes into defining and analyzing, parametrically, one or more alternative approaches to achieving thermal control of the spacecraft. The thermal requirements of all elements of the vehicle must be considered; housekeeping electronics, payload electronics, batteries, sensors, propulsion, antennas, etc., for all mission phases from prelaunch testing through on-orbit operations. For many of these elements it may be quickly obvious that standard thermal control techniques involving finishes, small heaters, or multilayer insulation will be sufficient. These elements are of little concern at this point in a program, and a very rough estimate of the types of finishes, weights, and heater powers based on experience with other programs is usually quite sufficient.

The real effort in the concept phase centers around elements of the thermal control system that have significant system-level impacts due to size, weight, power requirements, or development complexity. Examples of these elements include heat-pipe systems for high-capacity batteries, deployable radiators for rejecting unusually large amounts of waste heat, cooling high-power-density electronics, cryogenic coolers, etc. For each "significant" element a possible thermal-design approach may be identified, and each approach analyzed parametrically to determine the relative merits of each in terms of performance, weight, volume, radiator area, heater power, etc. For example, an IR-sensor design operating at cryogenic temperatures might be analyzed as a function of focal-plane temperature or parasitic heat-leak rates. The bottom line for these analyses is usually thermal-subsystem weight, volume, and power requirements.

There is often a temptation at this stage to jump directly to a point design rather than performing parametric analyses. This is generally unwise, however, since the input parameters upon which the thermal design is based will usually change quickly with time, and more importantly, the parametric analyses allow the design team greater insight into the impact of system requirements on the weight and complexity of the thermal control subsystem.

From the parametric analyses performed by each of the subsystem engineers, a baseline spacecraft design is synthesized by the study team. Ideally, this design represents an optimum balance between the competing requirements of different subsystems and overall system constraints such as weight, volume, reliability, and cost. The thermal engineer is responsible for specifying a baseline thermal control system design and providing a preliminary assessment of its performance capabilities and characteristics, weight, power requirements, and any relevant issues or areas of concern.

The final task during this phase is to assess the subsystem state of the art through a review of all required thermal control technologies. Most of these technologies will be mature and flight-demonstrated. Others, however, may require significant development effort before they are flight-ready. A critical input at the concept definition phase is identification of any research and development that is required to bring needed technologies to maturity, along with an assessment of what work is currently in progress in the industry or required in the future to meet program needs. Useful sources of information on the current state of the art for various technologies include The Aerospace Corporation Engineering Group and Laboratory Operations technical specialists, the Air Force Phillips Laboratory in Albuquerque, NASA, recent conference papers, and the on-line databases available through many libraries. A particularly useful way to characterize the state of the art in a given technology is the NASA 9-point scale of technical maturity, shown in Table 1.

Table 1. NASA Civil Space Technology Readiness Levels

Basic Technology Research	Level 1	Basic Principles Observed and Reported
Research to Prove Feasibility	Level 2	Technology Concept and/or Application Formulated
	Level 3	Analytical & Experimental Critical Function and/or Characteristic Proof-of-Concept
Technology Development	Level 4	Component and/or Breadboard Validation in Laboratory Environment
Technology Demonstration	Level 5	Component and/or Breadboard Validation in Relevant Environment
	Level 6	System/Subsystem Model or Prototype Demonstration in a Relevant Environment (Ground or Space)
System/Subsystem Development	Level 7	System Prototype Demonstration in a Space Environment
System Test, Launch, and Operations	Level 8	Actual System Completed and "Flight Qualified" Through Test and Demonstration (Ground or Space)
	Level 9	Actual System "Flight Proven" through Successful Mission Operations

At the onset of a Concept Definition study it is incumbent upon the technical specialist to plan the depth and breadth of support to be consistent with the study schedule, amount of funding available, and the level of effort of the other technical disciplines. Understand the overall study milestones and gear your effort accordingly. Be careful that your projected effort does not drive the study schedule. If such is the case, advise and warn the study leader in advance. If available funding is not sufficient to provide meaningful support to the study, this issue should be raised with the study-team leader. Any issues that cannot be adequately addressed due to funding or schedule limitations must be identified and documented.

The scope and detail of supporting thermal analyses should be tailored to fit the schedule and available funding. In general, a large number of parametric analysis cases with a small model are of greater value to a concept study than are detailed analyses with large models. Scaling existing designs from other programs can also be an efficient way of answering study needs without conducting time-consuming analyses and "reinventing the wheel." In other cases, back-of-the-envelope calculations using Earth heating and view factor tables and/or very simple thermal math models are sufficient to parametrically characterize a design. If, however, more extensive analyses are absolutely necessary, be sure to budget adequate time to account for machine turnaround time and re-analysis due to inadvertent errors. If large computer charges are anticipated, the study leader should be made aware of this in advance.

The type of documentation required will vary from one concept study to another. Some team leaders may expect weekly viewgraph presentations to the study team, while others may ask for periodic written reports. In any case, one should have a clear understanding of what inputs are expected and the format in which they should be presented.

As a final note, be sure that you understand and convey to the study leader your commitments and their respective priorities prior to joining the study team. Since many engineers work in matrix organizations and support more than one program, complete or dedicated support during key phases of the study may be out-prioritized. There is nothing that a program office hates more than to have a change in support personnel in the middle of a study. One way to preclude this from happening is to keep your supervisor updated such that he/she can step in if this situation arises.

The Validation Phase

Once the concept definition phase studies have been completed, the design concept, supporting trade studies, and predicted system performance are reviewed by officials and a decision is made on whether or not to proceed with the project. If a "go ahead" is given, the program enters a validation phase in which the customer team must generate a description of the system that they will ask a contractor (or contractors) to build. This is done by refining the concept phase studies and determining what technologies and capabilities can realistically be achieved given the cost and schedule constraints of the program. A system requirements review is then held to reach agreement between customer and contractor personnel as to what the top-level requirements will be. Once these are established, special studies and tests are performed to address any critical technology questions. As this effort matures, a system design review is often held to complete the validation of the system design. The end result of this phase will be a Request For Proposal (RFP), which the customer issues to industry.

Contractors who wish to bid on the program will be given an RFP package that includes basic program information such as schedules, instructions for submitting a proposal, and government points of contact. Of greater interest to the technical specialists, however, are the Statement of Work (SOW), system and subsystem design specifications, applicable specifications and compliance documents, and the Contract Data Requirements List (CDRL).

For all practical purposes, the SOW is the top-level technical document from the proposal phase onward. The basic contract that is awarded to the winning contractor at the completion of the source-selection phase (to be discussed in the next section) actually takes legal precedence over the SOW, but since it usually contains little technical information it is not of much interest to the technical specialist.

The SOW core document contains numbered paragraphs that define what the contractor shall do and under what ground rules and assumptions. The SOW is basically a list of *tasks*, such as performing thermal analyses, trade studies, tests, etc. (Specifications for the design, on the other hand, are usually contained in ancillary SOW documents, which will be discussed later.) Since the SOW is part of the legal contract, wording is extremely important and all tasks are identified and clearly specified as things the contractor *must* do.

Included in various SOW attachments and appendices will be the Contract Data Requirements List (CDRL), a list of applicable specifications and compliance documents, and system/subsystem design specifications, if any.

The CDRL is basically a list of reports, meeting minutes, Interface Control Documents (ICDs), drawings, and documentation that the contractor is required to deliver to the customer. The list is intended to include only major items of documentation and does not include the memos and small reports that will be routinely transmitted informally at the technical level.

Other common specifications and applicable documents include:

MIL-STD-1540B	"Test Requirements for Space Vehicles"
MIL-STD-24236	(Rev C) "General Specification for Metallic and Bimetallic Thermostat Switches"
ICD 2-19001	"Shuttle Orbiter/Cargo Standard Interfaces"
JSC 07700	Vol. XIV "Space Shuttle System Payload Accommodations"
ICD-A-81200	"IUS/Spacecraft ICD"
ICD-D-E0001	"STS/IUS Interface Control Document"

The final area of the SOW is the system/subsystem design specification. Specifications for the design of each subsystem, including the thermal control subsystem, may or may not be included depending upon the degree of control that the customer wishes to exercise over the contractor's design. In the event that a thermal control subsystem specification is included, the key items that the thermal engineer should expect to find in the specification are:

- A requirement to keep all component temperatures within allowable limits during all mission phases from prelaunch to end of life, considering worst-case variations in power dissipations, environments, operating modes, and contamination/degradation.

- Emphasis on the simplest, most reliable flight-demonstrated technology with no single-point failures.

- For military programs, mandatory compliance with MIL-STD-1540B. For NASA and commercial communications-satellite programs, compliance with analysis margin and test requirements.

- A design that is predictable by thermal analyses and verifiable by ground test.

- Identification of key tradeoff studies required.

- Identification of key development and final verification tests.

Proposal Evaluation

Once an RFP has been issued, any contractor who is interested may usually submit a proposal. A typical proposal consists of an executive summary, a technical proposal, a management proposal, and a cost proposal, all limited in length to a certain number of pages, as specified in the RFP.

To evaluate the technical proposal(s), a team of customer personnel is assembled, representing a wide range of technical disciplines. If you are placed on such a team, the source selection takes priority over all other assignments. Because of the competitive, and sometimes contentious, nature of the contract-award process, no information, notes, or documentation concerning the evaluations or the content of any of the offerer's proposals may be taken outside of the proposal evaluation office area or discussed with anyone who is not a member of the source-selection team. Furthermore, each evaluator is expected to provide their own technical evaluation of the proposals free from management concurrence. Your findings are not reported to your management, even if they are members of the same source-selection team. The objective here is to provide a fair, unbiased evaluation of the proposals submitted.

The proposal-evaluation process usually begins with an overview briefing to all of the evaluators by the customer program managers. From this briefing and from reading the actual RFP each evaluator must become thoroughly familiar with the RFP requirements, the evaluation criteria, and evaluation procedures. Familiarity with these is critical since each proposal must be evaluated against the RFP requirements and evaluation criteria and not against the other proposals. At no time can Contractor A's proposal be compared to Contractor B's; all proposals are evaluated separately for compliance with RFP requirements only.

The actual review of the proposal documents will occur in an office area set aside for that activity, and will be accessible only to members of the evaluation team. Although you will be asked to evaluate only these areas of the proposals related to your area of expertise (thermal control), you should read the entire technical proposal, and sometimes the executive summary, to ensure that issues affecting other subsystems are properly addressed. If the spacecraft requires large deployable radiators, for example, the impact of that on vehicle dynamics, sensor fields of view, and launch packaging should be covered in other areas of the proposal. Similarly, the need for large heaters may affect the sizing of the electrical-power subsystem.

Any thermal control issue affecting other spacecraft subsystems should be discussed with the team members evaluating these subsystems to make sure that they are aware of the impacts. To maintain propriety and provide documentation of the evaluation process, a large amount of paperwork is unfortunately required.

Evaluation of proposals must be conducted in light of the specialists' knowledge of what is required to analyze, design, test, and build a spacecraft thermal control system. Neither the SOW nor the proposal attempt to specify every detailed task that must be performed by the contractor, but the reviewer must ensure that what the contractor has written demonstrates a sound understanding of what is required to develop the thermal control system and ensure successful completion of the project, on time and on budget. An overly vague proposal that does not reflect a sound grasp of the requirements and that does not commit to specific tasks should be viewed with a great deal of concern.

The proposal should discuss the proposed thermal-design approach and testing that the contractor will do to verify the thermal control system. The proposal should also show that the contractor understands and has made a preliminary evaluation of all environments, operating modes, unusual thermal requirements, and potential areas of concern. Analysis and test schedules and estimated manpower levels should also be presented in enough detail to show that the contractor has made a realistic estimate of the thermal subsystem cost and development schedule. Particular attention should be paid to any newer exotic-technology items, such as capillary pumped loops, which may require extensive development effort.

The reviewer should also verify that all values presented by the contractor are reasonable and consistent. For instance, if the power subsystem puts out 10,000 watts and the thermal design is based on 2,000 watts of heat dissipation, there is clearly something wrong. Similarly, simple hand calculations can be performed to determine if estimated radiator sizes are approximately correct. Simple sanity checks like these are useful at this stage of a program to ensure that the contractor has a handle on the thermal control task.

All findings are documented and are used to create a final briefing to customer management, which summarizes the strong and weak points and risk assessment of the entire team. You will find that proposal-evaluation team leaders will be coming back to you for further explanation of significant issues that you have raised. It may take several conversations with a team leader before he/she understands your concerns completely, especially if the leader does not have a thermal background. Some extra patience is sometimes required at this stage, but an accurate understanding of this situation must be passed along to ensure a fair and accurate evaluation.

At the conclusion of the source selection, one or more contractors will be selected to continue into the full-scale development phase. There may be one prime contractor, two or more prime contractors who will compete until a down-selection to one prime contractor is made in the program, or two or more associate contractors developing different parts of the satellite. Generally speaking, the larger the program, the greater the number of contractors involved.

The Full-Scale Development Phase

Contract Award Through PDR

During the period from contract award through Preliminary Design Review (PDR), customer personnel establish working relationships with the contractor and begin engineering development of the spacecraft and related ground systems. The emphasis during this phase is at the system level and the type of effort is similar to that during the concept definition phase, only to a greater level of detail. The spacecraft design is still fairly flexible at this point and significant changes in configuration, payloads, and subsystem designs (including thermal) should be expected. For this reason, it is important to continue the types of trade-off studies and parametric analyses that were conducted during the concept definition phase. This assists the systems engineering staff and program management in defining an optimal baseline spacecraft design before the PDR.

At this stage, it is also important to identify any potential technical or development problems with the thermal or system designs under consideration. Thinking ahead is crucial, since it is far easier to change a design now than it is later when program "inertia" makes fundamental changes nearly impossible. You will have the greatest possible leverage on a program at this stage, and forward thinking here can save a great deal of grief and frustration in later phases of the program. If you see a possible problem now, don't put off addressing it, as you may come to regret this later.

Meetings with specialists in other spacecraft subsystems, program managers, and the customer will increase significantly. These meetings may seem like time-consuming affairs that take you away from your "real work," but it should be borne in mind that it is through these meetings that the results of your studies and analyses are funneled into the program. You may perform endless analyses and write volumes of memos, but if you do not present your results at these meetings and press for changes that you think are required, your work will tend to get lost or ignored by program managers and customer personnel who may be preoccupied with their own activities and ideas. Sometimes it takes a great deal of persistence to get your point of view to be recognized and to effect substantive changes to a program, so don't give up prematurely. Walking away from an issue and writing a "Pearl Harbor" memo may be great for "I told you so's" later on, but it doesn't do much to bring a program to a successful completion.

During this phase, key requirements must be identified and a preliminary thermal design selected and documented. Among the requirements that should be addressed are:

- Range of mission orbits
- Normal attitude(s) of satellite
- Launch phase configurations and attitudes
- Ground cooling needs
- Autonomy requirements
- Attitudes during stressed/failure modes
- Temperature limits and reliability requirements
- Equipment power dissipations and operating modes
- Thermal-distortion budgets
- Launch-system interfaces
- Interfaces with other subsystems, such as
    - Payloads
    - Propulsion
    - Attitude control
    - Electrical power
    - Structures
    - Telemetry, tracking, and command
    - Computer and data handling
- Contamination control

- Special thermal control requirements for batteries, crystal oscillators, sensors, etc.

To support the selection of a preliminary thermal design, the contractor is expected to provide results of tradeoff studies addressing issues such as the degree of active versus passive control, solid-state heater controllers versus thermostats, deployable versus fixed radiators, refrigerators versus stored cryogenics versus cryogenic radiators, hardware or material trades to minimize weight, etc. Clearly not all of these are applicable to any given satellite, but they are representative of the types of trades that are expected.

Once a preliminary thermal design is established it should be well documented prior to PDR. This documentation should provide a complete description of the design, including key assumptions, radiator areas, insulation requirements, thermal finishes and their assumed optical properties, heater sizes and locations, heat sinks, heat-pipe types, sizes, and locations, flight instrumentation list, refrigerator descriptions and power requirements, etc. In addition, any thermal development and acceptance/qualification test plans should be addressed, a schedule for all thermal analysis and testing should be presented, and any potential problem areas discussed. By the time a preliminary design is identified, analysis should exist to back up the design choices. This analysis, although preliminary in nature, should address all of the issues listed earlier and be well documented.

PDR to CDR

The period from PDR to Critical Design Review (CDR) is when most of the design and analysis work takes place. Starting from PDR itself, the design and supporting analysis and development efforts should be paced against the standards to which they will be judged at program completion. Although a greater deal of work remains to be done and substantial uncertainties may exist, the contractor should at all times be able to demonstrate to the customer that they are on the right track to deliver a high-quality product. Table 2 shows an extensive list of tasks and criteria that the contractor should eventually be able to meet. Progress against this list should be closely monitored during this period, with the goal of being able to answer all items by CDR.

During this period, close contact should be maintained between the customer and the contractor thermal people. Regular Technical Interchange Meetings (TIMs), either formal or informal, should be scheduled to discuss progress and any concerns you may have. Face-to-face meetings at the contractor's office. where data and reports are readily available, are far superior to telephone discussions and should be scheduled on a regular basis, even if the program is going smoothly.

Table 2. Thermal Control Subsystem (TCS) Evaluation Criteria

The TCS design emphasizes the following features:

- No single-point failures possible
- Reliable
- Flight-proven
- Predictable by thermal analyses
- Verifiable by ground test
- Provides adequate thermal margin
    - Passively controlled components: 11-deg C beyond worst-case predictions made by a test correlated thermal model
    - Passively controlled components where a temperature margin is not feasible: a rational, well-documented equivalent of 11-deg C
    - Actively controlled components: Control authority of at least 25 percent that can be shown to be equivalent to 11-deg C margin specified for passively controlled components
- Meets satellite life requirement
- Insensitive to the space environment
    - Vacuum
    - Natural and ultraviolet radiation
    - Contamination
    - Temperature cycling
    - Micrometeoroids and manmade debris
    - Electrostatic charge accumulation
- Insensitive to the ground and launch environment
    - Vibration
    - Acoustic Noise
    - Venting
    - Handling and Storage
    - Contamination
- Takes into account the maximum range of component power dissipations
- Considers the maximum range of orbital thermal environments
    - Operational hot orbit
    - Operational cold orbit
    - Eclipsing orbit (if different than cold)
    - On-orbit maneuvering
    - Failure mode and recovery
    - Initial outgassing orbit attitude
- Takes into account the maximum range of other important mission environments such as:
    - Prelaunch
    - Launch
    - Transfer Orbit

- TCS hardware including MLI, paints, coatings, adhesives, conductive thermal compounds, thermal shorts (braids), isolators, thermal doublers, heat pipes, fasteners, tapes, etc. meet or exceed the NASA outgassing criteria.
    - Weight loss no greater than 1.0 percent
    - CVCM less than 0.1 percent
- Autonomous
- Fault tolerant
- Allows for proper venting and outgassing by well defined paths for all spacecraft parts, subsystems, and payloads
- Can be readily integrated
- Imposes the minimum amount of operational restrictions on the satellite and launch vehicle
- Allows for growth

<u>Hardware development programs in support of the TCS are of sufficient depth and breadth to reduce the risk of not having flight-qualified hardware when needed</u>.

- Heat pipes
- Phase Change Materials (PCM)
- High capacity Constant Conductance Heat Pipes (CCHPs)
- High capacity Variable Conductance Heat Pipes (VCHPs)
- High fin effectiveness metal matrix radiators
- Heat plane materials and heat pipes for electronic equipment

<u>Completion of Key Tradeoffs</u>

- Extent of passive versus active thermal control
- Distributed versus centralized thermal control using an on-board computer
- Solid-state temperature control versus bimetallic thermostats
- Selection of location and configuration of radiators
- Selection of heat pipes
- Extent of ground cooling required
- Transfer-orbit battery requirements from upper stage

<u>Demonstration that the TCS Design Meets Requirements (Analyses)</u>

- Geometric model of selected configuration
- Thermal mathematical models
- Thermal analysis results for orbital worst-case hot and cold environments
- Thermal analysis results showing sensitivity to assumed beginning of life (BOL) and end of life (EOL) thermal properties
- Analysis results showing adequate margin for both hot and cold cases
- Documentation and substantiation of key assumptions

- Thermal analysis results for worst-case hot and cold environments corresponding to prelaunch, launch, and transfer orbit

System, Subsystem, and Component Level Testing

- Plans and commitment to perform thermal balance (T/B) testing to validate the proposed design
- Development tests
- Heat-pipe performance tests
- System-level qualification TB test

Extensive detailed thermal analysis of the spacecraft and all of its components under worst-case hot and cold conditions must be performed during this period. The description of the Hubble Space Telescope thermal design in Chapter III illustrates the number of components that must be analyzed. The report summarizing the results of the thermal analyses conducted for that program is over 500 pages long, which gives the reader some idea of the amount of analysis that may be required. By CDR, a thermal design must be firmly established and all supporting analyses and development tests of critical components completed.

CDR to Launch

The period following CDR is generally devoted to making any design changes dictated by the outcome of the CDR, conducting subsystem development tests, building the satellite, and testing it. The work during this period becomes less analytical and more hardware-oriented. Final drawings must be made by the designer and signed off by the technical specialists, thermal control system hardware must be specified in detail and manufactured or purchased, and thermal balance and thermal vacuum tests must be planned and executed (see Chapter IX). Although the level of effort for the thermal engineer is generally less then during the PDR and CDR period, there is still a great deal of work to be done. Attending to all the nut and bolt details and completing all of the required documentation and test planning can seem like an endless task.

The most important activity in this phase, and perhaps the most important single event in the entire program for the thermal engineer, is the thermal-balance test. Despite the sophistication of today's analytical techniques and the maturity of satellite thermal technology, there are frequent errors in the analysis and design of spacecraft thermal control subsystems. The thermal-balance test is the one chance the designer has to catch these errors before it is too late. In some sense, the test is even more important than all of the work that came before it, in that it may uncover a problem that would have caused the loss of a mission costing hundreds of millions of dollars if the satellite had been launched as-is. Errors large enough to represent a "fatal" flaw in the satellite design do occur in thermal control subsystems, although this is not common. A more likely outcome of the test is finding that some adjustments must be made to the design or to the mathematical models. In any event, the thermal balance test is *the* critical verification of the thermal design, and a conclusive test is to be considered mandatory to the success of the program.

The Operational Phase

Once the satellite has been built and successfully tested, it is delivered to the customer for launch and operation. During this phase, the thermal engineer will support launch rehearsals as well as the actual launch, and will typically provide an assessment of the performance of the thermal control subsystem on orbit. If there are any anomalous performances or failures in the thermal subsystem, the thermal engineer will typically be called upon to perform analyses, assessments, or tests to determine what caused the problem and what corrective action should be taken. If there are to be additional spacecraft built in the future, then design changes may need to be investigated.

Assuming that the thermal subsystem performs as expected, the biggest activity during this phase will be launch support. There is tremendous variation from program to program in the level of support expected from thermal engineers during launch. Some small programs may have no thermal support, while other programs may have 24-hour coverage by several thermal engineers for up to two weeks. Most programs will require a launch-site inspection of the satellite configuration and functional verification of commandable thermal components, such as heaters. Technical assistance is also generally required of the thermal engineer at the satellite control room from launch until the satellite is stabilized in its operational orbit and initial deployments of solar arrays, antennas, and other appendages are complete. This is generally a period lasting from a couple of days to a week, and may involve one thermal engineer who comes in for critical events or several engineers on shifts to provide around-the-clock coverage. The principle activities are monitoring temperatures, heater status, and other telemetry to insure that the thermal design is functioning normally, and also to provide recommendations for corrective action if there is a malfunction of the thermal or other vehicle subsystems.

THERMAL DESIGN/ANALYSIS PROCESS OVERVIEW

The thermal-design process is a combination of design selection and supporting analysis. The selection of a viable thermal-design approach may become almost intuitive for a thermal engineer who has worked a number of programs. Detailed thermal analyses are, however, always required to verify and refine the design. Experience minimizes the number of time-consuming analysis iterations required to close in on a final design.

A wide range of thermal control hardware and techniques are available to achieve thermal control, from simple surface finishes to complex refrigeration systems (see Chapters 4, 7, and 8). The spacecraft system requirements to minimize weight, cost, and test complexity while maximizing reliability are usually served best by keeping the thermal design as simple as possible and by avoiding the use of active components. A design that relies only on surface finishes and insulation blankets will be lighter, far less expensive to build, more reliable, and easier to test than a design involving heat pipes, louvers, or refrigerators. Therefore, although active components will sometimes be required, they should be avoided wherever possible.

Before starting the design-analysis process, the overall effort must be planned. It is important to keep in mind that the goal is to provide a reliable thermal control system at minimum cost for the satellite or component in question. All of the analysis, design, and testing activities are only tools to be used to reach that goal, and any unnecessary expenditure of time or money on these processes should be avoided. This means that the design should be no more complex than is required to do the job, the thermal math models should have the minimum number of nodes needed to verify the design, and, if any difficulties are encountered in analysis, design or testing, the engineer should ask if there may be a simpler alternate route to reach the final goal of a viable thermal control system. It is easy to get bogged down in an overly complex analysis or design and some experience is required to know what is the optimum trade between detail/complexity and practicality in the thermal engineering process.

The first step in the thermal-design process is a clear understanding of the objective(s) and any ground rules/constraints. The objective might be to develop a complete thermal control system for a new satellite, to predict temperatures for an existing satellite in a new attitude, to modify the thermal design of a component in response to changes in component design, etc. Understanding the objective may be more difficult than it would appear and may require a number of meetings with program managers and other subsystem specialists to properly understand what is required. Once an objective is established, project ground rules and constraints must also be considered, since these will effect the thermal design effort. These factors include such issues as how much manpower is available and what the priority of this project is relative to other tasks or considerations. These factors will play a major role in deciding how the effort will be structured.

Once the objectives and ground rules are understood, an approach to problem resolution must be found. The approach may be to do an analysis, perform tests, do hand calculations, adapt the thermal design from a thermally similar device, or a combination of these activities. It is important to evaluate each of these approaches to determine what is meaningful to problem solution. Some problems, for instance, may be intractable to analyze in detail and a more meaningful solution may involve simple hand calculations followed by a good test. Identifying the approach must also involve consideration of schedule, budget, and any risks, such as reliance on new or unproven technologies or analysis software.

Once a technically sound approach to the design effort is established, a preliminary schedule and cost estimate should be made. The engineer should put together an outline of tasks required to support the job, which should include major milestones, criteria for determining if objectives are met, manpower level, and a clear definition of what is expected from whom and when. While this outline may adequately be handled mentally for a simple task, it will quickly grow to memo size for even a small thermal-design effort. This top-level plan is invaluable in keeping the effort on track as the process unfolds. Milestones for a typical design analysis might be:

- Establish thermal design requirement/constraints
- Obtain system design information/details
- Determine heat dissipation and weights
- Identify environmental and boundary conditions
- Define operating modes
- Identify any long lead time materials or components
- Identify development test requirements
- Form matrix of computer runs to be made
- Math model definition (number and nocation of nodes)
- Obtain thermal property data
- Construct math model(s)
- Debug model
- Make production runs (number of runs, cost)
- Conduct development tests
- Data reduction/review
- Finalize design
- Design analysis documentation/presentation
- Test planning

- Test support
- Test data evaluation/design changes
- Test documentation

If the estimated time and cost for the effort is not consistent with program requirements, the thermal engineer must either find a simpler/lower cost approach to doing the thermal design, or renegotiate funding or schedule. Planning up front and fighting any necessary battles with program management early will avoid immense headaches later in the project. As the effort unfolds, cost and schedule should be monitored regularly for measuring performance and any program slips or changes should be incorporated into the thermal plan.

Once a plan that meets technical and program requirements is established, the work of performing the design analysis begins. The first step here is to establish working relationships with all of the individuals who provide needed inputs or receive results. These will typically be the lead engineers responsible for the other spacecraft subsystems and payloads, such as Propulsion, Battery, Payloads, Attitude Control, etc. It is important to coordinate with these individuals to establish objectives, understand requirements of their subsystems, determine impacts on the thermal design, etc. Failure to communicate regularly may result in wasted time analyzing an out-of-date design.

In preparation for the design effort, a fair amount of data and information about the system must be gathered. This data would typically include drawings and sketches of the hardware, estimated heat dissipation and weights of components, definition of orbit and attitude, understanding of thermal environments from prelaunch through end-of-life, operating modes of the spacecraft, and thermal property data for materials that may be used. This information is needed to identify a preliminary thermal-design approach and to construct the thermal math models.

Before the thermal analysis can begin, a thermal-design approach must be identified. This is usually done by a combination of experience and simple hand calculations to determine if a given approach is viable (this will be discussed in more detail later). It is important at this point to consider all factors including cost, practicality, analyzability, reliability, and testability.

Using the data and design approach discussed above, the analyst constructs the thermal models, which consist of a geometric math model (GMM) for calculating radiation interchange factors, and a thermal math model (TMM) for predicting temperatures. The GMM is a mathematical representation of the physical surfaces of the satellite or component and is used to calculate the radiation couplings between all surfaces in the model, as well as heating rates to each surface from external flux sources such as solar, Earth IR, and albedo radiation. The TMM is usually a lumped-parameter representation of the thermal capacitance of each node and thermal conduction terms between nodes, and is directly analogous to an electrical RC network. These models are constructed using a combination of computer-aided design (CAD) technologies and hand calculations, and will be discussed in detail later in this chapter.

The completed (and debugged) thermal model is run to predict hardware temperatures under worst-case hot and cold conditions. A number of runs may be required to determine what exactly is the worst-case combination of factors, such as orbit beta angle, operating mode, vehicle attitude, surface properties, etc., and a number of parametric runs may be required to close-in on optimum sizing of radiators, heaters, and so on. In addition, many analyses will have to be rerun to reflect design changes or updates to analysis inputs, such as box-power dissipations, that will occur as the vehicle design matures. Periodic review meetings with management and other program personnel are required to ensure that the analysis reflects the current system design and will provide the results needed for other subsystem design efforts. Peer review is also a good way of uncovering the errors that inevitably occur in any analysis before they can do any harm.

The final and sometimes most-tedious step is documentation. The thermal design analysis report(s) should include a complete description of the final thermal design, an in-depth discussion of all significant math model inputs and assumptions, a listing of the thermal models (in an attachment), predicted temperatures and margins for all components and heater powers for worst-case conditions and operating modes, and a discussion of any significant concerns or recommendations. In preparing such a report it is important to first review all of the work performed and to have a good grasp of it. A critical appraisal of all results to ensure that they are valid, complete, and consistent must be made. The report itself must be written to the needs of those to whom it is addressed, as well as to provide a record for the future reference of the analyst. The report should conclude with a concise summary of why the task was done, how it was done, what was found, and what should be done as a result.

FUNDAMENTALS OF THERMAL MODELING[1]

Thermal Math Modeling as a Cognitive Process

A brief introduction to the rudimentary techniques of thermal modeling, coupled with a simple understanding of the various basic heat-transfer

1 This section reprinted from the "Thermal Network Modeling Handbook" prepared by TRW under NASA contract 9-10435.

mechanisms, are the prerequisites for a beginning thermal math modeler. Properly applied, the body of concepts, principles, and techniques applicable to thermal math modeling constitutes a valid engineering tool that can be applied to the solution of real engineering problems. A good lumped-parameter representation of a thermal system requires, in addition to the basic principles and techniques, an elusive mixture of experience (with real systems, both physical and model) and engineering judgment to transfer the end product into an accurate, versatile, and cost effective thermal math model.

Generally, the problems encountered in developing a thermal math model reduce to an overall object of achieving the greatest accuracy for the least cost. Cost factors are rather well defined, and fall into two classes: (1) the cost of developing the model, and (2) the cost of using the model. Development costs can be based almost solely on the actual engineering manpower required to do the job within the constraints of time and budget. However, the potential costs involved in using a model are often not as obvious nor as linear.

The problem of achieving accuracy in a math model, while subject to cost constraints, varies greatly from one thermal math model to another. General accuracy requirements may be as straightforward as "temperature accuracy shall be compatible with thermocouple A/D converter quantization error." On the other hand, accuracy levels might be indirectly indicated by requiring that a model must "be sufficiently detailed to permit meaningful parametric analyses with respect to insulation thickness variations in increments of 1/4 inch." Clearly, there is going to be a great deal of engineering judgment involved in developing a model that is "sufficiently detailed" to be "meaningful."

Succeeding sections of this report will present many of the basic principles and techniques involved in thermal math modeling. Experience, of course, can only be acquired from hands-on familiarity with real thermal systems and participation in the modeling and analysis thereof. Engineering judgment can probably be described more accurately as the result of abstracting from the body of unique familiar information, a general understanding that can be extended to guide the investigation and comprehension of new and unfamiliar areas. As such, engineering judgment cannot be presented in a table, or a figure, or even an entire book.

Network Solution

Two systems are said to be analogous when they have similar equations and boundary conditions, and the equations describing the behavior of one system can be transformed into the equations for the other by simply changing symbols of the variables. Thermal and electrical systems, are two such analogous systems, as shown in Table 3.

Table 3. Thermal-Electrical System Analogy

Quantity	Thermal System	Electrical System
Potential	T	E
Flow	$\dot{Q}$	I
Resistance	R	R
Conductance	G	$\frac{1}{R}$
Capacitance	C	C
Ohm's Law	$\dot{Q} = GT$	$I = \frac{E}{R}$

The analogy between thermal and electrical systems allows the engineer to utilize the widely known basic laws such as Ohm's Law and Kirchhoff's Laws which is used for balancing networks. Numerical techniques used to solve the partial differential equations describing such systems have been conveniently adapted to computer solutions, thus enabling the engineer to readily compute temperature distributions and gradients of complex physical thermal networks.

Thermal-analyzer computer programs have been developed that require the user to define a thermal network of the system analogous to an electrical circuit. The network components are input into the computer and preprogrammed routines perform the transient or steady state solutions. This section discusses the development of a thermal-network and the numerical techniques for solving this network.

Nodes

In order to develop a thermal network and apply numerical techniques to its solution, it is necessary to subdivide the thermal system into a number of finite subvolumes called nodes. The thermal properties of each node are considered to be concentrated at the central nodal point of each subvolume. Each node represents two thermal-network elements, a temperature (potential) and a capacitance (thermal mass) as shown in Figure 1.

The temperature, T, assigned to a node represents the average mass temperature of the subvolume. The capacitance, C, assigned to a node is computed from the thermophysical properties of the subvolume material evaluated at the temperature of the node, and is assumed to be concentrated at the nodal center of the subvolume. Because a node represents a lumping of parameters to a single point in space, the temperature distribution through the subvolume implied by the nodal temperature is linear, as shown in Figure 2c, and not a step function, as illustrated in Figure 2b.

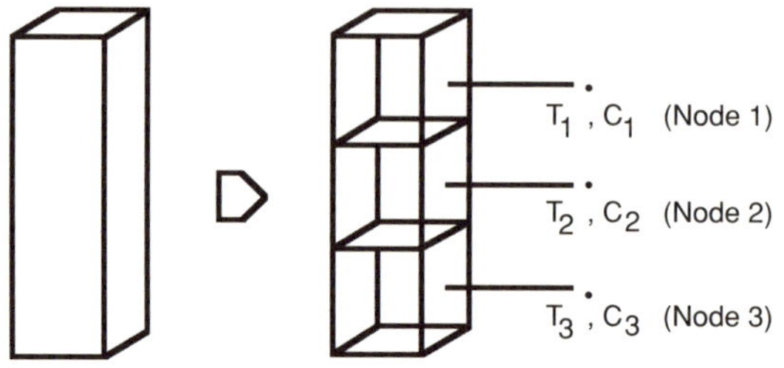

Figure 1. Nodalization

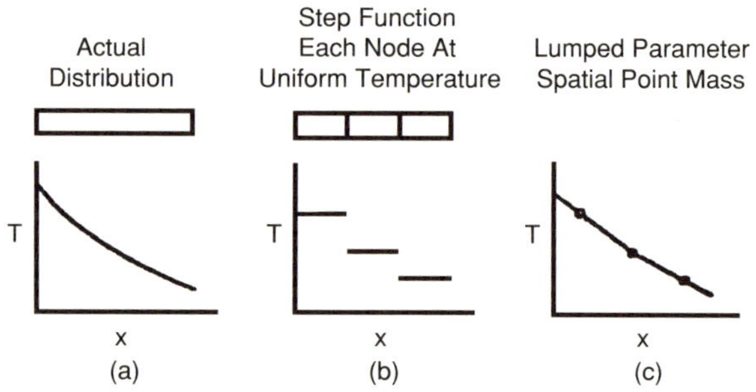

Figure 2. Temperature distributions

In a homogeneous material, the temperature at a point other than the nodal point may be approximated by interpolation between adjacent nodal points where the temperatures are known.

The error introduced by dividing a system into finite-sized nodes, rather than volume dx^3 where dx approaches zero, is dependent on numerous considerations: material thermal properties, boundary conditions, node size, node center placement, and time increment of transient calculations. The techniques for proper nodalization to minimize the error will be discussed in a later section.

To this point only nodes that represent subvolumes with a finite thermal mass (capacitance) have been discussed. In many instances, two other types of nodes are required to define a thermal network. They are nodes having a zero capacitance or an infinite capacitance. Thermal analyzers such as SINDA usually give the three types of nodes particular names, as follows:

Nodal Type	Name
Finite Thermal Mass	Diffusion
Zero Thermal Mass	Arithmetic
Infinite Thermal Mass	Boundary

The diffusion node has a finite capacitance. It is used to represent normal material, the temperature of which can change due to heat flow in or out of the nodes. It is characterized by a gain or loss of potential energy, which depends on the capacitance value, the net heat flow into the node, and the time over which the heat is flowing. Mathematically, a diffusion node is defined by the expression:

$$\Sigma\dot{Q} - \frac{C\Delta T}{t} = 0.$$

The arithmetic node (zero capacitance) is a physically unreal quantity; however, its effective use with numerical solutions can often be helpful in interpreting results in such applications as surface temperatures, bondline temperatures, and node-coupling temperatures. It also finds use in representing thermal system elements that have small capacitance values in comparison to the large majority of the other nodes in the system, which results in computer-run-time reduction with minor changes in overall accuracy. Examples of these could be small components such as bolts, films, or fillets; gaseous contents of small ducts or tubes; and low-mass insulations. Arithmetic nodes should be few in number when contrasted to the total number of nodes in the network. The temperature of an arithmetic node responds instantaneously to its surroundings. Mathematically, an arithmetic node is defined by the expression:

$$\Sigma\dot{Q} = 0.$$

The boundary node (infinite capacitance) is used to represent a boundary or sink whose temperature is set and will not change no matter how much heat flows into or out of it from other nodes in the model. Common uses are: deep-space sink temperature, recovery temperature, and lunar-surface temperature. In addition, a boundary node may be used to represent thermal-system components such as the bulk propellant in a large tank that has a very large thermal mass (capacitance) relative to the other nodes. Mathematically a boundary node is defined as:

$$T = \text{Constant}.$$

The placement of the diffusion-node centers and the choice of node shapes is dependent on several factors: the points where temperatures are desired, the expected temperature distribution, physical reasonableness, and the ease of computation. The actual size of the node is dependent on other considerations:

accuracy desired, structural design, computer storage capabilities, and computer time required. Each factor, however, embodies other considerations. For example, to anticipate the expected temperature distribution one must draw heavily on engineering judgment as to the effects of the expected boundary conditions and associated material properties.

In general, the shape of a diffusion node is chosen to be a simple geometric figure, the areas and volumes of which can be easily calculated. Irregularly shaped structural members may be approximated by simple shapes by employing assumptions that are consistent with the desired results. In some cases, nodal divisions are decided first, with the node-center locations thus defined as a consequence. In these cases, nodal edges will usually lie along structural edges, and structural members will be divided in a symmetric and equal fashion. In other cases, output requirements will dictate the locations of node centers, with the nodal edges assigned as a consequence. These two approaches are illustrated in Figure 3. In the first case, Figure 3(a), it is desired to prepare a general model of the structure, but in the second case, Figure 3(b), it is desired to model the response of the two thermocouples located on the bond line between the two members.

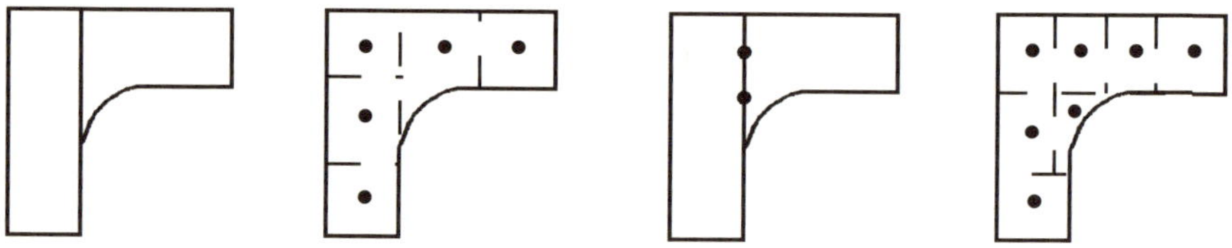

(a) General Node Boundaries

(b) Node Centers at Thermocouple Locations

Figure 3. Alternate nodalization methods

The above example alludes to a general desirability for rectangularly shaped nodes. This is true for the simple reason that it is easy to compute the areas and volumes required for the input calculations. Such simple nodal shapes are in keeping with current engineering practice. By contrast, Dusinberre suggested that nodalization be performed in such a manner that the paths of heat flow assume a triangular pattern, as shown in Figure 4(a). The only drawback to this theoretically sound approach is that it remains for the engineer to compute the volumes of the irregular polygonal nodes that are the consequence of such a tact, as shown in Figure 4(b).

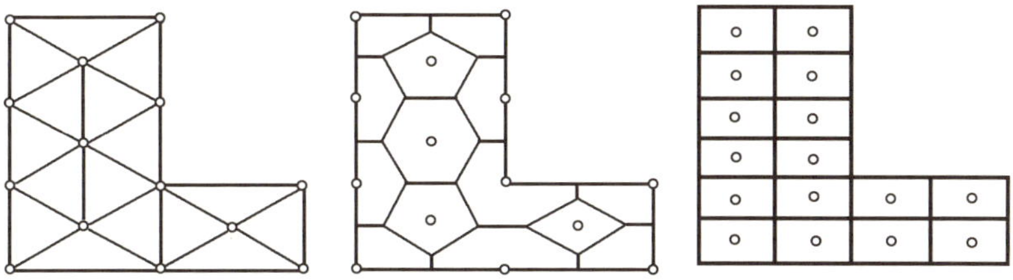

(a) Triangular Heat Flow Paths (b) Nodes Shapes Resulting from (a) (c) Simple Rectangular Nodalization

Figure 4. Polygonal nodalization vs rectangular nodalization

Note how much simpler the rectangular nodalization approach is, as indicated in Figure 4(c). As might be expected, to achieve the same simplicity of calculation, circular structures are nodalized in pie-shaped wedges, annular rings, or a combination of the two, as shown in Figure 5.

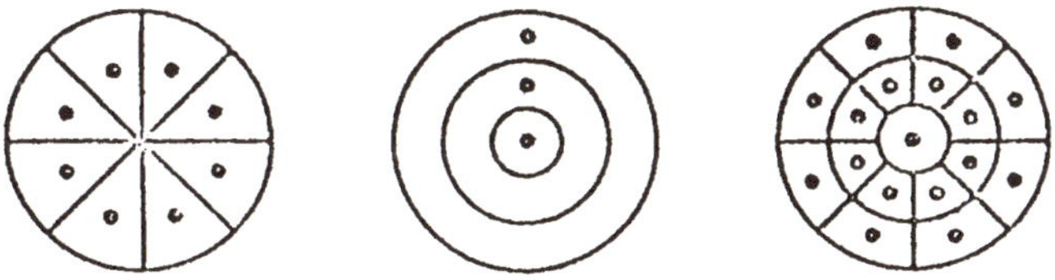

(a) (b) (c)

Figure 5. Nodalizaton of circular elements

Boundary nodes are used to define points, lines, or surfaces of constant temperature in one- two-, or three-dimensional models, respectively. The physical location of a boundary node is determined solely by the conduction paths connected to it. A single boundary node may be used to model *all* boundaries that are at the same temperature. This point is illustrated in Figure 6, which shows that the indicated boundary node will suffice as a model of the entire, constant-temperature edge of the structure (in this case, at 100-deg F).

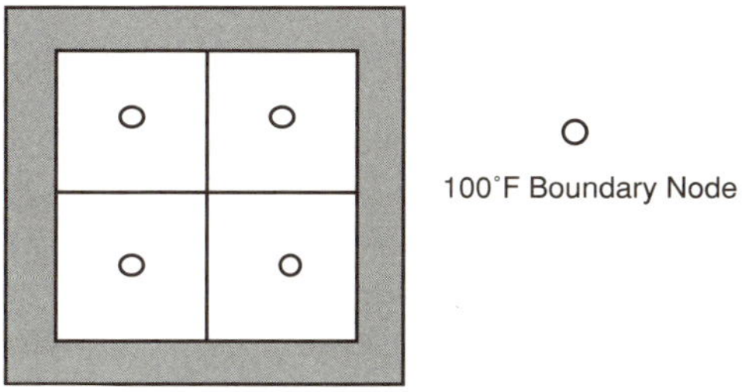

Figure 6. Sample boundary node

Arithmetic nodes have a number of uses that are consequences of the fact that such nodes serve as an engineering model of the proverbial "wafer of thickness dx, where dx approaches zero." A typical application lies in the modeling of exterior surfaces of reentry vehicles, which are often subjected to severe, rapidly changing boundary conditions. In the physical system, the surface temperature remains very close to radiation equilibrium with the surface heating rate, indicating that this system can be accurately simulated by the use of a surface arithmetic node. This application is illustrated in Figure 7.

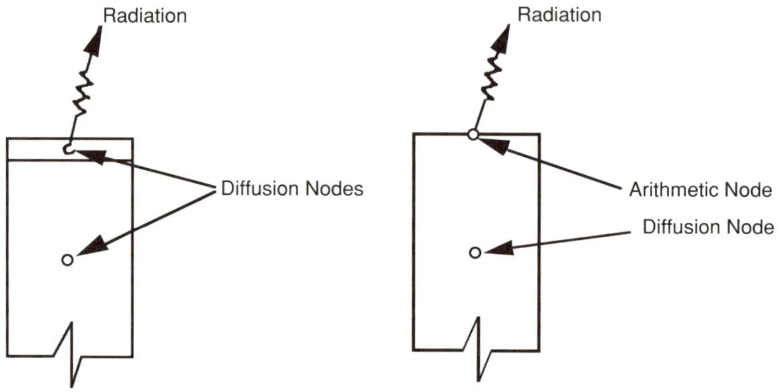

Figure 7. Use of arithmetic nodes to model surfaces

The case where heat flows from a surface by conduction is usually one in which two structures are bonded together and a bondline temperature is sought. When the structures are homogeneous, a bondline temperature may be established by simple linear interpolation between the nearest node centers. When the materials are dissimilar, it is more appropriate to use an arithmetic node at the bondline, leaving to the computer the process of performing a conductance-weighted average of the adjoining diffusion node temperature, which, in essence, is the result of finding the steady-state (heat in = heat out) temperature for an arithmetic node.

Arithmetic nodes may also be used advantageously in place of diffusion nodes that have a capacitance that is small when compared to great majority of nodes in the system. This often occurs when modeling a small quantity of gas in a tube or other enclosure, or when modeling small structural parts, such as wires, bolts, fillets, films, and sheets, where detailed temperatures are desired (which preclude lumping such items along with larger nearby nodes). The correct use of arithmetic nodes in these cases generally results in a considerable saving of computer time when the model is processed.

In developing a thermal network, computations with respect to nodes are generally limited to calculating the capacitance of diffusion nodes. The following formula is used:

$$C = \rho * V * C_p,$$

where

C	=	thermal capacitance - Btu/°F
ρ	=	density - LB/FT^3
V	=	volume - FT^3
C_p	=	specific heat - Btu/LB°F

The specific heat (C_p) and the density (ρ) of materials may vary with temperature. The necessity to utilize temperature-dependent properties for analysis depends on the degree with which the properties vary and the temperature range over which the capacitance of the material will be calculated. Most thermal analyzers can accommodate temperature-varying thermal properties.

The use of arithmetic nodes may also require some computations. Replacement of small-capacitance diffusion nodes with an arithmetic node must be preceded by computations to verify that the capacitance-conductor effects are such that the node in question will essentially reach steady-state temperatures during the time step required by the larger nodes. The use of an arithmetic node to predict surface temperatures where surface radiation or very high heating rates are involved requires careful analysis to ensure the stability of the arithmetic node. Stability criteria and solution techniques are discussed later. From this section, it can be seen that solution techniques using linearized "last-pass" temperature values may require the use of analyzer control constants to restrict the maximum node temperature change or computation time step. The engineer must further be cautioned against using coupled arithmetic nodes without a complete understanding of the implications and required analyzer control constants used to ensure a valid solution.

Conductors

Conductors are the thermal-math-modeling network elements that are used to represent the heat-flow paths through which energy is transferred from one node to another node. Figure 8 illustrates the element node temperatures (T), capacitances (C), and conductors (G) that comprise a thermal network.

The three processes by which heat flows from a region of higher temperature to a region of lower temperature are conduction, convection, and radiation. Conduction is the process by which heat flows within a medium or between different mediums in direct physical contact. The energy is transmitted by molecular communication. Figure 9 illustrates conduction conductors.

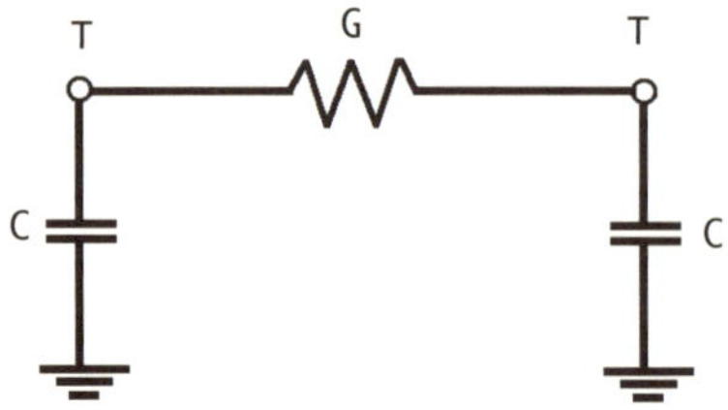

Figure 8. Thermal network elements

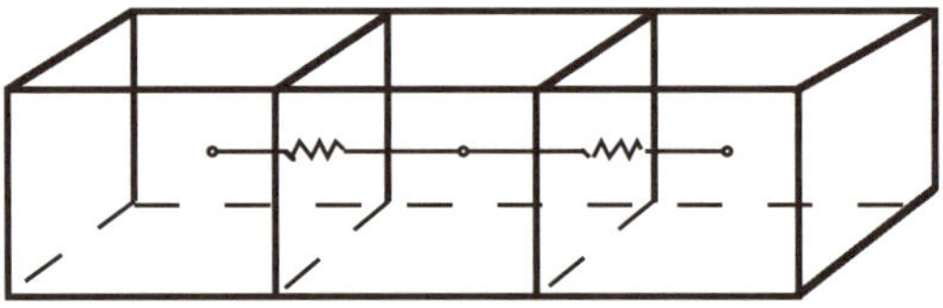

Figure 9. Conduction conductor

Convection is the process of energy transport by combined action of heat conduction, energy storage, and mixing motion. Heat will flow by conduction from a surface to adjacent particles of fluid; then the fluid particles will move to a region of lower temperature where they will mix with, and transfer a part of their energy to, other fluid particles. The energy is actually stored in the fluid particles and is carried as a result of their mass motion. Figure 10 illustrates the convection conductor.

Figure 10. Convection conductor

Conductors that represent conduction or convection paths are referred to as linear conductors because the heat-flow rate is a function of the temperature difference between nodal temperatures to the first power.

$$\dot{Q} = G_{ij} (T_i - T_j).$$

Radiation is the process by which heat flows between two bodies when the bodies are separated in space. Energy is transferred through electromagetic wave phenomena. Radiation conductors are termed non-linear because the heat flow between two surfaces by radiation is a function of the difference of the fourth powers of the surface temperatures:

$$\dot{Q} = G_{ij} (T_i^4 - T_j^4).$$

Figure 11 illustrates the radiation conductor.

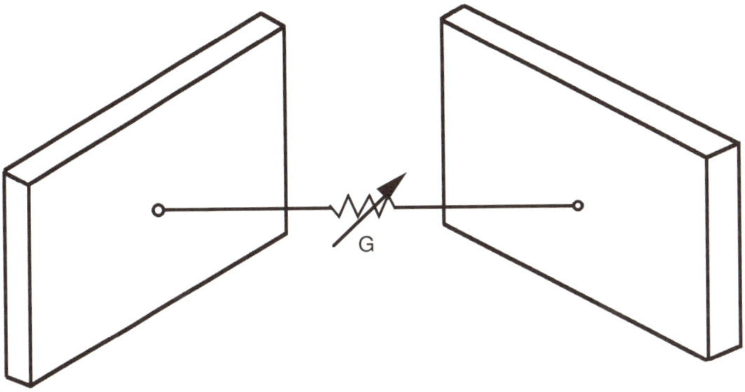

Figure 11. Radiation conductor

Fluid-flow thermal systems may also be simulated by thermal modeling. Energy stored in the thermal mass (capacitance) of a fluid lump (node) is transferred from one point to another by the movement of the fluid mass. This type of conductor is generally referred to as a one-way or mass-flow conductor, and is illustrated in Figure 12. The mass-flow conductor is linear and actually asymmetric, since upstream nodes are unaffected by what happens downstream.

$$\dot{Q} = G_{ij} (T_i - T_j)$$

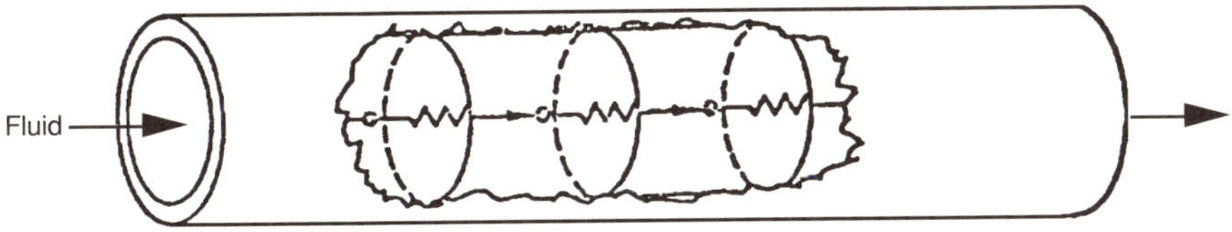

Figure 12. Mass-flow conductors

Computational Methods - Conduction Conductors

Conduction conductors for rectangular nodes are computed from the equation:

$$G = \frac{kA}{L},$$

where

G	=	thermal conductance - Btu/HR-°F
k	=	thermal conductivity - Btu/HR-FT-°F
A	=	cross sectional area through which heat flows - FT^2
L	=	length between adjoining nodes - FT

(The English units are shown, but other consistent units could be used.)

The thermal conductivity (k) of materials may vary with temperature or other influencing factors within the system; the crossectional area through which the heat flows (A) and length between node centers (L) are determined by the size and shape of the adjoining nodes. As with the capacitance calculations, the necessity to use temperature-dependent properties depends on the degree with which the conductivity changes over the temperature range expected during the analysis.

Computational Methods - Rectangular Nodes

The length, L, of the heat-flow path, used for conduction-conductance calculations for rectangular nodes, is the distance between node centers, and the area, A, to be used is the crossectional area perpendicular to the line joining the node centers. The convention is depicted in Figure 13.

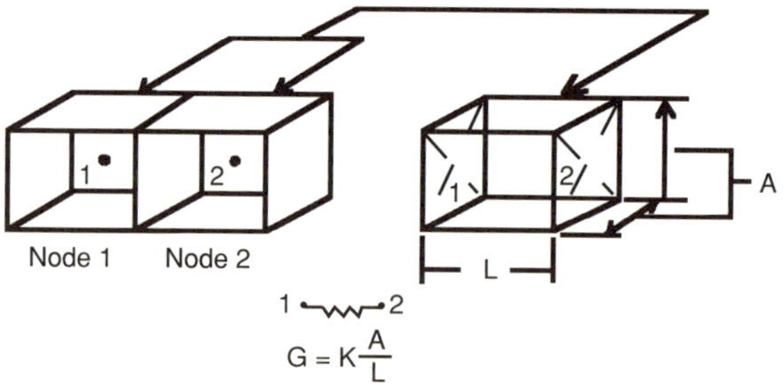

Figure 13. Simple conductor representing a heat-flow path through material

Computational Methods - Circular Sections

For conductors between nodes that are circular sections, the conventions shown in Figure 14 should be used:

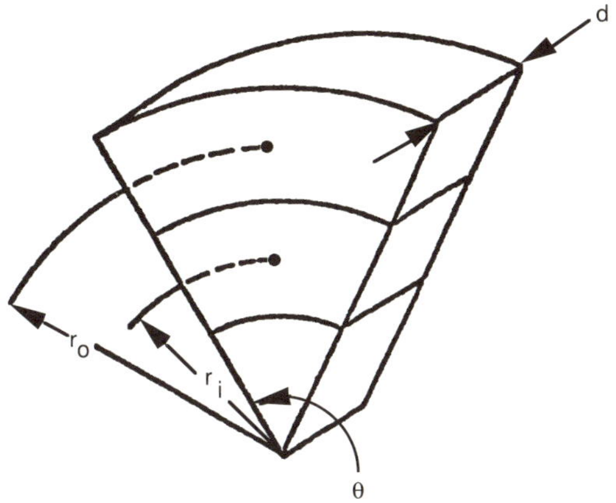

$$\frac{A}{L} = \frac{\theta d}{\ln(r_o/r_i)},$$

where

θ = radians
d = FT
r_o = consistent units with r_i
r_i = consistent units with r_o

Figure 14. Area and length equivalents for circular section nodes

Two or more parallel conduction paths between nodes may be summed to create one conductor value by the following equation:

$$G_T = G_1 + G_2 + \ldots G_n.$$

This may be helpful in computing an equivalent conductor between two nodes, as illustrated in Figure 15.

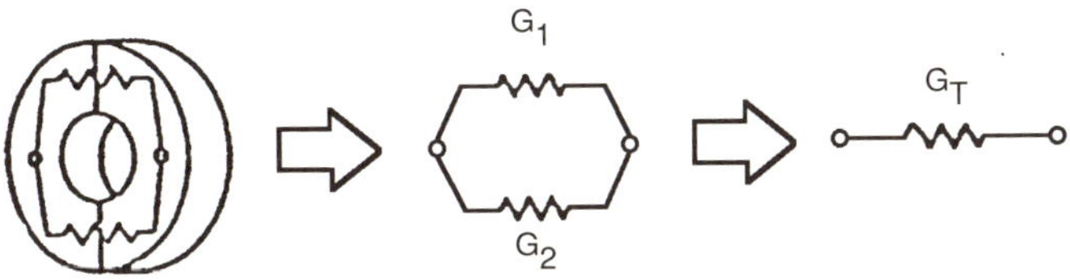

Figure 15. Parallel conductor flow paths

Computation Methods - Series Conductors

Two or more series conduction paths between nodes may be combined to create one conductor value by the following equation:

$$\frac{1}{G_T} = \frac{1}{G_1} + \frac{1}{G_2} + \ldots \qquad G_T = \frac{1}{\frac{1}{G_1} + \frac{1}{G_2} + \ldots \frac{1}{G_n}} .$$

This may be helpful in computing the conductors between two dissimilar-shaped or dissimilar-material nodes, as shown in Figure 16.

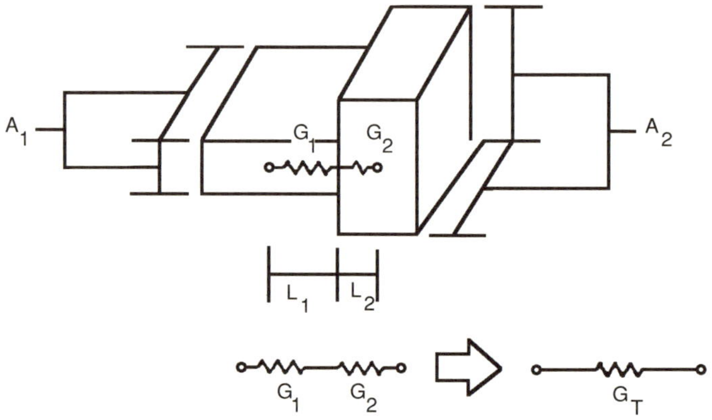

$$G_T = \frac{1}{\frac{1}{G_1} + \frac{1}{G_2}} = \frac{G_1 G_2}{G_1 + G_2}$$

Figure 16. Series conductor paths

Computational Methods - Convection Conductors

Convection conductors are computed from the expression:

$$G = hA,$$

where

G	=	thermal conductance - Btu/HR-°F
h	=	convective heat transfer coefficient - Btu/HR-FT^2-°F
A	=	surface area in contact with the fluid - FT^2

(Again, English units are used as an example.)

G is the product of the average-unit thermal convective conductance h (convective heat-transfer or film coefficient) and the nodal surface area A in contact with the fluid. However, h is a complicated function of fluid flow, the thermal properties of the fluid medium, and the geometry of the system.

Since the convective process of heat transfer is so closely linked to fluid motion, it is first required to establish whether the fluid flow is laminar or turbulent. In *laminar flow,* the fluid moves in layers and the fluid particles follow a smooth and continuous path. Heat is transferred only by molecular conduction within the fluid as well as at the interface between the fluid and the surface. In *turbulent flow,* the path of the fluid particles is irregular, and although the general trend of the motion is in one direction, eddies or mixing currents exist. In addition to the conduction mechanism being modified, increased heat transfer occurs in turbulent flow when energy is carried by fluid particles across flow streamlines and mixes with other particles of the fluid.

In addition to knowing whether the fluid motion is laminar or turbulent, it is necessary to know the process by which the motion was induced. When the heat flows between the fluid and the surface as a result of fluid motion caused by differences in fluid density resulting from temperature gradients in the fluid, the heat transfer mechanism is called *free* or *natural convection.* When the motion is caused by some external agency, such as a pump or blower, the heat transfer mechanism is called *forced convection.*

The following table illustrates typical values of average heat-transfer coefficients encountered in engineering practice.

Table 4. Order of Magnitude of Convective Heat Transfer Coefficients

Convective Medium	Convective Heat Transfer Coefficient h - Btu/HR-FT^2-deg F
Air, Free Convection	.2-2.
Air, Forced Convection	5-50
Oil, Forced Convection	10-30
Water, Forced Convection	50-2000
Water, Boiling	500-10000
Steam, Condensing	1000-20000

It should be remembered that the predicted values for h are only approximate. The accuracy of the heat-transfer coefficient calculated from any available equation or graph may be no better than 30 percent.

Computational Methods - Radiation Conductors

Most thermal analyzer computer programs linearize the radiation term prior to performing the heat balance at each time step. This operation simply amounts to factoring $(T_i^4 - T_j^4)$ into the following components: $(T_i^3 + T_i T_j^2 + T_i^2 T_j + T_j^3)(T_i - T_j)$, the term $(T_i^3 + T_i T_j^2 + T_i^2 T_j + T_j^3)$ is evaluated by the computer each time-step using the current values of T_i and T_j. This quantity is then multiplied by the input value of the radiation conductor, thus reducing the radiation equation to a linear form. The thermal engineer need only be concerned with the input value of the radiation conductor, which takes the following form:

$G_{ij} = \sigma \varepsilon_i F_{i-j} A_i$ for radiation to a black body
$G_{ij} = \sigma \mathcal{F}_{i-j} A_i$ for radiation between grey surfaces,

where

G_{ij}	=	input value for radiation conductors - Btu/HR-°R^4
σ	=	Stephan-Boltzmann constant = $.1713 \times 10^{-8}$ - Btu/HR-FT^2-°R^4
ε_i	=	emittance of surface i - dimensionless
F_{i-j}	=	geometric configuration factor from surface i to surface j - dimensionless
A_i	=	area of surface i - FT^2
$\mathcal{F}_{i-j}$	=	grey body radiation factor - dimensionless

(Again, other units could be used as long as they are consistent.)

The emittance, ε, is a measure of how well a body can radiate energy as compared with a black body. Emittance is the ratio of the total emissive power of a real surface at temperature T to the total emissive power of a black surface at the same temperature. The emittances of various surfaces are a function of the material, surface condition, and temperature of a body. The surface of a body, and therefore the emittance, may be altered by polishing, roughing, painting, etc. The values of ε for most common materials and surface conditions have been measured at various temperatures and are presented in Chapter IV, Appendix A, and in many reference manuals. It is left to the engineer to determine the value of emittance to be used and whether the variation of ε with temperature is significant over the temperature range expected for the surface.

The geometric-shape (configuration) factor from surface i to surface j, F_{i-j}, is the fraction of total radiated energy from surface i that is directly incident on surface j, assuming surface i is emitting energy diffusely. F_{j-i} would be the fraction of total radiant energy from surface j that is intercepted by surface i. The configuration factors for finite regions of diffuse areas are related by

$$A_i F_{i-j} = A_j F_{j-i} \ .$$

The configuration factor, F_{i-j}, is a function of the geometry of the system only. Several computer programs have been developed to compute the shape factors between surfaces with complex geometries, and will be discussed later. Form factors between some surfaces with simple geometries, can be hand computed. Hand-calculated view factors can be used for preliminary analysis or to check the results of view factors generated by computer programs.

Ref. 10 presents configuration factors for various simple geometries. The use of these figures and configuration-factor algebra will allow the engineer to determine form factors for many simple radiation problems.

The gray-body shape factor $\mathcal{F}_{i-j}$ is the product of the geometric shape factor F_{i-j} and a factor that allows for the departure of the surface from black-body conditions. For radiation enclosures, the $\mathcal{F}_{i-j}$ factors are generally evaluated with a computer program. The input for the program is the $A_i F_{i-j}$ values for every surface of the enclosure to every other surface and the emittance and area for each surface. Simplified equations for $\mathcal{F}_{i-j}$ exist for two-component gray enclosures.

Infinite parallel flat plates: $F_{1-2} = F_{2-1} = 1$

$$\mathcal{F}_{i-j} = \frac{1}{\left(\frac{1}{\varepsilon_1} + \frac{1}{\varepsilon_2} - 1\right)} \ .$$

Concentric cylinders of infinite height or concentric spheres:

$$F_{1-2} = 1, F_{2-1} \neq 0$$

$$\mathcal{F}_{1-2} = \frac{1}{\frac{1}{\varepsilon_1} + \frac{A_1}{A_2}\left(\frac{1}{\varepsilon_2} - 1\right)} \ .$$

For "non-enclosed" surfaces, an effective emittance, ε_{eff}, between the surfaces may be used to compute the gray-body form factor with the following equation:

$$\mathcal{F}_{i\text{-}j} = \varepsilon_{eff\ i}\, F_{i\text{-}j} \quad .$$

The effective emittance is a function of the emittances of the two surfaces and the configuration factors (F) between them. The error induced with use of ε_{eff} is the result of neglecting secondary reflections from surfaces other than the two for which the effective emittance was determined. By reducing Hottel's method for two flat surfaces with emissivities of ε_1, ε_2, in a black enclosure, the following equation can be constructed:

$$\varepsilon_{eff} = \frac{\varepsilon_1\, \varepsilon_2}{1 - F_{1-2}\, F_{2-1}\,(1-\varepsilon_1)(1-\varepsilon_2)} \quad .$$

The following examples of configuration factor algebra should be helpful:

$A_1F_{1\text{-}3} = A_3F_{3\text{-}1}$
$A_1F_{1\text{-}34} = A_1F_{1\text{-}3} + A_1F_{1\text{-}4}$
$A_{12}F_{12\text{-}34} = A_1\, F_{1\text{-}34} + A_2F_{2\text{-}34}$
$A_{12}F_{12\text{-}34} = A_1F_{1\text{-}3} + A_1F_{1\text{-}4} + A_2F_{2\text{-}3} + A_2F_{2\text{-}4}$
$A_1F_{1\text{-}4} = A_3F_{3\text{-}2}$ (symmetrically positioned)

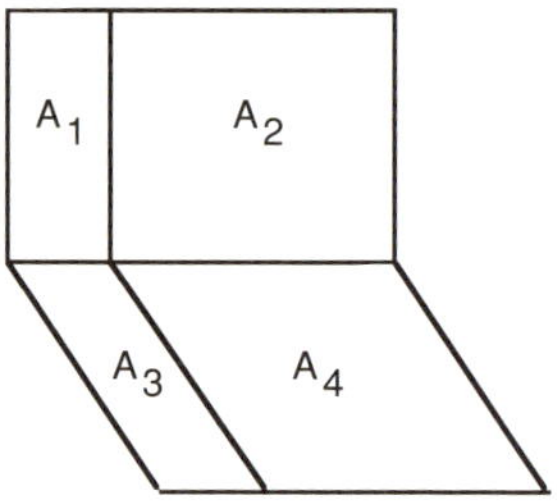

Energy Sources Or Sinks

Energy sources or sinks, Q, are modeling elements that allow the impression of positive or negative heating rates on the nodes of a thermal network, independent of conductor paths to the node.

Common engineering applications of heat sources in thermal models are:

- Solar and planetary heating
- Aerodynamic heating
- Avionic cold plate heat loads
- Change-of-state latent energy
- Thermal control heaters

Common applications for heat sinks are:

- Change-of-state latent energy
- Radiator heat rejection
- Aerodynamic cooling

Heating rates may be impressed on diffusion (finite capacitance) or arithmetic (zero capacitance) nodes. Most thermal analyzers provide a separate entry block for entering heating or cooling rates. For example, the SINDA computer program uses the SOURCE DATA BLOCK for such entries. In the usual case, heating rates are not considered when computing the time steps for transient analysis, and large heating rates on low capacitance nodes may create instability in the network solution. Also, the impression of large heat sources on arithmetic nodes with radiation (non-linear) conductors attached often causes large erroneous temperature oscillations in the arithmetic and adjoining nodes. Both of these difficulties can be avoided with the use of the program control constants incorporated in most thermal network analyzers. These control constants are the time-step multiplication factor and the maximum temperature changed allowed.

THERMAL DESIGN ANALYSIS EXAMPLE - POAM

The Polar Ozone and Aerosol Monitor (POAM) sensor will be used as an example of how a thermal analysis is conducted. The POAM sensor is used to measure the concentrations of ozone and aerosols in the upper atmosphere in the Earth's polar regions. The experiment was funded by the U.S. Strategic Defense Initiative Office, administered by the Office of Naval Research, and was flown on the French SPOT Earth resources satellite.

The sensor measures the concentrations of ozone and aerosols by observing the attenuation of sunlight as it passes through the atmosphere during sunrise and sunset events as the satellite circles the Earth in a polar, sun-synchronous orbit, as shown in Figure 17. The sensor actually contains nine small telescopes, each of which has a filter and a sensor, which measure the intensity of sunlight in nine very narrow wavebands. By observing the intensity of the sunlight as the sun sets or rises, concentrations can be measured at different heights in the atmosphere to a resolution of about 1 km. These measurements are used to support research into the depletion of the protective ozone layer in the upper atmosphere, as well as other atmospheric studies.

Physical Configuration

The POAM sensor is shown in Figure 18. It consists of a rectangular base with four mounting feet, called the azimuth housing, and a dome-shaped enclosure containing the telescope assembly. The dome is attached to a short shaft that rides on a pair of bearings in the azimuth housing. The only physical connections between these two assemblies are the bearings and a small cable bundle that runs down the center of the hollow shaft, which is not shown in the figure. The telescope assembly is similarly connected to the dome only through a pair of bearings and a few small wires. Since heat conduction across ball bearings and along fine wires is relatively weak, the telescope, dome, and azimuth housing are only rather weakly coupled together thermally.

The entire sensor assembly is mounted to an exterior face of the host SPOT satellite, as shown in Figure 19. The mounting is accomplished through the use of a bracket, as shown in the figure. The satellite itself is placed in a 822-km, 98.738-deg inclined circular sun-synchronous orbit with a period of 100 minutes and a range of beta angles of 14.5 to 29.8 deg. At the sunrise event, the sensor makes one minute of observations and then rotates the dome about 130 deg in azimuth to be in the proper position for the sunset event that occurs about 40 minutes later. Only small rotations of the telescope on its elevation bearings are required to track the sun during each observation. A typical observation sequence is illustrated in Figure 20.

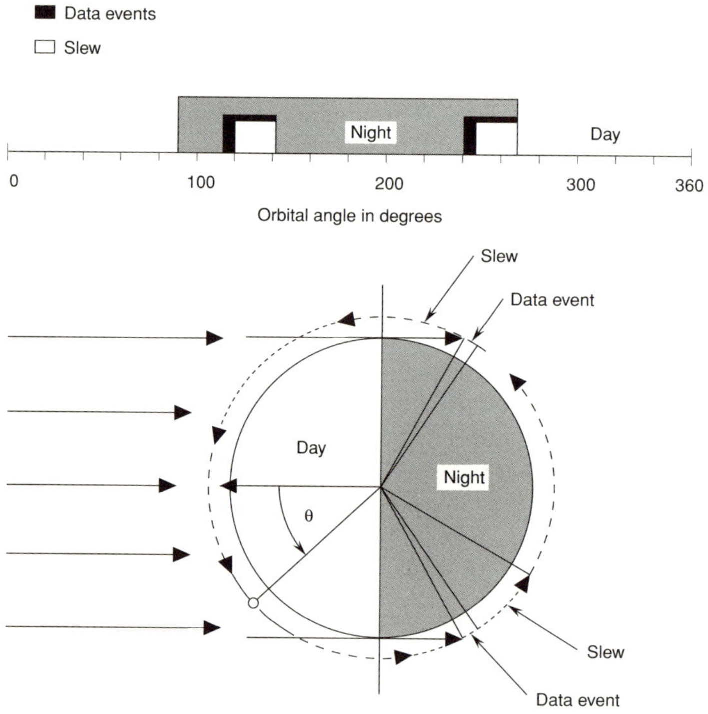

Figure 17. POAM data events

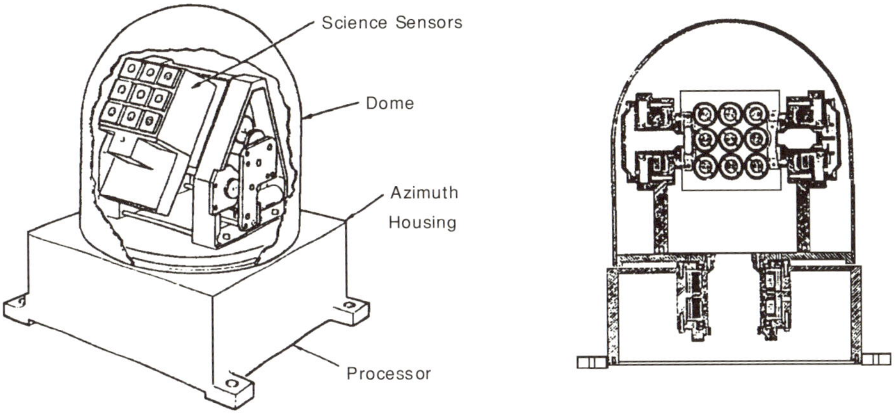

Figure 18. POAM sensor (TTC)

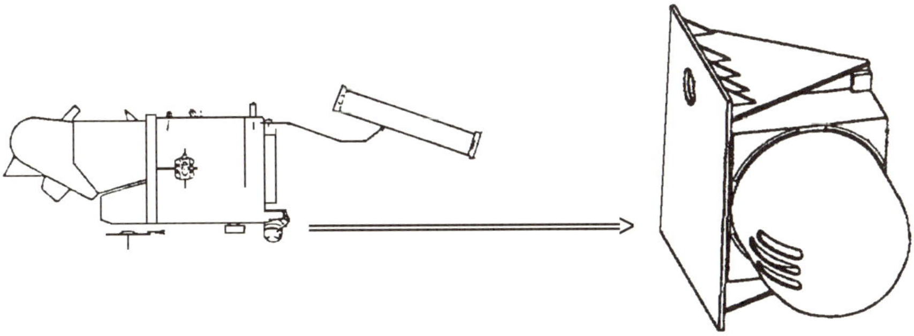

Figure 19. POAM on SPOT spacecraft (TTC, Matra-Marconi)

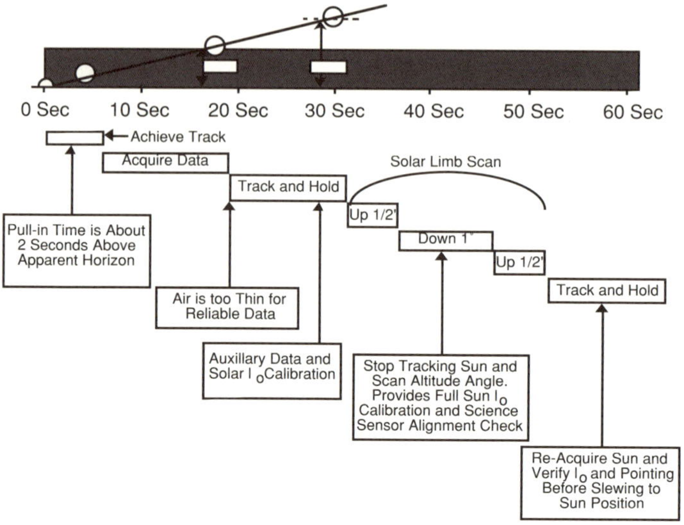

Figure 20. POAM observation sequence

Thermal Design Requirements

The thermal-design requirements for POAM are driven by both the instrument and the host spacecraft, and are listed in Table 5. From the POAM point of view, the instrument must be maintained between -10- and +50-deg C while operating, and between -30- and +50-deg C while not operating or at turn-on. From the spacecraft point of view, certain mission-related requirements must be met, including: the ability to go for 3 1/2 hours after launch with no power supplied to the instrument; to survive the spacecraft safe-mode condition in which only survival-heater power will be available to the instrument; to be able to withstand a 2-orbit (approximately 3 1/2 hours) transition from safe mode to normal operating mode, during which time neither electronics nor survival-heater

Table 5. Thermal Design Requirements

- Optical head case temperature: -10 to +50 deg C
- Survival/turn-on limit: -30 deg C
- Telecope detector rate or temperature change: < .07 deg C/min during observation period
- Must survive for 3 1/2 hours after launch without power
- Must survive for 2 revs without power returning from safe to nominal operating mode
- Maintain 10 deg C margins on predicted temperatures, 25 percent margin on heater power
- Limit conduction between optical head and bracket to less than .07 W/K

power will be available; and, to limit conduction between the instrument and the spacecraft mounting bracket to less than 0.07 w/deg K. It was further agreed by all organizations involved that an uncertainty margin of 10-deg C would be applied to all temperature predictions and any heaters would be sized to provide either 10-deg C margin to lower temperature limits or a 25 percent excess capacity at the lower temperature limit.

Conceptual Design

The first step in the design process is to identify the factors that will drive the design. This includes the previously discussed design requirements levied by the instrument designers and the satellite, as well as the instrument heat dissipation and range of external environments.

The instrument heat dissipation varies around the orbit due to the operation of motors during telescope slewing. At the conceptual design phase of this program, the electrical-power draw for the instrument (which is all converted to heat since there is no significant energy output) was estimated to be no greater than that shown in Figure 21. Since there could also be periods of several orbits or longer during which there would be no observations and no operation of the drive motors, the minimum power draw was assumed to be a constant 4.4 watts. Most of this heat is dissipated in the azimuth housing, with only a small portion dissipated in the telescope.

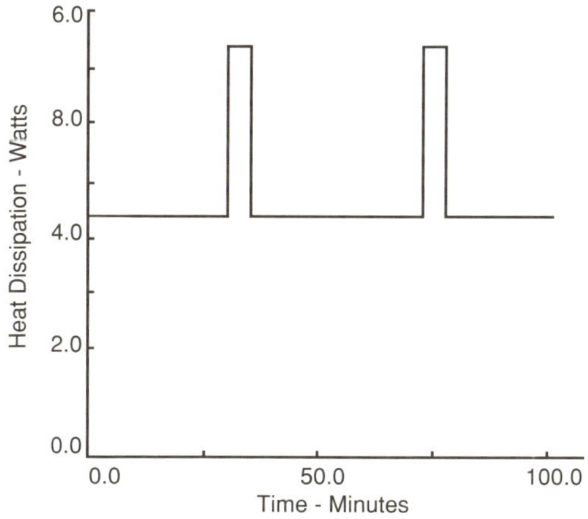

Figure 21. Instrument heat dissipation

The instrument is also exposed to solar, Earth IR, and albedo environmental heating fluxes. Because the satellite is Earth-facing in a sun-synchronous orbit, the sun position relative to the vehicle forms a cone as the satellite goes around the

Earth, as shown in Figure 22. This cone has an elevation angle that equals the orbit beta angle, 14.5 to 29.8 deg. Eclipse time ranges from 32.5 to 34.6 minutes, as can be calculated from the equations found in Chapter II. Since the spacecraft is Earth-facing, the instrument is always pointing with the dome facing straight down at the Earth, and Earth-IR loads can therefore be assumed to be constant around the orbit. Albedo loads will, of course, vary around the orbit, but the narrow range of beta angle ensures that orbit average albedo loads will not change a great deal over time.

Given the estimated heat dissipation, and the requirement that the instrument be conductively isolated from the satellite, some amount of radiator area will be required to reject the instrument waste heat to space. Most of this heat dissipation is in the base, which is pretty much covered on three sides by its mounting bracket. The side facing away from the spacecraft is the only one with a fairly clear view to space, although it does have a small view to the spacecraft solar array. To determine if this side would have sufficient area to reject the waste heat, a simple calculation can be performed. The maximum orbit-average internal heat plus the maximum orbit-average environmental heat flux must not exceed the energy radiated from the surface at the maximum allowable temperature;

$$\dot{Q}_{ELECTRONIC} + \dot{Q}_{ENVIRONMENTAL} = A\varepsilon\sigma T^4 \ . \qquad (1)$$

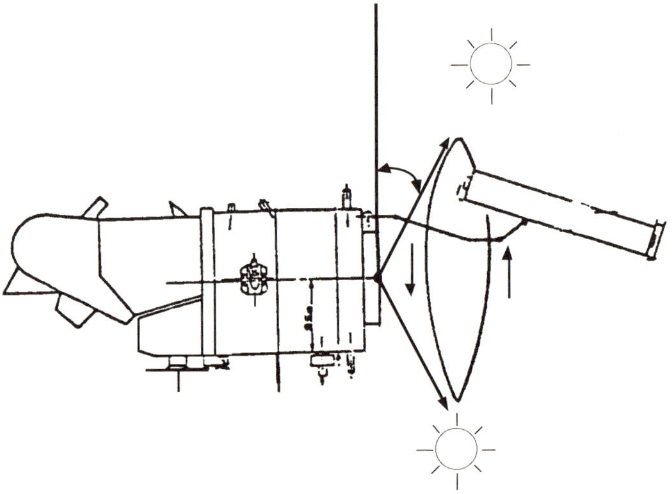

Figure 22. Solar illumination angles

(The small amount of heat backload from the spacecraft solar array may be neglected for this preliminary evaluation. Also, the mass of the instrument, 25 pounds, gives it large heat capacity relative to the heat pulses during motor operations. This insures that the temperature will not vary too much from the orbit average, making these orbit-average calculations reasonably accurate.)

The maximum orbit-average internal heat dissipation can be calculated in a straightforward manner from Figure 21 to be 4.7 watts (16.0 Btu/hr). The worst-case solar heating for this surface would occur when the sun is at its maximum elevation angle above the surface, which is 29.8 deg, as shown in Figure 22. The orbit-average solar load is given by:

$$\dot{Q}_{solar} = (\sin 29.8^\circ)\, S\, \alpha\, (\%\text{ of orbit in sunlight}), \qquad (2)$$

where

S in the solar constant
α = absorptance of surface

Assuming a 5-mil silvered Teflon radiator-surface finish, the end-of-life absorptance would be approximately .18 after 3 years in low Earth orbit. This is based on a beginning-of-life α of .05, per Table 1, and a degradation of .09, per Figure 8 in the Thermal Surface Finishes section of Chapter IV. The maximum solar constant is 444 Btu/hr ft^2 (Chapter II), and the percent sunlight time is

$$\frac{\text{orbit period-eclipse time}}{\text{orbit period}} = \frac{100\text{ min} - 32.5\text{ min}}{100\text{ min}} = 67.5\%.$$

Substituting these values in equation (2) gives a maximum orbit average absorbed solar load of 26.8 Btu/hr ft^2.

The orbit-average Earth IR load can be calculated using Figure 23, which is applicable to flat, unblocked surfaces. The Earth-pointing orientation of the satellite means that the instrument radiator surface remains perpendicular to the Earth all around the orbit, so Earth-IR heating will not change. The ρ angle for Figure 23 is therefore 90 deg and the altitude is 510 statute miles (822 kilometers). At the intersection of ρ=90 deg and h=500 miles, find F_E=.22, project F_E=.22 to the line labeled q_E, then project this point horizontally to the q_E scale to read q_E=.0042 Btu/ft^2/sec = 15.1 Btu/hr ft^2. The heat absorbed is the incident value times the emissivity of the surface, or (15.1)(.78) = 11.8 Btu/hr ft^2 for 5-mil silvered Teflon.

Albedo loads can be calculated in a similar fashion using Figure 24; however, it is necessary to calculate the value for several points since albedo changes as the satellite goes around the orbit. Using Figure 24, the altitude and ρ angles are the same as for the Earth-IR calculation, i.e., 510 miles and 90 deg . At the intersection of ρ=90 de and h=510, find F_R=.22. Since the satellite is in an

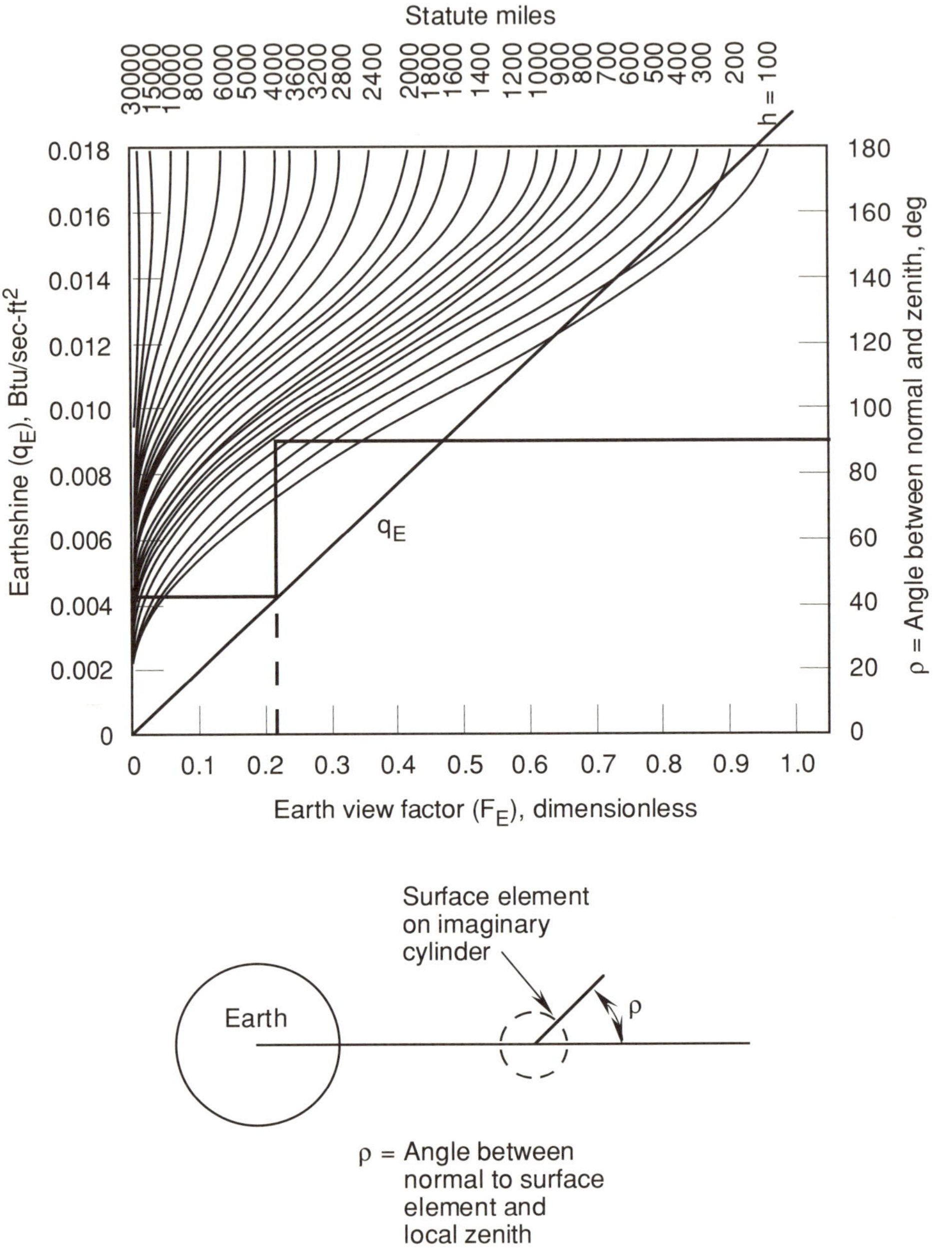

Figure 23. Incident Earthshine irradiation on a surface element in an Earth orbit (courtesy of Lockheed)

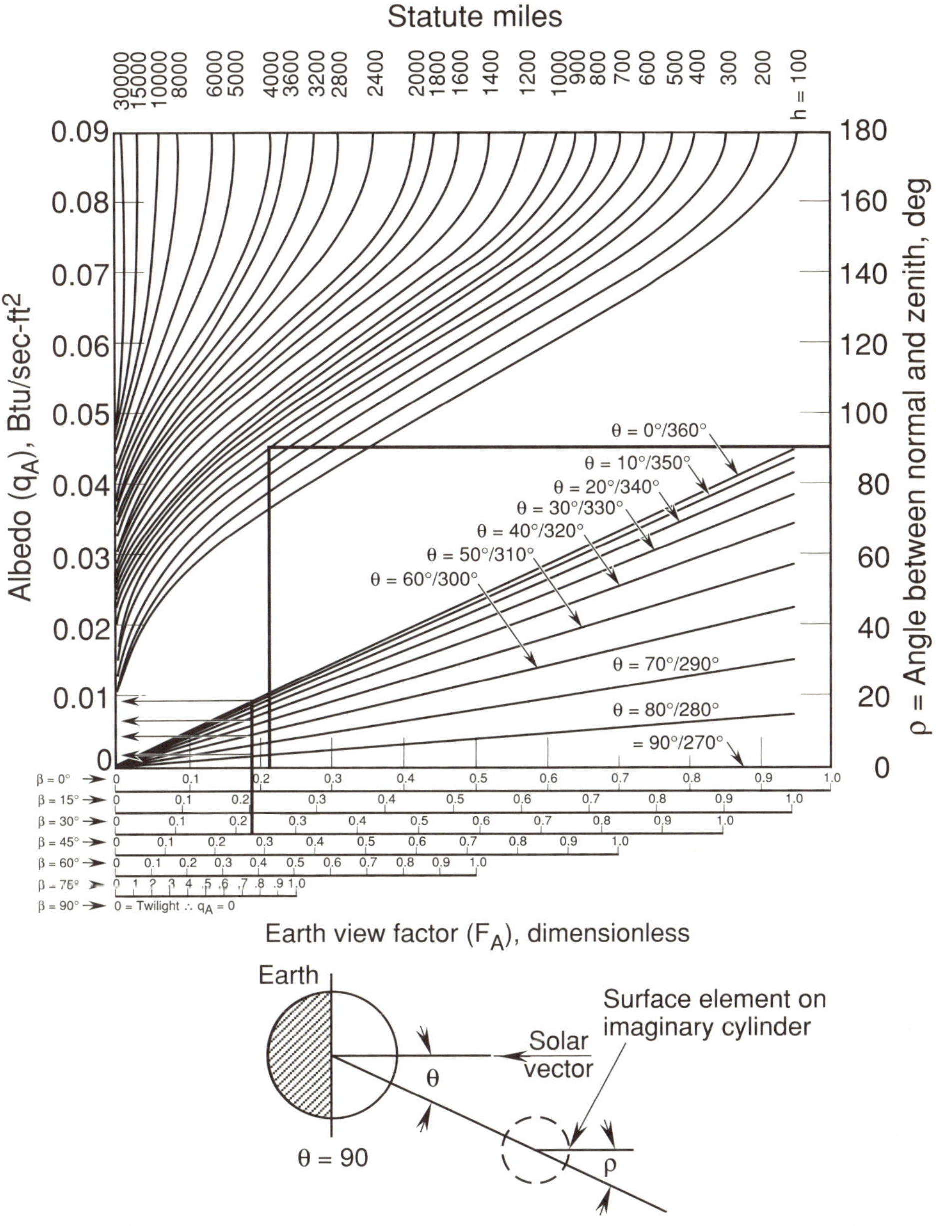

ρ = Angle between normal to surface element and local zenith

θ = Orbit angle, measured from sub solar point to local zenith

Figure 24. Incident albedo irradiation on a surface element in an Earth orbit (Note: Earth reflectance assumed to be 0.38. Albedo plotted is an approximation, with the largest error near the terminator (θ = 90)) (courtesy of Lockheed)

orbit with a beta angle of almost 30 deg, shift down to the scale labeled β=30 deg and draw a vertical line at F_R=.22. The intersections of the line with the family of lines labeled θ=X deg gives the incident albedo flux for various points around the orbit at position angles of θ deg, measured from the closest approach to the subsolar point. Figure 24 gives the values for the half of an orbit, on the sunlit side of the Earth. If these values are averaged and then divided by two to account for the dark half of the orbit which experiences no albedo load, this will give the orbit-average incident albedo of 10.7 Btu/hr ft^2. Multiplying this by the silvered Teflon solar absorptivity of .18 gives an orbit-average absorbed-albedo load of 1.9 Btu/hr ft^2. The albedo load is therefore a fairly small contributor compared to the solar (26.8 Btu/hr ft^2) and Earth IR (11.8 Btu/hr ft^2) heat loads.

Substituting the orbit-average electronics heat, solar IR, and albedo loads into Eq. (1) and solving for the radiator area at a temperature of 40-deg C=564-deg R gives:

$$\dot{Q}_{ELECT} + \dot{Q}_{ENVIR.} = A\varepsilon\sigma T^4$$

$$16.0 + [26.8 + 11.8 + 1.9]\,A = A\,(.78)(.1714\times10^{-8})(564)^4$$

$$A = .169\ \text{ft}^2 = 24.3\ \text{in}^2\ .$$

Since the face of the azimuth housing has an area of 31.5 in^2, there is adequate area for a radiator. (Note: A radiating temperature of 40-deg C was selected to allow for the required 10-deg C margin between analysis and the maximum allowable instrument temperature limit of 50-deg C.)

The minimum temperature of the instrument under cold-case conditions using the radiator size calculated above was determined in the same manner. Solar, albedo, and Earth IR heating for the cold-case orbit (β=14.5 deg, summer) and electronics waste heat without motor operations were summed and, using an area of 24.3 in^2, Eq. (1) was solved for T. This gave a cold-case temperature of 71-deg F (22-deg C). Based on these preliminary hot-and cold-case calculations, the entire side of the azimuth housing (31.5 in^2), rather than the 24.3 in^2 calculated above, could be used as a radiator to bring the average temperature down a little. Lower operating temperatures generally increase the life and reliability of electronic components.

Based on the preliminary radiator sizing and the requirements listed in Table 5, the thermal-design concept shown in Figure 25 was identified. The side of the azimuth housing facing away from the spacecraft would be covered with 5-mil silvered Teflon, and would serve as the primary radiator. All of the other surfaces of the azimuth housing and dome would be covered with MLI insulation

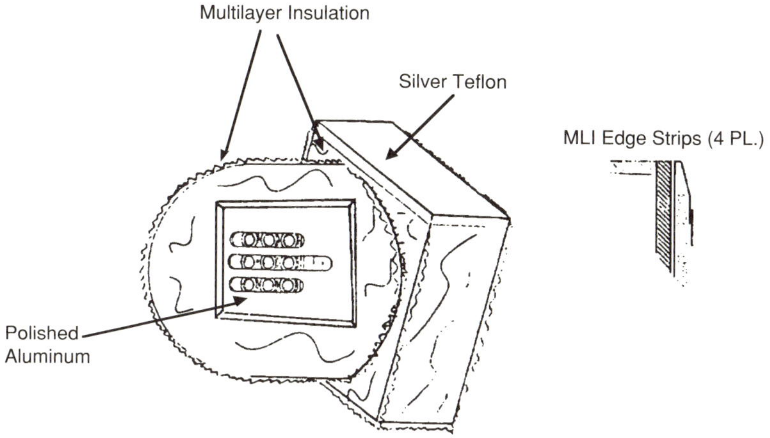

Figure 25. POAM thermal design

blankets to minimize the loss of heat through these surfaces and to essentially eliminate radiative thermal interactions between the instrument and the spacecraft. A small "window" would be made in the MLI covering the dome to allow the telescopes a view out. The surface of the aluminum dome exposed in the window area, however, would be polished to provide a low absorptance and emittance to minimize both radiative heat loss and energy absorbed from incident environmental heat fluxes. Because of their poor conductive-heat transfer, the bearings between the telescope and the dome and between dome and azimuth housing tend to thermally isolate these components. Therefore, to tie them together as much as possible radiatively, the telescope external surface and the dome internal surface are given a black high-emittance finish. The bottom of the dome and the top of the azimuth housing are also painted black to maximize the radiative coupling in the interface. Plastic isolators would also be placed under each of the mounting feet to meet the requirement of limiting conductive-heat transfer between the instrument and the spacecraft-supplied support bracket.

Detailed Design Analysis

Once a design concept is identified, a detailed analysis must be conducted to fine tune the design and predict instrument temperatures under the entire range of flight conditions. This involves identifying analysis cases to be run and constructing a geometric math model (GMM) and a thermal math model (TMM) of the instrument. For the POAM program, a thermal analysis of the overall instrument was conducted, with separate additional analyses performed of the individual circuit cards and telescope photo detectors. This discussion will be limited to the instrument-level analysis.

The GMM and TMM serve different purposes. The GMM is a mathematical representation of the physical surfaces, of the instrument and is used to calculate blackbody radiation couplings between surfaces as well as heating rates due to environmental fluxes. The TMM is most often a lumped-parameter network representation of the thermal mass and conduction and radiation couplings of the instrument, and is used to predict instrument temperatures. The radiation interchange couplings and environmental heat fluxes calculated by the GMM are used in constructing the TMM. Both the GMM and TMM are constructed and executed using industry-standard codes. The most common GMM codes are TRASYS and NEVADA (Ref. 11), and the most common TMM code is SINDA (Refs. 4 and 5). Other commercially available codes do exist, however, and some large companies have their own internally developed codes, which they use instead of those mentioned above. These codes will be discussed in detail in later sections of this Chapter.

Analysis Cases

Based on the instrument operating modes and thermal-design requirements discussed earlier, four significant thermal-design analysis cases were identified, as shown in Table 6. Normal on-orbit operations are bounded by the hot and cold operating cases. The response of the instrument to launch and a potential spacecraft "safe mode" condition were also analyzed.

Table 6. Design Environments/Assumptions

Hot Operating	Cold Operating	Safe Mode	Launch/Ascent
β = 29.8 deg, winter	β = 14.5, Summer	N/A	β = 14.5, Summer
EOL α = .17	BOL α = 0.08	N/A	BOL α = .08
MLI ε^* = 0.01	MLI ε^* = 0.05	MLI ε^* = 0.05	MLI ε^* = .05
Tel. Sees Sun Twice Per Rev	Tel. Does Not See Sun	Tel. Does Not See Sun	Tel. Does Not See Sun
10 Min/Rev Motor Ops (3 W)	No Motor Ops	No Motor Ops	No Motor Ops
Hot Spacecraft	Cold Spacecraft	Cold Spacecraft	Cold Spacecraft
Radiator to Solar Array View Varies Around Rev	Radiator to Solar Array View Varies Around Rev	Fixed Radiator View to Solar Array	Radiator to Solar Array View Varies Around Rev
4.4 W Electronics Heat	4.4 W Electronics Heat	No Power (Heaters Only)	No Power, No Heaters
Earth IR = 74 Btu/hr ft^2	Earth IR = 66.	No. IR	Earth IR = 66
Albedo = 0.42	Albedo = 0.34	No Albedo	Albedo = 0.34

- Orbit 822 km altitude, 98.738 inclination, 337.5 deg ascending node

The hot operating case conditions include maximum solar heating, which occurs at the highest beta angle with the winter solar constant, maximum Earth IR and albedo, maximum (end of life) solar absorptance on the external surface finishes, good insulation-blanket performance, maximum motor operations, telescopes looking at the sun twice per orbit, maximum spacecraft temperatures, and maximum electronics waste heat. Cold operating case conditions include minimum solar loads, minimum (beginning of life) solar absorptances, poor insulation-blanket performance, no motor operations, telescope stopped in a position where it does not see the sun, cold spacecraft temperatures, minimum Earth IR and albedo, and minimum electronic waste heat (which happens to be the same as the maximum heat since it is constant for this instrument).

During safe mode, the spacecraft turns and points constantly at the sun instead of the Earth and the POAM instrument is turned off, although some power is available to run heaters, if required. Since the instrument is off and shadowed from the sun by the spacecraft, this is a cold-case condition. It is therefore also assumed that there is no Earth IR or albedo heating for conservatism.

During the launch phase, POAM is turned off. While sitting on the launch pad the instrument will be at approximately the same temperature as the purge gas inside the booster fairing (15-deg C), since it has no internal heat dissipation. For the first three minutes after lift-off the booster fairing is in place and experiences a large temperature rise. Because of the large thermal mass of POAM and the brief duration of this phase, the thermal effect on the instrument is negligible. This is followed, however, by a 1/2-hour period during which the spacecraft attitude is not controlled and the POAM radiator may see the sun, Earth, and/or deep space. Since the instrument is not powered, this is a cold case, and it was therefore assumed that the radiator was facing deep space with no environmental heat fluxes incident on any surface. After 1/2 hour the spacecraft is stabilized in its normal attitude, but POAM is still not powered. The point of this analysis case is to determine how long the instrument can go after launch with no power without violating its lower survival temperature limit of -30-deg C.

GMM Construction

The GMM of the POAM mounted on the host spacecraft was constructed using the NEVADA code. The model, shown in Figure 26, consists of a simple representation of the spacecraft, POAM, and the support bracket. It was constructed on a CAD-like system using rectangular, circular, hemispherical, and cylindrical surface elements available in the NEVADA package, and each surface was assigned the appropriate absorptance, emittance, and specularity. Details about how these models are constructed using NEVADA can be found in Ref. 11.

The GMM was then run using NEVADA to calculate the radiation interchange factors between all surfaces. NEVADA also outputs a radiation-conductor block that may be merged directly into the SINDA TMM. This block of conductors will be discussed later.

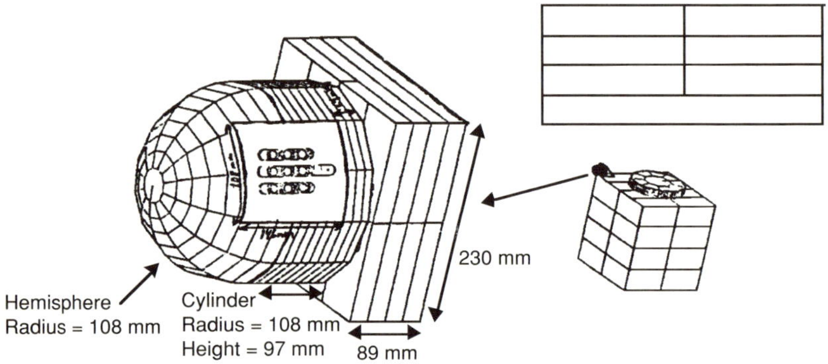

Figure 26. POAM geometry

The GMM was then placed mathematically into the proper orbit and attitude using another section of the NEVADA software package. Solar, Earth IR, and albedo heat loads absorbed on each surface were calculated for a dozen points around the orbit for both the hot case (maximum beta angle, winter, maximum absorptance) and the cold case (minimum beta angle, summer, minimum absorptance) orbits. These heat rates are also output by NEVADA in arrays that can be merged directly into the TMM.

TMM Construction

The TMM consists of nodes representing parts of the instrument, diffusion and radiation conductors between nodes, blocks of arrays, and constants for storing inputs such as environmental heating rates calculated by NEVADA and logic blocks for controlling the execution of the program. A listing of the POAM TMM in SINDA format is shown in Table 7.

Table 7. POAM TMM Listing

```
BCD 3THERMAL LPCS
END
BCD 3NODE DATA
    101,50.,.085         AZIMUTH HOUSING NODES
    3,50.,.13
    4,50.,.13
    5,50.,.13
    6,50.,.13
    7,50.,.13
    8,50.,.13
    9,50.,.54
    10,50.,.54
    11,50.,.089
    12,50.,1.6           DOME NODE
    112,50.,0.61         TELESCOPE NODE
    22,50.,-1.           DOME MLI NODE
    23,50.,-1.           AZIMUTH HOUSING NODE
    -997,150.,0.         SPACECRAFT SOLAR ARRAY BOUNDARY NODE
    -998,80.,0.          SUPPORT BRACKET BOUNDARY NODE
    -999,-460.,0.        SPACE BOUNDARY NODE (TEMPERATURE=ABSOLUTE ZERO)
END
BCD 3CONDUCTOR DATA
    1,3,101,1.04       DIFFUSION CONDUCTORS IN AZIMUTH HOUSING
    2,8,101,1.04
    3,101,9,3.13
    4,3,4,1.
    5,4,5,1.
    6,5,6,1.
    7,6,7,1.
    8,7,8,1.
    9,9,10,3.48
    10,9,8,1.15
    11,9,3,1.15
    12,10,4,1.15
    13,10,5,1.15
    14,10,6,1.15
    15,10,7,1.15
    16,11,101,.12
    17,11,3,.063
    18,11,4,.063
    19,11,5,.063
    20,11,6,.063
    21,11,7,.063
    22,11,8,.063
    23,3,998,.033      CONDUCTION ACROSS ISOLATORS TO SPACECRAFT
    24,4,998,.033
    25,7,998,.033
    26,8,998,.033

                                          RADIATION CONDUCTORS
    -101,101,999,1.61E-10       FROM RADIATOR TO SPACE
    -102,101,997,4.14E-11       FROM RADIATOR TO SOLAR ARRAY
    -103,12,999,7.48E-11        FROM DOME WINDOW TO SPACE
    -104,12,997,8.31E-12        FROM DOME WINDOW TO SOLAR ARRAY
    -105,112,999,4.97E-12       FROM TELESCOPE TO SPACE
    -106,112,12,1.23E-9         FROM TELESCOPE TO DOME (INTERNAL)
    -107,22,999,1.49E-9         FROM DOME MLI OUTER LAYER TO SPACE
    -108,22,997,0.166E-9        FROM DOME MLI OUTER LAYER TO SOLAR ARRAY
    -109,22,12,1.34E-11         FROM DOME THROUGH MLI TO OUTER LAYER
    -110,12,9,1.42E-10          FROM BOTOM OF DOME TO AZIMUTH HOUSING
    -111,12,10,1.42E-10             "       "      "       "       "
    -112,23,9,0.72E-12          FROM AZIMUTH HOUSING THROUGH MLI TO OUTER LAYER
    -113,23,10,0.72E-12             "       "      "       "       "
    -114,3,998,9.37E-13         FROM AZIMUTH HOUSING THROUGH MLI TO OUTER LAYER
    -115,4,998,9.37E-13
    -116,5,998,9.37E-13
    -117,6,998,9.37E-13
    -118,7,998,9.37E-13
    -119,8,998,9.37E-13
    -120,11,998,0.45E-11
    -123,23,22,1.13E-10         FROM DOME MLI TO AZIMUTH HOUSING MLI (EXTERNAL)
    -124,23,999,3.6E-10         FROM AZIMUTH HOUSING MLI OUTER LAYER TO SPACE
END
```

Table 7. (continued)

```
        BCD 3CONSTANTS DATA
            TIMEO=Ø.,ARLXCA=.1,DRLXCA=.1,NLOOP=5ØØØ      PROGRAM CONTROL CONSTANTS
            NDIM=5ØØØ,BALENG=.Ø5,CSGFAC=1.                  "       "        "
            1,Ø.,2,Ø.,3,Ø.,4,Ø.,5,Ø.                      CONSTANT STORAGE LOCATIONS
            6,Ø.,7,Ø.,8,Ø.,9,Ø.,1Ø,Ø.                              ||
            11,Ø.,12,Ø.,13,Ø.,14,Ø.,15,Ø.                          ||
            16,Ø.,17,Ø.,18,Ø.,19,Ø.                                \/
        END
        BCD 3ARRAY DATA
C         AZIMUTH HOUSING RADIATOR AREA*VIEW TO SPACE VS. TIME FOR ONE ORBIT
            1,Ø.,.Ø945,.21,.Ø917,.42,.116,.63,.1Ø8,.84,.111
             1.Ø5,.111,1.26,.Ø917,1.47,.Ø945,1.68,.Ø945,END
C         AZIMUTH HOUSING RADIATOR AREA*VIEW TO SOLAR ARRAY VS. TIME FOR ONE ORBIT
            2,Ø.,.Ø245,.21,.Ø273,.42,.ØØ27,.63,.Ø1Ø5,.84,.Ø1Ø5,1.Ø5,.ØØ8
              1.26,.Ø273,1.47,.Ø245,1.68,.Ø245,END
C         SOLAR ARRAY TEMPERATURE VS. TIME FOR ONE ORBIT
            3,Ø.,176.,.653,133.,.822,-112.,1.212,-126.,1.213,133.
              1.68,176.,END
C         ENVIRONMENTAL HEATING ON TELESCOPE BODY VS. TIME FOR ONE ORBIT
            4,Ø.,1.17,.Ø47,.79,.327,.56,.374,.19,.512,.19,.513,5.26
              .56,5.34,.6Ø6,5.44,.653,5.88,.654,.19,1.212,.19
              1.213,5.85,1.4,5.Ø2,1.54,3.74,1.68,1.17,END
C         ENVIRONMENTAL HEATING ON DOME IN WINDOW AREA VS. TIME FOR ONE ORBIT
            5,Ø.,2.5,.Ø4,2.5,.186,4.82,.42,5.9,.513,6.3
              .653,4.8,.654,1.5,1.212,1.5,1.213,4.8,1.245,5.
              1.32,2.Ø,1.68,2.5,END
C         ENVIRONMENTAL HEATING IN GAP BETWEEN DOME AND AZIMUTH HOUSING VS. TIME
            6,Ø.,1.6,.467,1.4,.56,4.3,.653,4.2,.654,1.,1.212,1.
              1.213,4.2,1.3Ø7,4.3,1.4,1.4,1.68,1.6,END
C         ENVIRONMENTAL HEATING ON OUTER LAYER OF DOME MLI
            7,Ø.,58.,.Ø9,48.,.19,48.,.28,58.,.37,7Ø.,.47,83.
             .56,89.,.653,95.,.654,24.,1.212,24.,1.213,96.
             1.26,96.,1.44,83.,1.68,58.,END
C         AZIMUTH HOUSING ELECTRONICS WASTE HEAT
            8,Ø.,15.,1.68,15.,END
C         AZIMUTH MOTOR WASTE HEAT
            9,Ø.,Ø.,.652,Ø.,.653,1Ø.23,.736,1Ø.23,.737,Ø.,1.244,Ø.,1.245,1Ø.23
             1.32,1Ø.23,1.321,Ø.,1.68,Ø.,END
C         SOLAR HEAT FLUX PER SQ. IN. INCIDENT ON AZIMUTH HOUSING RADIATOR VS. TIME
            11,Ø.,1.53,.653,1.53,.654,Ø.,1.212,Ø.,1.213,1.53,1.68,1.53,END
C         ALBEDO HEAT FLUX PER SQ. IN. INCIDENT ON AZIMUTH HOUSING RADIATOR VS. TIME
            12,Ø.,.19,.187,.15,.373,Ø.,1.3Ø7,Ø.,1.493,.15,1.68,.19,END
C         EARTH IR FLUX PER SQ. IN. INCIDENT ON AZIMUTH HOUSING RADIATOR VS. TIME
            13,Ø.,.1Ø7,1.68,.1Ø7,END
        END
        BCD 3EXECUTION
M         TIMEND=1ØØ.              PROGRAM CONTROL CONSTANTS
M         OUTPUT=1Ø.                   "       "        "
            ATSDUF                     "       "        "

        END
        BCD 3VARIABLES 1
M           IF(TIMEO.GT.98.3) OUTPUT=1./6Ø.
            D11CYL(1.68,TIMEO,A1,XK1)          INTERPOLATES ARRAYS AND PLACES CURRENT
            D11CYL(1.68,TIMEO,A2,XK2)             VALUES IN CONSTANT LOCATIONS
            D11CYL(1.68,TIMEO,A3,XK3)                     ||
            D11CYL(1.68,TIMEO,A4,XK4)                     ||
            D11CYL(1.68,TIMEO,A5,XK5)                     ||
            D11CYL(1.68,TIMEO,A6,XK6)                     ||
            D11CYL(1.68,TIMEO,A7,XK7)                     ||
            D11CYL(1.68,TIMEO,A8,XK8)                     ||
            D11CYL(1.68,TIMEO,A9,XK9)                     ||
            D11CYL(1.68,TIMEO,A11,XK11)                   ||
            D11CYL(1.68,TIMEO,A12,XK12)                   \/
            D11CYL(1.68,TIMEO,A13,XK13)
```

Table 7. (continued)

```
M       T997=XK3                                   SETS CURRENT SOLAR ARRAY TEMPERATURE

M       G1Ø1=(XK1*8.4)*(3Ø.6*.78)*(.1714E-8)/144. ADJUSTS RADIATOR RELATIVE VIEW TO
M       G1Ø2=(XK2*8.4)*(3Ø.6*.78)*(.1714E-8)/144. SPACE AND SOLAR ARRAY AS ARRAY TURNS

M         Q23=32.             ENVIRONMENTAL HEATING ON AZIMUTH HOUSING MLI OUTER LAYER
M         Q22=XK7                   "          "    "  DOME MLI OUTER LAYER
M         Q1Ø1=3Ø.6*(.15*(XK11+XK12)+.78*XK13)  "   "  AZIMUTH HOUSING RADIATOR
M         Q3=XK8/4.           AZIMUTH HOUSING ELECTRONICS WASTE HEAT
M         Q4=Q3                     "        "        "       "
M         Q7=Q3                     "        "        "       "
M         Q8=Q3                     "        "        "       "
M         Q112=XK4        ENVIRONMENTAL HEAT ON TELESCOPE
M         Q12=XK5+XK6/2.      ENVIRONMENTAL HEATING ON DOME
M         Q9=XK9/2.+XK6/4.    ENVIRONMENTAL AND AZIMUTH DRIVE MOTOR WASTE HEAT
M         Q1Ø=Q9                    "          "         "         "
      END
      BCD 3VARIABLES 2
      END
      BCD 3OUTPUT CALLS
          TPRINT
      END
      BCD 3END OF DATA
```

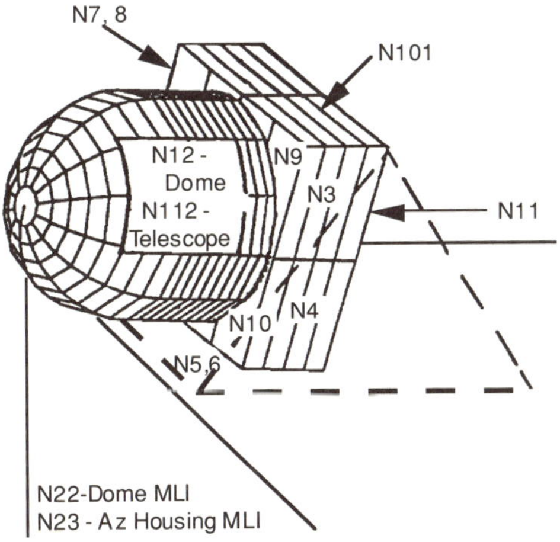

Figure 27. POAM TMM nodes

The nodalization scheme chosen for the TMM is shown in Figure 27. Since the base, dome, and telescope are all constructed of thick (.1- to .2-inch thick) aluminum, they can each be assumed to be fairly isothermal and a minimum number of nodes are required. The entire telescope is therefore modeled as one node, as is the dome assembly. The location of the radiator on one side of the azimuth housing, however, means that some temperature gradient could exist between it and heat-dissipating elements on the opposite face. The azimuth housing was therefore modeled using ten nodes, as shown in Figure 27. The MLI covering the dome and azimuth housing were represented by one node each. Temperatures of the spacecraft solar array, mounting bracket, and the outer layer of the spacecraft MLI were all supplied by the spacecraft engineers and were therefore put in the POAM TMM as boundary-driver nodes.

Capacitances of each node were hand calculated either by using a weight found in a mass-properties report for the instrument and multiplying by the specific heat, or by calculating the volume of material and multiplying by the density and specific heat. Arithmetic (zero capacitance) nodes were used to represent MLI blankets because they are extremely light and respond almost as if they had zero mass. The boundary nodes do not require a capacitance since they are treated by the program as a constant-temperature (infinite capacitance) sink.

Diffusion conductors were calculated in a straightforward manner using the $k\frac{A}{\ell}$ relationship as discussed in the "Fundamentals of Thermal Modeling" section. One exception to this was the conduction between the azimuth housing and its rear cover, which is held in place with screws. The contact conduction between these nodes was based on screw-conduction terms found in the "Mountings and Interfaces" section of Chapter IV. The other exception to this was the conduction across the azimuth and elevation bearings. Since bearing conduction is so uncertain, two cases were run to bound the problem: zero conductivity at one extreme and a conduction equal to 100 Btu/hr ft^2 over the entire area of the bearing race at the other extreme. Both of these values were used for each of the four design analysis cases for POAM, and the value that resulted in the most extreme temperatures was chosen.

Conduction heat transfer between the instrument and its mounting bracket had to be limited to less than .07 w/deg C per spacecraft requirements. To accomplish this, the fiberglass isolators shown in Figure 28 were designed for installation under the four mounting feet. The calculations of the conductance across these isolators are shown in the figure.

Radiation conductors generated by NEVADA were merged into the TMM. The conductors through the MLI blankets, however, were calculated manually. The heat leak through MLI can be modeled as an effective emittance, ε^*, as discussed in the "Multilayer Insulation and Barriers" section of Chapter IV. A range of effective emittances were used for this analysis, since it is hard to predict the exact performance of an insulation blanket before it is built and tested. For the hot case, a value of .01 was chosen, while .05 was used in the cold case. An ε^* of .05 is rather high, but it is justified in this case because the blankets are small and therefore more susceptible to the heat-leak effects of edges and attachments. For each node covered with MLI, a radiation conductor was calculated as (Area)(ε^*)(σ) between the instrument node and the node representing the outer layer of the MLI blanket. The radiation couplings from the outside face of the MLI blanket to space and to the spacecraft were previously calculated by the NEVADA model and were already merged into the TMM.

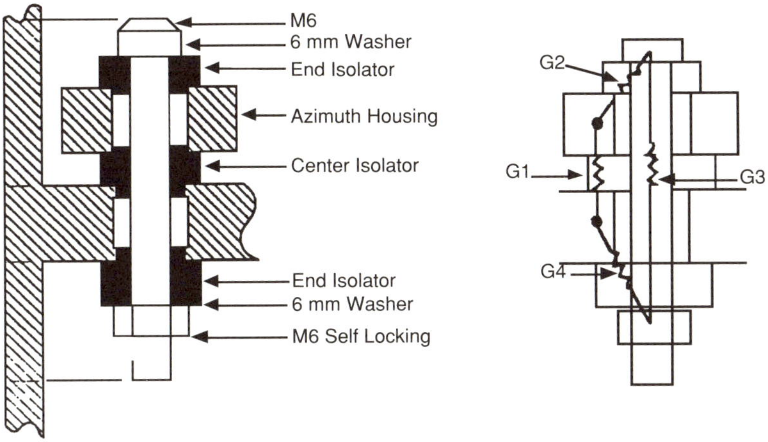

$$G_{THROUGH\ BOLT} = \frac{1}{\frac{1}{G_2} + \frac{1}{G_3} + \frac{1}{G_4}}$$

$$G_{TOTAL} = G_{THROUGH\ BOLT} + G1$$

Figure 28. POAM mounting isolator

The complete hot case TMM is shown in Table 7. The first "block" contains the node data. Each node is given an integer number, initial temperature, and capacitance. Arithmetic (zero capacitance) nodes are represented in SINDA by assigning them a negative capacitance value, and boundary (infinite capacitance) nodes are represented by giving them a negative node number, as can be seen in the table. The next block contains the conductor data. Each conductor input contains an integer-conductor number, the nodes that the conductor connects together, and a conductor value. Radiation conductors are given negative conductor numbers in SINDA. The next block contains the user and SINDA data constants. In this case, there are a number of program control constants as well as ten constant-storage locations, which will be discussed later. The next block contains array data. In this case there are arrays giving time-varying environmental-heat fluxes previously calculated by NEVADA, time-varying electronic-waste-heat rates, and time-varying radiation conductors between the POAM radiator node and the rotating spacecraft solar array (this was also previously calculated by NEVADA and input manually into the TMM). The next three blocks control the execution of the program. The second of these, "Variables 1," is used in this case to specify how much heat is on each node as well as what the radiator coupling is from the POAM radiator to the solar array at any given time. This block is accessed before the start of each time step as the program calculates the change of POAM temperatures with time. The final block specifies the data to be output by the program. In this case, temperatures and impressed heat rates for each node are requested.

Predicted Temperatures

The file shown in Table 7 is executed by the SINDA program and temperatures are calculated. Similar files were constructed for the cold-operating, safe-mode, and launch-ascent cases. Predicted temperatures for these conditions are shown in Figures 29 through 31. Comparison of these results to the requirements of Table 5 shows that all requirements are met with adequate (10-deg C or greater) margin. For the safe-mode case a heater was required to maintain the instrument above its lower survival temperature.

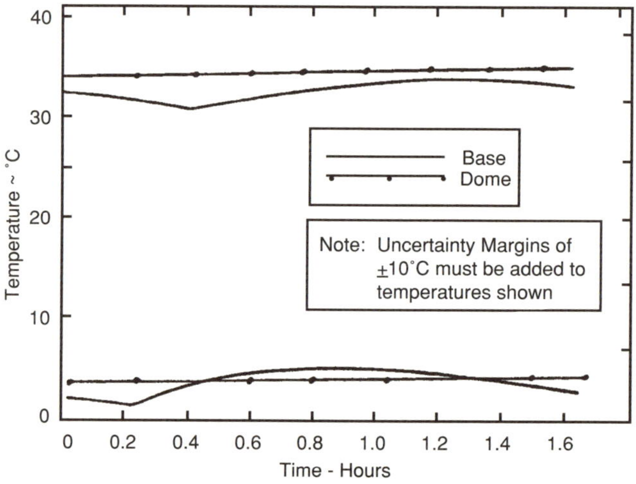

Figure 29. Hot and cold operating temperatures

Thermal Balance Test

The thermal analysis described above was verified by a thermal balance test. This was conducted during spacecraft-level thermal vacuum testing in Toulouse, France. The POAM instrument was installed on the spacecraft in the flight configuration. Hot- and cold-case test phases were planned. Due to limitations associated with the spacecraft, these were not precise representations of the flight hot and cold cases, but they were close and provided two good conditions to which the thermal math models could be checked and correlated.

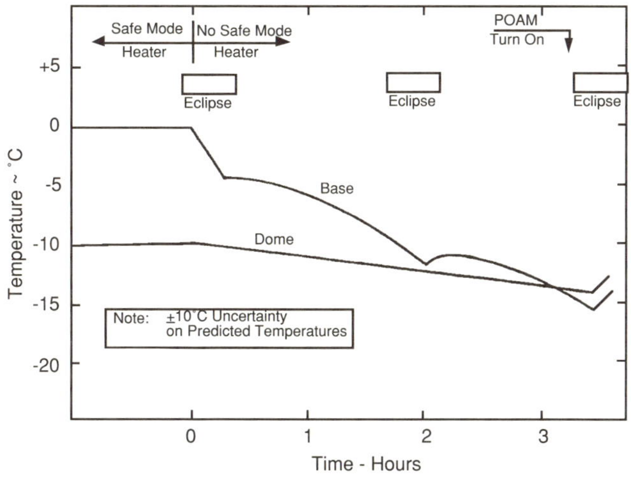

Figure 30. Safe-mode temperatures

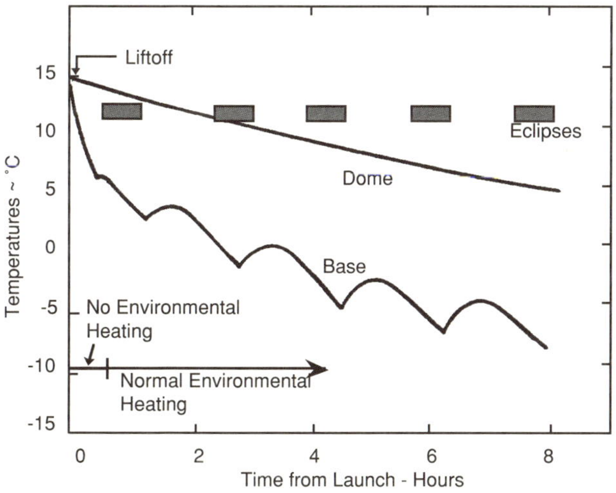

Figure 31. Launch/Ascent temperatures

The temperature instrumentation used is shown in Figure 32 and the hot- and cold-case steady-state temperatures are shown in Figure 33. As can be seen from this data, the azimuth housing is nearly isothermal with only a 2-deg to 3-deg C variation around the box. The dome assembly, however, ran approximately 15-deg C cooler than the azimuth housing. This temperature difference indicates that there is little conductive coupling between the dome and azimuth housing, as was assumed in the design analysis for conservatism. The dome runs cooler because there is no electronics waste heat dissipated in the telescopes and no sunlight shining into the telescope aperture in the test chamber. Looking at the sun twice per rev on orbit will cause the dome temperature to rise closer to that of the base during flight operations.

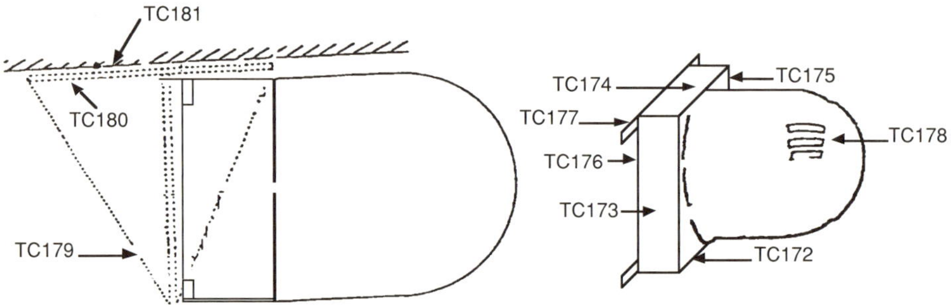

Figure 32. Test thermocouples

The optical-head thermal model was run using the as-run test environment. In comparing the model predictions to test data it was discovered that the conductance value through the thermal isolators between the optical head and its support bracket was low by approximately 40 percent. The lower conductance value reflected an earlier isolator design that used titanium rather than stainless-steel bolts. Further comparisons also revealed that heat losses through the gap between the dome and azimuth housing, which had been neglected in the analysis, had a noticeable impact on both the optical-head temperature and the temperature difference between the dome and azimuth housing. After corrections for these two effects, the thermal math model and test data agreed within $\pm$3 deg C, as shown in Table 8.

Table 8. TMM Correlation to Test Data

	Cold Case			Hot Case		
	Test	Model	Δ	Test	Model	Δ
Dome	-11 deg	-8 deg	+3 deg	9 deg	8 deg	-1 deg
Azimuth Housing	6 deg	8 deg	+2 deg	25 deg	23 deg	-2 deg
Fin	7 deg	9 deg	+2 deg	24 deg	24 deg	0 deg

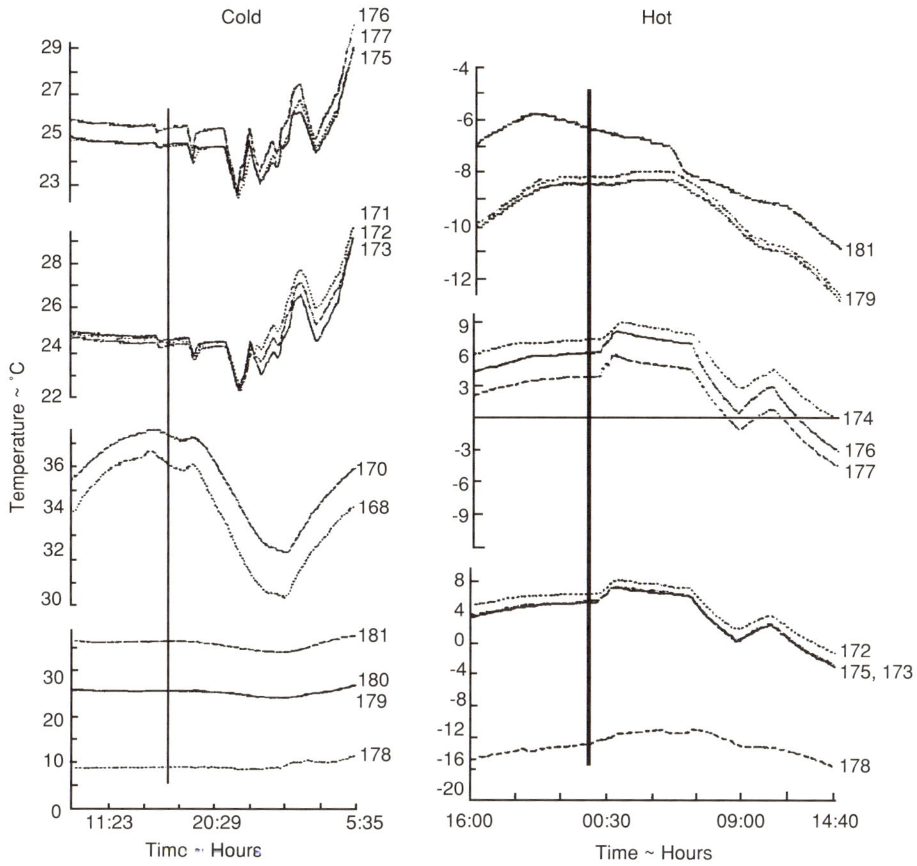

Figure 33. Thermal balance test temperatures

MARGINS

Even the best thermal analyses are subject to uncertainties. Despite our best efforts and the sophistication of today's analytical codes and computer workstations, flight experience teaches us that our predicted temperatures are not always precisely accurate. Some of the inaccuracies are due to factors that are known to be uncertain, such as contact conductances and the performance of insulation blankets. Some uncertainties are just due to the simplifications inherent in the analytical techniques. Some are due to errors. In any event, our understanding of these uncertainties is not yet sufficient to eliminate them from the analysis process.

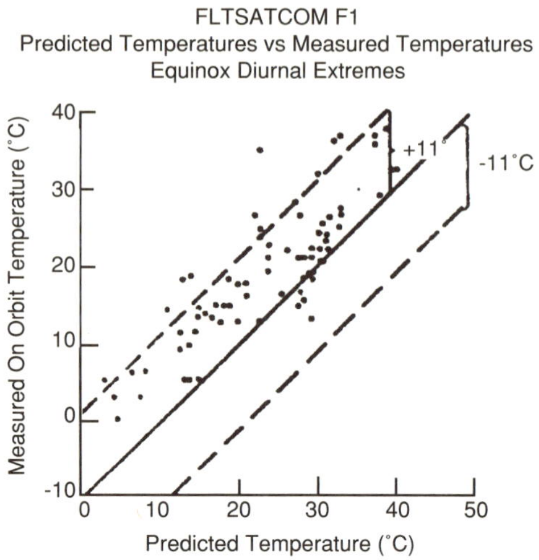

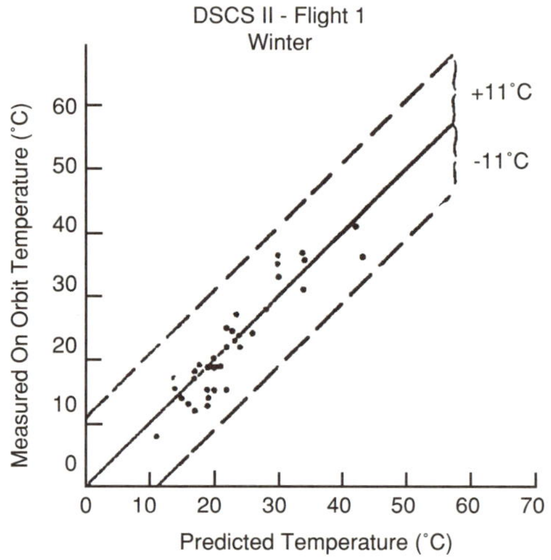

Figure 34. Thermal model accuracy assessment

When one compares temperatures predicted by analysis with what actually happens in flight, there are significant dispersions. Figure 34 shows such a comparison for two satellites; Fltdatcom and DSCS II. A study of a number of satellite programs conducted by Starck (ref. 7 of Chapter IX) concluded that an 11-deg C margin was required to provide "two-sigma" (95 percent) confidence that flight temperatures would be within limits (see Table 9). This study is the basis of the MIL-STD-1540B analytical uncertainty margin of 11-deg C. It is important to note that this margin is applied to predictions made by analytical models that have been correlated to thermal-balance test data. For an uncorrelated model, the uncertainty jumps to 17-deg C. It should also be noted, however, that very large discrepancies (40- or 50-deg C) do occur now and then. A thermal-balance test is needed to catch these large, potentially mission-threatening, errors before the satellite is launched. Simply using the 17-deg C margin and forgoing a thermal balance test could be a costly mistake.

Table 9. Temperature Uncertainty Margin Based on Space Flight Data Base

Standard Deviation	Percent of Confidence	Temperature Uncertainty (deg C)	
		Unverified Analytical Predictions	Predictions Verified by Testing
1.0	68 percent	8.3	5.6
1.4	85 percent	12.2	7.8
2.0	95 percent	16.7	11.8
3.0	99 percent	25.0	16.7

Unlike military programs, NASA and commercial-satellite procurement agencies do not have a specification on uncertainty margins for thermal analysis. An informal survey of NASA and commercial-satellite programs showed that 5-deg C was the most common margin used, although significantly different margins were used on some programs. A summary of margins typically used on commercial programs is shown in Table 10.

Table 10. Commercial Satellite Component Temperature Ranges

	Thermal Analysis Range	Acceptance Range	Qualification Range
Hughes	+5 to 55-deg C	0 to 60-deg C	-5 to 65-deg C
Martin Marietta	-5 to +50-deg C	-10 to +55-deg C	-15 to 60-deg C
Loral	+5 to 55-deg C	0 to 60-deg C	-5 to 65-deg C

Note: Temperture ranges for many commerical programs are self-imposed by contractor and are not contractually required by the customer.

Recommended Margins

For components that have no thermal control or have passive thermal control only, an uncertainty margin of at least 11-deg C should be included in all cases in determining the maximum or minimum expected flight temperature. This 11-deg C thermal margin is applied to the temperature predictions made by the thermal math models after correlation to the thermal-balance test. This implies that even larger thermal margins are required at the beginning of a program to accommodate changes that typically evolve from preliminary design to final product. The suggested margin during the design phase is 17-deg C which can be reduced to 11-deg C after the thermal balance test.

For cryogenic systems operating below approximately minus 70-deg C, the heat load margins shown in Table 11 are recommended in lieu of the 11-deg C temperature margin.

Table 11. Thermal Uncertainty Margins for Cryogenic Systems

Milestone	Heat Load Margin (percent)
Program go-ahead	50
PDR	45
CDR	35
Qualification	30
FDR/Flight Acceptance	25

Constant conductance heat pipes are considered passive thermal control elements and should use the 11-deg C margin discussed above. In addition, the heat-transport capability of the pipe should be at least 50 percent greater than that required for the maximum heat load at the maximum expected flight temperature.

Self-regulating heaters that use resistance elements that exhibit a large variation in resistance with temperature (such as "auto trace" or positive-temperature coefficient thermistors) are considered passive devices, and a margin of 11-deg C is required.

For thermal designs in that temperatures are actively controlled by variable-conductance heat pipes, louvers, heat pumps, expendable coolant systems, or refrigerators, a heat-load margin of 25 percent may be used in lieu of the 11-deg C specified above at the worst-case hot and/or cold extreme design conditions. Similarly, for thermostatically or proportionally controlled heaters, a 25 percent heater capacity margin may be used in lieu of the 11-deg C at the cold end, which translates into a duty cycle of no more than 80 percent at the minimum expected flight temperature under worst-case cold conditions.

A discussion of how these margins relate to test temperatures of spacecraft components can be found in Chapter IX.

THERMAL MATH MODEL COMPUTER CODES (SINDA)

Solving the general heat-transfer equation is the objective of all the thermal analysis codes in the spacecraft industry. The general partial, differential equation of heat conduction with source terms for a stationary heterogeneous, anisotropic solid is

$$\rho C_p \frac{\partial T}{\partial t} = \nabla \cdot (K \cdot \nabla T) + Q(T, t) \text{ (Energy rate per unit volume)}, \tag{3}$$

where ρ = density (lbs/ft ^{3}),

C_p = specific heat $\left(\frac{\text{Btu}}{\text{lb} - {}^\circ\text{F}}\right)$,

∇ = gradient operator (1/ft) ,

K = conductivity tensor $\left(\frac{\text{Btu}}{\text{hr. Ft. }^\circ\text{F}}\right)$,

T = temperature (∞F),

t = time (hr),

Q = source term $\left(\frac{\text{Btu}}{\text{hr. Ft.}^3}\right)$

(Note: A consistent set of units is required.)

Eq. 3 is a parabolic differential equation in which the Fourier conduction law ($q = - K \bullet \Delta T$) is used. Although temperature (T) is a scalar that can vary with position, i.e., T=(x,y,z,t), the heat flow depends on the temperature gradient in a particular direction and is therefore a vector quantity. The intent of this summary is not to derive heat-transfer equations or provide a detailed discussion of SINDA applications, but as an overview of the theory and available software. The textbooks, by Frank Kreith in Ref. 1, J. P. Holman in Ref. 2, and M. S. Carslaw and J. C. Jaeger in Ref. 3, provide a good review of heat-transfer theory. The SINDA manuals (Refs. 4 and 5) can be consulted for building detailed thermal models. Most of the aerospace companies in the spacecraft industry use finite-difference numerical techniques to solve Eq. (3) for various heat-transfer problems with appropriate boundary conditions. Generally, each company has its own finite difference thermal analyzer and some version of SINDA. Most of these versions are either SINDA87 written by J. Gaski (Ref. 4) or SINDA85/FLUINT (Ref. 5) developed by Martin Marietta for NASA Johnson Space Center (JSC). Some companies still use the original version of SINDA developed by J. Gaski in 1966, which is referred to as CINDA (Chrysler Improved Numerical Differencing Analyzer).

SINDA is a thermal analysis code that consists of a preprocessor and execution library. By following certain rules, the preprocessor reads the SINDA input data file developed by the analyst and constructs a FORTRAN executable. The analyst selects various SINDA subroutines from the SINDA library to obtain temperatures. SINDA allows the user to include the necessary FORTRAN logic to solve a specific heat-transfer problem. The FORTRAN code can be added into any one of the SINDA operation blocks. The Gaski version has a one-dimensional incompressible-fluid thermal-analysis capability for evaluating pumped-fluid-heat-transfer networks.

SINDA85 represents a significant evolution from the previous SINDA-type codes. SINDA85 has fluid-network analysis capability for incompressible, compressible, two-phase flow, etc., and also allows the analyst to build a thermal model from separate submodels. Both are very powerful features.

These codes determine the solution to a finite-difference model that approximates the physical object. The nodes or subvolumes are assumed to be isothermal and to have constant physical properties within the node. Some heat-transfer books refer to finite-difference node meshes as lumped parameter representations. The nodes are then interconnected by conduction and/or radiation. The governing partial differential equation is converted into a system of finite-difference equations by constructing a finite-difference mesh. The basis for making this step is the Taylor series approximation. A three-dimensional Cartesian coordinate system is assumed for this discussion. From Figure 35, which shows typical one- and two-dimensional finite-difference meshes, the Taylor series about x_0 for T(x) is written for the one-dimensional mesh

$$T(x_o + \Delta_x) = T(x_o) + \left.\frac{\partial T}{\partial x}\right|_{x=x_o} \bullet \Delta x + \left.\frac{\partial^2 T}{\partial x^2}\right|_{x=x_o} \bullet \Delta x^2/2! + \left.\frac{\partial^3 T}{\partial x^3}\right|_{x=x_o} \bullet \Delta x^3/3! + --- .$$

From this approximation, the first and second derivatives can be derived

$$\left.\frac{\partial T}{\partial x}\right|_{x=x_o} = \frac{T(x_o+\Delta x) - T(x_o)}{\Delta x} + 0(\Delta x)$$

$$\left.\frac{\partial^2 T}{\partial x^2}\right|_{x=x_o} = \frac{\dfrac{T(x_o+\Delta x) - T(x_o)}{\Delta x} - \dfrac{T(x_o) - T(x_o-\Delta x)}{\Delta x}}{\Delta x} + 0\left(\Delta x^2\right) ,$$

where the 0 (Δx) and 0 (Δx^2) are a means of expressing the order of the truncation error associated with the approximation. Eq. 3 can be written for a heterogeneous, anisotropic solid, the conductivity of which in each of the three principal directions is a function of temperature;

$$\rho C_p \frac{\partial T}{\partial t} = \frac{\partial}{\partial x}\left[k_x(T)\frac{\partial T}{\partial x}\right] + \frac{\partial}{\partial y}\left[k_y(T)\frac{\partial T}{\partial y}\right] + \frac{\partial}{\partial z}\left[k_z(T)\frac{\partial T}{\partial z}\right] + Q(T,t) \quad . \qquad (4)$$

The x-partial derivative $\frac{\partial}{\partial_x}\left[k_x(T)\frac{\partial T}{\partial_x}\right]$ can be written as

$$\left[k_x\left(\delta^+\right)\bullet\left(\frac{T_{n+1}(x+\Delta x, y, z, t) - T_n(x, y, z, t)}{\Delta x}\right) - k_x\left(\delta^-\right)\bullet\left(\frac{T(x, y, z, t)\,n - T_{n-1}(x-\Delta x, y, z, t)}{\Delta x}\right)\right] /\Delta x \ , \qquad (5)$$

where n is the node number about which the Taylor series is applied,

$$\delta^{\pm} = \left[T_n(x, y, z, t) + T_j(x \pm \Delta x, y, z, t)\right] /2$$

or

$$\delta = T_n(x, y, z, t) \ ,$$

j is the adjacent node, x, y, z are the spacial coordinates of n, and t is the time. By multiplying Eq. (5) with the volume $\Delta x \bullet A$ where $A = \Delta y \bullet \Delta z$, there results

$$A \bullet k_x\left(\delta^+\right) \bullet \frac{\left(T_{n+1} - T_n\right)}{\Delta x} - A \bullet k_x\left(\delta^-\right) \bullet \frac{\left(T_n - T_{n-1}\right)}{\Delta x}\,, \qquad (6)$$

where T_i is shorthand for T_i (x, y, z, t) and i = n.

Let the coefficient $A \bullet k_x\,(\delta^{\pm})\,/\,\Delta x$ be defined as the parameter G, the conductance. Hence, Eq. (6) becomes

$$G_{n+1} \bullet \left(T_{n+1} - T_n\right) - G_{n,n-1} \bullet \left(T_n - T_{n-1}\right)\,, \qquad (7)$$

where $G_{n+1,n} = \dfrac{k(\delta^+) \bullet A}{\Delta x}$ and $G_{n+1,n} = \dfrac{k(\delta^-) \bullet A}{\Delta x}$. Similar expressions can be written for the other terms $\dfrac{\partial}{\partial y}\left[k_y\,(T)\dfrac{\partial T}{\partial y}\right]$ and $\dfrac{\partial}{\partial z}\left[k_z\,(T)\dfrac{\partial T}{\partial z}\right]$ in Eq. (4).

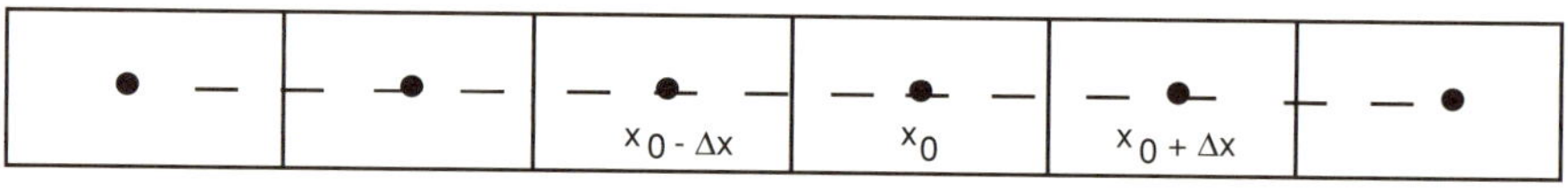

1 - D Mesh Uniform Spacing

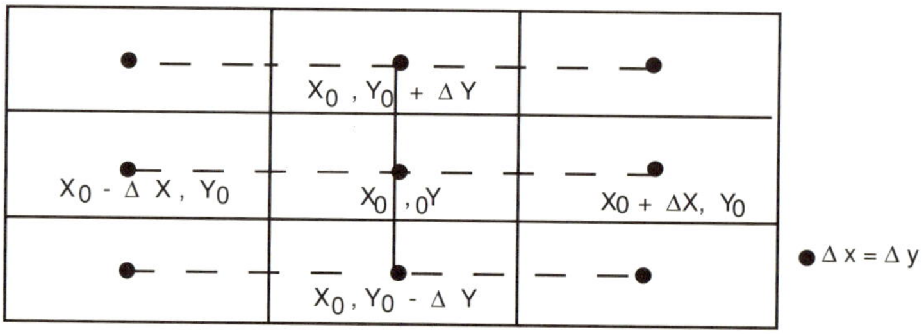

2 - D Mesh Uniform Spacing

Figure 35. Finite-difference method (FDM)

The conductance G is placed by the analyst in the conduction block of SINDA. Hence, through the Taylor series approximation, a partial differential equation has been converted into a set of finite-difference equations that can now be solved numerically. The source term in Eq. (4), Q (t), is how external and internal radiation, convection, and heat sources are added to the difference equation. The radiation term is typically written as

$$\sigma A \mathscr{F}_{n,n+1} \left(T_n^{\ 4} - T_{n+1}^{\ 4} \right) , \tag{8}$$

where σ is the Stefan-Boltzmann constant, A is the surface area of the radiating surface, and $\mathscr{F}n$, n + 1 represents the net radiation exchanged between two real surfaces, including all possible reflection paths. From the SINDA perspective $A\mathscr{F}_{n,n+1}$ is just another conductance, except it is a radiation coefficient. In SINDA, radiation conductances are distinguished from convection and conduction coefficients by a minus sign, i.e., - G denotes a radiation conductor and G indicates a normal conductor (conduction or convection).

In building a thermal model the analyst decides how many nodes to use, how to distribute them, and how to connect them either by radiation, conduction, or convection. The resulting model network produces a system of finite-difference equations with either constant or variable coefficients. The number of equations to be solved depends on the number of nodes selected by the user in the thermal model minus any boundary nodes, which have a prescribed temperature history. For example, space is considered a boundary node and is set at 0-deg K or -460-deg F, depending on the units used.

In order to convert the finite difference equations to a set of algebraic equations that are then solved within SINDA, the time derivative must be approximated just like the spacial derivatives. The $\frac{\partial T}{\partial t}$ in Eq. (4) can be approximated as follows :

$$T\left(t^{*} + \Delta t\right) = T\left(t^{*}\right) + \theta \bullet \left.\frac{\partial T}{\partial t}\right|_{t^{*} + \Delta t} \bullet \Delta t + (1 - \theta) \bullet \left.\frac{\partial T}{\partial t}\right|_{t^{*}} \bullet \Delta t , \tag{9}$$

where θ is a variable-weighted implicit factor. Multiplying Eq. (4) by the volume $(\Delta x \bullet A)$, the coefficient for $\frac{\partial T}{\partial t}$ becomes

$$C_n = \rho \bullet Cp \bullet \Delta x \bullet A ,$$

where C_n denotes the capacitance of node n and A is the crossectional area $\Delta y \bullet \Delta z$. By combing Eqs. (7), (8) and (9), Eq. (4) becomes

$$C_n \frac{[T_n (t + \Delta t) - T_n (t)]}{\Delta t} = \theta \bullet \left[\sum_{j=1}^{N} G_{jn} \left(T_j - T_n\right) + \right.$$

$$\sigma \bullet \sum_{j=1}^{N} \mathcal{F}_{jn} A_n \left(T_j^4 - T_n^4\right) + Q_n \left(T_n, t\right)\Big]_{t^* + \Delta t} +$$

$$(1-\theta) \bullet \left[\sum_{j=1}^{N} G_{jn} \left(T_j - T_n\right) + \right.$$

$$\sigma \bullet \sum_{j=1}^{N} \mathcal{F}_{jn} A_n \left(T_j^4 - T_n^4\right) + Q_n \left(T_n, t\right)\Big]_{t^*} . \tag{10}$$

This equation contains the parameter θ, which can be adjusted along with the finite-difference mesh size and time step to yield various finite-difference approximations with different local truncation errors. The values $\theta = 0$, 1/2, and 1 yield the forward-explicit, Crank-Nicolson, and backward-implicit approximations.

Selecting a particular finite-difference mesh scheme and evaluating the coefficients in Eq. (10) yields a system of n algebraic equations where n is the number of finite-difference nodes. Note n does not include boundary nodes. If $\theta = 0$, each equation is explicit and has only one unknown temperature, T_n. If $\theta > 0$, a system of algebraic equations exists and must be solved by either iterative techniques, matrix inversion schemes, or decomposition procedures. Typically the system of equations is written as

$$T_{new} = [A] \bullet T_{old} \quad , \tag{11}$$

where [A] is an n x n matrix and T is an n x 1 or column matrix. For thermal models of ten or more finite-difference nodes, [A] is typically a sparse matrix because each node is normally connected to a small subset of the total number of nodes in the model. For most heat-transfer problems, [A] is not banded because of radiation interchange between the nodes. Consequently the efficient solvers for tridiagonal matrices are not generally useful.

Three types of errors can occur with the application of finite-difference to heat-transfer problems. The first error is the truncation error, which is the difference between the differential equation and the approximating difference equations (Refs. 6, 7, and 8). This can be illustrated for the one-dimensional heat-transfer equation with constant conductivity. Let

$$F_{pde}(T) = \left(\frac{\partial T}{\partial t} - k \bullet \frac{\partial^2 T}{\partial x^2} \right) \quad \text{(partial differential equation)}$$

$$F_{fd}\left(T_i\right) = \left(\frac{T_{t^* + \Delta t,\, x^*} - T_{t^*,\, x^*}}{\Delta t}\right) -$$

$$k\left(\frac{T_{x^* + \Delta x,\, t^*} - 2T_{x^*,\, t^*}\; T_{x^* - \Delta x,\, t^*}}{\Delta x^2}\right) \text{(finite-difference equation),}$$

then $[F_{fd}(T_i) - F_{pde}(T)]$ represents the truncation error at each node. T_i refers to the temperature at three successive nodes $x^* + \Delta x$, x^*, and $x^* - \Delta x$, and t^* designates a discrete time. The temperature T in the analytical solution is a continuous function. The truncation error is determined from the finite-difference node spacing (mesh size) and the size of the time step. As the number of finite-difference nodes is increased and the time step decreased, the error associated with the Taylor series approximation (truncation) decreases and approaches zero in the limit. In this case the truncation errors approach zero and the difference equation is said to be consistent with the partial differential equation. However, as the number of nodes in the network expands, the corresponding number of difference equations to be solved increases. This, in turn, increases computer-execution time. From the viewpoint of algebraic simplicity, the analyst would prefer the coarsest network possible. The best thermal model is a compromise between node size and computational cost. There are no specific rules for selecting a network size, only insight and experience for choosing the optimum size. One way to judge the truncation errors introduced by too coarse a network is to estimate the truncation error as the calculation proceeds. The second potential error source is related to the stability of the numerical solution. If the effect of errors tends to diminish as the numerical solution progresses, the solution is stable and converges. However, if the errors tend to grow with time, then the solution becomes unstable and diverges. The third error source is the result of computer round-off errors during numerical calculations. This is the difference between the exact numerical answer and the actual numerical answer. Round-off error is a real problem with 16-bit computers, somewhat of a problem with 32-bit machines, and generally not a problem with 64-bit computers. Hence, the numerical temperature, T_{num}, is given as

$$T_{num} = T_{ex} + \underbrace{\left(T_{num} - T_{exn}\right)}_{\text{round off}} + \underbrace{\left(T_{exn} - T_{ex}\right)}_{\text{discrete}}, \qquad (12)$$

where T_{ex} is the analytical solution and T_{exn} is the exact numerical solution. The discrete error is the combination of the truncation and stability errors. As discussed previously, these errors are directly coupled to the mesh size and time step assumed by the analyst. The truncation error for a uniform mesh (See Figure

35) is typically 0 (Δx^2) i.e., of second order. However, for a non-uniform mesh (See Figure 36), the truncation error becomes 0 (Δx), i.e., first order. Hence, a non-uniform finite-difference mesh reduces the order of the truncation error and hence decreases the accuracy of the approximation. Most spacecraft thermal models are not uniform. However, if sufficient thermal nodes are used, the numerical answers will be reasonably accurate.

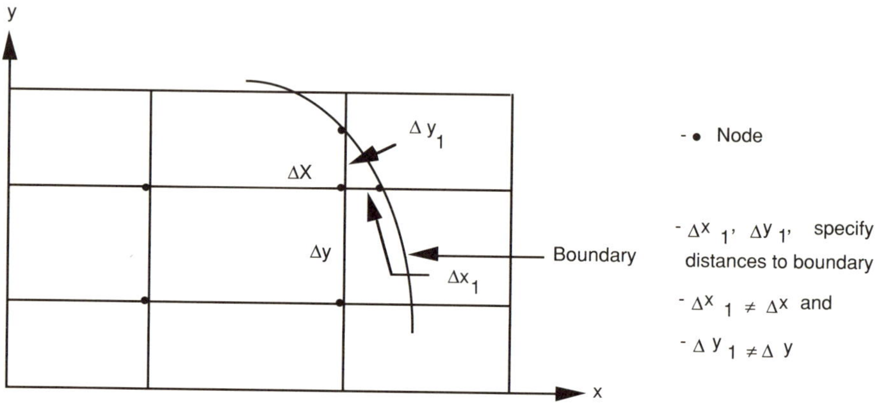

Figure 36. FDM two-dimensional non-uniform mesh

The forward-differencing expression is obtained from Eq. (9) by setting θ = 0; hence,

$$T\left(t^* + \Delta t\right) = T\left(t^*\right) + \left.\frac{\partial T}{\partial t}\right|_{t^*} \bullet \Delta t \quad , \qquad (13)$$

and is shown in Figure 37. This method requires that the calculation of T_i at $t^* + \Delta t$ be based on T_i that are known at t^* the previous time. This is illustrated by setting $\theta = 0$ in Eq. (10). The forward-differencing assumption is explicit and the solution can be unstable if the time step, Δt, is too large. The criteria for stability is determined by calculating the minimum value

$$\tau_n = \frac{C_n}{\sum_j G_{nj}}$$

for each finite-difference node, where j is the sum over all nodes connected to n by conduction and radiation. Convection would be included in a prelaunch environment. The thermal capacitance of the node is C_n and the G_{nj}'s are the conductance values between adjacent nodes. If radiation occurs between two nodes, the value is linearized to obtain

$$G_{nj} = \sigma \mathcal{F}_{nj} A_j \left(T_n^2 - T_j^2\right)\left(T_n, T_j\right) . \tag{14}$$

In SINDA, τ is called CSGMIN. CSGMIN represents the smallest time constant in the thermal network at each time step. It can change from time step to time step. CSGMIN includes the effect of boundary conditions if the node that has the smallest τ is connected to any boundary nodes. The solution process will remain stable if the time step Δt is always less than CSGMIN. In SINDA $\Delta t = 0.95*$ CSGMIN/CSGFAC is always used, with CSGFAC defaulted to 1.0.

The forward differencing equation has one unknown node temperature at $t^* + \Delta t$,with all the other temperatures known at t^* (see Eq. 10 with $\theta = 0$). Any radiation terms are approximated by Eq. (14). Although this explicit equation is simple to solve, the time step Δt is limited by the stability criteria for the node with the smallest time constant. Hence, in using this technique the analyst is trading off simplicity versus potentially many small time steps. This can cause excessive computer-execution times and significantly increase costs on central computers, or completely consume the CPU on a local workstation. In applying the forward-differencing equations, the analyst does not have to specify the convergence criteria and a time step, since this can be conveniently computed from the specified thermal data.

Another technique used to solve heat-transfer equations is backward differencing. In this case the heat balance is written in terms of the unknown temperatures at $t^* + \Delta t$,

$$T\left(t^* + \Delta t\right) = T\left(t^*\right) + \left.\frac{\partial T}{\partial t}\right|_{t^* + \Delta t} \bullet \Delta t . \tag{15}$$

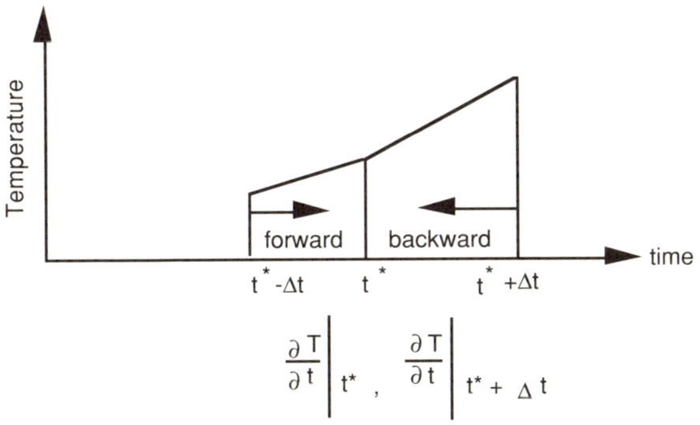

Figure 37. Explicit and implicit derivatives

This is obtained by setting $\theta = 1$ in Eq. (9). Figure 37 illustrates the backward slope $\frac{\partial T}{\partial t}$. This approach yields a system of n equations, where n is the total number of finite-difference nodes whose temperatures are calculated at each time step. Boundary nodes are excluded. This formulation is called implicit. The minimum time constant CSGMIN is still calculated in SINDA for implicit methods. Since implicit methods are unconditionally stable, the time step Δt can exceed CSGMIN. However, if the time step selected is too large, although stable, the truncation error can become significant. When using an implicit method, the analyst must specify the time step. The user should always compare the specific time step to CSGMIN. If the selected time step is five to ten times CSGMIN, it is probably too large. This is obviously dependent on the problem being solved.

Iterative schemes are typically used to solve systems of equations. Such techniques require a convergence criteria. For transient problems, the SINDA constants DRLXCA and ARLXCA must be specified to use the implicit schemes. The reason two constants are required is because SINDA allows both diffusion and arithmetic nodes. Diffusion nodes have mass, whereas arithmetic nodes do not. DRLXCA is the convergence criteria for diffusion nodes and ARLXCA is the convergence criteria for arithmetic nodes.

The advantage of backward differencing rests with the ability to vary the time step. During periods of rapidly varying boundary conditions, the time step can be reduced. Similarly, during periods of slowly changing boundary conditions, the time step can be increased. Typically, implicit numerical schemes are faster than the explicit-forward method because of the large time steps allowed. However, the larger the time step, the more iterations required to achieve a solution. Each iteration is essentially equivalent to a time step. Hence, the actual implicit time step is approximately the specified Δt divided by the number of iterations required to achieve a solution. The user needs to compare this modified time step to CSGMIN to verify that the specified time step is providing the increased computational speed expected over the explicit method. For some problems the implicit scheme may not be any faster than the explicit method.

The following stability criteria are associated with Eq. (10):

$\theta = 0$, $\Delta t <$ CSGMIN.

$\theta < 1/2$, conditionally stable, i.e., CSGMIN $< \Delta t <$ CSGMIN $(1+\Delta)$ where $\Delta \to \infty$ as $\theta \to 1/2$.

$1/2 \leq \theta \leq 1$, unconditionally stable for any Δt.

All the SINDA codes allow three types of nodes: diffusion (with mass), arithmetic (no mass), and boundary (specified temperature). These definitions are particularly useful when solving equations whose time constants vary by several orders of magnitude or higher. If arithmetic nodes were not allowed, the algebraic system of equations would be very stiff. This class of problem (stiff equations) can only be solved with implicit techniques. However, when the CSGMIN is very small for some diffusion nodes, they can be converted to arithmetic nodes. This helps make the equations less stiff and improves the computational efficiency without sacrificing accuracy. The temperature of an arithmetic node is obtained by noting that the sum of heat flow into the node is zero. Typically, MLI blankets are modeled as arithmetic nodes because they are light and respond instantaneously to the environment. Another example occurs when the analyst places arithmetic nodes between the interface of two surfaces in order to obtain the interface temperature.

In summary, the finite-difference approximations to the partial differential heat-transfer equation discussed in the preceding sections required that continuous variables be quantized. Spatial variables were quantized as nodes and connected by conductors while time was divided into discrete steps, Δt. For finite-difference nodes of size $\Delta v = \Delta x \bullet \Delta y \bullet \Delta z$, the time step and spatial dimensions are related through the CSGMIN definition. Since the finite-difference solution approaches the exact solution as ΔV and Δt approach zero, the logical question is, what limits ΔV or Δt? The answer is cost and computer storage (model size). Collectively these constraints limit ΔV to a non-zero minimum value. A small ΔV means a large number of nodes and conductors, and the computer memory must contain enough space to hold all of the parameters (capacitance, temperature, conductance, etc.) associated with the model. A large model is expensive to execute on a computer and expensive in engineering man-hours to develop. The analyst has to use engineering judgment to decide how much detail is sufficient to determine the thermal response of the physical system being analyzed. The time step is chosen consistent with CSGMIN for the forward-explicit method. If CSGMIN is too small, this can greatly increase the computational costs. The implicit methods allow time steps greater than CSGMIN. In this case the analyst needs to determine the largest acceptable Δt that will not impact the accuracy of the calculated temperatures. For each thermal problem the analyst is faced with the task of developing a model and selecting a solution technique that will yield accurate stable temperatures for the least cost.

Numerous other approaches to formulating and solving finite-difference equations are available. The Gaski 1987 SINDA has several transient and steady-state solutions. This approach provides the analyst with the flexibility to select a solution subroutine that will work. As noted previously, the forward-explicit method is limited to time steps less than CSGMIN. However, two other explicit schemes, one by Saul'yev and the other by Dufort-Frankel, allow time steps greater than CSGMIN because these methods are unconditionally stable. The Saul'yev alternating-direction explicit method is implemented in the Gaski SINDA/1987 as the execution subroutine SNADE, and the modified Dufort-Frankel technique is called SNDUFR in Gaski SINDA/1987 (Ref. 4). The original Dufort-Frankel method used the following approximations in time,

$$\frac{\partial T}{\partial t} = \frac{1}{2}\left(T_n\Big|_{t^* + \Delta t} - T_n\Big|_{t^* - \Delta t} \right) , \tag{16}$$

and in space,

$$T_n\Big|_{t^*} = \frac{1}{2}\left(T_{n+1}\Big|_{t^*} - T_{n-1}\Big|_{t^*} \right) . \tag{17}$$

These approximations produced a finite-difference technique whose truncation error was $0\,(\Delta t^2 , \Delta x^2)$ accurate as compared to the forward-differencing scheme, which is $0\,(\Delta t + \Delta x^2)$ accurate. The central-difference approximation in Eq. (16), which is $0\,(\Delta t^2)$ accurate, causes a term

$$\frac{\partial^2 T}{\partial t^2} \bullet \frac{\Delta t^2}{\Delta x^2}$$

to appear in the truncation error. Depending on the relationship between the spacing of the nodes and the time step, the resulting equation can be hyperbolic instead of parabolic. The old SINDA codes used the original Dufort-Frankel approximation, which generated temperatures that were off by 10 to 20-deg F. This was a direct result of the term

$$\frac{\partial^2 T}{\partial t^2} \bullet \frac{\Delta t^2}{\Delta x^2} .$$

By using the Euler approximation in time, which is normally used, we have

$$\frac{\partial T}{\partial^2} = T_n\Big|_{t^* + \Delta t} - T_n\Big|_{t^*} . \tag{18}$$

The modified Dufort-Frankel method uses Equations (17) and (18), and produces accurate solutions for time steps greater than CSGMIN. The truncation error is $0\,(\Delta t, \Delta x^2)$. The $\frac{\partial^2 T}{\partial t^2} \bullet \frac{\Delta t^2}{\Delta x^2}$ term is higher order and is not a factor with this approximation. One final note; the Richardson explicit method used the time approximation in Eq. (16) with the standard approximation in the spatial derivative. This method was unconditionally unstable. The application of Eq. (16) as an approximation in time has generally not been successful.

The most common implicit scheme is the Crank-Nicolson (CN) method. Their approximation is obtained from Eq. (10) with $\theta = 1/2$. The Crank-Nicolson technique is simply an application of the trapezoid rule. The method has a truncation error $O(\Delta t^2, \Delta x^2)$. The basic CN equation for one dimension can be derived from adding:

$$\frac{T_{t^* + \Delta t/2} - T_{t^*}}{\Delta t/2} = \left[\frac{T_{n+1} - 2T_n + T_{n-1}}{(\Delta x^2)}\right]_{t^*} \quad \text{Explicit Forward,} \qquad (19)$$

and

$$\frac{T_{t^* + \Delta t} - T_{t^* + \Delta t/2}}{\Delta t/2} = \left[\frac{T_{n+1} - 2T_n + T_{n-1}}{(\Delta x)^2}\right]_{t^* + \Delta t} \quad \text{Implicit Backward,}$$

to yield

$$T_{t^* + \Delta t} - T_{t^*} = \frac{\Delta t}{2\Delta x^2} \bullet \left[\left[T_{n+1} - 2T_n + T_{n-1}\right]_{t^* + \Delta t} + \left[T_{n+1} - 2T_n + T_{n-1}\right]_{t^*}\right]. \qquad (20)$$

This approximation is not limited to one dimension, which was selected to illustrate the method. The CN approximation can also be obtained by simply adding the classic explicit and implicit methods. Because the approximation is centered about $t^* + \Delta t/2$, this accounts for the increased accuracy in time over the classic-explicit and backward-implicit methods, which are centered at t^* and $t^* + \Delta t$, respectively. FWDBCK in the Gaski SINDA/1987 uses the CN method. The SNTSM method in SINDA/1987 (Ref. 4) is a Taylor series method with the weighted average approximation (Eq. 9) and automatic time step selection. Table 12 lists the various transient and steady-state execution subroutines used in SINDA/1987. The SINDA/FLUINT uses only four execution subroutines.

Several steady-state subroutines are used in SINDA/1987. Table 12 contains the available subroutines. For most reasonably sized thermal models, SCROUT, which is based on the Choleski method, is the best selection for steady-state solutions. The steady-state convergence criteria for the iterative methods is based on first meeting a global-temperature relaxation error, next a system-energy balance criteria, and finally a nodal-energy balance error. The user specifies the minimum acceptable criteria. Typically, a T_{i+1} - T_i delta difference of less than .01-deg F (if deg F is the unit being used), a system-energy balance error of 1 percent and a nodal energy balance error of .5 percent are used. Note i is the iteration count. The analyst can adjust the error constants in SINDA by specifying values for specific user constants. The reason three steady-state criteria are used is that temperature relaxation can signal false convergence for some problems. Thus by specifying system- and nodal-energy balance criteria, convergence to the correct answer is assured even if the temperature relaxation is misleading.

Table 12. Typical SINDA Execution Subroutines for FDM

- Transient:

SNFRDL		Explicit forward differencing
FWDBKL	(FWDBCK, FDBKCD)	Quadratic or linear equation, implicit finite differencing, successive point iteration
SNADE		Alternating direction explicit finite differencing
SNDUFR		Modified Dufort-Frankel explicit finite differencing
ATSDUF		Automatic time step selection, like SNDUFR otherwise
ATSFBK		Automatic time step selection, like FWDBCK otherwise
SNTSM	(SNTSM1) (SNTSM3) (SNTSM4)	Taylor series with weighted average, automatic time step selection

- Steady State

STDSTL		Quadratic or linear equation with successive point iteration
SNHOSS	(SNHOSD)	Taylor series, explicit
SNSOSS		Taylor series, implicit
SNDSNR		Newton-Rhapson with Gauss-Jordan reduction
SCROUT		Matrix decomposition, elimination method
SNSOR	(SNSOR1)	Successive over-relaxation

The finite-element method provides a second approach to numerically solving heat-transfer problems (Ref. 9). The finite-element mesh schemes are the real strength of this technique. Each finite-element model normally has hundreds of elements. Two approaches are used to develop a solution with finite-element. One is called the method of weighted residuals (MWR) and the other is the Ritz variational method. The most widely used finite-element technique is the Galerkin approach, which is one of four MWRs. Finite element seeks an explicit expression for the temperatures, $\tilde{T}$, in terms of known functions that on the average satisfy the governing differential equations and the boundary conditions exactly on an element. The $\tilde{T}$ is the finite-element approximation to the actual temperature, T. The form used for $\tilde{T}$ is

$$\tilde{T}(t, a_i) = \sum_{i=o}^{N} a_i \varnothing_i (t) , \tag{21}$$

where the a_i are referred to as degrees of freedom (DOF), N is the total number of DOF, and the $\varnothing_i$ (t) are called by various authors as trial, basis, shape, interpolation or coordinate functions. Typically the $\varnothing_i$ are assumed to be powers if x, sine, cosine, etc., on the element. This approach parallels the analytical technique of finding a function or set of functions that solves the differential equation and also satisfies the prescribed boundary conditions. An example of a basis function is

$$\varnothing_i = 0, \quad t = a$$

$$\varnothing_i = \frac{a - t}{a - b}, \quad a < t < b \tag{22}$$

$$\varnothing_i = 1, \quad t = b \quad .$$

Hence $\varnothing_i$ is a linear function whose value varies from 0 to 1. A bar element has an element node at each end, and a triangle element has a node at each corner (Figure 38). The a_i are specified at each element node. For a thermal problem, a_i equals T_1, where the T_i are the element-node temperatures. The essence of the method is to obtain a set of algebraic equations for the element-node temperatures T_i which form a column vector called $\tilde{T}$. The temperatures between element nodes are found by applying the basis function between those nodes. For example Eq. (22) for $\varnothing_i$ would be used to find the temperature between element nodes a and b. For a reasonable finite-element mesh, $\tilde{T}$ should approach T, the exact temperature solution. Basis functions used in Eq. (21) can be linear (Eq. 22), quadratic, cubic, or quartic. Examples of element shapes are:

One-Dimensional,	bar.
Two-Dimensional,	Triangular, rectangular, quadrilaterals.
Three-Dimensional,	Hexahedrals, pentahedrals, tetrahedrals.

To improve the accuracy of the finite-element method either a smaller mesh (more elements) is used or higher-order basis functions (increased DOF) on the elements are needed. This FEM technique does not provide the analyst with an error estimate like finite difference, which is based on Taylor series expansions. Hence, the analyst either repeats the problem with a smaller mesh or, based on experience, develops a finite-element mesh that appears to provide an acceptable solution. This leads to detailed-mesh structures, since the associated errors are not easily calculated and the analyst does not want to solve the same problem twice for two different mesh sizes in order to establish a convergence criteria.

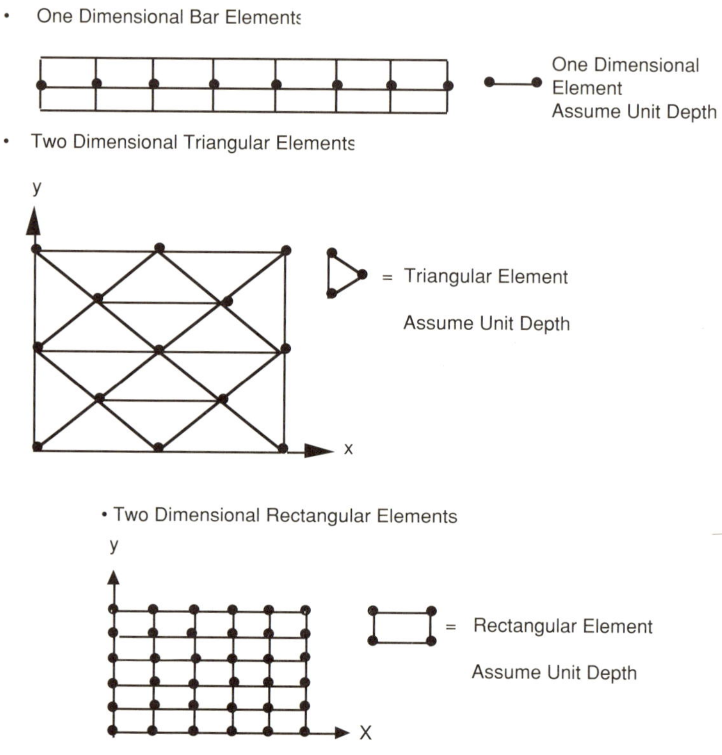

Figure 38. Finite-element mesh (FEM) subdivision

There are many finite-element codes available in the aerospace industry that can be used for thermal analysis. These include NASTRAN, ADINAT, ABAQUS, ANSYS, COSMOS, and TOPAZ. All these codes were developed to perform structural analysis. Every major aerospace company has at least one of these codes.

In summary, either finite difference or finite-element can be used to solve heat-transfer problems. Finite difference is based on Taylor series approximations to develop the algebraic equations that are solved numerically to find a set of temperatures. Each finite-difference node is located at the center of mass and is assumed isothermal throughout the volume occupied. The error associated with the calculation can be estimated. The finite-element method is based on using elements that are one-, two- or three-dimensional, depending on the problem being solved. Combinations of elements can also be used. Each element has element nodes at its corners. The values, for example temperatures, are usually specified or calculated at element nodes. Variations within the element are calculated by using interpolation (basis) functions within the element. Hence the properties and temperature can vary across the element. The Galerkin method of weighted residuals is normally used to develop the algebraic equations that determine the element-node temperatures.

The finite-difference method is excellent for building spacecraft-system models. The method is compatible with the basic surface primitives, e.g., cones, cylinders, spheres, etc., used to describe spacecraft surfaces in the radiation codes. Heat-transfer problems that are primarily driven by radiation can be easily solved with this method. The finite-difference mesh does not have to be uniform; however, the truncation error decreases from $O(\Delta x^2)$ to $O(\Delta x)$. The accuracy of the method is judged by the truncation error from the Taylor series expansions. This assumes that the analyst is using an inherently stable integration scheme and round-off error is small. This may not always be the case. The truncation error can be reduced with a smaller finite-difference mesh and smaller time steps. It is very difficult to compare the accuracy of finite-difference and finite-element methods unless an exact solution is available. This is never the case for nonlinear problems, which is typical for radiation-dominated thermal analyses. The dominant mode of heat transfer for spacecraft is radiation. Thermal models developed with this method can have three types of isothermal nodes: diffusion, arithmetic, and boundary. The arithmetic node, which is a zero-mass node, can be used to avoid stiff equations that always have a large spread in the time constants. In many spacecraft models one thermal node represents an electronic box. This is easily accommodated by finite-difference, but not by finite-element, schemes. In order to overcome the mesh-generation problem for finite-difference techniques, several aerospace companies have used finite-element mesh generators like PATRAN to build a mesh and then convert that mesh to finite-difference for the finite-difference analysis codes like SINDA. The resulting temperatures are then returned to the finite-element mesh-generation code for display. Without a finite-difference pre- and post-processor, this method has a serious disadvantage at building thermal models and displaying the results.

The finite-element method is extensively used in structural analysis. The method is excellent for solving thermal/stress problems. Normally the structural model requires significantly more detail than the equivalent thermal model. Hence the structural characteristics will typically drive the size of the thermal analysis for a combined thermal-stress analysis. Applications of finite-element techniques to the thermal analysis of circuit boards, traveling wave tubes and rocket nozzles are common. The real strength of finite-element techniques are the mesh generation schemes. These techniques can easily handle irregular surface shapes and the interface between two different mesh schemes. Because of the long time application of finite-element methods to structural problems, several excellent commercial mesh-generation packages are available. These include PATRAN from PDA, GEOMOD (IDEAS-SDRC), and ANVIL-5000 from MCS. The pre- and post-processing capabilities of these codes are excellent. Currently, no commercial finite-difference mesh-generation/post-processing package exists of comparable capability. This is why finite-element mesh-generating schemes have been used to develop and post-process finite-difference temperature results. The finite-element codes have the equivalent to diffusion (nodes with mass) and boundary nodes. They do not allow arithmetic (zero mass) nodes. Because of this the resulting algebraic equations can be very stiff and lead to excessive computational costs. Also, the finite-element codes cannot use just one node for an electronic-box simulation, as with finite-difference. For typical thermal analysis, finite-element models will always be larger than necessary. This is driven by the requirement that each element face must share a complete interface with another element. This is also driven by the lack of knowing the error associated with the calculations. Hence, the analyst tends to construct smaller meshes than may be necessary. Typically, curved surfaces like cones and cylinders require far more finite-element surfaces to describe the shape than is needed for finite-difference. One node of 360 deg may be all the analyst really needs. Such a representation is not possible with finite-elements. The Monte-Carlo radiation codes recognize and use the actual surface description for a cone and cylinder. This provides radiation-interchange factors that are correct. If these surfaces are approximated by flat surfaces or polynomial fits, this can impact the accuracy of the interchange factors and unnecessarily increase the cost and complexity of obtaining them. The combination of increased surface numbers and resulting interchange factors can significantly impact the execution time of the thermal model. This increased detail forced by the method of solution will normally not add any additional real information.

Most finite-difference codes like SINDA allow the analyst to include extensive use of user logic, e.g., FORTRAN subroutines, in the thermal model. The finite-element codes like NASTRAN are far more restrictive in this area.

Implicit-solution schemes are best for transient finite-element analysis. This is mainly driven by the fact that the algebraic equations being solved can be very stiff. Many finite-element solution schemes are most efficient with banded matrices; however, with radiation the matrices are not conveniently banded.

The only way to check the accuracy of the finite-element codes is to run the problem again with a smaller mesh size or high-order elements. This is obviously not an inexpensive procedure for determining the error. Typically, error calculations are not made within the finite-element codes.

An approach used to construct a spacecraft thermal model is shown in Figure 39. The various codes used to complete this process at The Aerospace Corporation are listed in Table 13. In the 1960s and 1970s most analysts developed thermal models by hand. This was often a very time-consuming task. In the 1980s with the development of the minicomputers, e.g., VAXs and workstations such as the SUN, Hewlett-Packard, and high-end PCs, the time required to build a thermal model could be greatly reduced through the interactive use of software codes that aid the analyst in model construction, debugging, and execution.

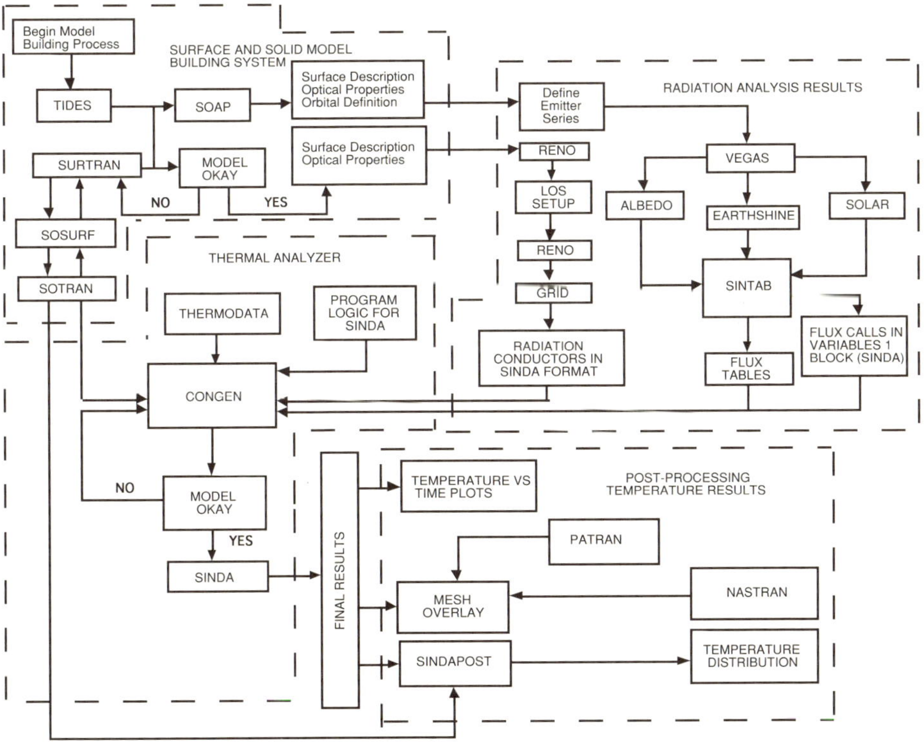

Figure 39. Thermal analysis system flow

Table 13. Thermal Analysis Computer Codes used by The Aerospace Corporation

Component	Function
• SURTRAN	Construct, verify surface geometric radiation model
• SOAP	Verify radiation model and orbit parameters
• SOTRAN	Construct solid-conduction model
• SOSURF	Convert radiation-surface model to conduction-solid model & vice versa
• SINDAPOST	Graphical post-processing for SINDA
• TIDES	Convert surface geometric models between radiation codes
• MESH OVERLAY	Passes temperatures from finite-difference to finite-element meshes
• CONGEN	Output SINDA input deck
• NEVADA	Calculate radiation interchange factors and absorbed fluxes
• SINTAB	Formulate NEVADA to SINDA tables
• SINDA	Thermal analyzer code
• THERMODATA	Contains thermophysical properties in a database

Current efforts in the industry are directed at developing integrated software platforms that allow the analyst to generate complete TMM and GMM models, execute them, and display results in a user-friendly, menu-driven environment, on high-end workstations. They are:

1. Integrated Thermal Analysis Platform (ITAP) by The Aerospace Corporation (in use).

2. Thermal Synthesizer System (TSS) by Lockheed for NASA/JSC (in use).

3. Graphics Enhanced Thermal Analysis Environment (GRETAE) by Martin Marietta (under development).

The thrust of these systems is to facilitate the ability of the analyst to build thermal models in a fast and efficient manner. The goal is to let the computer perform as many of the calculations as possible, so that the analyst can think more about the physics of the problem and less about the mechanics of the calculations. The workstations presently available allow the incorporation of all aspects of the thermal-model building process (Figure 39). This relieves the user of the need to interface with different computer systems to build, analyze, and post-process the results of a thermal model. These workstations provide the analyst with enough computer power to support graphics, analysis packages, and window-display systems, all integrated into powerful analysis platforms. Because the workstation platforms are constantly being improved, the software systems developed for specific workstation applications need to be portable in order to reduce redevelopment costs because of hardware obsolescence.

In addition to these high-end platforms being developed by major aerospace organizations, there are several commercially available thermal-modeling codes available for use on PCs and other machines. Although not as highly integrated as the high-end platforms, they nonetheless dramatically enhance analysis productivity. Such codes include:

- SINDA/FLUINT, SINAPS (Cullimore and Ring Technologies, Inc.)
- Integrated Thermal Analysis System (ITAS, Analytix Corp.)
- SINDA/G with pre- and post-processors (Network Analysis Associates, Inc.)
- Pre-SINDA (MRJ, Inc.)
- SSPTA/386 (Swales and Associates, Inc.)

REFERENCES

1. F. Kreith, *Principles of Heat Transfer*, Intext Educational Publishers, New York, 3rd Edition, 1976.

2. J. P. Holman, *Heat Transfer*, McGraw-Hill Inc., 4th Edition, 1976.

3. H. S. Carslaw and J. C. Jaeger, *Conduction of Heat in Solids*, Oxford University Press, 2nd Edition, 1959.

4. J. D. Gaski, *SINDA 1987/ANSI*, Network Analysis Associates, Inc., POB 8007, Fountain Valley, CA, 1992.

5. B. A. Cullimore, *SINDA '85/FLUINT System Improved Numerical Differencing Analyzer and Fluid Integrator*, Version 2.3, Martin Marietta, 1990.

6. L. Lapidus and G. F. Pinder, *Numerical Solution of Partial Differential Equations in Science and Engineering*, John Wiley and Sons, 1982.

7. J. H. Ferziger, *Numerical Methods for Engineering Applications*, John Wiley and Sons, 1981.

8. G. D. Smith, *Numerical Solution of Partial Differential Equations*, Oxford University Press, 1978.

9. D. S. Burnett, *Finite Element Analysis*, Addison - Wesley Publishing Co., 1987.

10. D. C. Hamilton, W. R. Morgan, Radiant Interchange Configuration Factors, National Advisory Committee for Aeronautics Technical Note 2836, Dec. 1952.

11. *NEVADA User's Manual,* Turner Associates Consultants, Incline Village, NV.

12. G. M. Dusinberre, *Heat Transfer Calculations by Finite Differences*, International Textbook Company, 1961.

Chapter VI

Space Shuttle Integration

Gwynne E. Gurevich
The Aerospace Corporation

INTRODUCTION

The function of the flight- and cargo-integration effort is to ensure that the total payload complement for a given flight is compatible in form, fit, and function, and that all of the associated flight-design parameters and crew activities are within the National Space Transportation System (NSTS) capabilities.* This effort includes assessment by flight- and ground-systems engineering, safety, and all elements of the NSTS operations. The Cargo Integration Review (CIR) is the major cargo-related review of this assessment effort and the time at which customer concurrence is obtained. A typical schedule of events leading to the CIR, subsequent reviews, and, ultimately, flight is outlined in Figures 1 and 2.

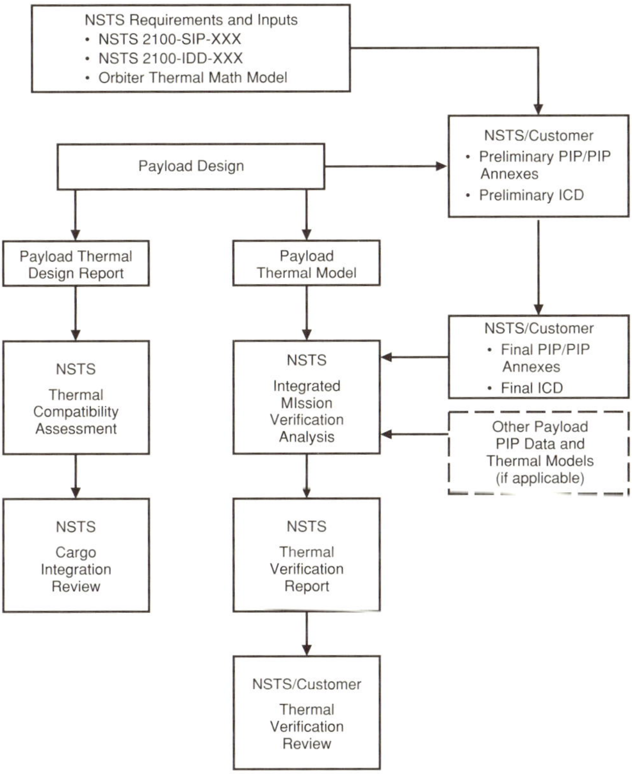

Figure 1. NSTS/payload thermal integration

* Figure 1 describes the Payload/NSTS thermal integration process in flow-chart form.

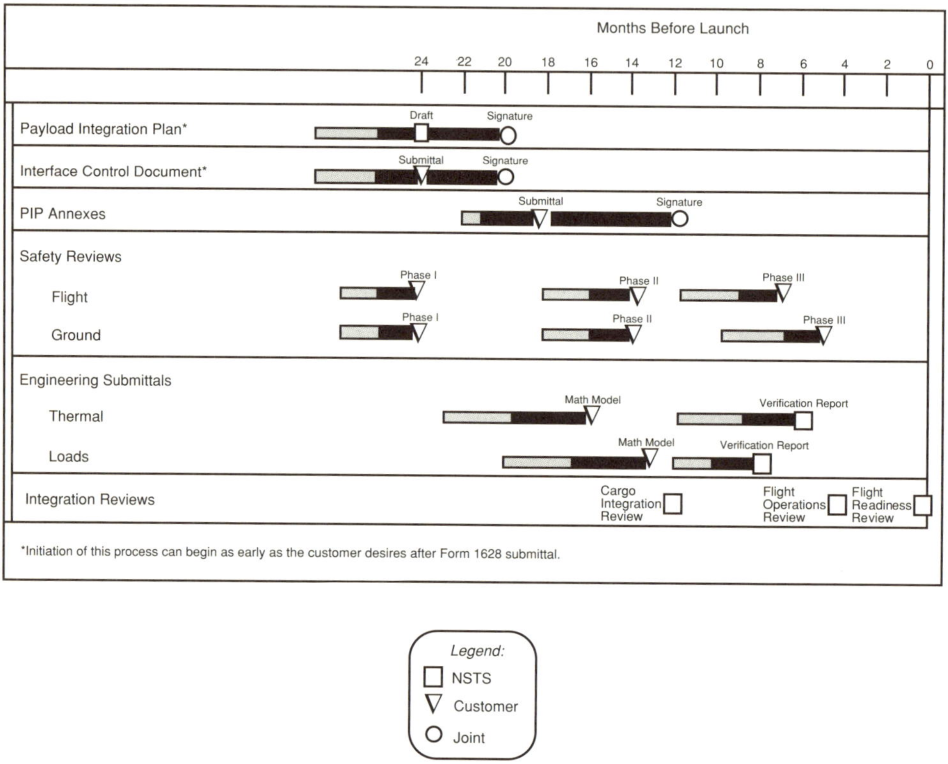

Figure 2. Space Shuttle payload integration timeline

The following assessments are required prior to formal reviews to ensure an adequate NSTS understanding of the cargo and flight requirements and the ability to support such requirements.

- Crew-activities assessment
- Flight operations and support assessment
- Payload operations control center/Mission control center
- Network assessment
- Training assessment
- Ground-operations assessment
- Conceptual flight-profile assessment
- Human use
- Engineering-compatibility assessment
- Interface-verification status
- Safety assessment

ENGINEERING COMPATIBILITY

The purpose of the engineering-compatibility assessment is to verify the compatibility of the integrated Space Transportation System (STS) flight hardware, software, and engineering flight products with current mission requirements, cargo requirements, and Orbiter accommodations.

This is accomplished by teams with specialization in the following areas: avionics, structures, electromagnetic compatibility, thermal, flight preparation, interface verification, reliability and certification, and cargo interfaces. The individual payload assessment begins with the process of developing the Payload Integration Plan (PIP), the payload-unique interface control documents and drawings, the Orbiter Crew Compartment Annex (Annex 6), the Command and Data Annex (Annex 4), and the Extravehicular Activity Annex (EVA) (Annex 11). The individual payloads' physical and functional requirements are integrated by NASA into an engineering data package.

The thermal review must assess both the active and passive thermal control requirements of the integrating hardware against the STS and cargo thermal capabilities and requirements, as defined in the Shuttle Orbiter/Cargo Standard Interfaces, ICD 2-19001. This document contains the Orbiter vehicle attitude hold-time constraints, Orbiter deorbit and entry-preparation constraints, typical temperature ranges for the cargo bay wall/liner, entry-air inlet conditions, typical pre-launch and post-landing environments, typical Remote Manipulator System (RMS) thermal interfaces, Orbiter surface materials and their optical properties, and the vent/purge and active cooling systems capabilities and parameters. This information should be reviewed and checked against the payload requirements to ensure that no payload requirement conflicts with the Orbiter's capability.

In addition to determining the payload-to-Orbiter compatibility, the compatibility of the payload with the stated mission objectives must be assessed. This is a more complicated task, as the mission objectives and companion payload's requirements and limitations may not be well defined. However, once payload thermal compatibility with the Orbiter and mission is determined, a compatibility statement must be signed by both the contractor/payload thermal representative and the NASA thermal-engineering team leader. This is typically done at the CIR. Examples of both the active and passive thermal-compatibility statements are shown on Figures 3 and 4.

Because any thermal assessment requires temperature predictions, accurate thermal models must exist. Accuracy is determined by comparing model predictions to thermal-balance test or flight data. Discrepancies should be worked so that precise temperatures can be predicted. Margin should be added to allow for any analysis uncertainties.

Many analyses must be completed to adequately assess the payload's compatibility with the Orbiter and mission. Analyses should be completed to determine the operational and survival thermal capability of the payload in various standard and mission-specific attitudes, i.e., how long can an attitude be maintained before payload components begin to exceed their temperature limits. A thermal-recovery analysis should also be performed that defines the length of time required for the payload components to warm up (recover from a cold attitude) or cool down (recover from a hot attitude). This information is required by NASA to aid nominal and contingency mission planning and is placed in the Payload Integration Plan (PIP); see Figure 5 for a sample chart. As the payload-bay thermal environment for each attitude is affected by the beta angle (sun angle to orbit plane), a sensitivity analysis should be performed to determine the thermal effects of beta-angle variation. This information should also be incorporated in the PIP.

Payloads that share flights with other payloads and utilize standard accommodations must incorporate into the payload design a minimum thermal capability common to all users of a particular flight. To ensure this mixed-cargo thermal compatibility, the NSTS has defined a set of on-orbit Orbiter attitude requirements with which, as a minimum, all payloads must be compatible. A description of these attitudes is given in NSTS 07700, Volume XIV, Appendix 2: "System Description and Design Data, Thermal." It should be recognized that the Orbiter has, in general, greater attitude-hold capability than is required for most missions' payloads. However, some payload requirements do exceed the capability of the Orbiter. Payload attitude-hold requirements may not exceed those defined for the Orbiter, to ensure there are no Orbiter temperature-limit violations and that the heat-rejection requirement imposed by the Orbiter systems, crew, and payloads are met.

Payload surfaces or elements that may be located near one or more payload-bay floodlights should be analyzed to determine if a temperature violation could result from the normal operation of a floodlight. Also, analysis of floodlight circuitry in the payload bay indicates that a single failure could result in the activation of a floodlight in the failed-on position. Temperature-sensitive components such as electrical components, propellant tanks, batteries, and fuel lines should be analyzed for this condition to ensure no safety hazard is presented.

STS __________ CARGO INTEGRATION REVIEW
COMPATIBILITY STATEMENT

The NASA Team Leader and the contractor/payload representative have assessed the compatibility of the integrating hardware and software design against the STS and cargo requirements in the

TR/Thermal Systems Engineering - Passive as of ____________________
(system or other) (date)

The engineering assessments listed below, with the exception of (A) open items, and (B) open DN's/ECR's are found to be compatible.

☐ Verify compatibility of IH with cargo design

☐ Verify compatibility of IH with mission objectives

☐ Verify compatibility of IH with thermal constraints (max-min average temperatures, max-min entrapment temperatures, and predicted versus ICD temperatures)

☐ Status ICD TBD's, TBR's, and PIRN's

☐

☐

☐

(A) Open Items

(B) Open DN's/ECR's

________________________________ ________________________________
Contractor/Payload Representative Date NASA Team Leader Date

Figure 3. Passive thermal control compatibility statement

STS __________ CARGO INTEGRATION REVIEW
COMPATIBILITY STATEMENT

The NASA Team Leader and the contractor/payload representative have assessed the compatibility of the integrating hardware and software design against the STS and cargo requirements in the

TR/Thermal - Active ______________________ as of ____________________
(system or other) (date)

The engineering assessments listed below, with the exception of (A) open items, and (B) open DN's/ECR's are found to be compatible.

- ☐ Verify compatibility of Orbiter active thermal systems with payload heat loads
- ☐ Verify compatibility of Orbiter gas supply system with payload requirements
- ☐
- ☐

(A) Open Items

(B) Open DN's/ECR's

______________________________ ______________________________
Contractor/Payload Representative Date NASA Team Leader Date

Figure 4. Active thermal compatibility statement

Ascent and re-entry analyses should be performed both for normal and contingency scenarios. Ascent analyses show that most payloads have the same temperature upon the opening of the payload bay doors as at liftoff. It has been found that during entry and post-landing phases, the temperature environment is

MICB CHANGE REQUEST/DIRECTIVE NASA-Johnson Space Center	Page: Date: 9/12/89
Number P17559-38	CHANGE TITLE Thermal Environment

4.2.3.1. Thermal Environment - (cont):

IS:

The payload design and operation shall be compatible with the following attitude conditions. The Orbiter will normally be oriented in one of the attitudes contained in Table 4-1. The payload will be designed to allow deep-space excursions that Include a 35-min inertial measurement unit (IMU) alignment occuring approximately every 12 hr. The table specifies the payload constraints and recovery times for these excursions, so that repeat of the attitudes can be planned.

Table 4-1.-Attitude Requirement/Capabilities

Attitudes	Time Constraint	Nominal recovery attitude	Time	Preferred operational recovery attitude	Time
+ZLV	Continuous	N/A	N/A		
PTC	Continuous	N/A	N/A		
+Z Solar	30 min	+ZLV	TBD		
+Z Space	TBD	TBD	TBD		
-XLV*	TBD	+ZLV	TBD	TBD	TBD
-XLV**	TBD	+ZLV	TBD	TBD	TBD
-XLV***	TBD	+ZLV	TBD	TBD	TBD
±XSI	TBD	+ZLV	TBD	TBD	TBD
+YLV, XPOP***	TBD	+ZLV	TBD	TBD	TBD

* Nose down right wing velocity vector and rotated approximately 30°North of orbit plane.
** Nose down left wing in velocity vector and rotated approximately 30°South of orbit plane.
*** Nose down wing perpendicular to orbit plane and bay on RAM.
**** Wing down, nose perpendicular to orbit plane and bay on RAM.

2. Revise 1st sentence of 3rd para as follows:

WAS:

In the event of an anomaly, the STS will observe the attitude constraints of either Table 4-1a or Table 4-1b, as appropriate, to the extend possible.

IS:

In the event of an anomaly, the STS will observe the attitude constraints of Table 4-1 to the extend possible.

3. Add the following paragraph to the end of section:

The payload must be designed to be safe with any cargo bay flood light failed on. (Reference para. 6.1.6 of ICD2-19001 for floodlight characteristics.) If floodlight operation impacts mission success, operational constraints and appropriate safeguards will be negotiated between the NSTS and customer and will be documented in the Flight Operations Support Annex, Annex 3.

Figure 5. Payload Integration Plan tailoring

influenced by the pre-entry condition, entry heating and subsequent heat conduction, ground purge, and weather conditions at the landing site. Generally, maximum temperature is reached after landing as a result of heat soak-back through the Orbiter structure and air entering the payload through the vent doors.

Thermally sensitive payload surfaces that may be located near a payload-bay vent should be analyzed to determined the impact of being exposed to hot entry air after the vent doors are opened. The entry-air temperature declines rapidly from approximately 400-deg F to 100-deg F in 60 seconds at the vent-door opening.

Post-landing and ferry-flight analyses should be performed to ensure that components do not exceed their temperature limits during this portion of the mission. When the Shuttle lands at Edwards Air Force Base, the payload (cargo) typically remains aboard the Orbiter, which is flown or ferried to the launch site for deintegration. During ferry-flight operations, payloads within the bay are exposed to ambient conditions that are not controlled or monitored. Payloads are normally not powered, heated, or cooled. Payloads with unique requirements should specify the requirements in the PIP. The temperature in the bay during ferry flight (each flight segment is limited to 4 hours) could range from 35-deg F to 80-deg F. To compensate for an adverse environment, conditioned air provided by the Orbiter purge system while on the ground can be used to bias the payload-bay environment. The period on the ground while enroute to selected Air Force bases or NASA facilities may range from a few hours to more than 24. Here the payload bay temperature may range from 10-deg F to 125-deg F, caused by diurnal and seasonal variations and geographic location. Stops can be made to provide conditioned air. This requirement must be specified in the PIP.

Other analyses, depending on the thermal interfaces involved, that should be completed are:

1. Heat-rejection analysis for payloads utilizing the cabin air in the middeck and aft flight deck.

2. Heat-rejection analysis for payloads utilizing the payload heat exchanger.

3. Heat-rejection analysis for payloads utilizing the spigot system.

4. Payload-grapple fixture-end effector interface analysis for payloads utilizing the Remote Manipulator System (RMS).

A realistic mission-timeline analysis should include the complete sequence of attitudes defined for the mission, which may be quite complicated. Merely analyzing the payloads' response to representative attitudes may not be sufficient, especially if preliminary, simplified analyses predict temperature-limit exceedances for payload components.

THE CARGO INTEGRATION REVIEW

The engineering-compatibility work should be completed prior to the CIR. This review is a 4-day session held approximately 11.5 months prior to the subject flight. A dry run (CIRD) of the briefings will be completed one month prior to the CIR. A data package will then be sent to the customer. The first two days of the CIR will be devoted to team reviews of the engineering detailed in the package, and identification of any discrepancies. The third day will be a preboard review of all discrepancies, issues, and recommendations. This review includes Kennedy Space Center, Orbiter, and payload-engineering members. The fourth day will be a CIR Board review of assessment summaries, unresolved discrepancies/issues, and recommendations. The Board, chaired by the Deputy Director of the National STS Program, is responsible for the direction, conduct, and authorization of flight production. The following is a typical agenda:

- Introduction
- Flight overview
- Flight planning
    - Flight-design assessment
    - Crew-activities overview and assessment
- Flight-operations support
- Ground data systems - MCC/POCC requirements/implementation
- Systems assessment
- Training
- Ground operations - payload processing
- Engineering compatibility summary
- Safety (ground and flight)
- Summary/actions

Engineering-compatibility concerns and issues brought out from the CIR should be worked and closed at subsequent status meetings, such as the Payload Operation Working Group Meetings (POWG).

SAFETY

In addition to the design considerations associated with completion of payload mission objectives, special considerations associated with STS safety and missions compatibility are required.

The customer is responsible for investigating the effect of unplanned events that may occur to ensure that no thermal-limit violations exist that could endanger the crew or compromise the flight during any mission phase. There are two types of considerations: design requirements for contingency operation and analyses-defining limitations during contingency operations.

1. Contingency Design Requirements

- Payloads must be designed to be thermally compatible with abort during any mission phase:

  During powered ascent, abort can occur as either a return to launch site (RTLS) or abort to an alternate landing site, such as a transatlantic landing site. On-orbit aborts can occur prior to or subsequent to payload bay door opening. Prior to door opening, abort-once-around (AOA) presents the minimum orbit time, while the maximum time depends on the orbit inclination. The payload bay doors are normally opened 1 to 1 1/4 hours after liftoff; however, customers must design for a maximum time for door opening of 3 hours. If the doors are not opened by 3 hours, an abort will be declared and landing will occur by liftoff plus 6.5 hours for 28.5-deg inclination missions, or 11.5 hours for 57-deg inclination. Following the 3-hour abort time, special Orbiter contingency operations may be required necessitating curtailment of standard payload services i.e., power, cooling, etc. Following payload bay door opening, aborts can occur at anytime; therefore, payloads must be compatible with an abort from the worst hot or cold condition that could be encountered for that particular mission.

- Payloads must be designed to not present a hazard to the Orbiter for flights ending at contingency landing sites; i.e., where ground services (such as payload-bay purge or active cooling) are not available.

- Payloads using Orbiter-provided heat-rejection provisions must be designed to not present a hazard to the Orbiter for reduced or loss of heat rejection.

- Payloads using Orbiter-provided electrical energy for thermal control must not present a hazard in the event of loss of power.

2. Contingency Analysis Defining Limitations

Data must be provided to the NSTS to support contingency planning:

- Long-term off-nominal exposure to worst hot or cold mission environments must be analyzed and temperature limits affecting safety must be identified.

- For deployable payloads, limitations associated with delay in the deployment sequence or restow of erectable spacecraft (if applicable), and delayed deployment must be identified and thermal recovery periods defined.

Additional contingencies may exist due to payload-peculiar characteristics, and these contingencies as well as the contingencies noted above will be defined and documented in the applicable PIP or PIP annex.

Also, payload operational constraints associated with implementation of payload objectives should be established by conducting appropriate thermal analyses of the payload design.

Safety assessments of the mission design and configuration for cargo is conducted in three safety-assessment activities.

a. Payloads are assessed for compliance with requirements as specified in NHB 1700.7 ("Safety Requirements for Payloads using the NSTS").

b. NSTS cargo-integration hardware is assessed for compliance with requirements as specified in NHB 5300.4 ("Safety, Reliability, Maintainability, and Quality Provisions for the Space Shuttle Program").

c. The plan for an Integrated Cargo Hazard Assessment (ICHA) is presented at the CIR for review and approval. A final report will be presented to the Payload Safety Panel and to the MICB, and will be available prior to the FRR.

The status of all three assessments will be presented at the CIR. The final results of these assessments, along with the safety assessments of other NSTS elements, are used to develop a NSTS Mission Safety Assessment (MSA).

Chapter VII

Heat Pipes and Capillary Pumped Loops

Robert C. Prager
The Aerospace Corporation

OVERVIEW

The heat pipe uses a closed two-phase liquid-flow cycle to transport relatively large quantities of heat from one location to another without the use of electrical power. The heat pipe can be used to create isothermal surfaces, can be used as a thermal "transformer" to change the flux density of the heat flow, and can be used in various ways as thermal control devices. One-way (diode) heat pipes have been tested and flown, as have variable-conductance heat pipes (VCHPs), which maintain a constant-temperature evaporator surface under varying load conditions. Since the driving mechanism is capillary pumping, a relatively weak force, heat pipes may be susceptible to severe performance degradation when operating in a gravity field, and planning is needed to facilitate ground testing.

WHY A HEAT PIPE WORKS

Consider a simple horizontal heat pipe in equilibrium with an isothermal environment. The liquid in the wick and the vapor in the vapor space are at saturation. If heat is applied to the evaporator, raising its temperature, liquid in the wick evaporates (removing some of the added heat), which "depresses the meniscus" in the evaporator since less liquid is present there. This process also raises the local vapor pressure, since it must be in saturation with the heated liquid in the wick.

The difference between the increased curvature of the meniscus in the evaporator wick and the unchanged meniscus in the condenser wick causes a difference in capillary pressure sufficient to pull liquid from the condenser wick toward the evaporator. This replenishes the liquid in the evaporator wick. At the same time, heated vapor flows from the evaporator to the condenser, which is at a lower pressure. When this vapor comes in contact with cooler condenser surfaces, it condenses. This cycle is shown schematically in Figure 1.

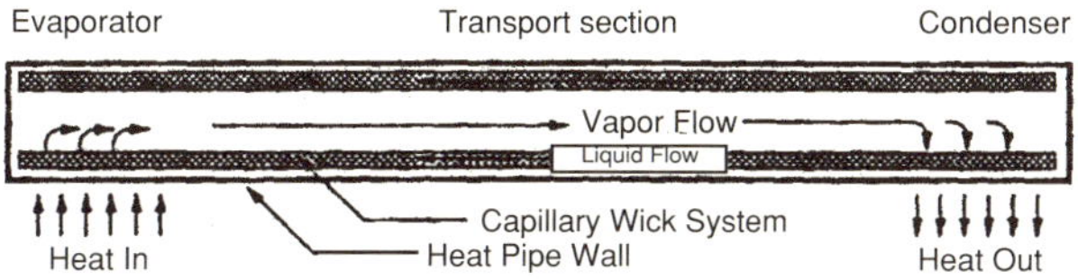

Figure 1. Heat-pipe schematic

Since the latent heat of vaporization of most heat-pipe working fluids is high, only small amounts of fluid need to flow to transport significant quantities of heat.

The driving mechanism, the temperature difference between the evaporator and condenser wall, is also small.

CONSTANT-CONDUCTANCE HEAT PIPE

Elements of this most basic heat pipe consist of a working fluid, a wick structure, and an envelope. This type of heat pipe is used to move heat from one location to another (possibly changing its flux density in the process) or to isothermalize a surface. It need not be shaped like a "pipe" — large flat plates several feet across have been built and tested for special applications. Constant-conductance heat pipes are often categorized according to the type of wick structure used.

Groove Wicks. The simplest design, consisting of axial grooves in the wall of extruded aluminum tubing. Grooves can be formed in tubes of other materials such as copper (by swaging) or even refractory metals (by deposition), but aluminum is by far the most common. This class of wick is very susceptible to gravity effects during ground testing, but is relatively inexpensive to produce and very consistent. Heat-transfer capability is moderate, but sufficient for many applications. Grooves are typically rectangular or trapezoidal in shape, but more complex shapes such as the "teardrop" or "keyhole" have been extruded with difficulty (see Figure 2).

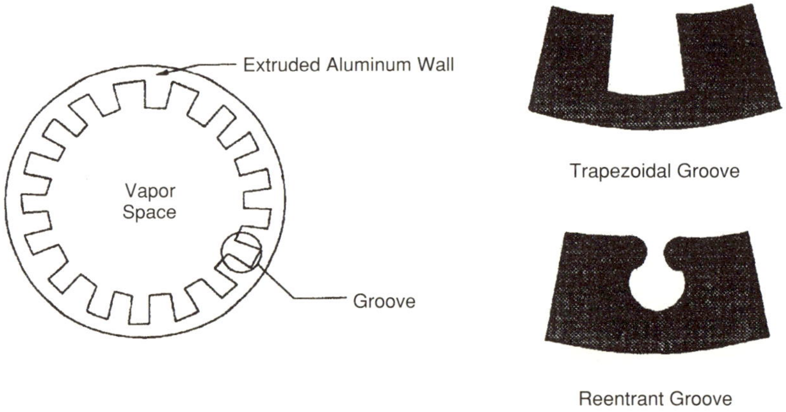

Figure 2. Grooved heat pipe

"Monogroove" Designs. These high-capacity designs, consisting typically of one large, teardrop-shaped groove connected to a vapor space (see Figure 3), can be considered to be an extension of the basic-groove concept. Unlike a heat pipe with many smaller grooves of the same total area, the monogroove's large groove provides relatively unrestricted longitudinal flow. Liquid is distributed on the evaporator wall by means of a secondary wick consisting of small circumferential grooves or screen. This design has shown very high capacity during ground testing, but difficulties were encountered during early Shuttle testing. Later experiments were more successful. As of this date (early 1993) no monogroove heat pipe has ever been used on a production spacecraft.

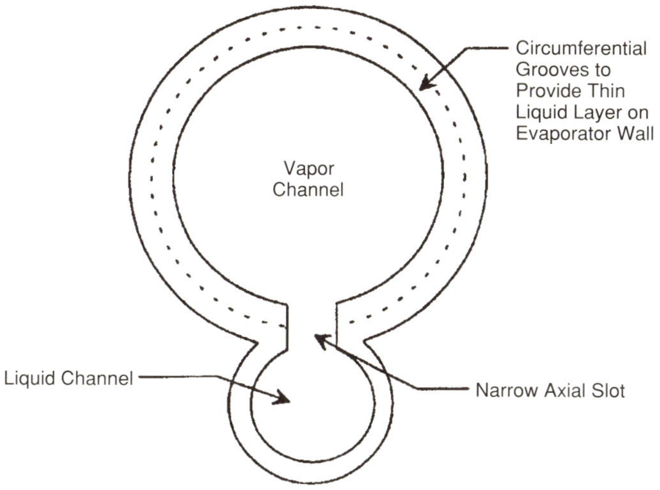

Figure 3. Monogroove heat pipe

Composite Wicks. The simplest wick in this class (and the oldest heat-pipe wick) consists of several layers of screen fastened to the inside wall of the heat pipe. More capacity can be obtained by using more layers of screen, to increase the wick flow area — at the cost of increasing the heat pipe temperature difference due to temperature drop to conduct through the thick saturated wick. To overcome this penalty, some heat-pipe manufacturers separate the wick into two parts, the portion that spreads the fluid circumferentially about the wall of the evaporator, and the portion that carries the fluid down the length of the heat pipe. The former is kept as thin as possible and can consist of circumferential grooves cut in the wall of the heat pipe or of a single layer of screen or metal mesh bonded to the wall. The latter is held off the wall by means of legs or straps, or makes contact with the wall in only a few places. This type of wick has capacities similar to the axially grooved heat pipe, but has much more capability when tilted. Since the wick must be assembled of relatively fragile materials, care is required in building such a pipe, and no two supposedly identical pipes will perform in exactly the same manner. Some sample wick designs of this type are shown in Figure 4.

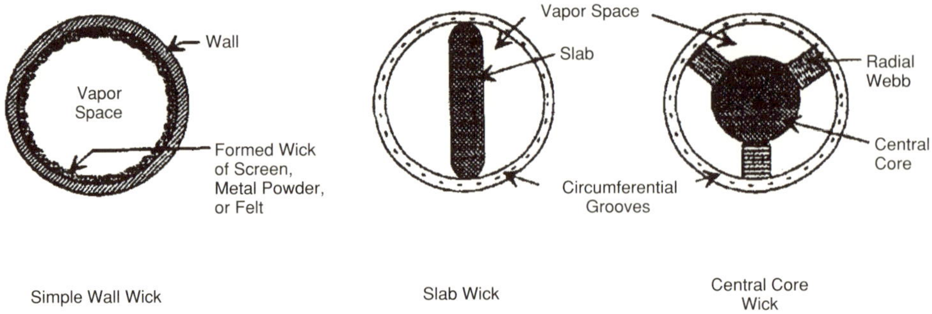

Figure 4. Composite wicks

Artery and Tunnel Wicks. This class of heat pipe is based on the composite wick, but provides one or more relatively unrestricted liquid-flow paths in parallel with the longitudinal wick. These paths will fill with fluid in space due to minimum surface-energy considerations, and greatly reduce the viscous pressure drop in the heat pipe, thereby increasing capacity. When properly designed, these arteries will fill as the heat pipes operate in a gravity field. Wicks in this class can be blocked by bubbles of noncondensable gas in the arteries (see Abhat et al. and Saaski), but they are attractive because of their large heat-transfer capability in a small envelope. If the liquid in the artery remains subcooled when it reaches the evaporator, bubble formation can be avoided. A number of mechanical schemes have been proposed and tested to prevent bubbles from blocking the arteries of variable-conductance heat pipes (VCHPs) (see Eninger). These pipes are particularly prone to bubble formation since the liquid in the artery contains dissolved control gas, which tends to come out of solution as the liquid warms during its transit of the pipe from condenser to evaporator.

Crossections of some of these wick structures are shown in Figure 5.

DIODE HEAT PIPES

A constant-conductance heat pipe can be modified so that heat-pipe operation occurs normally in one direction, but the pipe ceases operation when an attempt is made to transfer heat in the other, "wrong" direction, resulting in a diode action. Even when blocked, however, some heat is transferred, if only by conduction down the pipe wick and wall. This heat leak is particularly significant in cryogenic systems. Common concepts are the liquid-trap, liquid-blockage, and gas-blockage diodes.

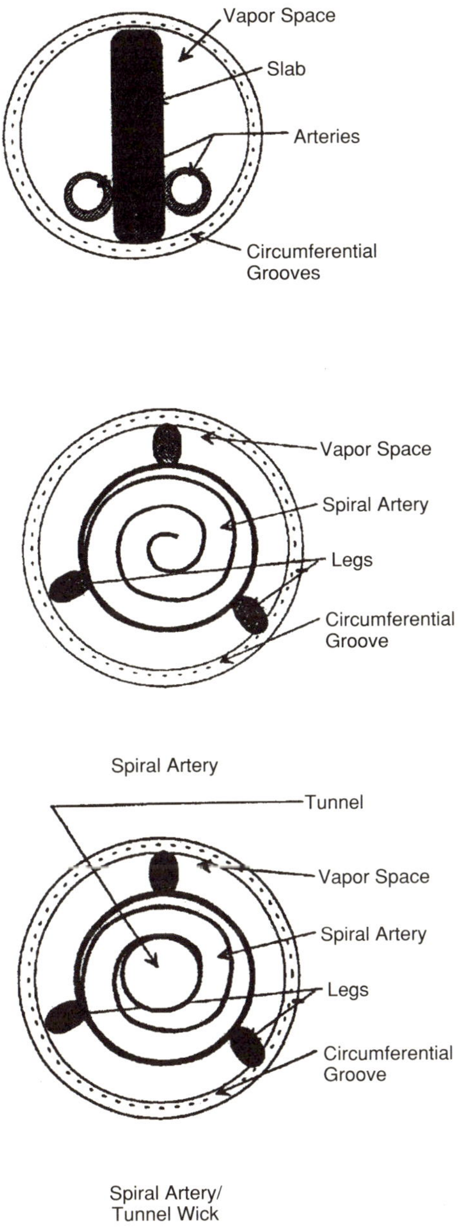

Figure 5. Artery and tunnel wicks

Liquid-Trap Diode. This is the most common type of heat-pipe diode. This heat pipe has a wicked reservoir at the evaporator end designed so that it is heated by the same environment that heats the evaporator. Although the envelopes are connected, the reservoir wick is not connected to the rest of the heat pipe. When, during normal operation, heat is applied to the evaporator and reservoir, heat is transferred from the evaporator to the condenser as in the constant-conductance heat pipe, and any fluid in the reservoir wick evaporates and joins the vapor flow to the condenser. (The reservoir wick should be dry during normal operation.) When this pipe is reversed, and the evaporator and reservoir become cooler than the condenser, some of the hot vapor coming from the condenser condenses in the reservoir, and is lost to the rest of the heat pipe. Sufficient liquid is tied up in the reservoir to cause the pipe to dry out. "Shutoff" is neither instantaneous nor complete. A schematic of this type of diode operation is shown in Figure 6.

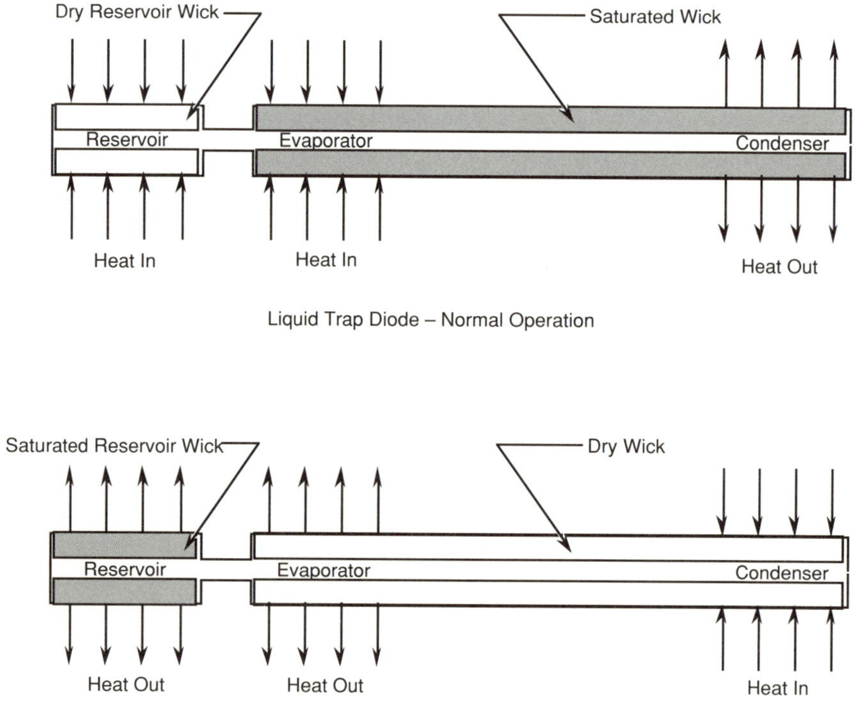

Liquid Trap Diode – Normal Operation

Liquid Trap Diode – Reverse (Shut Off) Operation

Figure 6. Liquid-trap diode heat pipe

Liquid-Blockage Diode. At the condenser end, this heat pipe has a wicked-reservoir end cooled by the same environment that cools the condenser. Its wick is not in contact with that of the remainder of the heat pipe, and it is normally full of working fluid — in effect, it traps a large fluid slug. When the pipe is reversed, the fluid slug travels to the normal evaporator end, where it completely fills the evaporator vapor space (and that of a large portion of the transport section), preventing condensation. Optimum design of the wick structure and vapor space must be compromised to control the liquid slug during shutoff, and such control requires maintaining close tolerances during the manufacturing process. A schematic of this type of diode operation is shown in Figure 7. Proper control of the fluid (and therefore operation as a diode) in a gravity field requires maintaining the gap between the evaporator wall and the blocking plug at a dimension that will fill with liquid if it is available.

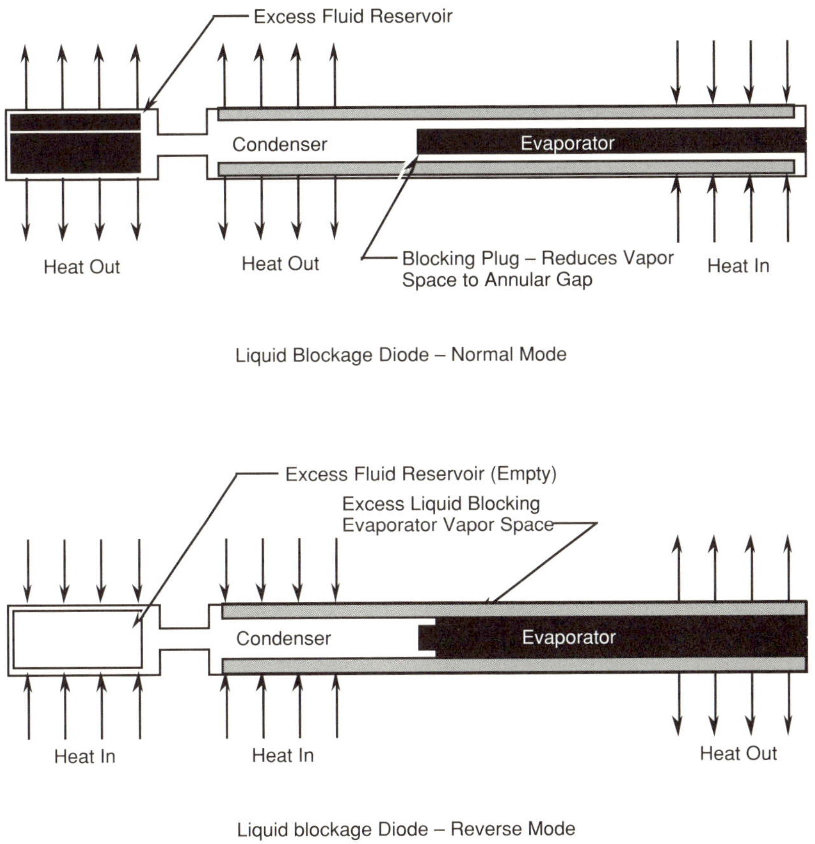

Figure 7. Liquid-blockage diode heat pipe

Gas-Blockage Diode. The gas-blockage diode is similar in design to the liquid-blockage diode, except that the reservoir can be unwicked and contains a non-condensable gas. When the pipe is reversed, the gas flows to the evaporator and, as above, completely fills the vapor space, preventing condensation. However, as the temperature rises, the gas slug can be compressed to the point where the heat pipe will start working again. Furthermore, convection within the gas slug may be a significant heat-leak component. A schematic of this type of diode operation is shown in Figure 8.

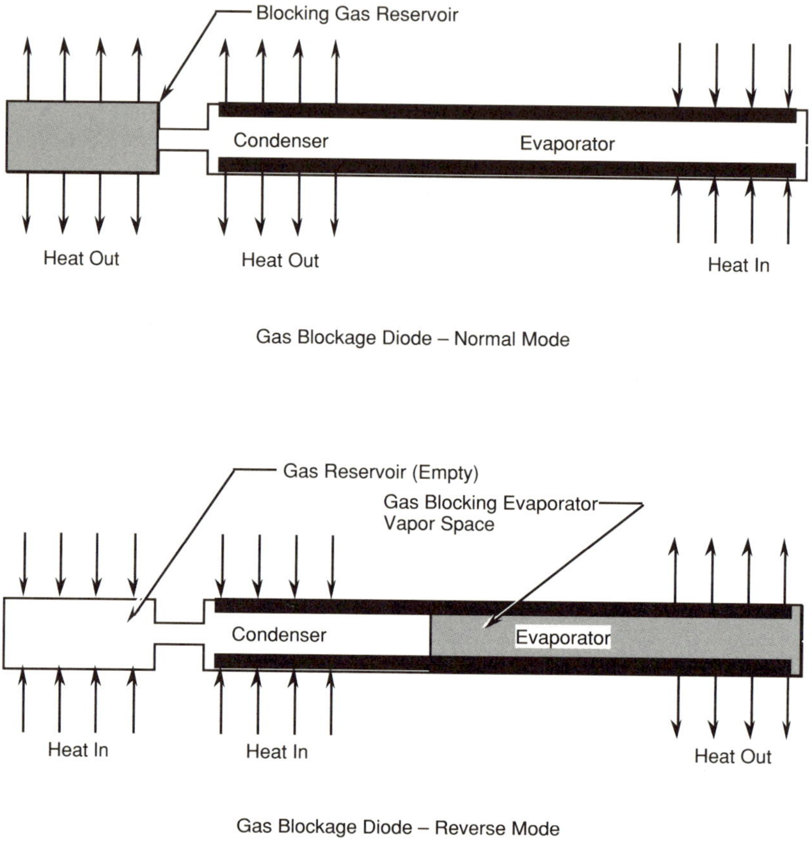

Figure 8. Gas-blockage diode heat pipe

Other diodes. Any heat pipe with a wick with a finer pore size in the evaporator than in the condenser or the adiabatic section will show some signs of diode operation, because its capacity will be different depending upon which direction it is trying to move heat. The most extreme case is that of a heat pipe in which there is no wick in the condenser (see the capillary pumped loop, below), as the pipe will dry out quickly and shut off if heat is applied there.

VARIABLE-CONDUCTANCE HEAT PIPES (VCHPs)

These heat pipes use a gas reservoir, connected to the end of the condenser, that is filled with a non-condensable gas to control the operating area of the condenser based on the evaporator temperature. (In effect, in a typical spacecraft application, the active-radiator area becomes a function of the electronic box cold-plate temperature, with increasing box temperatures leading to increased radiator areas.) Although complicated models of the gas front exist, the gas front may be considered an impermeable floating piston. If the temperature at the cold plate rises, the vapor in the evaporator (at the saturation pressure of the liquid in the evaporator) rises rapidly. The pressure of the mixture of control gas and vapor in the reservoir must rise to compensate, so the "gas-front-as-piston" will move further into the condenser, decreasing the volume of control gas. This opens up more of the condenser area to heat-pipe operation. This is shown schematically in Figure 9.

A number of schemes have been flown, which differ mainly in the treatment of the reservoir (some have wicks, some are kept hot or cold by exposure to different environments, and some become elements of what is arguably an active thermal control system by means of heaters connected via feedback control to sensors at the evaporator or payload). Sufficient control gas is usually present in the reservoir to enable these pipes to function as gas diodes if the heat pipe is reversed.

CAPILLARY PUMPED LOOPS

The capillary pumped loop (CPL) is a two-phase heat-transfer loop that uses a capillary structure to return liquid to the evaporator. The basic concept is still under development, although several systems have been flown as Shuttle experiments. It shows promise when high heat-transfer rates must be sustained over relatively long distances. A schematic of the loop in operation is shown in Figure 10. As heat is applied, vapor is vented from the evaporator into the condenser duct, where it begins to condense on the wall. Liquid film on the wall is carried along by the vapor flow, with the thermodynamic quality (the ratio of vapor to total mass) of the flow dropping. At a certain point in the loop, surface-tension forces will predominate, and the flow will consist of slugs of liquid mixed with collapsing bubbles of vapor. By the time it reaches the liquid side of the evaporator, pulled by capillary forces within the evaporator, the stream will consist solely of subcooled liquid.

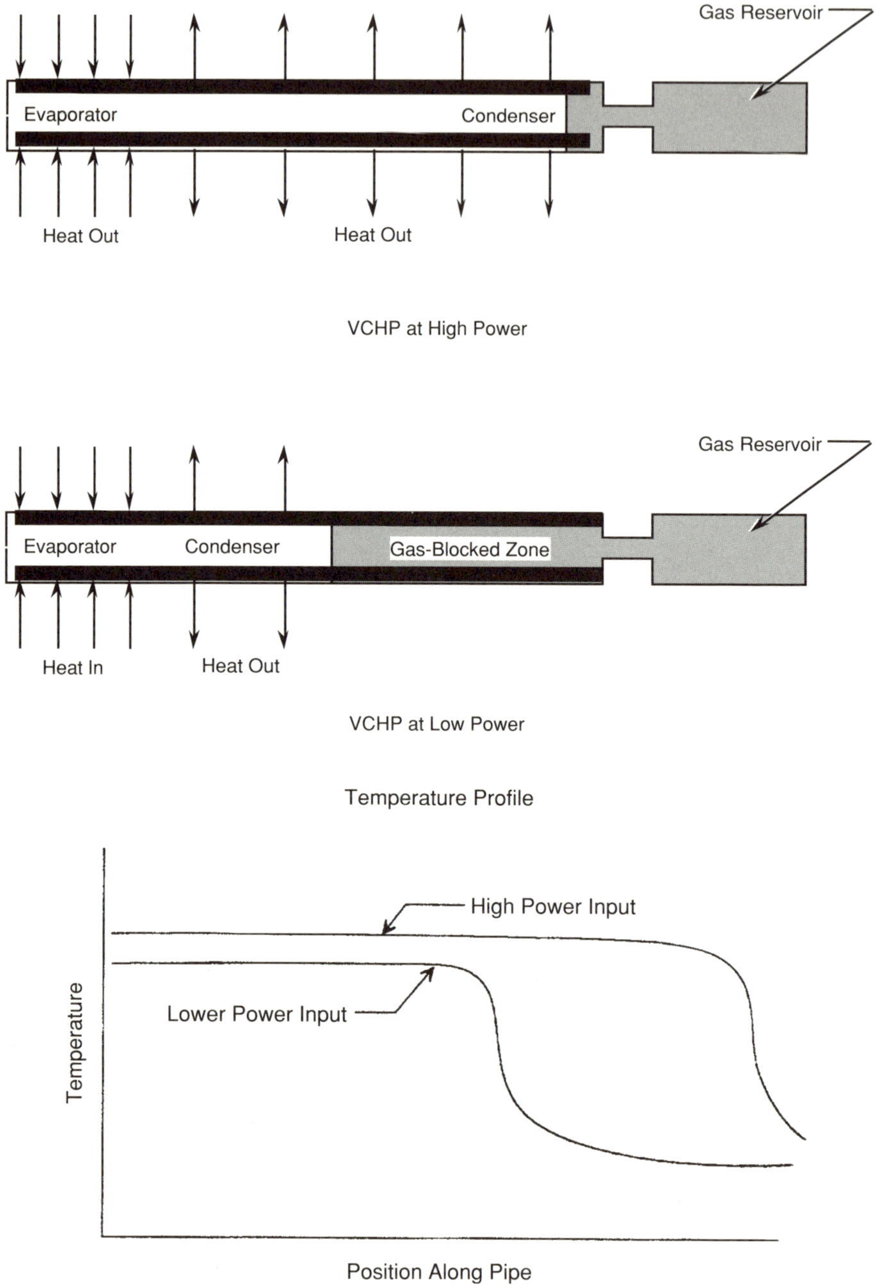

Figure 9. VCHP operation

This device is essentially a heat pipe with the return flow of condensate to the evaporator in a separate unwicked tube. The only wick in the CPL is in the evaporator, where it distributes liquid from the supply tube over the actual evaporator surface.

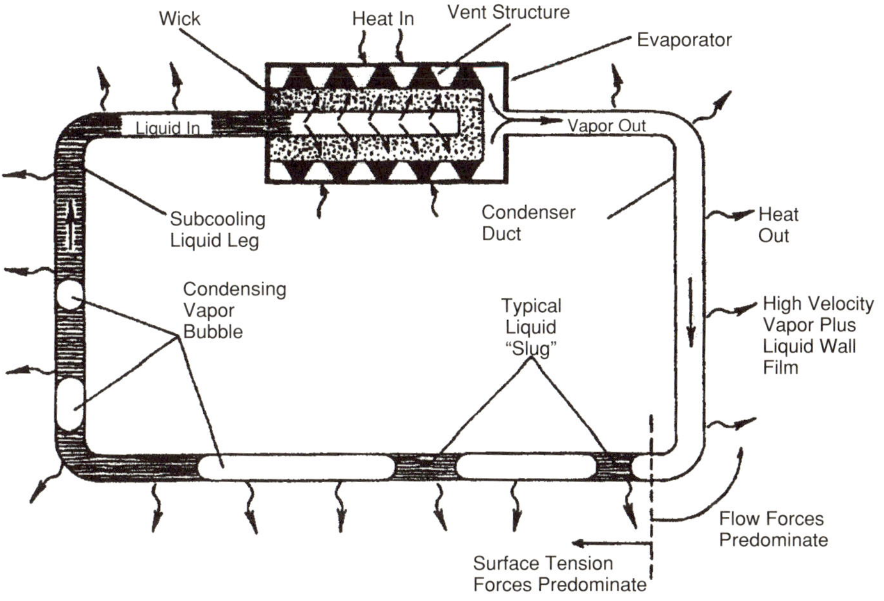

Figure 10. Capillary pumped loop schematic

HYBRID (MECHANICALLY-ASSISTED) SYSTEMS

These concepts are essentially extensions of the capillary pumped loop. They abandon any claim to be passive thermal control systems by the addition of small pumps to force liquid flow. Because they are two-phase systems, only small quantities of the working fluid need to be carried to the evaporation site in the liquid phase to transport large amounts of heat energy. Several have been proposed for use on the Space Station, and a number of prototypes have been built and tested.

ANALYSIS

Heat-Pipe Capacity (Capillary Pumping Limit). Return flow of liquid from the condenser to the evaporator is caused by differences in the capillary pressure between the evaporator and condenser. The capillary pressure acting on a liquid surface is inversely proportional to the radius of curvature of the fluid surface at the liquid/vapor interface in the wick. For purposes of the analysis, the liquid surface in the condenser is usually assumed to be flat, so that the radius of curvature (and hence the capillary force) is zero. As liquid evaporates, the meniscus in the evaporator depresses, causing a difference in capillary pressure between the evaporator and condenser surfaces. (see Figure 11). This difference in pressure pulls liquid through the wick from the condenser to the evaporator in an attempt to restore equilibrium.

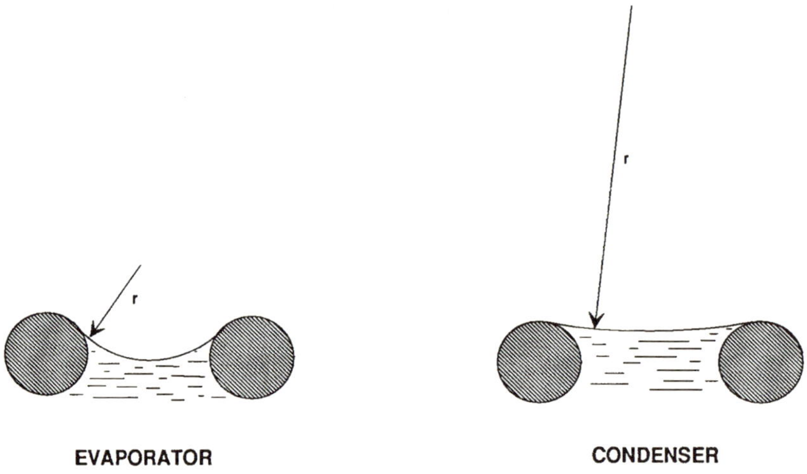

Figure 11. Depression of meniscus

A heat pipe "dries out" when the flow through the wick caused by this pressure difference is insufficient to supply liquid at the same rate at which working fluid is being vaporized in the evaporator.

This point is calculated by balancing the pressure drops in the system:

$$\Delta P_{CAPILLARY} - \Delta P_{GRAVITY} = \Delta P_{LIQUID} + \Delta P_{VAPOR},$$

where

$\Delta P_{CAPILLARY}$ (capillary pressure rise) is the maximum possible difference in capillary pressure between the evaporator and the condenser. This term is a function of the surface tension (which depends on the choice of working fluid and the temperature) and the wick pore size (which depends upon the wick material and type of wick).

$\Delta P_{GRAVITY}$ (gravity head loss) is the "head loss" that must be overcome by capillary pressure to sustain fluid in the evaporator. In addition to gravity, this term can result from other accelerations, such as those on a spinning spacecraft.

ΔP_{LIQUID} (liquid pressure drop) is the pressure loss due to viscous flow through the wick. This term is simple for an axial groove wick, but it can become extremely complicated for a composite artery wick, where viscous pressure losses in flow-through complicated structures of layered screens, metal felt, or sintered powder must be modeled. Expressions for these losses usually contain empirical constants, which is one of the reasons why performance testing of each pipe is usually necessary.

ΔP_{VAPOR} (vapor pressure drop) is the pressure loss due to vapor flow from the evaporator to the condenser. This term is usually small unless the vapor velocity is high due to constricted vapor space or if the vapor density is very low.

The exact equation will depend upon the wick design used. Many formulations are given in the references.

Thermodynamic Considerations. If operation near the freezing point is needed (as would be the case for water at typical room temperatures), for almost any cryogenic liquid, or for liquid metals at startup, high vapor velocities and large vapor-pressure drops will be encountered, since the vapor density and pressure are very low. These large pressure drops cause their own temperature drops in the pipe (since saturation temperature is a function of pressure). In some cases, the pressure drop in the vapor required to support the calculated heat-pipe capacity would result in a negative vapor pressure in the condenser, an obvious impossibility. Under similar low density conditions, choked flow (the "sonic limit") has been observed in liquid-metal heat pipes. Although not a true limit, the operating temperature of the heat pipe rises so thermal equilibrium can be established, which may cause the temperature to rise beyond the desired range. In short, it is not a good idea to design a heat pipe that must run in a temperature regime where its working fluid has a very low vapor pressure.

If the relative velocity of liquid and vapor is high enough (as measured by the Weber number), liquid can be pulled out of the wick and returned to the condenser as droplets entrained in the vapor. This phenomenon (the "entrainment limit") was first observed in liquid-metal heat pipes where the droplets could be heard to "ping" against the end cap. It is an operating limit in that, to support a given rate of heat transfer from the evaporator, an excess of liquid must be pulled through the wick, since not all of the liquid will reach the evaporator.

The "boiling limit" or "heat flux limit" is concerned with the flux density of the thermal load on the evaporator. Even if the heat-pipe wick could theoretically return the liquid from the condenser required by the heat load, if the load is concentrated in too small an area, nucleate boiling can occur in the evaporator wick. The creation of bubbles in an otherwise filled wick reduces the area of the wick available for fluid flow, and hence reduces the capacity of the wick.

MATERIALS

Working Fluid. The choice of working fluid is usually governed by the temperatures of the desired operating range. A heat-pipe working fluid can be used effectively between a temperature somewhat above its triple point and below its critical temperature. If the triple point is approached too closely, temperature drops in the vapor flow increase (see the discussion above, "Thermodynamic Considerations"). As the critical point is approached, the distinction between liquid and vapor blurs, and the surface tension drops to zero. (The pressure that must be contained by the envelope also increases significantly.) The triple points and critical temperatures of several heat-pipe working fluids are given in Table 1.

Table 1. Heat Pipe Working Fluids

Fluid	Melting Point		Boiling Point		Critical Temp	
	deg K	deg F	deg K	deg F	deg K	deg F
Hydrogen	14.0	-434.4	20.4	-423.0	33.0	-400.3
Neon	24.5	-415.6	27.1	-410.9	44.4	-379.8
Oxygen	54.3	-361.8	90.2	-297.3	154.8	-181.1
Nitrogen	63.1	-346.0	77.3	-320.4	126.2	-232.4
Ethane	89.9	-297.8	184.5	-127.6	305.5	90.2
Methane	90.7	-296.4	111.4	-259.2	190.5	-116.8
Methanol	175.2	-144.3	337.9	148.5	513.2	464.1
Acetone	180.0	-135.7	329.4	133.2	508.2	455.1
Ammonia	195.5	-107.8	239.8	-28.0	405.6	270.4
Water	273.2	32.0	373.2	212.0	647.3	705.4
Potassium	336.4	145.8	1032.2	1398.3	2250.0	3590.0
Sodium	371.0	208.1	1152.2	1614.3	2500.0	4040.0
Lithium	453.7	357.0	1615.0	2447.0	3800.0	6380.0

Data from Brennan and Kroliczek, *Heat Pipe Design Handbook.*

Two parameters have been developed as an aid in comparing the relative performance of heat-pipe working fluids. The "zero-g Figure of merit" is given by

$$\frac{\sigma\rho\lambda}{\mu} ,$$

where σ is the surface tension
ρ is the liquid density
λ is the latent heat of vaporization, and
μ is the dynamic viscosity.

This parameter neglects vapor flow entirely, but for most applications, vapor flow is not the limiting factor. This group of fluid properties appears in the heat-pipe capacity equation. A second parameter, the "one-g Figure of merit" or "wicking height factor," compares the relative sensitivity to gravity effects of working fluids:

$$\frac{\sigma}{\rho},$$

where the parameters are as defined above. It is a relative measure of how high a given wick structure will be able to pump a working fluid in a gravity field (or due to inertia effects, as in a spinning spacecraft).

COMPATIBILITY

Since a heat pipe is a completely sealed container, any chemical reactions between the working fluid and the wall or wick material can be disastrous. None of the reaction products can escape, and any material that is consumed cannot be replaced. Certain combinations of materials such as ammonia and copper are known to react quickly with one another, and hence are not likely to be chosen even by a novice.

However, combinations of materials that are traditional and acceptable in the chemical-process industry (such as water and stainless steel, or water and nickel) have been demonstrated to react with one another, causing noncondensable gas. In general, the cryogenic working fluids up through ammonia can be used with either stainless steel or aluminum (although there is some evidence that ammonia reacts slowly with aluminum, and the combination of ammonia, aluminum [such as a wall material], and stainless steel [such as would be found in a typical wick material] can react more quickly with one another).

Methanol works well with stainless steel, but reacts with aluminum. Water appears to work well with copper, and possibly monel, but not 304 or 316 stainless steel or nickel. Some short-term success has been achieved with carbon steel, but these pipes appear to be generating hydrogen gas, which diffuses through the pipe wall, and some reaction must be taking place internally.

Materials available for higher-temperature (liquid metal) heat pipes must hold together at these higher temperatures, and be inert to some very corrosive working fluids. This area is still under investigation.

TESTING

During Fabrication. The heat-pipe envelope will be checked for leaks during the fabrication process, usually with a helium mass-spectrometer leak detector. However, once the pipe is sealed at the fill tube, the integrity of this seal is open to question. Although some chemical tests have been used (see Edelstein), the most thorough seems to be checking for the presence of working fluid outside the heat pipe when placed in an evacuated chamber.

Performance of each heat pipe as a function of tilt should be measured at some typical operating temperature(s) to determine whether the wick functions properly. Testing at a low temperature will show whether non-condensable gas is present. (At high temperatures, the non-condensable gas can be compressed into a thin plug so that it isn't detectable using thermocouples mounted on the heat pipe.)

If the heat pipe is to be installed in a spacecraft in a position where it will be tested vertically (with gravity assist) during system-level testing, such as thermal vacuum or thermal balance, the heat pipe must be tested in the same orientation with a similar heat load before installation. In this way, the performance of the heat pipe that will be seen in the vacuum chamber will be known before the test is performed. This will help to avoid unpleasant surprises and scrambling for logical explanations at a time when the heat pipe can't be reached without breaking vacuum and tearing open the spacecraft.

If the heat pipe is to be curved in three dimensions so that it can't be tested in a single plane, some manufacturers build a test pipe with the same number of curves in the wick, but with all of the curves in a single plane. In this way, the wick performance to be expected in space can be characterized.

After Integration into the System. After integration of a heat pipe into a system, the heat pipe should be verified to determine whether any deterioration took place during the integration procedure, and also to verify the performance of the integrated thermal control system.

HEAT PIPE APPLICATIONS/PERFORMANCE

The most obvious application of a heat pipe is one requiring physical separation of the heat source and sink. If a heat pipe is used, it is not necessary to mount all hardware to be cooled directly on radiator panels, or to use relatively inefficient conductive couplings. (This requirement tends to occur when cooling boxes that must be kept close to each other for more efficient electrical or microwave design.) By the same token, heaters need not be mounted directly on hardware to be heated if a heat pipe is employed.

A closely-related class of application is that of the thermal transformer. A small high-powered box can be mounted on one side of a radiator with integral heat pipes; the heat generated will be spread and dissipated at a much lower flux density over the entire surface of the radiator. This approach also permits more efficient use of available "real estate" — the area available for a radiator is seldom centered symmetrically about the heat source, facing the optimal direction.

Heat pipes have been used to reduce temperature gradients in structures to minimize thermal distortion. The telescope tube of the NASA Orbiting Astronomical Observatory had three ring-shaped heat pipes to minimize circumferential temperature gradients. The ammonia heat pipes worked throughout the eight years of mission life.

The diode heat pipe was first proposed as a means of connecting a device to two radiator panels on opposite sides of a spacecraft, with the understanding that at least one of the radiators would be free of any direct solar load at all times during the orbit. The diodes would couple the device to the cold radiator, while preventing heat from leaking back into the system from the radiator in the sun. This type of thermal design problem — in which heat from a temporarily warm radiator or from a failed refrigerator must be kept from leaking back into the system — is an obvious application for a diode heat pipe.

The variable-conductance heat pipe (VCHP) can be used to control the amount of active radiator area, providing reasonably good temperature control without the use of heaters. This is particularly attractive if electrical power is limited, and this type of design has been flown on a number of satellite experiments. However, if the application requires maintaining a box or baseplate at virtually a constant temperature, feedback control (at the expense of some heater power) may be employed. A sensor on the baseplate of the device to be controlled can be routed to an on-board computer, and whenever the temperature drops below the desirable range, heaters on the VCHP reservoirs are activated, causing the control gas to expand and block off more of the radiator area. If the temperature rises above the range desired, power to the reservoir heaters is reduced, increasing the active radiator area. This concept usually requires less power than using heaters directly on the box or system to be controlled.

The use of flexible heat pipes or rotatable joints in heat pipes to cool devices on rotating or gimbaled platforms has been proposed, but flexible heat pipes tend to have too much resistance to motion, and rotating joints in heat-pipe walls leak under extreme conditions. These areas are still under active investigation.

REFERENCES

The following references are general, covering a broad range of topics concerning heat-pipe design and applications, as well as lists of more specialized references.

Papers concerning new developments in heat-pipe design and analysis, and discussing new applications or the results of tests or experiments are usually presented at the AIAA Thermophysics Conference each summer. In addition, volumes of proceedings from the International Heat Pipe Conference held every four years can be found in technical libraries. While heat-pipe papers are presented at other conferences and technical meetings, they tend to be concentrated here.

P. Brennan, and E. J. Kroliczek, *Heat Pipe Design Handbook,* B&K Engineering, Towson, MD 1979. (A final report prepared for NASA Goddard, Greenbelt, MD under contract NAS5-23406).

S. W. Chi, *Heat Pipe Theory and Practice: A Source Book,* Hemisphere Publishing Co., McGraw-Hill, New York 1976.

P. Dunn, and D. A. Reay, *Heat Pipes*, 2nd ed., Pergamon Press, Oxford 1978.

A list of more specialized reports follows:

Historical:

R. S. Gaugler, U.S. Patent Application, December 21, 1942. Published U.S. Patent No. 2350348, June 6, 1944. (First American description of a heat pipe.)

G. M. Grover, T. P. Cotter, and G. F. Erickson, "Structures of Very High Thermal Conductance," *J. Appl. Phys.* 35, p. 1990 (1964). (First modern report of a heat pipe.)

T. P. Cotter, "Theory of Heat Pipes," Los Alamos Scientific Laboratory Report LA-3246-MS, February 1965. (Discussion of the operating principles of a heat pipe.)

Fabrication:

F. Edelstein, "Heat Pipe Manufacturing Study," Grumman Aerospace Corp., Final Report to NASA, NAS5-23156, August 1974.

VCHPs:

B. D. Marcus, *Theory and Design of Variable Conductance Heat Pipes*, TRW, NASA Contractor Report CR-2018, April 1972. (A good early survey of variable conductance heat pipe design and application, containing much general heat pipe information.)

Gas Bubbles in Arteries:

J. E. Eninger, "Menisci Coalescence as a Mechanism for Venting Noncondensable Gas from Heat-Pipe Arteries," AIAA Paper 74-748, July 1974.

A. Abhat, M. Groll, and M. Hage, "Investigation of Bubble Formation in Arteries of Gas-Controlled Heat Pipes," AIAA Paper 75-655, May 1975.

E. W. Saaski, "Arterial Gas Occlusions in Operating Heat Pipes," AIAA Paper 75-657, May 1975.

Thermal Diodes:

A. Basiulis, "Unidirectional Heat Pipes to Control TWT Temperature in Synchronous Orbit," Symposium on Thermodynamics and Thermophysics of Space Flight, Palo Alto, March 1970. (First published description of liquid-trap diode.)

B. Swerdling, R. Kosson, M. Urkowitz, and J. Kirkpatrick, "Development of a Thermal Diode Heat Pipe for the Advanced Thermal Control Flight Experiment (ATFE)," AIAA Paper 72-260, April 1972. (A successful use of a liquid-blockage diode in space.)

Grooved Heat Pipes:

P. J. Brennan, E. J. Kroliczek, H. Jen, and R. MacIntosh, "Axially Grooved Heat Pipes - 1976," AIAA Paper 77-747, June 1977. (A good general survey paper on grooved-heat-pipe technology.)

Monogroove and other High Capacity Designs:

J. Alario, R. Haslett, and R. Kosson, "The Monogroove High Performance Heat Pipe," AIAA Paper 81-1156, June 1981.

E. J. Kroliczek, J. Ku, and S. Ollendorf, "Design, Development, and Test of a Capillary Pump Loop Heat Pipe," AIAA Paper 84-1720, June 1984.

G. P. Peterson, "Analytical Comparison of Three External Artery Heat Pipes," AIAA Paper 87-1616, June 1987.

Chapter VIII

Cryogenic Systems

Martin Donabedian
The Aerospace Corporation

INTRODUCTION

Some spacecraft components, such as IR telescopes and their focal plane detectors, low-noise amplifiers, and super-conducting devices, require cooling to low temperatures ranging from 200-deg K down to within a couple degrees of absolute zero. A number of technologies exist to provide this cooling, with the technology of choice depending on the desired temperature level, the amount of heat to be removed at the temperature, and the required operating life.

Figure 1 provides a generalized overview of which technologies are usually employed in each temperature/heat-load regime, assuming normal spacecraft mission durations on the order of one year or longer. The chart was constructed by laying out data points from over 60 systems either fabricated and flown, tested, or having a preliminary design proposed.

Radiators can be used, theoretically, down to about 60-deg K under ideal conditions, but below about 100-deg K the rejection capability falls dramatically due to the T^4 nature of radiation-heat transfer, and the overall feasibility is highly dependent on the spacecraft orbit, orientation, and the attitude-control limitations. This lower radiating capability leads to prohibitively large and heavy radiators, which may also be too sensitive to environmental loads and heat leaks through support structures and insulation to be effective.

At higher temperatures, beginning at about 150-deg K thermoelectric coolers are attractive for small cooling requirements. For temperatures in this same range, but with higher total heat-rejection requirements, radiators combined with heat pipes or pumped-liquid or - gas loops may be required to transport the heat and spread it out over a relatively large radiator.

The use of stored expendable systems provides a reliable and relatively simple method of cooling over a wide range of temperatures from about 1.5-deg K (superfluid helium) to about 150-deg K (solid ammonia) These systems rely on the boiling or sublimation of a low-temperature fluid or solid to absorb the device waste heat and reject it overboard in the vented gas.

As the heat-rejection requirement increases at low temperature, or if a long mission duration is required even at low heat rates, then the weight of stored cryogenic required would become very large, and active refrigerators would become a more attractive option. Reliability, operating life, power, and consideration for vibration control are key issues. Several development programs offer the potential for use of these systems over a wide range of cooling requirements for extended durations. The following sections provide a more detailed discussion of each of these cryogenic-cooling technologies.

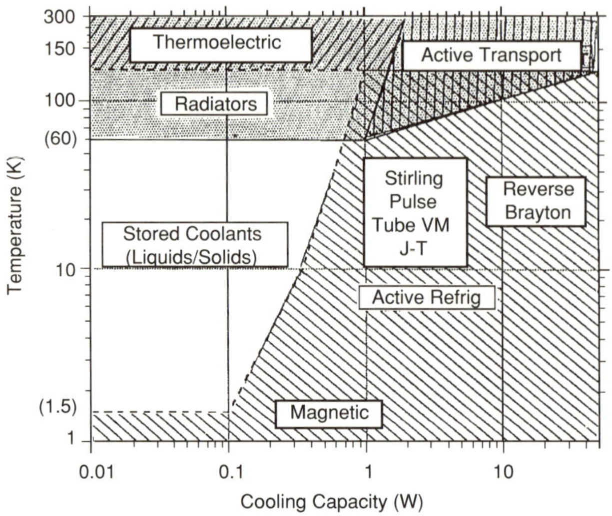

Figure 1. Cryogenic-cooling methods

STORED-CRYOGEN COOLING SYSTEMS

System Concepts. Stored-cryogen expendable-coolant systems use either cryogenic liquids in either the subcritical or supercritical state, cryogenic solids, or high-pressure gas combined with a Joule-Thompson (J-T) expansion valve system to provide cooling of spacecraft components.

The advantages of these systems are simplicity, reliability, relative economy, and negligible power requirements. In most cases, the technology is developed. The basic disadvantages of cryogenic storage systems are their limited life due to parasitic heat leakage and the high weight and volume penalty for extended durations of use. Although high-pressure-gas storage systems with J-T valves can overcome the long-storage limitation, the penalties associated with the storage of high-pressure gas and poor J-T expansion efficiencies generally make system-weights prohibitive as operating time increases. Stored gas J-T systems are thus applicable only to special cases of short-term or intermittent cooling requirements. A more complete discussion of these systems can be found in the IR Handbook (Ref. 1).

Operating Temperature Ranges. The operating temperatures attainable with expendable coolants using the heat of vaporization or sublimation are illustrated in Figure 2. The limits for each coolant are based on a minimum, as defined by the solid phase at 0.10-mm Hg pressure (arbitrary limit selected by the author) and on a maximum defined by the critical point. Cooling can be provided from essentially near absolute zero to over 300-deg K. Liquid helium (He^4), which has a normal boiling point of 4.2-deg K at one-atmosphere pressure, can actually be used near absolute zero at very low pressures, since it does not freeze under its own vapor pressure and must be compressed to about 25 atmospheres before freezing. Liquid helium actually exists in three fluid states with different temperature ranges: superfluid (less than 2.18-deg K), normal (2.18- to 5.2-deg K), and supercritical (above 5.2-deg K). Several operational systems developed with NASA funding have successfully operated in the superfluid ranges. The Infrared Astronomical System (IRAS) (Ref. 1) operated for a year at 1.8-deg K, while the Cosmic Background Explorer (COBE) (Ref. 2), operating at 1.5-deg K, was launched in 1989 and provided data for about 10 months. Both of these were built by Ball Aerospace Systems Division, Boulder, Colorado and utilized a porous plug in the vent system for liquid/vapor phase separation in the microgravity environment.

The primary limitation of the liquid-storage-type systems are the complex tank design to minimize boil-off, phase separation of subcritical storage in the space environment, and the large weight and volume penalty for extended mission time.

An alternative to this is the use of stored cryogens in the solid state, which provides a higher heat content than the liquid, higher density, and simpler system design. The concept is schematically shown in Figure 3. Limitations of solid systems include restrictions on detector mounting, specialized filling procedures, the need for a metal form or mesh to reduce temperature gradients as the solid dissipates, and more complex ground-hold considerations. The operating pressure of the cryogen must be controlled to a relatively small value, as indicated in Table 1. A second alternative to liquid storage is a J-T cooler that utilizes the expansion of a high-pressure gas (i.e., 2000 to 6000 psia or higher) through a J-T valve, which results in cooling of the gas and eventually formation of liquid to be used at the sensor-cooling point.

The J-T effect involves the ratio of temperature change to pressure change of an actual gas in the process of throttling or expansion (during a constant enthalpy process), without doing work or transferring heat. Under normal pressure and temperature conditions, a perfect gas provides no cooling effect or temperature change for a throttling process. However, in actual gases under conditions of high pressure and/or low temperatures, molecular forces cause a change in internal energy when the gas expands. The change in internal energy during the expansion process results in cooling of the gas. The cooled, expanded gas is passed back over the incoming gas to cool itself. This results in regenerative cooling. The process continues until liquid begins to form at the orifice to produce a bath of liquid at the cooling temperature of the gas. For certain gases, this effect occurs only below a specific inversion temperature. Helium (45-deg K), hydrogen

(204-deg K) and neon (250-deg K), for example, require precooling to the indicated temperature before the J-T expansion cooling effect occurs. Most other gases, such as nitrogen, argon, and air, have inversion points well above room temperature, and no special precooling is required.

The J-T cooler (Figure 4) consists of a finned tube in the form of a coil, an orifice and orifice cap, and an outer shield or coil. The finned tube is made of very small-inside-diameter tubing to provide the large ratio of surface area to volume necessary for effective heat exchange. For a fixed-orifice cryostat, the flow will vary with pressure. Thus there is only one pressure that will provide just the desired refrigeration, as illustrated in Figure 5. Traditional J-T systems have suffered because of the inefficient matching of the pressure to the desired refrigeration. In recent years, self-regulating (or variable-orifice-size) cryostats have increased the capability for correct matching, or for compensation for changes in heat load or the gradual tendency of clogging. This also allows for high flow during cool down and nominal flow during steady-state operation.

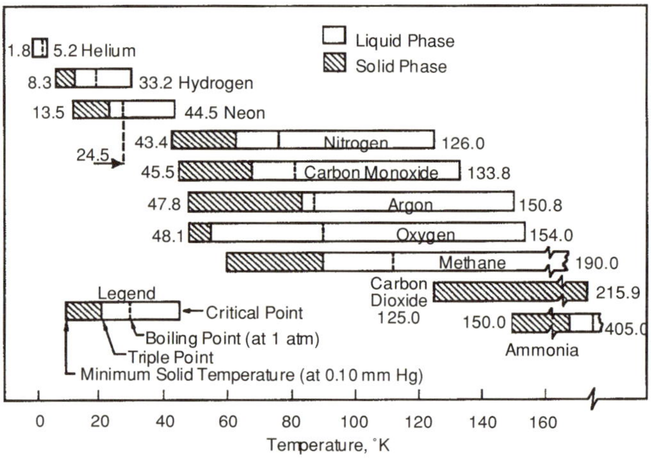

Figure 2. Operating temperature ranges for expendable cryogens

<u>Summary of Flight Systems</u>. Fluids stored at cryogenic temperatures in a liquid or supercritical state have been used as atmosphere constituents for manned systems, reactants for fuel cells or chemical lasers, or for cooling of various infrared or other low-temperature sensors (Figure 6). Tanks for storage of fluids at cryogenic temperatures are very complex. Typically they utilize one or more vapor-cooled shields, separated by high-efficiency MLI blankets and low-thermal conductivity supports (usually composite materials) to minimize the parasitic heat transfer from the external environment to the stored fluid. The state of the art in low heat-leak design for space operations is represented by the IRAS and COBE systems, which are shown in some detail on Figures 7 and 8. The IRAS operated for 10 months

with liquid helium (superfluid) at 1.8-deg K, while the COBE system, launched in November of 1989, operated successfully at about 1.6-deg K and was designed for a 14-month life.

Table 1. Properties of Selected Cryogens

Symbol	*Cryogen* Name	*Heat of Sublimation* (Btu lb^{-1})	(J g^{-1})	*Density of Solid at Melting Point* (lb ft^{-3})	kg m^{-3})	*Operating Temperature Range (K)* 0.10 mm Hg	Triple Point*
NH_3	Ammonia	739.0	1718.9	51.3	821.8	150.0	195.4
CO_2	Carbon dioxide	246.6	573.6	97.5	1562.0	125.0	215.7
CH_4	Methane	244.5	568.7	31.1	498.2	59.8	90.4
O_2	Oxygen	97.5	226.8	81.3	1302.4	48.1	54.9
A	Argon	79.8	185.61	107.0	1714.0	47.8	83.6
CO	Carbon monoxide	126.0	293.1	58.0	929.2	45.5	68.0
N_2	Nitrogen	96.6	224.7	63.8	1022.1	43.4	63.4
N_e	Neon	45.5	105.8	89.8	1438.6	13.5	24.5
H_2	Hydrogen	218.5	508.2	5.02	80.4	8.3	13.98

*Corresponds to highest temperature at which solid phase can exist.

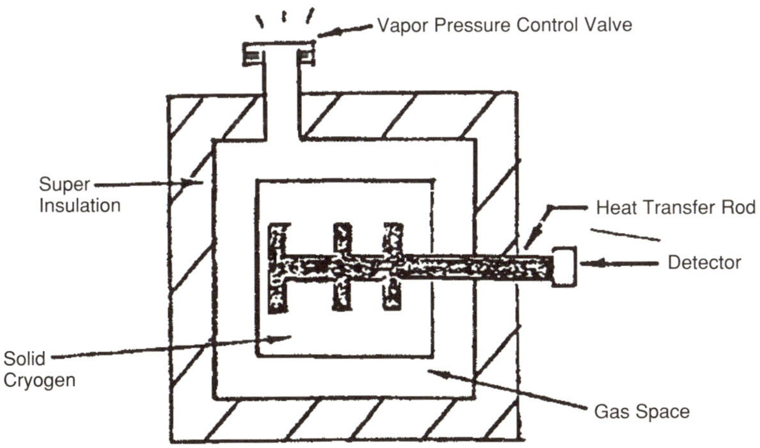

Figure 3. Typical solid-cryogen IR detector cooling system

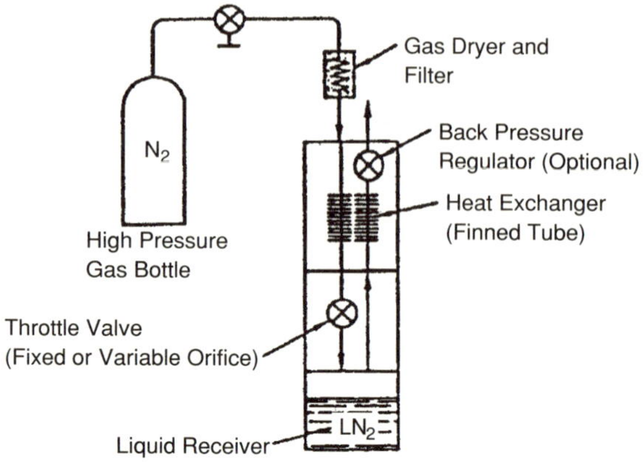

Figure 4. J-T Cooling-system schematic

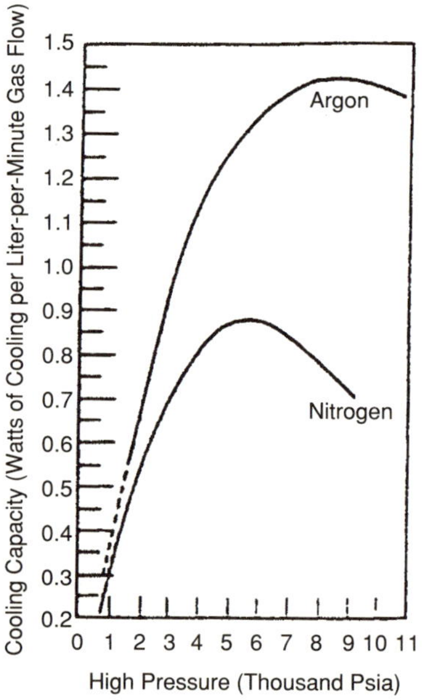

Figure 5. Cooling capacity versus inlet pressure for an ideal J-T cooler (300-deg K gas temperature). Gas flow is standard liters per minutes at 1 atm and 273-deg K. The liquid is at 1-atm pressure.

As of December 1992, six solid coolers (4 different designs) have operated successfully in space, some of which are shown in Table 2. Two of these configurations are depicted on Figures 9 and 10. Several other experimental and/or development units have been fabricated and tested. The Teal Ruby sensor (Figure 11), which was completed and ground-tested but never flown, provided a concept demonstration of a two-stage (neon/ammonia) cooler operating at a focal plane temperature of 15.8-deg K. The final thermal-vacuum test demonstrated an equivalent on-orbit life of about 10.5 months. The lowest-temperature solid cooler, using solid hydrogen at 9.5-deg K, is currently being developed for the SPIRIT III sensor to be flown on the Midcourse Surveillance Spacecraft (MSX), illustrated in Figure 12. The predicted on-orbit life of this demonstration concept is 25 months, based on the estimated loads defined at the PDR in June of 1989. The Spirit III design was based on the original CLAES concept, developed by LPARL under NASA/ESA sponsorship and subsequently modified to a 2-stage neon-CO_2 cooler, which was flown on the UARS in August of 1991 (Figure 13).

Storage Life		Fluid	Use
Short Term (Weeks/Months)			
Gemini		O_2, H_2	ECS, Power
Apollo/LEM		O_2, H_2, Helium	ECS, Power
MOL		O_2, H_2	ECS, Power
Shuttle		O_2, H_2	ECS, Power
ELMS (Up to 1 month)		HE (SC, 10 K)	Sensor Cooling
Intermediate (Up to 1 Year)			
HEAD	1 Yr (Develop only)	HE (Liquid, 4 K)	Sensor Cooling
OTTA	6 Mo (Develop)	O_2, N_2	Storage
HTTA	6 Mo (Develop)	H_2, HE	Storage
IRAS	1 Yr (1983)	HE (Liquid, 2.4 K)	IR Telescope
HEAO (B,C)	10 Mo (1978, 1979)	Solid CH_4 NH_3 (85 K)	Spectrometer
NIMBUS	8 Mo (1975, 1978)	Solid CH_4 NH_3 (60 K)	Radiometer
SESP-72-2	7 Mo (1972)	Solid CO_2 (125 K)	Spectrometer
Long Term (> 1 Year)			
Teal Ruby	1 to 1-1/2 Yr (Grd. Testing)	Solid Neon, CH_4 (18 K)	Detector Cooling
R&D Program	3 Yr (Engr. Test)	Solid CH_4, NH_3, HE	Technology
SB Laser	5-7 Yr (Studies)	H_2, D_2, O_2, NE_3, HE	Laser Reactants
ASAT MV	5-7 Yr (Studies)	HE (10 K)	Detector Cooling
COBE	1-1/2 Yr (1990)	HE (1.6 K)	Astronomy
CLAES-DM	2 Yr (Dev. Model)	Solid H_2 (10 K)	IR Telescope
Spirit III	3 Yr (Sched. 1992)	Solid H_2 (10 K)	IR Telescope
CLAES	1.25 Yr (UARS, 1991)	Solid Neon/CO_2 (14 K)	IR Telescope

Figure 6. Cryogenic storage systems

Table 3 includes data on representative J-T coolers. The use of helium, hydrogen, argon, and nitrogen gas enables the units to provide cooling from approximately 4.2- to 87.4-deg K at capacities of 0.50 to 10.0 W. J-T units have been fabricated for general-purpose commercial use, military missile-system applications (e.g., Sidewinder and Falcon), and for use in space. Air Products Model AC-2 was used in the Mariner program to provide 30 minutes of cooling at approximately 23-deg K. Advanced concept demonstrations in recent years have included the use of mixed gases to improve cooling capacity of N_2 very-rapid cool-down units, and various configurations of self-regulating demand-flow types.

Self-regulating demand-flow J-T cryostats are made by SBRC (Santa Barbara, CA), Hymatic Engineering (Worchester, UK), General Pneumatics Corp. (Scottsdale, AZ), and APD Cryogenics Inc. (Allentown, PA). A schematic of an SBRC/APD type of self-regulating unit is shown on Figure 14.

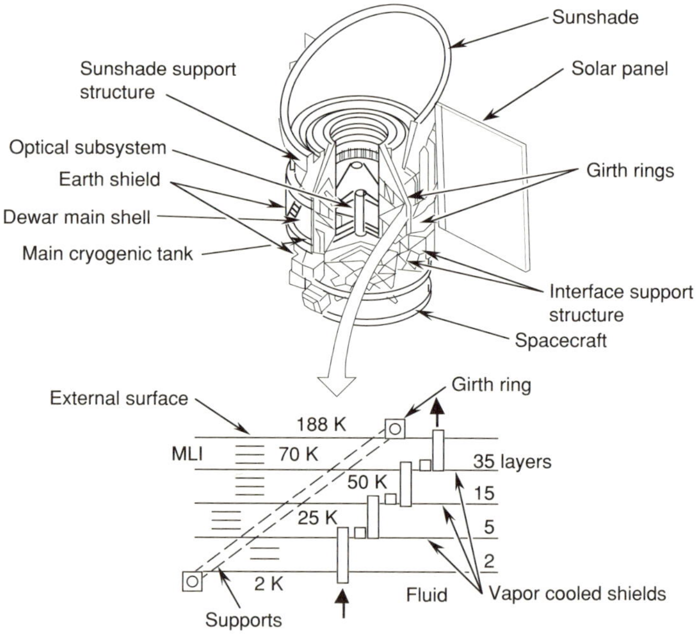

Figure 7. Infrared Astronomical Satellite (IRAS) (Ball)

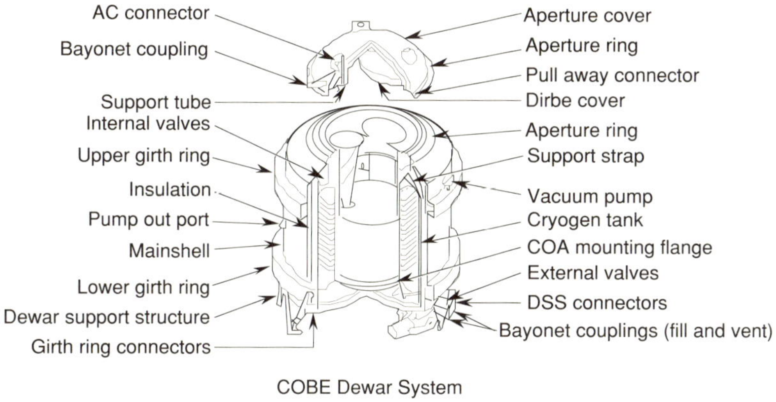

Figure 8. COSMIC background explorer (COBE) (Ball)

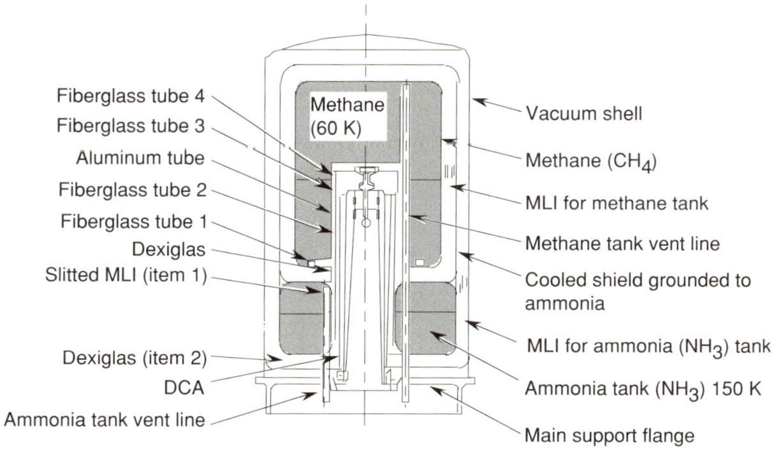

Figure 9. NIMBUS-F IR detector solid cooler (Lockheed)

Table 2. Solid Cryogenic Cooler Experience

Parameter	Operational Programs			CLAES	Experimental		Development	
	Nimbus - F/G	SESP 72-1	HEAO - B/C		3-Yr Dev. Cooler	Long Life Cooler	Teal Ruby	Spirit III (MSX)
Experiment or Purpose	Limb Radiance Measurement	Gamma Ray Detector	X-Ray Spectrometer	Limb Spectrometer	Lab Demon-stration	Flight Prototype	LWIR Exp.	LWIR Exp.
Operational Date	1975, 1978	1972	1978, 1979	1991	1976 (test)	1987 (fabr.)	Canceled (1989)	1992 (est.)
Exper. Temp (°K)	65	125	85	15	63	65	15	
Cooler Type/Cryogens	Dual Stage Methane/Ammonia	Single CO_2	Dual Stage Methane/Ammonia	Dual Stage Neon/CO_2	Dual Stage Methane/NH_3	Dual Stage Methane/NH_3	Dual Stage Neon/CH_4	Solid Hydrogen
Heat Loads (mW)								
Experiment	90	156	150	192 (FPA)	542	610	400	400
Parasitic	430 (148 for CH_4)	311	1050 (425 for CH_4)	1500 (Aperture) Ukn. (Parasitic)	445	608 (284 for CH_4)	(100 for Neon)	217
Operating Life (mo.)								
Design	12	12	12	22	36	36 (req.)	18	-
Actual	7.7	8	11.8	-	29 (test)	36 (calc.)	12 (est.)	25 (est.)
Weight (lbs)								
Total	53	48	170	2265	450	540	350	897**
Dry	27	18	-	793	480	250	151	172
Diameter (inches)	13	14	32	49	25	33	32	56
Length (inches)	34.5	16	36	56	48	43	35	68
Surface Area (sq ft)	11.6	7.1	36	-	32.3	42.2	35.2	54
FOM (Btu/hr-ft^2)*	0.043	0.149	0.041	-	0.017	-	0.010	0.0137
Manufacturer	Lockheed	Lockheed	BASD	Lockheed	Lockheed	Lockheed	Lockheed/ Rockwell	Lockheed

* Figure of Merit (FOM) = Q/A which is based on parasitic load to cold stage divided by total surface area
** Sensor Weight = 438 Lbs additional

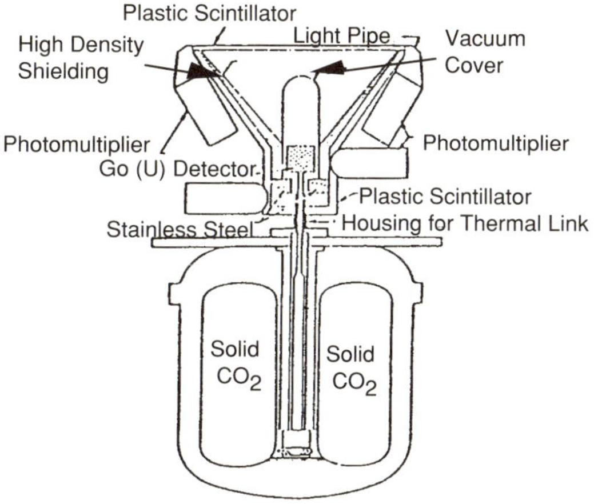

Figure 10. SESPP 72-1 Gamma-ray detector solid cooler (Lockheed)

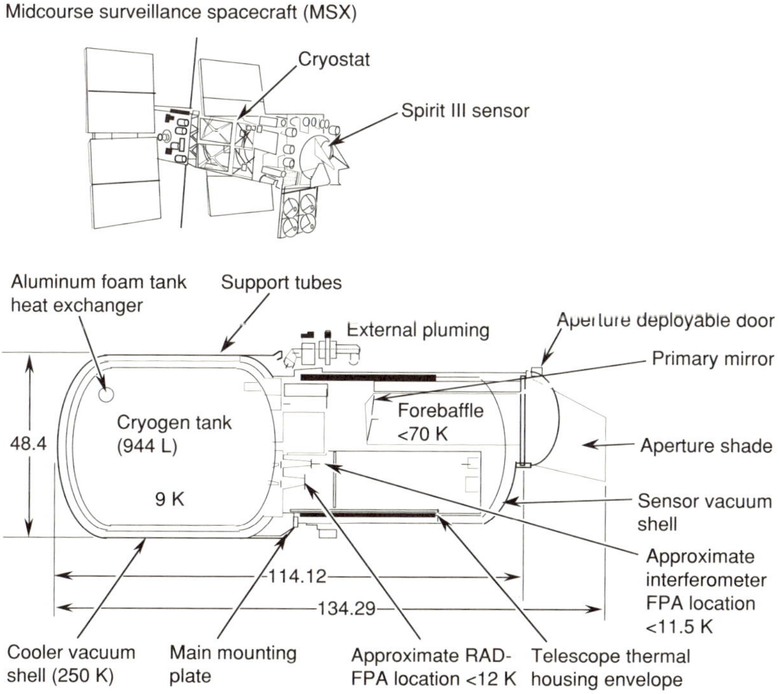

Figure 11. Spirit III sensor (Lockheed)

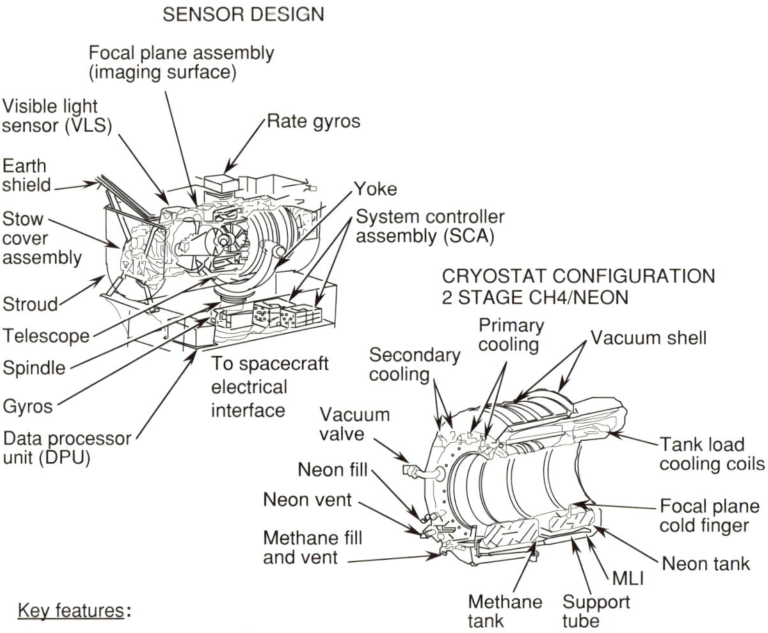

Key features:

Neon	99	lb
Methane	68	lb
On-orbit life	10.3	mo
Primary temp	15	K
Secondary temp	75	K
Dry weight	145	lb

Figure 12. Teal Ruby sensor (Rockwell)

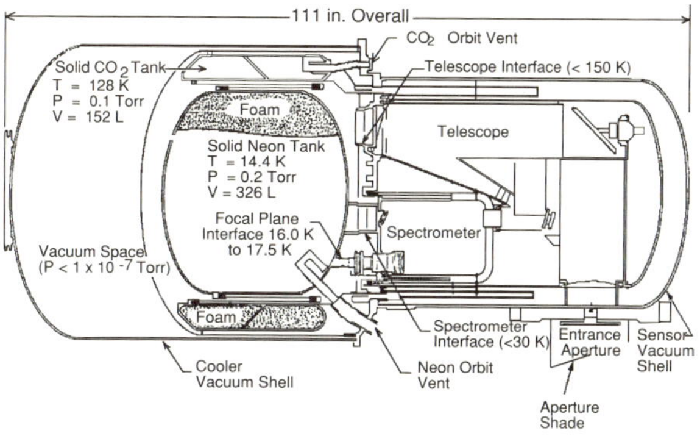

Figure 13. CLAES (2-stage version flown on UARS) (Lockheed)

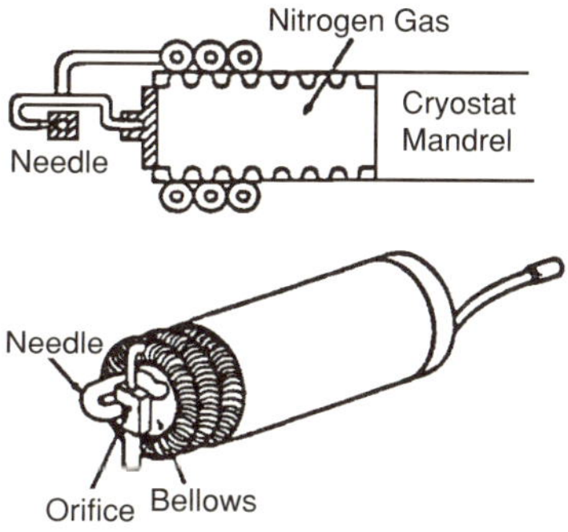

Figure 14. Schematic of an SBRC/APD-type self-regulating J-T cryostat (Walker, Ref. 34)

Performance Analysis Techniques. Design and analysis of cryogenic storage dewars for liquids or solids requires development of detailed thermal models and extensive parametric analyses and optimization. However, simplified computer programs ideal for concept validation and feasibility analysis have been developed for liquids (Refs. 4 & 6) and for solids (Ref. 5). The method of analysis revolves around the concept of a heat-leak parameter (Q/A), which was first introduced in 1971 (Ref. 7) and later used to characterize tanks in general (Ref. 1).

Table 3. Representative Joule-Thomson Coolers

Manufacturer	*Air Products*					*Hymatic*			*SBRC*			
Model	*AC-1*	*AC-2*	*AC-2L*	*AC-3L*	*VF2020*	*MAC8*	*MAC227*	*MAC215*	*42902-1-F*	*42902-2-F*	*9174*	*9186*
Nominal Cooling capacity load (W)	7.0	4.0	6.0	4.0	0.20	10.0	1.0	2.0	0.20	0.20	0.25	0.19
temp. (K)	80.0	23.0	22.0	22.0	77.0	80.0	82.0	21.00	87.0	77.0	77.0	77.0
Working fluid primary	N_2	N_2, H_2	N_2,LN_2	He,H_2,LN_2	N_2	N_2	N_2	H_2,Air	A	N_2	N_2	N_2
Gas consumption rate at nominal capacity ($1\ min^{-1}$) 2000 psia	22.6	26.0 (N_2) and 22.6 (H_2)	27.2(H_2) and 0.008 (LN_2)	36.0 (He) and 24.0(H_2) .008 (LN_2)	-	14.0 (4000 psia)	-	-	1.2	0.8	14.0	9.5
Minimum operating pressure (psi)	1160	1160	1160	-	-	1030	1030	1764	600	1000	1000	900
Cool-down time	5 min	10 min	10 min	40 min	20 sec (2000 psi)	30 sec	-	3 min	20 sec (2000 psi)	35 sec (2000 psi)	2 min (2000 psi)	1.5 (2000 psi)
Orifice characteristics type and diameter (in.)	Fixed	Fixed	Fixed	Fixed	Variable 0.326	Fixed 0.285	Variable	Fixed	Variable 0.21 max.	Variable 0.21 max	Fixed 0.325	Fixed 0.326
Overall dimensions (in.) length	0.5	3.625	17.0	23.0	2.55	1.7	1.8	2.5	2.75	2.75	2.5	2.5
diameter	3.5	7.75	6.0	6.0	0.875	0.3	-	1.0	0.25	0.25	0.1875	0.1875
Primary use	Lab.	Lab.	Lab.	Lab.	Airborne	Missile	Missile	-	Missile	Missile	Lab.	Lab.
Remarks	-	-	Incl. LN_2 precooler	Incl. LN_2 precooler	Cooling rate variable from 0.20 to 2.0 W		Freon 13 or Freon 14 are acceptable	-	Demand Flow	Demand flow	Open-loop	Closed-loop

Technology Development. With the existing technology in high-efficiency MLI blankets, multiple vapor-cooled shields, low-thermal-conductivity structural support materials, and passive thermal control of the outer-shell temperature, parasitic heat leak to the stored fluid can be reduced to values equal or less than 0.01 w/m^2 for liquid helium, 0.05 w/m^2 for liquid hydrogen, and 0.10 w/m^2 for liquid oxygen or nitrogen (see Ref. 6).

The limiting life factor for long-life storage systems is the heat leak through the structural support members, plumbing, electrical wires, and other miscellaneous heat shorts. To reduce these heat loads, retractable supports to be activated after launch have been proposed, but a flight-qualified design is yet to be built. For the near term, the use of hybrid concepts utilizing a mechanical refrigerator to reduce the external shell temperature provides a practical alternative. Mechanical refrigerator concepts and performance characteristics are provided later in this chapter.

CRYOGENIC RADIATORS

Concepts. In spaceborne applications, temperatures as low as about 60-deg K can be achieved by use of a suitably designed radiant cooler radiating into space. The low effective sink temperature of deep space provides an ideal environment for passive radiant cooling of infrared detectors and related devices to the temperatures indicated. This approach involves no moving parts, provides inherently long life, and requires no power.

The effective temperature of deep space is approximately 4-deg K. One or more detectors mounted to a suitably sized cold plate of high emissivity can radiate to this sink. The high vacuum of orbital altitudes minimizes the effect of convective heating. The cold plate must be shielded (with a cone, for example) against heat from direct sunlight, and in the case of near-Earth orbits, against the heat inputs from thermal emission and reflected sunlight from the Earth and its atmosphere (Figures 15 and 16). Furthermore, the cold plate must be thermally shielded from the parent spacecraft. These considerations usually result in a passive cooler design that is tailored to a particular spacecraft system.

The type of orbit (e.g., near polar, equatorial), orbit altitude, heat load, temperature, orientation of the spacecraft relative to the Earth or sun, and the location of the radiator all significantly influence radiator design. Ideally, the radiant-cooler patch (i.e., the detector mounting surface) is large enough so that thermal inputs, e.g., Joule heat, lead conduction, and radiative input through the optics, produced by the attachment of detectors or other components are small compared to the total power radiated by the patch at its equilibrium temperature. This permits flexibility in the optical and electrical design.

Table 4. Characteristics of Selected Radiant Coolers

No.	Inst. of Exp.	Sponsor	Flight or Program	Launch Date	Orbit	Altitude (nmi)	Temp (K)	Net Cooling Capacity Q(mW)	Cold Patch Area A (cm^2)	No. of Stages	Field of View (Deg.)	Overall Dimensions (cm)	Weight (lb)	Q/A (Net)		Q/A_{ideal} W/ft^2 (ε=1.0)	Fraction Parasitics
														($W\ ft^{-2}$)	($mWcm^{-2}$)		
1	High Resolution Infrared Radiometer (HRIR)	NASA/ ITT	NIMBUS I, II, III	1964-1969	Near polar, sun synchronous	~590	195	1.7	9.8	1	± 30	8 x 6 x 10	1.0	0.16	0.17	7.62	0.973
2	Filter Wedge Spectrometer (FWS)	NASA/ ITT	NIMBUS IV	1970	Near polar sun synchronous	~590	175	5.8	8.9	1	± 48 horz.	21 x 12 x 13	2.0	0.605	0.65	4.94	0.844
3	Very High Resolution Radiometer (VHRR)	NASA/ RCA	ITOS	1972	Circular, polar	-	97	1.4	85.0	3	-	20 x 18 x 15	3.3	0.0153	0.01	0.476	
4	High Resolution Radiometer (WHR)	USAF/ SAMSO/ ADL	DMSP	1972-1974	Near polar, sun synchronous	450	100	10.0	95.0	2	101	30 x 16 x 12	3.5	0.098	0.10	0.527	0.763
5	Surface Composition Mapping Radiometer (SCMR)	NASA/ ITT	NIMBUS V	1972	Near polar sun synchronous	500	110	4.7	23.0	2	104 horiz. 76 vert.	17 x 25 x 13	9.0	0.189	0.20	0.771	0.686
6	Visible IR Spin-Scan Radiometer (VISSR)	NASA/ SBRC	SMS-1	1974	Synchronous, geostationary	-	80	2.0	45.0	2	130 circ.	Dia.=45 L=22	8.0	0.041	0.04	0.22	0.813
7	High Resolution Infrared Radiation Sounder (HIRS)	NASA/ ITT	NIMBUS F(VT)	1975	Near polar, sun synchronous	~620	120	8.0	23.0	2	+52 hortz. +50, -28 vert.	18 x 13 x 25	9.0	0.32	0.35	1.09	0.628
8	Multiple Spectral Scanner (MSS)	NASA/ SBRC	ERTS-1	1972	Near polar, sun synchronous	~195	95	1.0	64.0	2	100 horitz. 77 vert.	56 x56 x 26	16.0	0.0145	0.0156	0.434	0.967
9	RM 20A	USAF/ SAMSO (LMSC)	STP-72-2	-	Near polar (i-98°)	400	100	1.0 W	12 ft^2	1	-	Dia.=46 in. (Radiator only)	-	0.0825	0.088		
10	RM20B	USAF/ SAMSO (RI)	STP-72-2	-	Near polar (i=98°)	400	135	10.0 W	14 ft^2	1	-	65x31 in. (Radiator only)	35.0*	0.358 0.714	0.38 0.760	1.75	0.477
11	NIMS	NASA SBRC	Galileo	1985	Jupiter Fly-By	High	75	10.0	627	2	-	D=40 L=12	-	0.015	0.0159	0.166	0.914
12	CRTU	RI	Expr.	--	--	Sync. Orbit	70	5.0 W	86 ft^2	2	--	--	320	0.058	0.071	0.128	0.550
13	VGIR	JPL	Concept Demo	--	--	--	75	200	1780	1	--	14	14	0.104	0.112	0.169	0.39
14	IR Sensor	AF /AESC	DSP	--	Geosynch.	--	173	2.2 W	--	1	~180	NOT A SHIELDED RADIATOR					
15	Thermatic Mapper (TM)	NASA/ SBRC	Landsat	1983	Polar, sun synchronous	437	87	26	435	2	114 x 160	51 x 51 x 24	40	0.055	0.059	0.380	0.802
16	PMIRR	NASA/ SBRC	Mars Obsv.	1992	Mars Polar	--	76	39	435	3	114x160	60 x 60 x 29	24	0.083	0.0897	0.176	0.53
17	SWIRLS	NASA/ AEC	EOS	--	Polar	705	90	210	796	3	119 x 180	60 x 60 x 29	26	0.24	0.26	0.35	0.29

Flight Systems. A summary of selected cryogenic radiators that have flown, been developed for test, and/or are space qualified are shown in Table 4. Many of these are relatively small (i.e., 1- to 5-milliwatt load) units sponsored by NASA. Most of the higher capacity units (e.g., nos. 9, 10, 12, 13, and 14) were developed under USAF sponsorship. Schematics of several cryogenic radiators are illustrated in Figures 17 through 20. Performance characteristics of the RM 20A and RM 20B radiators (nos. 9 and 10) are summarized in Table 5. Additional data on the JPL/Martin Marietta advanced passive v-groove radiator concept, which was flown on the Mars Observer, can be found in Ref. 9, while design and performance details of the Rockwell CRTU concept can be found in Refs. 10 through 12.

Because of the sensitivity of the performance of cryogenic radiators to environmental conditions, it is difficult to establish a meaningful figure of merit (FOM) for them. Two parameters that can be used to relate the overall efficiency, however, are the net cooling capacity, Q/A, and the fraction of the total load due to parasitic heat leak (f_p).

The ideal heat rejection capacity of a radiator is given by

$$Q/A = \sigma \varepsilon T^4 \quad , \tag{1}$$

where

Q/A	=	heat rejected watts/ft^2
σ	=	Stefan-Boltzman constant, 0.527 x 10^{-8} W/ft^2-K^4
T	=	radiator temperature, deg Kelvin
ε	=	emissivity of the radiator surface

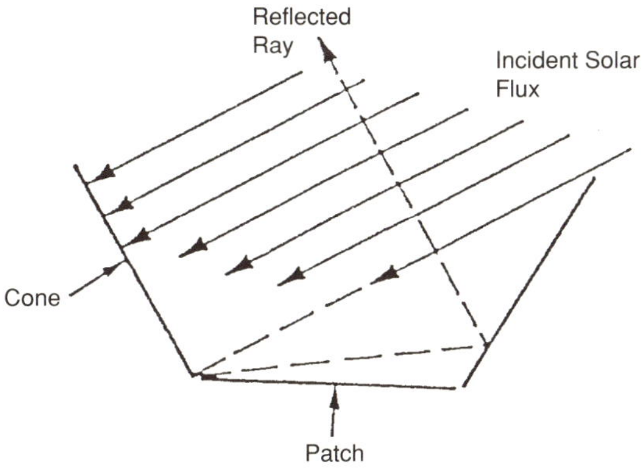

Figure 15. Cooler-cone solar-flux reflection

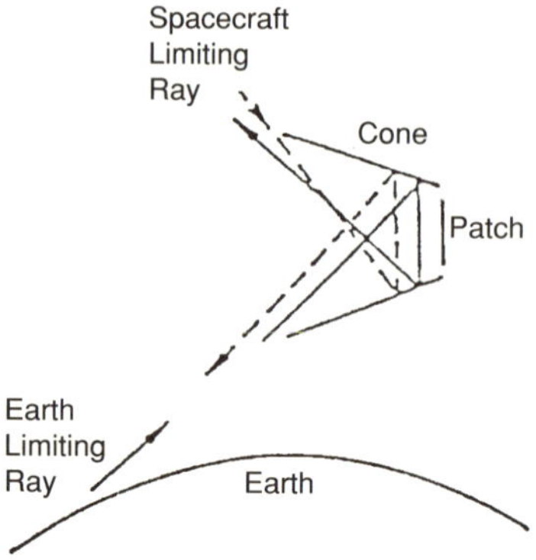

Figure 16. Cooler-cone Earth and spacecraft-flux reflection

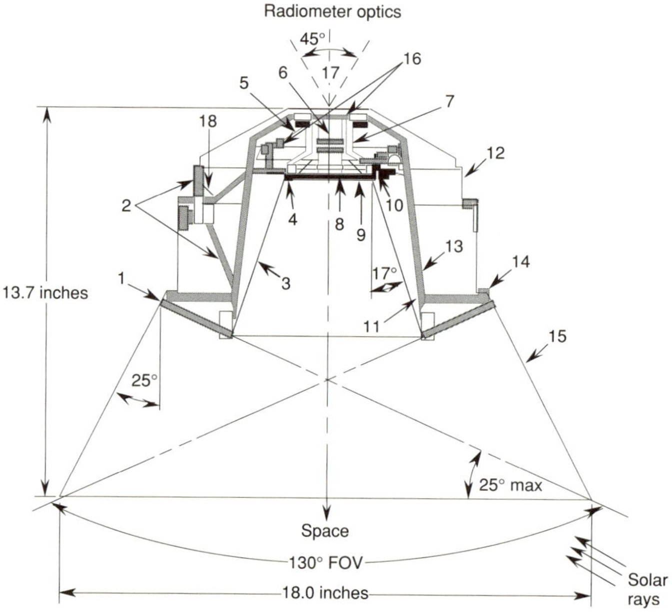

1. Intermediate stage radiator
2. Stage support bands
3. Intermediate stage shield
4. Cold stage heater
5. Cold stage support bands
6. Detector dewar assembly
7. Cold stage assembly
8. Cold stage damper
9. Cold stage radiator
10. Intermediate stage heater
11. Intermediate stage assembly
12. Ambient housing
13. Super insulation
14. Shield dampers
15. Sun shield
16. Warm and cold filters
17. Detector FOV
18. Cable assembly

Figure 17. Schematic of SBRC VISSR radiant cooler

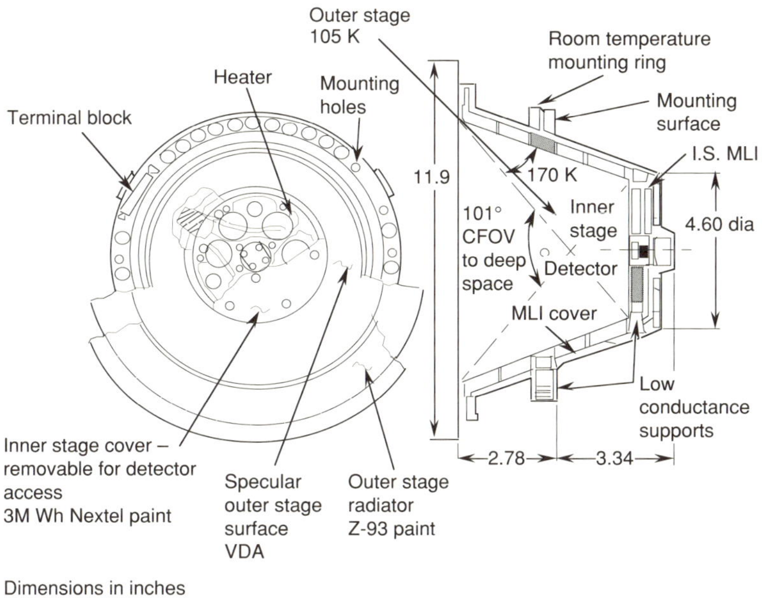

Figure 18. DMSP radiative cooler (Westinghouse/A. D. Little)

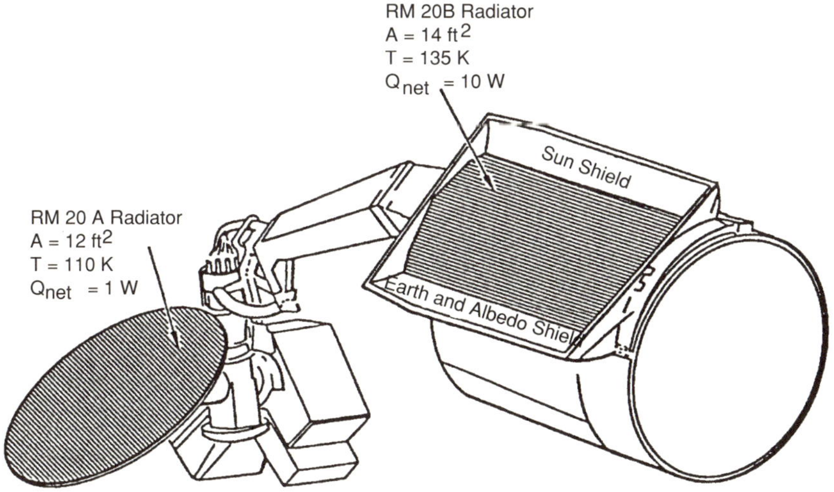

Figure 19. Radiator concepts (Rockwell, Lockheed)

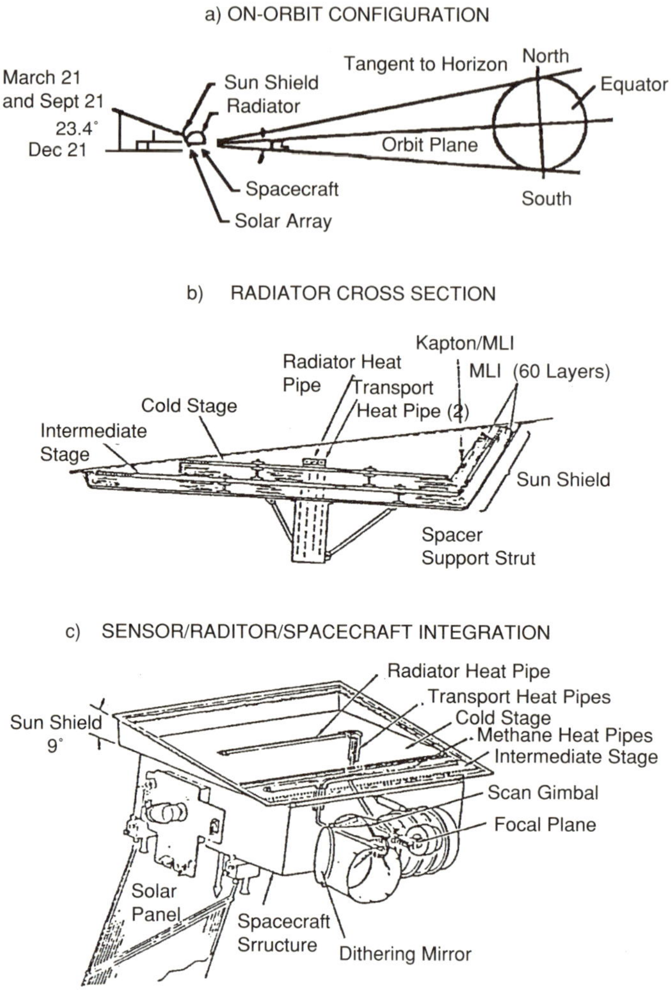

- Rockwell program under contract to Wright Patternson Air Force Base
- Two stage radiator for geosynchronous orbit
- Technological advance beyond Rockwell's RM20B radiator
- Uses shields, MLI, and low conductivity supports
- Cold Stage
    - 5 W at 70 K
    - Four oxygen heat pipes
    - 3M nextel black velvet paint
    - 74 ft^2
- Intermediate stage
    - 20 W at 140 K
    - 11 ft^2 exposed area
- Total weight 320 lbs
- Overall area = 86 ft^2

Figure 20. Cryogenic-radiator test unit (Rockwell)

The fraction of the total heat load due to parasitic heat leaks (f_p) can be expressed by

$$f_p = 1 - \frac{(Q/A)_{actual}}{(Q/A)_{ideal}} \quad , \qquad (2)$$

where the $(Q/A)_{ideal}$ is the theoretical maximum achievable. This provides a basis for comparison for different radiators. The radiator area required per watt of net heat rejection can then be expressed by combining equations (1) and (2) as

$$(A/Q)_{actual} = \frac{1.89 \times 10^8}{(1 - f_p)\, \varepsilon T^4} \quad . \qquad (3)$$

The values of f_p and actual A/Q for various radiators are presented in Figure 21.

Performance Analysis. Recent interest in multispectral sensors with relatively large heat loads over a range of cryogenic temperatures has established the need for improved methods of analysis and optimization. General expressions and a computer program for analysis and optimization of multistage radiators is presented in Reference 13. This was later modified and expanded (Reference 16) to include real hardware characteristics, to allow for parametric analysis and evaluation of contractor design proposals.

The principal of operation of a multistage radiator is based on the idea that each intermediate stage intercepts heat leakage through the insulation and supports from the stage below (or the spacecraft boundary), and rejects this heat to space. In this manner, each successive stage presents a colder boundary temperature to the stage above it. In general, for the same value of insulation effective emittance, one can achieve a lower cold-stage temperature the more stages one uses. Thus, a multistage radiator tends to approach closer and closer to the thermodynamic limit, given by the Stefan-Boltzman law, the more stages are utilized.

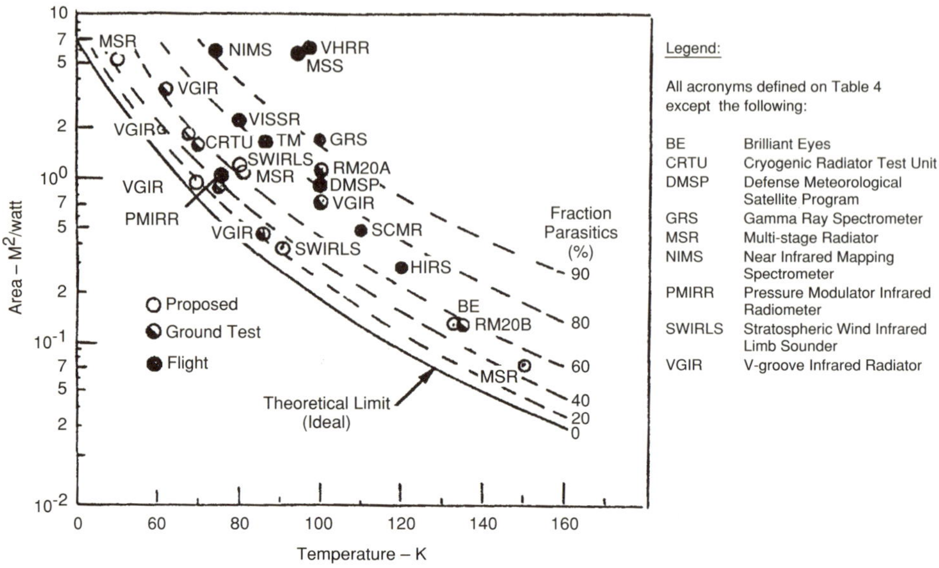

Figure 21. Comparative performance of cryogenic radiators

Table 5. RM20 A and B Cryogenic Performance

Summary Data	RM-20A Radiator	RM-20B Radiator
Manufacturer	Lockheed Missiles and Space Co.	Rockwell International
Type	Single Stage	Single Stage
Radiator Area	1.1 M^2	1.3 M^2
Dimensions	1.2 M Diameter	1.5 M × 87 M
Radiator Surface Coating	A-10 White Paint (Nextel)	Z-93 White Paint (IITRI)
Operating Temperature	100°K	138°K
Heat Dissipation	2 watts	12 watts
Parasitic Heat Load	3 watts	8 watts
Environmental Heat Load	3 watts	5 watts
Minimum Temp (Zero Load)	85°K	105°K
Number of Heat Pipes	2	10
Working Fluid	Methane	Freon 13, Freon 14, Methane
Type of Orbit	Sun Synchronous	Sun Synchronous
Development Status	Space Qualified	Space Qualified

It is important in these optimizations to use proper values for surface emittance and MLI-blanket performance (ε^* or k_{eff}) to obtain realistic designs. Data on surface emittances at low temperatures can be found in Reference 15. The performance of several MLI configurations are presented in Figures 22 and 23, extracted from Ref. 17. Additional data are available from Ref. 18.

REFRIGERATORS

Concepts. The primary considerations that differentiate space-system refrigerators from industrial or airborne systems are the extreme need for low weight and power, the requirement for extended reliability without maintenance or repair, and, generally, the need for minimum vibrational or other mechanical disturbances that significantly impact sensor pointing or jitter requirements. The practical refrigeration cycles for spaceborne use include the Stirling, Vuilleumier (VM), reversed Brayton/Claude, closed Joule-Thompson (J-T), and more recently (since 1980) the chemisorption compression/J-T expansion system under development by JPL and Aerojet, and the pulse tube, being developed by TRW and others (Ref. 39). A description and evaluation of the first four cycles and performance characteristics are covered in several publications available (Refs. 19 through 21, and 36), and will not be duplicated here. The fifth concept, sorption cryocoolers, is covered in Ref. 38.

Flight Systems. The limited spacecraft experience with mechanical refrigerators is summarized in Table 6. The first flights in 1970 and 1971 using Malaker Stirling Cycle units were basically commercial ground units converted for short-term space experiments. The longest operating system was the SEP-78-1 satellite. A set of 4 rhombic-drive Stirling units (Figure 24), developed by Philips NA, operated off and on for a period of about 5 years, although one failed early in life and the remaining units degraded significantly until the satellite was destroyed.

A VM-cycle refrigerator was flown as a two-stage unit on SESP 71-2 in 1972. The refrigerator operated successfully for approximately six months before the coolant loop failed.

The split-Stirling-cycle units flown on DSP as experiments were actually ground tactical coolers developed by Philips for NATO mobile ground units and marketed in the U.S. by Magnavox Corporation of Holland. A dual-piston version of similar design is now manufactured by Magnovox (Figure 25). These coolers were actually qualified for about 5000 hours of operation, but degraded rapidly during orbital operation. This was attributed to contamination of the working gas and subsequent degradation of the regenerators, and progressive mechanical imbalance and thermal degradation. The DSP installation is shown in Figure 26.

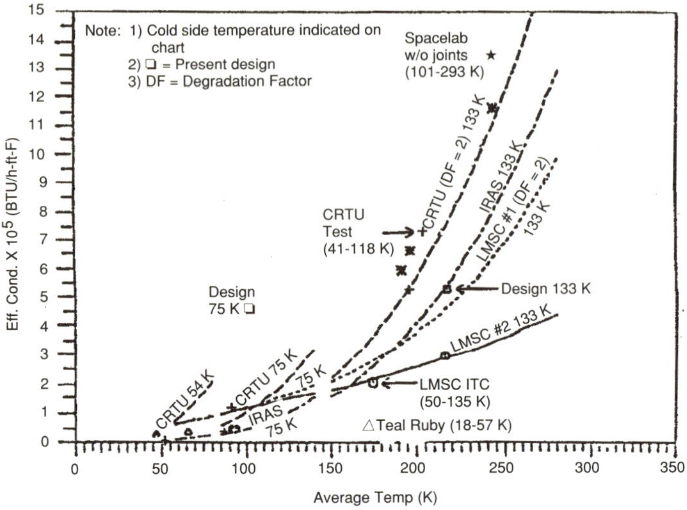

Figure 22. Cryogenic MLI effective conductivity

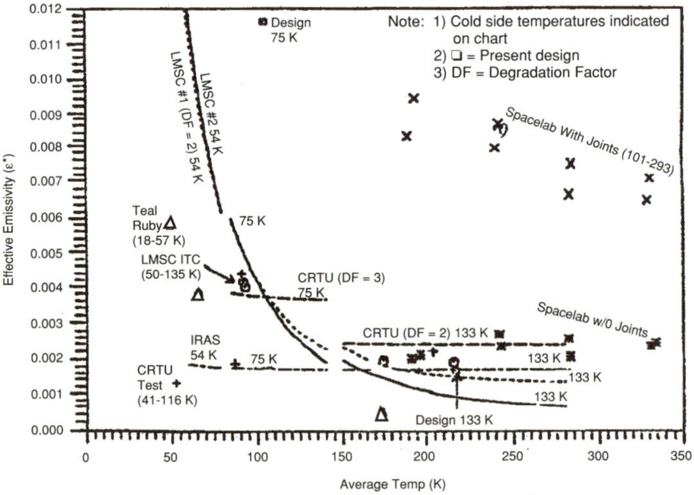

Figure 23. Cryogenic MLI effective emittance (ε*)

Table 6. Refrigerator Space Flight Experience

Program	RM-19 IR	S-192 S-193 Skylab Series	SESF 71-2 Celestial Mapping	P78-1 Gamma-Ray	ATMOS Spacelab 3	STS 61-C RCA IR Camera	3rd Color Expt.	ISAMS (UARS S/C)	ATSR FR/MATRA/ Spot S/C
Flight Date	1970	1971	1972	1979	1985	1986	1989	7/91	8/91
Sponsor	USAF	NASA	USAF	DARPA	NASA	NASA	USAF	ESA	ESA
Refrigerator	Malaker (2) 1-Stage Integral Stirling Commercial	Malaker (3) 1-Stage Integral Stirling Commercial	Hughes (1) 2-Stage Integral Vuilleimier	Phillips (4) 2-Stage Integral Stirling Rhombic Drive	CTI (1) 1-Stage Split Stirling	Cryodyn. (1) 1-Stage Integral Stirling	Philips (2) 1-Stage Split-Stirling	Oxford Split Stirling (2)	Oxford RAL Dual Comp. Split Stirling
Cooling	1.70 W @ 100 K	1.00 W @ 90 K	0.15 W @ 15 K 1.50 W @ 140 K	0.30 W @ 90 K 1.50 W @ 140 K	1.60 W 75 K	1.00 W @ 75 K		1.4 W @ 80 K	0.50 W @ 80 K
Power	40 W	45 W	540 W	30 W		30 W	58 W	80	45
Cooling System	Conduction	Conduction	Coolant Loop-Radiator	Conduction			Heat Pipe-Radiator	Cond.	--
Performance/ Status	2 Units operated for over 2 years with unknown duty cycle. 25 K degrad in 6 months due to contam.	10 Hrs over 90 days.	Coolant loop failed after 6 months, 490 flight and 689 grnd hrs.	8000 hr life test. 1 unit failed shortly in flight. Other 3 operated for 7 yrs at 50% D.C. with 30 K degrad.		6 day mission.	1 unit contamin. after 200 hrs. shut down, recovered, recontamin, 7-21 K degrad.	External contam 5 C/mo requires de-ice every 25 days.	Extr. contam 6 C/30 days.

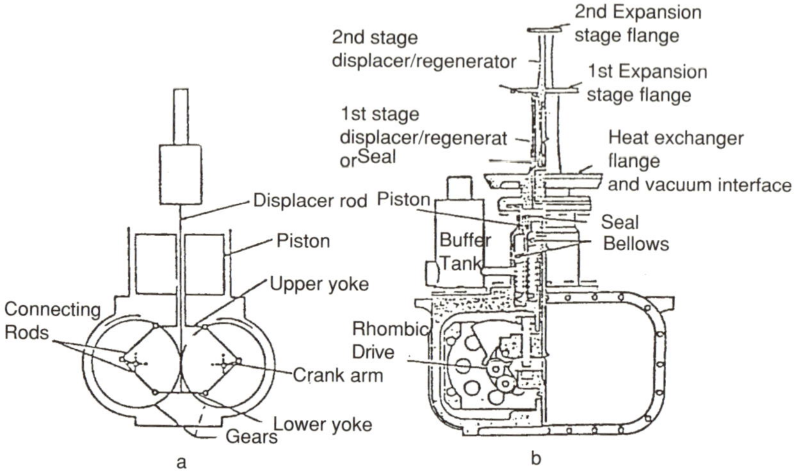

Figure 24. Phillips rhombic drive stirling

Several tactical coolers used by the DOD, with maintenance-free operation capability between 1000 and 2500 hours, are available and are included in Table 7. The Hughes (Electronic Dynamic Division) model 7044H is slated for several space experiments, including the CRYOHP (Ref. 35)

As of early 1992, the British Aerospace (BAE) split-Stirling cooler (Figure 27), developed from the original Oxford University concept, was the most advanced cooler available in terms of demonstrated life (26,000 hours). Variations of this concept were integrated by RAL/BAE and flown on experiments such as the ISAMS aboard the UARS satellite, and the ATSR experiment aboard the European Remote Sensing Satellite (ERS-1), both in 1991. Variations have also been developed by LMSC, in collaboration with Lucas (UK). A current specification sheet (for the BAE unit) is shown in Figure 28. Additional details can be found in Ref. 23. Several Aerospace studies have been made using this unit (Ref. 24), and off-design algorithms to predict performance over a range of conditions have been developed (Ref. 25).

Technology Development. As of March 1992, numerous programs to develop technology or qualify spaceborne coolers are being funded by AFSTC, NASA, DOE, NIST, U.S. Navy, ESA, and others, as well as several corporate internal research and development activities. All of these activities encompass several technologies, including integral and split-Stirling, reverse Brayton, sorption, or mechanical closed-cycle J-T systems. Although discussion of these programs and

Table 7. Short Life Tactical Coolers

Manufacturer	Magnavox	Magnavox	Cryodynamics	Hughes	Hughes	Hughes
Type	Split Stirling, Linear Drive	Split Stirling DC Motor/Crank	Stirling, Rotary Drive	Split Stirling, Rotary Drive	Split Stirling Linear Drive	Split Stirling
Model	MX-7051	MX-7045L	M15-S	7004H	7013H	7044H
Nominal Cooling Capacity	0.75 W @ 80 K	0.35 W @ 80°K	1.0 W @ 77°K	1.0 W 80°K	0.35 W @ 80°K	3.5 W @ 80 K
Weight (Lbs)	3.0	2.5	5	4	2.5	4
Size (Inches)	D=1.75, L=8	D=1.75, L=6	D=3, L=12	D=25, L=4.8	D=2, L=6	D=3, L=6
Power Requirement	16-30 Vac, 50 Hz	17-32 VDC	24 VDC	18-32 VDC	17-32 VDC	24-32 VDC
Power Input at Nominal Cooling Capacity (W)	60	30	29.5	50	30	85
Demonstrated Life (HRS)	7,000	5 000	1,000-3,000	1,000 HR (Quar) 1,500 HR (Demo)	4,000 2,500	>1,000 -1,500
Mean Time to Failure (HRS)	2,500 (Guaranteed) 25,000 (Est)	2,500 (Guaranteed)	Unknown	---	---	---
Status	Production --	Production	Production	Production -- U.S. Army	Production	Production
Prior Space Use	None	None	Skylab, Shuttle, RM-19 (USAF)	None - (Helicopers)	None	CRYOHP Experiment

Table 8. Cryocooler Development/Qualification Programs[1]

Program/ Sponsor	PSC/Phillips Lab	PSC/Phillips Lab	AIRS/ NASA JPL	EOS 80 K/ NASA GSFC	EOS 30 K/ NASA GSFC	BE AIT[2]/ BMDO	Turbo Brayton/ Phillips (SBIR)	COOLAR[3]
Refrig Design	1-Stage Stirling	1-Stage Stirling	1-Stage Stirling	1-Stage Stirling	2-Stage Stirling	1-Stage Pulse Tube	Reverse Brayton	Mech J-T 2-Stage
Vendors	LMSC/HAC	TRW	TBD	HAC/LMSC	Ball/Creare	TRW	Creare	Ball
Rqmt's	2 W @ 60 K Scalable 40 W/W 10 Yrs	1 W @ 150 K Balanced 15 W 10 Yrs	1 W @ 55 K 100 W 5.7 Yrs	0.8 W @ 80 K 30 Lbs 5.7 Yrs	0.3 W @ 30 K 80 W 40 Lbs 5.7 Yrs	0.25 W @ 70 K 20 W 4 Lbs 10 Yrs	5 W @ 65 K 200 W 55 Lbs 10 Yrs	3.5 W @ 65 K 300 W 150 Lbs 5 Yrs
Status	Phase 1 Prototype Fab	Phase 1 Prototype Fab	FM/RFP Dec 93	EM 4th Qtr/93	EM 1st Qtr/95	EM	BB Tested EM 4th Qtr/94	ABB Tests Completed
Schedule	Prototype 6/94 Perf & Life Tests-95	Prototype 2/94 Perf & Life Tests-4/94	EM 96 PF 98	TBD	TBD	Sys. Integ.	Life Test PL-94	EM-94

Notes:
1) Other Programs Include 10 K Sorption Periodic, 10 K Brayton (J-T), 35 K Pulse Tube
2) Advanced Interceptor Technology
3) Cryogenic On-Orbit Long Life Active Refrigeration

technologies are beyond the scope of this document, several of the programs, with near-term developments, as of this time period, are highlighted on Table 8. Ref. 37 provides a good overview of technology trends, including the pulse tube.

DESIGN AND TEST MARGINS FOR CRYOGENIC SYSTEMS

Most military spacecraft thermal control systems are designed to pass the test requirements of MIL-STD-1540B. These test requirements call for a 11-deg C temperature margin for passive systems (radiators) and a 25% heat load margin for active systems, which would apply to stored expendable systems and mechanical or chemical cryocoolers. Because the 11-deg C margin cannot be practically applied to very-low-temperature cryogenic systems, a study was made (Ref. 33) of the performance of NASA and DOD cryogenic systems developed during the period from 1970 to 1990 to establish guidelines for future systems.

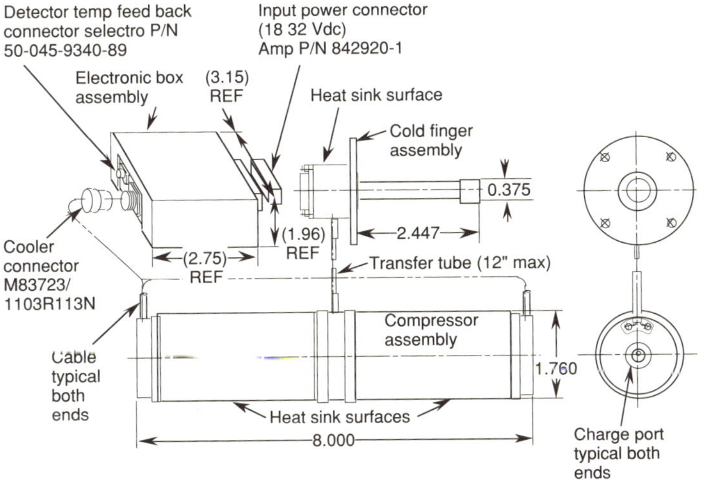

Parameter	MX7049
Refrigeration capacity at 77 K	1.75 watts
Cooldown time to 80 K	7 to 10 minutes with 1440 Joule thermal mass
Input power	100 watts (18-32 Vdc)
Ambient temperature	–54°C to +71°C
Vibration	
Axial	<0.25 lb
Radial	<0.25 lb
MTTF	≥2500

* Production Lots. This cooler is manufactured exclusively for Magnavox by Phillips Usis, BV of the Netherlands.

Figure 25. Phillips/Magnavox split-stirling tactical cryocooler

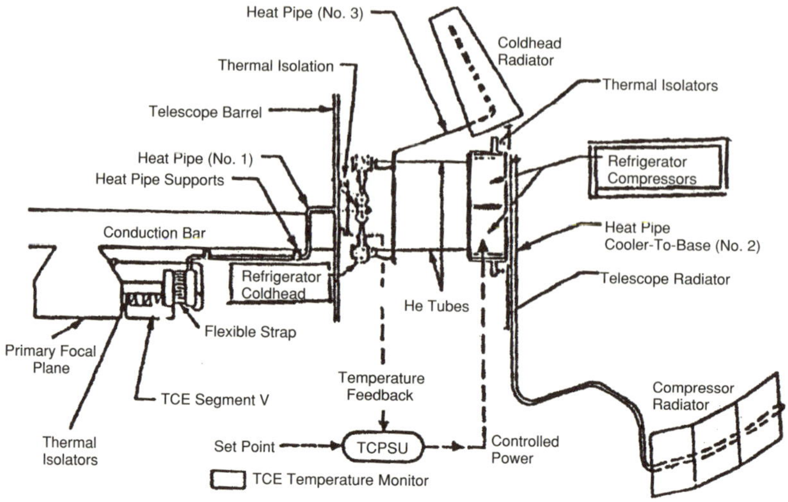

Figure 26. DSP third-color experiment cryocooler installation (Aerojet)

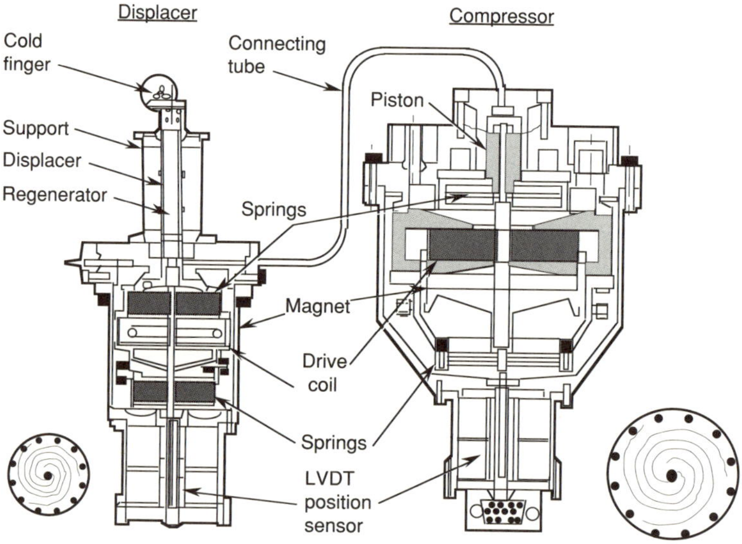

Figure 27. BAe 80-deg K Stirling cycle cooler

Specifications	Cold tip temperature:	80 K nominal		
	Cooling power at 80 K	800 mW nominal		
	Cold tip diameter:	10 mm (0.4 in)		
	Power supply:	28 V. ± 15V. 7 V nominal		
	Power consumption:	31 Watts without electronics nominal 40 Watts with electronics nominal		
	Momentum out of balance (uncompensated):	0.1 N see Peak	(Data derived from ISAMS cooler configuration by laser speckle photography)	
	Momentum compensation achieved by mounting compressor units back to back:	0.001 N see rms		
	Life time:	Approximately 24,000 hours have been accomplished to date without any performance degradation.		
	Weight (in 1 g):	Compressor	3.0 Kg	(6.6 lbs)
		Displacer	0.9 Kg	(2.0 lbs)
		Electronics	3.5 Kg	(7.7 lbs)
	Size:	Compressor	length	200 mm (8.0 in)
			diameter	75 mm (3.0 in)
		Displacer	length	190 mm (7.6 in)
			diameter	75 mm (3.0 in)
		Electronics	width	251 mm (9.9 in)
			length	254 mm (10.0 in)
			height	124 mm (4.9)
Environmental--Current Qualification Status	Temperature:	Storage -45°C to +85 C		
		Operation -20°C to +40°C (mechanical) -20°C to +40 C (electronics)		
	Vibration:	20-80 Hz increasing at 3 dB/Octave 80-350 Hz level at 0.24 g^2/Hz 350-2000 Hz decreasing .3 dB/Octave		

Figure 28. BAe 80-deg K cooler specifications

Conclusions from this research resulted in specific recommendations for design margins (which currently do not exist), and the application of the 11-deg C test margin for cryogenic systems.

Design Margins. Heat-load margins of 50 percent are recommended at program inception and are gradually reduced during the development cycle, reaching a minimum value of 25 percent at the time of flight-hardware acceptance testing, consistent with the test requirements of MIL-STD-1540B for active systems. This is illustrated in Figure 29.

Test Requirement Margins. The test-requirements margin of 25 percent is applicable to active systems, but the 11-deg C margin for passive systems requires adjustment below 203-deg K (-70-deg C). Algorithms developed in Ref. 33 show that the 11-deg C temperature margin and the 25 percent heat-load margin are equivalent at -70-deg C. Below that temperature, the 11-deg C temperature margin can be reduced as the design temperature is reduced, while maintaining an equivalent 25 percent heat-load test margin; this is illustrated in Figure 30 and Table 9. The temperature margins corresponding to the previously defined heat-load design margins of 50 percent are also defined in Table 9 and illustrated in Figure 30.

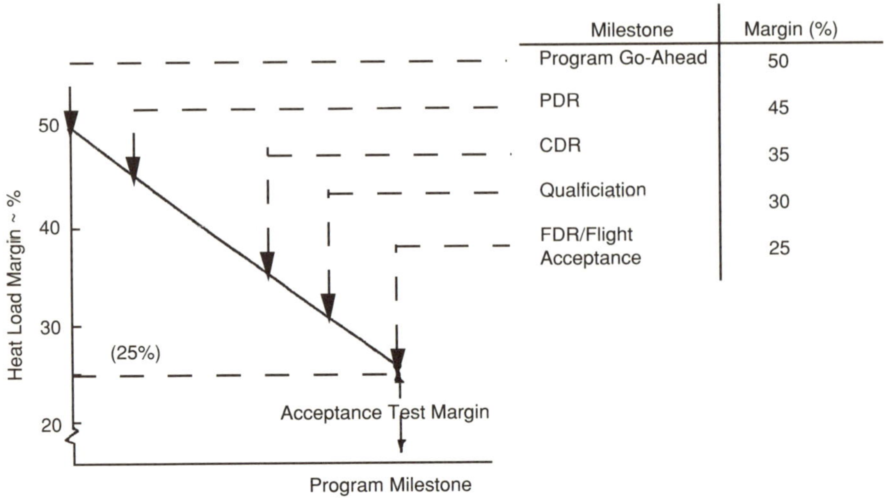

Figure 29. Recommended thermal margins for cryogenic systems

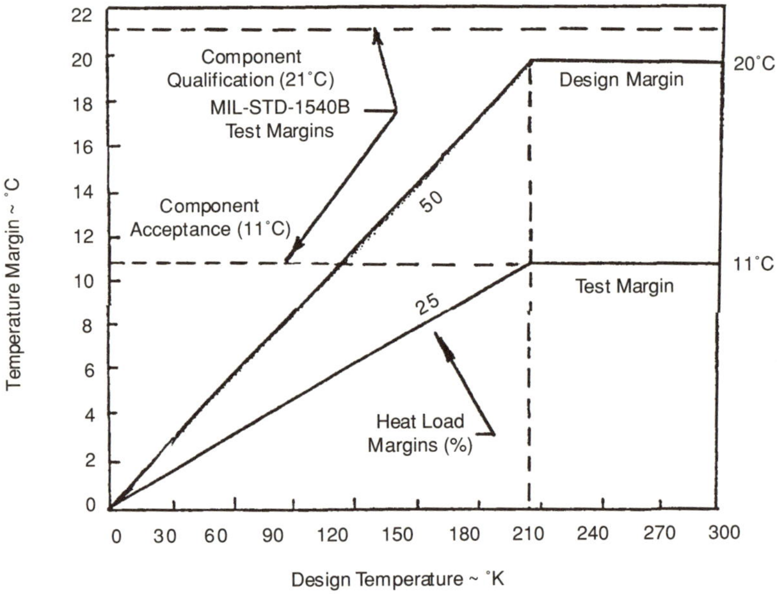

Figure 30. Proposed design and test-temperature margins vs design temperature

Table 9. Proposed Design and Test Margin for Passive Cryogenic Systems

Design Temperature		Test* Margin (25%)	Design Margin (50%)
(K)	(°C)	(°C)	(°C)
> 203	-70	11	20
185	-88	10	18
167	-106	9	16
149	-124	8	15
131	-142	7	13
113	-160	6	11
95	-178	5	9
77	-196	4	8
59	-214	3	6
41	-232	2	4
< 23	-250	1	2

*Proposed to be used used for systems below 203-deg K (-70-deg C) in lieu of the 11-deg C test margin specified in MIL-STD-1540B for passive systems.

REFERENCES

1. M. Donabedian, "Cooling Systems," *Military IR Handbook*, Chapter 15, Environmental Research Institute of Michigan, Rev. Ed., 1985.

2. A. R. Urbach, "The Infrared Astronomical Satellite (IRAS) Hardware Flight Performance," Proc. of the SPIE, Vol. 509, pp. 200-206, 1984.

3. R. A. Hopkins, "Measured Ground Performance of the Superfluid Helium Dewar for the Cosmic Background Explorer (COBE)," Proc. of the SPIE, Vol. 509, August 1984.

4. M. Donabedian, "Revised Computer Programs (CRYO2) for Cryogenic Fluid Storage," ATM No. 86(9975)-28, February 24, 1986.

5. M. Donabedian, "Revised Computer Programs (CRYO4) for Solid Hydrogen Cooler Analysis," ATM No. 89(4033-02)-6, September 8, 1989.

6. M. Donabedian, "A Computer Program for Analysis of Cryogenic Fluid Storage Systems," AIAA-87-1496, June 8, 1987.

7. M. Donabedian, "Survey of Cryogenic Cooling Techniques," TR-0073(3901-01)-1, October 30, 1972.

8. J. P. Wright and D. E. Wilson, "Cryogenic Test Unit," SSD-80-0176, Rockwell International, December 1980.

9. S. W. Petrick, et al., "Design, Fabrication and Testing of a V-Groove Radiator Mechanical Development Unit," AIAA-88-0558, January 11, 1988.

10. D. E. Wilson and J. P.Wright, "The Multistage Heat Pipe Radiator - An Advancement in Passive Cooling Technology," AIAA Paper No. 77-760, June 1977.

11. J. P. Wright and D. E. Wilson, "Fabrication and Test of a 70 K Heat Pipe Radiator," ASME Paper No. 78-ENAS-16, April 1978.

12. D. E. Wright and J. P. Wright, "Development and Testing of Two and Three Stage Heat Pipe Radiators," AIAA Paper No. 79-1060, June 1979.

13. L. Bledjian, "Computer Aided Optimization of Multi-Stage Radiators," ATM 89(4010-50)-3, March 20, 1989.

14. L. Bledjian, "Thermal Analysis in Support of the DMSP-5D2 Cooler Anomaly," ATM 80(5478-30)-4-6, May 29, 1980.

15. M. Donabedian, "Emittance of Selected Thermal Control Surfaces at Cryogenic Temperatures," ATM 90(9975)-10, December 15, 1989.

16. D. S. Glaister, "Multi-Stage Cryogenic Radiator Analysis Program" (ATM in progress).

17. D. S. Glaister, "Multi-Stage Radiator Parasitics Concern," ATM 90(9975)-32, April 13, 1990.

18. M. Donabedian, "MLI Performance Summary," ATM 85(9975)-12, November 8, 1985.

19. M. Donabedian, "Survey of Cryogenic Cooling Techniques," Aerospace Report TR-0073(3901-01)-1, October 30, 1972.

20. M. Donabedian, "Cooling Systems," *The Infrared Handbook*, Revised Edition, Chapter 15, 1985, Environmental Research Institute of Michigan.

21. A. L. Johnson, "Spacecraft Borne Long Life Cryogenic Refrigeration Status and Trends," paper presented at the 2nd Biennial Conference on Refrigeration for Cryogenic Sensors and Electronic Systems, December 7, 1982.

22. G. A. Klein, "Molecular Absorption Cryogenic Cooler for Liquid Hydrogen Propulsion Systems," Paper No. AIAA-82-0830, St. Louis, MO, June 7-11, 1982.

23. Werrett, "Development of the Stirling Cycle Coolers for the ISAMS Experiment," paper presented at 4th Cryocooler Conference, Eaton, MD, September 1986.

24. D. S. Glaister, "Expanded Cryogenic Hybrid Refrigerator-Radiator Point Design," ATM 90(9975)-26, March 8, 1990.

25. D. S. Glaister, "Updated British Aerospace Cryocooler Performance Algorithms," ATM 90(9975)-27, March 8, 1990.

26. A. L. Johnson, "Spacecraft Borne Cryogenic Cooling, Technology Status," November 1, 1989 (briefing).

27. A. L. Johnson, "Design Point Study for Mechanical Refrigerators," February 5, 1990 (briefing).

28. P. G. Wapato, "Application of the Turbo-Refrigerator to Long Term Cryogenic Storage," SAE Paper No. 820841, July 19, 1982.

29. D. S. Glaister, "Cryogenic Mechanical Cooling Systems," ATM 91(9975)-20, January 30, 1991.

30. MIL-STD-1540B (USAF), Test Requirements for Space Vehicles, Oct 12, 1982.

31. MIL-HDBK-340 (USAF), Application Guidelines for MIL-STD-1540B, July 1, 1985.

32. D. F. Gluck, "Space Vehicle Testing: Environments, Related Design and Analysis, Requirements and Practice," TOR-0086A(2902-08)-1, Sept 27, 1987.

33. M. Donabedian, "Thermal Uncertainty Margins for Cryogenic Sensor Systems," paper for AIAA 26th Thermophysics Conference, June 24-26, 1991, Honolulu, Hawaii.

34. G. Walker, *Miniature Refrigerators for Cryogenic Sensors and Cold Electronics*, Oxford University Press, NY, New York, 1989.

35. J. Beam P. J. Brennan, and M. Bello, "Design and Performance of a Cryogenic Heat Pipe Experiment (CRYOHP)," AIAA Paper No. 92-2867, presented at 27th Thermophysic Conference, Nashville, Tenn, July 6-8, 1992.

36. G. Walker, *Cryocoolers*, Plenum Press, 1983.

37. C. K. Chan, et al., "Overview of Cryocooler Technologies for Space-based Electronics and Sensors," *Adv. Cryo. Engr.,* Vol. 35, pp. 1239-1250.

38. J. A. Jones, et al., "Sorption Cooler Technology Development at JPL," *Cryogenics,* Vol. 30, March 1990, pp. 239-245.

Chapter IX

Thermal Testing

John W. Welch
The Aerospace Corporation

INTRODUCTION

To verify the thermal design and ensure successful operational use, space vehicles are subjected to extensive ground thermal testing. At the component level, these tests include thermal cycling, burn-in, and thermal vacuum. At the vehicle level, the primary thermal tests are thermal cycling, thermal vacuum, and thermal balance. In addition, developmental tests are performed as necessary. MIL-STD-1540A was written in 1974 to standardize the test requirements and establish a uniform set of definitions, environmental criteria, and test methods for military space vehicles, subsystems, and components. This document was in use until 1982, when MIL-STD-1540B (Ref. 1) was published. MIL-STD-1540B is oriented toward low-risk, long-life space vehicles and provides requirements for component, subsystem, and system-level testing. This includes orbiting vehicles, payloads that perform their mission while attached to recoverable launch vehicles, and airborne support equipment. No similar standards are available for NASA or commercial programs.

MIL-STD-1540B states that the test requirements should be tailored to the specific space program or project, considering design complexity, state of the art, mission criticality and acceptable risk. To aid in the classification of space programs and equipment, MIL-STD-343 (Ref. 2) defines four categories, considering such parameters as relative priority, assumed risk, level of national prestige, expected duration of flight life, complexity, use of redundancy, whether flight spares are available, cost, and criticality of launch dates. The four categories are: class A, high priority minimum risk; class B, high priority medium risk; class C, medium or higher risk effort; and class D, higher risk minimum cost effort. MIL-STD-343 reduces test requirements for class B, C, and D programs. The details of these modifications will be described later.

MIL-STD-343 also defines the different categories of space equipment. Consistent terminology is important because of its use in MIL-STD-1540B. The six levels of space equipment defined in MIL-STD-343 are, in ascending order of assembly: part, subassembly, component, subsystem, space experiment, and space vehicle. While MIL-STD-1540B establishes testing requirements for space vehicles, subsystems, and components, space experiments are often considered to be a lower level of assembly of a space vehicle, and are either tested at the space vehicle or subsystem level.

A component is a functional unit that is viewed as an entity for purposes of analysis, manufacturing, maintenance, or record-keeping. Examples are hydraulic actuators, valves, batteries, electrical harnesses, and individual electronic boxes such as transmitters, receivers, or multiplexers. A subsystem is an assembly of two or more components, including the support structure to which they are mounted and any interconnecting cables or tubing. A subsystem is composed of functionally related components that perform one or more prescribed functions. Typical space-vehicle subsystems are electrical power, attitude control, telemetry, thermal control, and propulsion. A space vehicle is a complete integrated set of subsystems and components capable of supporting an operational role in space. A space vehicle may be an orbiting vehicle, a major portion of an orbiting vehicle, or a payload that performs its mission while attached to a recoverable launch vehicle.

DESIGN ENVIRONMENTS

Paragraphs 3.8 and 3.9 of MIL-STD-1540B describe the design environments for a space vehicle and a vehicle component, respectively. The design environments for either are the composite of the various environmental stresses to which the vehicle or components are designed. The thermal stresses used in the design consider equipment operation, internal heating, eclipse conditions, space-vehicle orientation, environmental heating (including solar, Earthshine infrared, and albedo radiation), ascent heating, and degradation of thermal surfaces during the life of the mission. These stresses are used in the analytic modeling efforts to predict, over the life of the spacecraft, the worst-case hot and cold temperatures for the component, subsystem, and satellite. From these results, acceptance and qualification temperatures are derived.

Figure 1 illustrates how these levels are determined for components. The nominal extreme temperatures are generated by the thermal analytic models. These values must be found component by component, as a worst combination of conditions for one component may not prove to be worst for another. To these results, an uncertainty margin is added. This margin, which can be quite large at the beginning of a program (e.g., 20- to 40-deg C), is reduced as the design and analytic process progresses. Following successful correlation of the thermal analysis with thermal-balance test data, this uncertainty margin can be reduced to as little as ±11-deg C. If a component is heater controlled at the cold extreme, 25 percent excess heater-control authority is used in lieu of an 11-deg C temperature margin. These temperatures (called the maximum and minimum predicted temperatures) set the component acceptance-test levels, subject to the requirement that the mounting plate or case temperature be at least as cold as -24-deg C and at least as hot as +61-deg C. If the minimum predicted temperature is greater than -24-deg C (or if the maximum predicted temperature is less than +61-deg C) the acceptance temperature is then set to -24-deg C (or +61-deg C in the hot case). The specified temperature extremes are required to: (a) provide adequate environmental stress screening, (b) demonstrate component-survival capability, and (c) assure that temperature-insensitive and high-quality parts and materials are used in the design.

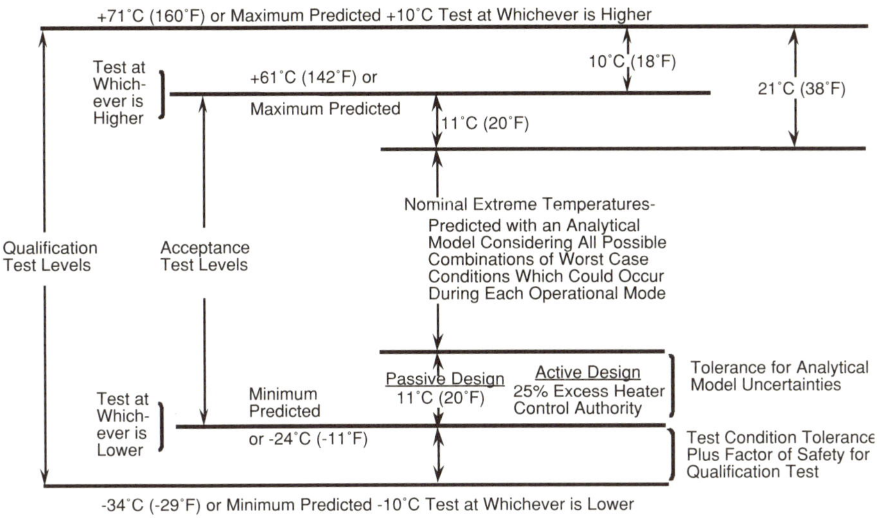

Figure 1. MIL-STD-1540B test requirements for components

Component qualification tests are conducted at temperatures 10-deg C colder (even if heaters are used for temperature control) and 10-deg C hotter than the acceptance-test temperatures, subject to the constraint that the mounting plate be at least as cold as -34-deg C and at least as hot as +71-deg C. Certain temperature-sensitive components are typically exempt from the design margins. Batteries, propellant valves, extremely accurate clocks, and inertial reference units are four such examples. Batteries are usually tightly controlled at the cold temperatures to increase life. Representative values from NiCd batteries are: operating, 0-deg C to +25-deg C, and survival/turn-on, -10- to +40-deg C. In survival/turn-on mode the component need not operate within specification, but must not experience any degradation when returned to the operating range. Most equipment have temperature excursions 20- to 50-deg C above and below room temperature. Components without active electronics that are mounted outboard, such as solar arrays and antennas, are usually designed to withstand wider temperature excursions, particularly at the cold end.

Thermal uncertainty associated with temperature predictions is reduced during the design-analysis-test process as the design becomes firm, as improved and more detailed analyses are conducted, and as developmental tests are completed. The thermal balance test and subsequent correlation of the analytic model to test data substantially reduces temperature-prediction uncertainty. Deviation between on-orbit temperature measurements and preflight temperature predictions is a measure of the final uncertainty associated with the analytic and test process.

The $\pm$11-deg C uncertainty margin is a result of extensive comparisons between preflight predictions and on-flight orbit-temperature measurements. A report by R. D. Stark, *Thermal Testing of Spacecraft*, summarizes much of the work. Among his results, a study of 20 critical spacecraft components showed that the thermal balance test and subsequent model correlation reduced the standard deviation between prediction and on-orbit measurement from 9-deg C to 5.5-deg C, and reduced the maximum deviation from 17-deg C to 11-deg C. As the intent of MIL-STD-1540B is to have a 95 percent confidence that design temperatures (maximum and minimum predicted temperatures) are never exceeded, the 11-deg C uncertainty should be used for predictions verified by the thermal-balance test, and margins greater than this should be used for unverified analytic predictions.

Although component- and spacecraft-level testing are the only two levels of assembly addressed in MIL-STD-1540B, as spacecraft size and complexity has grown and buildup time has lengthened, the need has developed for intermediate tests, between component and space-vehicle testing. The purpose of such tests can be environmental-stress screening, performance verification, and/or thermal balance. Such tests may be conducted on all or part of a subsystem. These tests allow use of smaller test facilities than required for the space-vehicle tests and make it easier to tailor the thermal environment to the specific requirements of the components under test. The uncertainty margin associated with this level of assembly depends upon whether the item tested is more thermally similar to a component or space vehicle. In most cases space-vehicle uncertainty margins should apply.

At the space-vehicle level, individual zones and temperature extremes associated with each zone are established. Test temperature extremes are based upon analytic temperature predictions for at least one component in each zone. Minimum and maximum predicted temperatures and qualification level temperatures are imposed in each region, ensuring that no component will exceed allowable temperature limits.

Employing Design Environments

As stated in the introduction, the purpose of thermal testing is to verify the thermal design and ensure successful operational use. This is accomplished by detecting flaws or defects in the thermal design, materials, or manufacturing process, and by verifying that the unit tested performs within operational specifications during the test. Environmental-stress screening is the process that subjects hardware to physical stresses and forces flaws that are not ordinarily apparent into observable failures. These flaws, when discovered, are repaired or equipment is replaced prior to usage in flight. Ideally, qualification tests expose design defects, while acceptance tests uncover defects in workmanship, parts, materials, and processes. Performance verification is achieved when the item tested operates within specifications when subjected to an extreme environment.

To aid in achieving effective ground testing, it is considered important to identify problems at the earliest practical point in time. Therefore, the test levels and techniques are designed to maximize test rigor at the lowest levels of assembly and lessen in severity as the level of assembly increases, in a funnel effect. In this fashion, problems can be identified in time for orderly solution and at a level of assembly that will minimize excessive tear down. While not specifically addressed in MIL-STD-1540B, it is assumed that adequate testing of parts will be performed and that high-quality parts will be utilized.

COMPONENT TESTING

As defined previously, a component is a functional unit made up of modules and assemblies that, in turn, are made up of piece parts. Although tests and screens are conducted at lower levels of assembly than components, the lowest level addressed in MIL-STD-1540B is the component level. The three environmental tests performed at the component level are thermal vacuum, thermal cycling, and burn-in. Functional tests, which are not considered to be environmental tests, are performed at temperature extremes during thermal cycling and thermal vacuum.

For various components, Table 1 calls out the tests that are required, optional, and not required at the qualification and acceptance levels. Note (2) for thermal vacuum acceptance testing deserves discussion. Most electronic boxes are unsealed, so this test would appear to be widely required. The second part of this note, however, implies that low-power RF equipment and all digital equipment do not require a thermal-vacuum-acceptance test. The Aerospace Corporation's Spacecraft Thermal Department believes that all electronic units should receive a thermal vacuum acceptance test.

Performance of moving mechanical assemblies can be very temperature sensitive. Binding of deployment mechanisms because of temperature or temperature gradients has occurred on-orbit. Temperature gradients can strongly influence friction in bearing assemblies. For these reasons, testing moving mechanical assemblies in vacuum at temperature extremes, with temperature gradients imposed (if applicable), is required.

Test plans for antennas are sometimes given inadequate attention. Antennas cannot be treated merely as "structure". While large power dissipations are not normally associated with antennas, proper design and workmanship must be demonstrated. Often, tests need to be conducted over wide temperature

Table 1. MIL-STD-1540B Component Test Baseline

Component Qualification Tests

Test	Reference Paragraph	Suggested Sequence	Electronic or Electrical Equipment	Antennas	Moving Mechanical Assembly	Solar Panel	Batteries	Valves	Fluid or Propulsion Equipment	Pressure Vessels	Thrusters	Thermal Equipment	Optical Equipment
Functional	6.4.1	1[1]	R	R	R	R	R	R	R	R	R	R	R
Thermal Vacuum	6.4.2	9	R	R	R	R	R	R	R	O	R	R	R
Thermal Cycling	6.4.3	8	R	O	O	O	O	O	O	—			—

Component Acceptance Tests

Test	Reference Paragraph	Suggested Sequence	Electronic or Electrical Equipment	Antennas	Moving Mechanical Assembly	Solar Panel	Batteries	Valves	Fluid or Propulsion Equipment	Pressure Vessels	Thrusters	Thermal Equipment	Optical Equipment
Functional	7.3.1	1[1]	R	R	R	R	R	R	R	R	R	R	R
Thermal Vacuum	7.3.2	7	R[2]	O	R	O	R	R	R	O	R	R	R
Thermal Cycling	7.3.3	6	R	O	O	O	O	O	O	—	—	—	—
Burn-in	7.3.9	8	O	—	O	—	—	R	—	—	R	—	—

Legend
R - Required
O - Optional Test
— - No Requirement

Notes
(1) Functional tests shall be conducted prior to and following environmental test
(2) Required only on unsealed units and on high power, RF equipment

extremes to simulate predicted on-orbit temperatures. Sometimes these tests need to be conducted with temperature gradients imposed.

Solar arrays may experience wide temperature excursions. Moreover, as they are thin and lightweight, they have small thermal capacitances and hence respond rapidly to varying environments. It is no longer a requirement of MIL-STD-1540B to conduct thermal-cycling tests at the maximum rate to be experienced during ascent or on-orbit. However, it may be prudent to test to these predicted rates. Solar panels may appear to be benign, but on several occasions flaws were discovered during thermal cycling at elevated temperatures. These defects would not have been apparent at room temperatures.

The performance and life of batteries can strongly depend on temperature. Additionally, recent designs of nickel-hydrogen batteries have employed numerous thermal interfaces. In battery testing, it is important to measure cell-to-cell temperature differences, characterize contact conductance at interfaces, and establish battery-power characteristics at temperature extremes.

Objective of Component Thermal Testing

The component-level thermal tests can be construed as having three functions: environmental-stress screening, demonstration of survival and turn-on capability, and performance verification. The intent of the screening function is to find faults in component design, workmanship, materials, and processes. Ideally, the qualification test should uncover design defects, while the acceptance test uncovers defects in workmanship, parts, materials, and processes. The intent of the survival and turn-on function is to demonstrate that the equipment can be soaked, started, and operated at cold and hot survival-temperature limits without experiencing permanent damage or performance degradation when returned to the operational temperature range. Finally, the tests verify that the component electronic performance is within specification.

With regard to screening and performance verification, thermal cycling and thermal vacuum tests have different roles. The thermal-cycling test is primarily a screening test; performance verification is secondary. The reverse is true for thermal vacuum testing.

Thermal Cycling Test

The requirements of component-level thermal-cycle testing are given in Table 2. As previously discussed, component-level thermal testing is performed at temperatures either based upon analytic predictions plus a design margin, or set at specified extremes, whichever is more severe. At acceptance level, either the maximum and minimum predicted temperatures (which include the 11-deg C uncertainty margin) are used, or cold and hot limits of -24-deg C to +61-deg C are used. For example, if a component has expected nominal extreme values of -18-deg C to +42-deg C, the component has minimum and maximum predicted temperatures of -29-deg C to +53-deg C, respectively. Because +53-deg C is less severe than +61-deg C, the acceptance-test temperatures for this component would be -29-deg C to +61-deg C. At qualification level, testing should be performed at temperatures either 10-deg C colder than the minimum predicted temperatures and 10-deg C hotter than the maximum predicted temperature, or at the specified extremes -34-deg C to +71-deg C. In the previous example, the qualification-test temperatures would be -39-deg C to +71-deg C.

Table 2. Component Thermal Cycle Test Parameters

Thermal Cycling Test Parameters	Qualification - Para. 6.4.3	Acceptance - Para. 7.3.3
Temperature Range (differential)	105-deg C	85-deg C minimum
Temperature Extremes	Minimum predicted with -10-deg C environmental design margin, to maximum predicted with +10-deg C environmental design margin, or at least -34-deg C to +71-deg C	Minimum predicted to maximum predicted, or at least -24-deg C to +61-deg C
Number of Cycles	3X acceptance (24 cycles minimum)	8 cycles minimum
Dwell	1-hour minimum at temperature extremes (each cycle)	1-hour minimum at temperature extremes (each cycle)

The minimum number of acceptance cycles performed during the test is eight. For qualification, three times the acceptance cycles, or a minimum of twenty-four, is required. A minimum dwell time of one hour at both temperature extremes on each cycle is required to allow the component to stabilize at the temperature extreme. The rate of transition between the two temperature extremes should be as rapid as possible. The stabilization criteria prior to dwell should be ±3-deg C/hour.

Although secondary in purpose, performance verification is accomplished in the thermal-cycle test by means of functional testing. Functional tests should be conducted at the high- and low-temperature levels during the first and last cycle, and after return of the component to ambient temperature. Although not specified in MIL-STD-1540B, abbreviated functional tests should be conducted at the high- and low-temperature levels on all cycles between the first and last. During these tests, parameter drift should be carefully monitored. Throughout the test, electrical components (including all redundant circuits) should be cycled through various operational modes and perceptive parameters monitored for failures and intermittents.

As previously stated, the primary purpose of thermal cycling is environmental-stress screening. The types of flaws uncovered during thermal cycling include defective solder joints, loose crimp connections, drift problems, broken wire bonds, bent connector pins, defective or contaminated parts, and thermal coefficient of expansion mismatch. The five parameters that appear to most greatly affect thermal-cycling screening effectiveness are: (1) number of cycles, (2) temperature extremes and range, (3) temperature transition rate of change, (4) operation/nonoperation profile, and (5) dwell at the temperature extreme.

Considerable work has been done on the subject of how many thermal cycles are required, especially at the unit level. Of particular importance are the 1972 Martin Marietta study sponsored by NASA and the more recent 1984 guidelines published by the Institute of Environmental Sciences (IES). Martin Marietta conducted an industry survey on the subject of thermal cycling as employed in the production acceptance testing of electronic units. Seven companies supplied test data relating failures to the number of temperature cycles. The curves of Figure 2 show that failure rate decreases with increasing number of temperature cycles, with more cycles required to reach the "knee" of the curve as component piece-part count increases.

The number of thermal cycles recommended in the Martin Marietta study for acceptance testing of electronic components is a function of part count: 100 parts, 1 cycle; 500 parts, 3 cycles; 2000 parts, 6 cycles; and 4000 parts, 10 cycles. For the nominal case, the IES guidelines recommend 12 thermal cycles at the unit level. The data used in making this recommendation show that while the knee in the failure rate versus number of cycles occurs by the fourteenth cycle for all components, a good portion of the failures have yet to occur. From this, it is apparent that the MIL-STD-1540B requirement of at least eight cycles for unit acceptance testing is well justified, and perhaps insufficient.

Comparison of MIL-STD-1540B unit acceptance testing (-24-deg C to +61-deg C) with the recommendations of Martin Marietta (-54-deg C to +55-deg C), and the nominal recommendations of IES for all units (-40-deg C to +70-deg C), show that MIL-STD-1540B is less stringent at the cold end, comparable at the hot end, and less stringent in terms of temperature range. It is concluded from this that the temperature extremes in the standard are within the design and performance capability and the experience base of electronic components. It is especially appropriate that these extremes be utilized in the screening program.

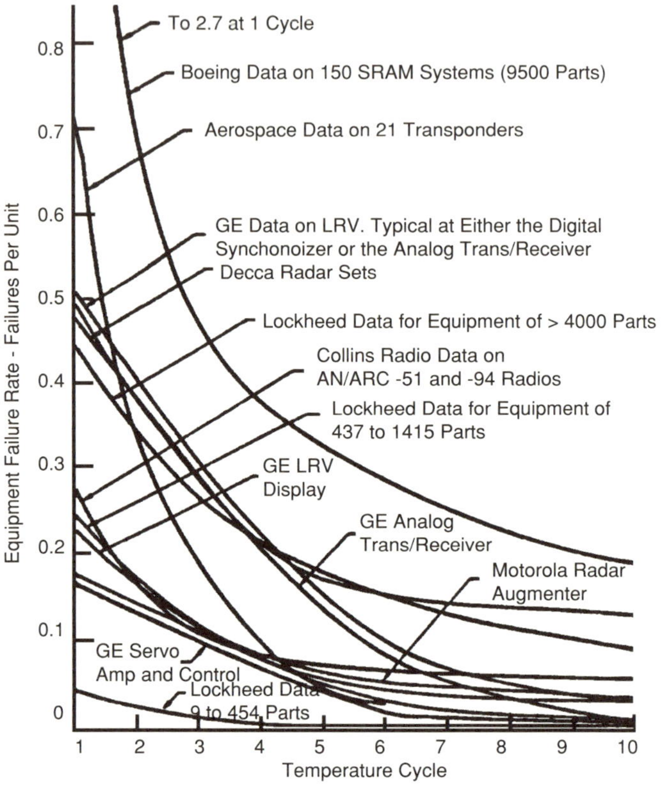

Figure 2. Industry data failures vs number of thermal cycles (Martin Marietta)

MIL-STD-1540B requirements for rate of change are given at representative location(s) on the unit, such as a mounting point on a baseplate of conduction-cooled designs or on the case of radiation-cooled designs. The requirement in MIL-STD-1540B is "an average rate of at least 1-deg C per minute." There is little data on the effect of different rates of change. However, faster transition rate of changes, at least as great as the "maximum rate to be experienced during ascent on-orbit or re-entry," should be encouraged.

Data in the IES guidelines support the MIL-STD-1540B position with regard to unit operating/nonoperating testing profiles: "After the component has stabilized (at specified low temperature) the unit shall be turned off and then cold started. With the component operating, the chamber temperature shall be increased to the upper temperature level. After the component has stabilized... the unit shall be turned off and hot started... and during... the test, electrical components, including all redundant circuits, shall be cycled through the various operational modes and perceptive parameters monitored for failures and intermittence to the maximum extent possible." With regard to both precipitating incipient failures and detecting these failures, thermal cycling is superior, with the unit operating as opposed to nonoperating. Quart and Wong report a doubling of failure rate with units operating compared to nonoperating. The IES guidelines recommend conducting unit thermal-cycling tests with power on. This recommendation appears to be well founded. Furthermore, cold and hot starts, at operational and survival limits, are effective stress screens, as well as demonstrating that the equipment is well designed and rugged enough to survive mission-derived extreme environments and subsequently, to perform within specifications over the narrower operational temperature range.

MIL-STD-1540B states that "each cycle shall have a one hour minimum dwell at the high and the low temperature levels during which the unit shall be turned off until temperature stabilizes and then turned on. The dwell time at the high and low levels shall be long enough to obtain internal thermal equilibrium." The requirement is necessary to ensure that the unit will be tested at the designated temperature extremes, and to provide an adequate screen.

Burn-In

A necessary adjunct to the screening program is burn-in testing during which the component is operated for a long period of time to precipitate failures. During burn-in, additional hours of operation beyond those accrued during thermal cycling are accumulated until 300 hours are reached; additional defects are precipitated, detected, and corrected; failure-free performance is demonstrated for the last 100 hours. This test, which is part of acceptance testing, is described as "modified thermal-cycling test" and includes the requirement that additional cycles, beyond those conducted during the thermal-cycling test, be conducted until a total of at least 18 cycles is reached. As shown in Table 3, the burn-in test parameter requirements closely resemble the thermal-cycle acceptance-test parameters.

It is the practice of some companies, at least for some units, to conduct burn-in at a constant elevated temperature. Burn-in at a constant elevated temperature aids in the detection of performance changes or drift of a parameter. Additionally, such a test tends to precipitate processes driven by Gibbs free-energy potential (e.g., corrosion, migration). If equipment is believed to be susceptible to problems of this type, the case can be made that more of the burn-in time should be spent at an elevated-temperature, even at the expense of thermal cycling. For other conditions, it is more effective to cycle the equipment, and the practice of constant elevated-temperature burn-in should be discouraged.

Table 3. Component Burn-In Test Parameters

Burn-In Test Parameters	Qualification	Acceptance - Para. 7.3.9
Temperature Range (differential)	No	85-deg C
Temperature Extremes	qual. test	Minimum predicted to maximum predicted or at least -24-deg C to +61-deg C
Number of Temperature Cycles	specified by	18 cycles minimum including thermal cycling test cycles
Total Operating Time	MIL-STD-1540B	300-hour minimum including thermal cycling time (or 100 cycles minimum for cycle-sensitive components)
Dwell		1-hour minimum at temperature extremes

Thermal Vacuum Test

The primary purpose of the thermal vacuum test is to verify the functional performance of the component, although the test is still effective at stress-screening the unit. Without the convective environment, temperatures and stresses will more closely simulate on-flight conditions than thermal cycle could. The test requirements are shown in Table 4.

The temperature range and extremes are identical to the thermal-cycle-parameter requirements. The number of thermal cycles, however, is considerably less, with a minimum of three required for qualification level and one for acceptance level. Because of the vacuum environment, a longer dwell, 12 hours, is necessary to ensure that the component has stabilized in temperature. The specified pressure of 10^{-4} Torr or less should be reduced to 10^{-5} or less to adequately simulate the vacuum environment.

To verify the operational performance during the test, functional tests should be conducted at high- and low-temperature levels during the first and last cycle, and after return of the component to ambient temperature and pressure. Abbreviated functional tests should be conducted at the high- and low-temperature levels on all cycles between the first and last. Throughout the test, electrical components (including all redundant circuits) should be cycled through various operational modes, and perceptive parameters monitored for failures and intermittents to the maximum extent possible.

Table 4. Component Thermal Vacuum Test Parameters

Thermal Vacuum Test Parameters	Qualification - Para. 6.4.2	Acceptance - Para. 7.3.2
Temperature Range (differential)	105-deg C	85-deg C
Temperature Extremes	Minimum predicted with -10-deg C environmental design margin, to maximum predicted with +10-deg C environmental design margin, or at least -34-deg C to +71-deg C	Minimum predicted to maximum predicted, or at least -24-deg C to +61-deg C
Number of Cycles	3 cycles minimum	1 cycle minimum
Dwell	12-hour minimum at temperature extremes	12-hour minimum at temperature extremes
Pressure	10^{-4} Torr or Less	10^{-4} Torr or less

Thermal-Test Facilities

A wide variety of test chambers are available for thermal-cycling tests. Nitrogen or humidity-controlled air is used to prevent water vapor condensation. During heating or cooling a rapid air or gas flow is required. A rapid rate of temperature change at the baseplate or case of the component of interest is often difficult to achieve. This is often the major technical challenge faced in thermal-cycling testing.

Thermal vacuum tests are divided for convenience into two categories; (1) those where conduction to a heat sink is the dominant mode of cooling, and (2) those where radiation to the surroundings dominates or where cooling is by both conduction and radiation. The former has proven to be the more likely occurrence. Conduction cooling is usually accomplished by torquing the component down onto a monolithic, nearly isothermal heat sink. This is not truly representative of actual component installation, which may for example have Delron inserts in an aluminum honeycomb with face sheets. However, it is usually acceptable for component testing and buy-off, provided the differences between test mounting and flight mounting are accounted for by analysis and verified by testing at the system or the subsystem level. Occasionally (e.g., where power density is very high) the component is mounted to a flight-like installation. In either case, the component is covered by an insulation blanket to assure that radiation is negligible.

Many components are cooled primarily by radiation or by both conduction and radiation. Such components include control-moment gyroscopes, inertial reference units, and accelerometers. Here, control of heat-loss paths should be such that radiation and conduction occur in the same proportion as calculated for the flight environment. This is necessary so that module and piece-part temperatures and component-temperature gradients duplicate those that occur in actual usage.

DEVELOPMENTAL AND SUBSYSTEM THERMAL TESTING

MIL-STD-1540B addressed thermal testing at the component and system (space vehicle) level. Besides these two levels, special tests may be necessary at levels of assembly different from these two. In addition, dedicated tests may be required to provide confidence in a new design or to aid the analysis. One such test is the developmental test. Developmental tests can provide early data to assist in the design or manufacturing process. Typically, a new design concept is experimentally assessed, or an old technique is evaluated in a new application. Often it is interesting to obtain data to aid the layup and handling of the insulation blanket or to assess the structural integrity of such blankets under the pressure differential of venting. Heat pipes usually receive extensive developmental tests, both at the pipe- and pallet-level of assembly. Such testing can include: evaluation of process variables, such as cleaning and bake-out techniques; evaluation of pipe performance, such as watt-inch heat-transport capability; or verification of fabrication techniques. Tests may be conducted to determine contact conductance along critical heat-flow paths or the conductance of thermal isolators. Engineering-model electronic units are sometimes instrumented and tested in a vacuum environment to verify the thermal analysis. Piece-part temperature predictions can also be verified.

Subsystem- and assembly-level tests can also be performed to provide environmental-stress screening, performance verification, or thermal balance. These tests allow use of smaller test facilities than required for the space vehicle test and make it easier to tailor the thermal environmental to the specific requirements of the components under test. Usually, configuration and leveling requirements can be more readily met in a subsystem, rather than in a space-vehicle test. Certainly the results are obtained in a more timely manner. In some cases, the subsystem tests suffice to demonstrate or prove some aspects of the design (e.g., thermal balance), when the test cannot be conducted in a meaningful way at the space-vehicle level.

SPACE VEHICLE THERMAL TESTS

Space-vehicle thermal tests are similar in purpose and structure to the component-level tests. Qualification tests are more demanding than acceptance test in that they require more cycles and wider temperature swings. Qualification tests are formal contractual demonstrations that the design, manufacturing, and assembly of hardware have resulted in conformance to specified requirements. The acceptance tests are required formal tests that are conducted to demonstrate acceptability of an item for delivery. They are intended to demonstrate performance to specified requirements and to act as quality-control screens to detect deficiencies in workmanship, materials, and quality. The three tests described in MIL-STD-1540B associated with space-vehicle level are thermal vacuum, thermal balance, and thermal cycling.

Thermal Vacuum Test

The thermal vacuum test is constituted primarily of system-level functional performance tests between and at temperature extremes. Emphasis is on component and subsystem interaction and interfaces, and on end-to-end system performance. Thermal functions that are verified during these tests include thermostat and heater actuation, heater duty cycle, louver operation, heat-pipe performance, and insulation effective emissivity. Table 5 summarizes the test requirements from MIL-STD-1540B. Temperature extremes are based upon worst-case analytic prediction for at least one component in each zone. Typically, the spacecraft will be divided into manageable zones and the test-temperature limits will be specified for each zone. A variety of components, often tested to different temperature extremes during component qualification and acceptance, must be accommodated during space-vehicle thermal vacuum testing. The approach taken is to drive as many components as possible (but at least one component per vehicle equipment zone) to their qualification- and acceptance-temperature extremes, with the constraint that no component should exceed its component-level test- temperature extremes. Temperatures must be monitored to avoid overstressing. The test temperatures are identical to the component-level test, except that the "at least as extreme" component requirement (-24-deg C to +61-deg C for acceptance, -34-deg C to +71-deg C for qualification) is not imposed at this higher level of assembly, and test temperatures for each zone are based upon the most extreme temperatures of components within the zones.

A minimum of eight cycles is required at the qualification level, and a minimum of four cycles is required at the acceptance level. MIL-STD-1540B allows that, at acceptance level, the four thermal vacuum cycles can be reduced to one if space-vehicle thermal cycling is performed. It is not prudent, however, to reduce the thermal vacuum cycles from four to one. This may be the only vacuum environment some components will be subjected to prior to flight. One cycle may be insufficient to accurately characterize the component behavior in a vacuum. Furthermore, the primary intents of thermal cycling (environmental-stress screening) and thermal vacuum (performance verification) are quite different. It is not justifiable to state that the thermal cycles performed provide the same level of performance verification as three cycles in thermal vacuum. To this extent, a minimum of four thermal vacuum cycles should be encouraged, even if thermal cycling is performed.

Table 5. Space Vehicle Thermal Vacuum Test Parameters

Thermal Vacuum Test Parameters	Qualification - Para. 6.2.7	Acceptance - Para. 7.1.7
Temperature Range and Extremes	Minimum predicted to maximum predicted temperature environments plus environment design margin of 10°C, for one component in each vehicle equipment area	Minimum predicted to maximum predicted temperature environments, for one component in each vehicle equipment area
Number of Cycles	Minimum of 8 cycles	Minimum of 4 cycles if thermal cycling not performed
Dwell	Minimum of 8 hours soak at each temperature extreme of each cycle	Minimum of 8 hours soak at each temperature extreme of each cycle
Pressure	10^{-4} Torr or less	10^{-4} Torr or less

An eight-hour dwell is required to bring the space vehicle to equilibrium prior to functional testing. For smaller satellites, this value can be reduced provided that analysis is shown demonstrating that a shorter dwell time will allow the satellite to equilibrate. Prior to the dwell, the temperature rate of change should be less than 3-deg C/hour.

Full functional tests are to be performed before and after the thermal vacuum test at ambient temperatures and pressure, and at the high- and low-temperature extremes for the first and last cycle. Abbreviated functional tests should be performed at both temperature extremes for the cycles between the first and last. Throughout the thermal vacuum test, the equipment should be on and functioning through different operational modes.

Thermal-Cycling Test

The thermal-cycling tests at the space-vehicle level, as in the component level, are primarily environmental screens to expose design, workmanship, material, and processing defects. Integrity of mounting, cabling, and connectors are verified. This test is optional at the spacecraft level. The test parameters are given in Table 6. The temperature extremes are not specified but must be within the maximum and minimum values and have a range of, at minimum, 70-deg C at qualification and 50-deg C at acceptance. A minimum of 50 qualification and 40 acceptance cycles are called out.

Table 6. Space Vehicle Thermal Cycle Test Parameters

Thermal Cycling Test Parameters	Qualification - Para. 6.2.9	Acceptance - Para 7.1.8
Temperature Range Differential	Maximum possible within constraints, with minimum of 70°C	Maximum possible within constraints, with minimum of 50°C
Temperature Extremes	Not Specified in Para. 6.2.9	No specified in Para. 7.1.8
Number of Cycles	No. of cycles = 125 percent of acceptance test = 50 minimum	40 minimum
Dwell	Duration not specified. On last cyclc only, at each temperature extreme, for functional test.	Duration not specified. On last cycle only, at each temperature extreme, for functional test.

Thermal-Balance Test

The thermal-balance test is comprised of dedicated thermal tests conducted during the thermal vacuum test to verify the thermal analytic models. Additionally, the functional capability of the thermal control hardware is demonstrated. A successful test and subsequent model correlation establishes the ability of the thermal control subsystem to maintain all payload and equipment within specified temperature limits for all mission phases. Although strictly a requirement of qualification-level testing, this test should be conducted for one-of-a-kind spacecraft, the lead vehicle of a series of spacecraft, a block change in a series of vehicles, upper stages, and sortie pallets designed to fly with the Shuttle.

The test usually involves several (but not all) mission phases and one or two vehicle configurations. In operational orbit, there may be several combinations of equipment operation and solar angles or solar angle versus time profiles that bear investigation. As vacuum-chamber time is expensive, judicious choice of cases should be made. The intent is not to drive all equipment to qualification-test limits, as this is done during the functional portion of the thermal vacuum test. The test should exercise all important internal heat-flow paths and external solar-flux absorptive and radiative surfaces. To provide a frame of reference for the operational orbits, the thermal-balance-test program may simulate as many as three or four hot cases that envelope the range of expected sun angles and equipment operation, both primary and redundant. This might include a transient case where angle to space vehicle varies with time. Hot cases tend to have high, but realistic, levels of equipment usage and absorbed solar energy. Typically, two cold operational cases are simulated: a cold steady-state condition and an eclipse simulation. Conditions for these tests involve minimal equipment usage, bus voltage, and solar heating. Sufficient data are needed to correlate the thermal math models, which are then used to predict component temperatures for all mission phases and power configurations. For on-orbit operational cases, the space vehicle is ideally in the fully deployed configuration. Deployable antennas are often stowed during the thermal-balance test. Large appendages such as solar arrays, booms, and antennas sometimes are not part of the tested configuration.

Nonoperational mission modes may also require simulation. Transfer-orbit, storage, and safe mode are the most important. A transfer-orbit simulation should utilize the stowed-vehicle configuration. A dedicated cold phase is recommended to verify and characterize heater operation.

Methods for simulation of environmental heat loads are sometimes divided into two categories: absorbed flux and incident flux. In the absorbed-flux method, analytically predicted absorbed fluxes are imposed on the space-vehicle surfaces. If long-wave radiation is employed, as with quartz lamps or heated plates, the emissivity of the absorbing surface must be known and calibration using absorbed-flux radiometers is required. If heaters are directly affixed to the space vehicle, heat losses must be well understood and refurbishment is required if the qualification space vehicle is also the flight vehicle.

The contractor should compare pretest temperature predictions with corresponding test data. As a guideline, those differences that fall outside a ±3-deg C band require either a good explanation or a model adjustment, depending on the size of the deviation. The ±3-deg C goal is especially important for temperature- sensitive or mission-critical components.

A variety of test-related factors contribute to a fairly large residual analytic uncertainty after completion of the thermal-balance test. Contractors often do not meet the $\pm$3-deg C correlation guidelines and allow correlation deviations as great as $\pm$6-deg C. Both incident-flux (solar simulation) and absorbed-flux methods of heat addition are imperfect. With the former, spectral matching is not exact, and reflection and reradiation sometimes occur from auxiliary equipment within the chamber. Hot-case conditions often cannot be properly simulated because critical surfaces have beginning-of-life (new) surface properties, whereas hot cases are dominated by end-of-life (degraded) properties. For example, paints, silverized Teflon, aluminized Kapton, and silverized fused-silica mirrors (OSRs) can experience solar-absorptance increases (due to contamination and environmental exposure) by factors of two to four for typical missions. The latter technique is heavily dependent on analysis for determining the intensity and distribution of solar flux. Measurement and calibration error also contribute. Model correlation to test data may not be effective if an incorrect heat-transfer mechanism is employed. Finally, some design changes that were made because of thermal-balance test results are not verified until the acceptance test of the first flight vehicle—and, sometimes, there is not a test validation.

Overall, the thermal-balance test has proved successful in correcting major thermal-modeling errors, in reducing the standard deviation between prediction and flight measurements, and in providing physical insight into heat-transfer mechanisms.

The thermal-balance test and portions of the thermal vacuum test serve to verify the design and performance of thermal control hardware. Primary and redundant heaters and thermostats are exercised, and 25 percent excess heater-control authority is demonstrated under cold-case conditions. Depending on the location of the heaters, this demonstration of heater-power adequacy can verify that insulation blankets are not "lossy," that surface finishes have the proper emissivity, and that conduction isolators perform properly. If heaters are not present, and for hot mission phases, the performance of these hardware elements is verified by temperature measurements. Temperature measurements made during simulated hot and cold mission phases characterize the performance of louvers and heat pipes. If the heat pipes are level (horizontal) within allowable tolerances during the thermal-balance test, then it is possible to verify the on-orbit performance of these heat pipes. If such leveling is not possible during the space-vehicle test, verification should be accomplished by a subsystem or pallet thermal-balance test.

The propulsion subsystem often presents especially difficult problems of thermal control. Many components are involved (e.g., valves, catalyst beds, thrusters) and there are long line runs. Insulation layup and closeouts are important, as are location of heaters and thermostats. The thermal-balance test allows determination of temperature and temperature gradients under flight conditions, and provides verification of heater and thermostat location.

Space-Vehicle Vacuum-Test Facilities

Thermal-vacuum-test facilities capable of handling the space vehicle can be quite large. The AEDC Mark 1 chamber in Manchester, Tennessee is a 42-foot-diameter, 82-foot-tall chamber housed in a ten-story building. The Lockheed Missile and Space Company Delta chamber is a horizontal stainless-steel cylinder with removable end caps measuring 80 feet in length, flange to flange, and 36 feet in diameter. A representative chamber is shown in Figure 3 and a list of facilities is shown in Appendix B.

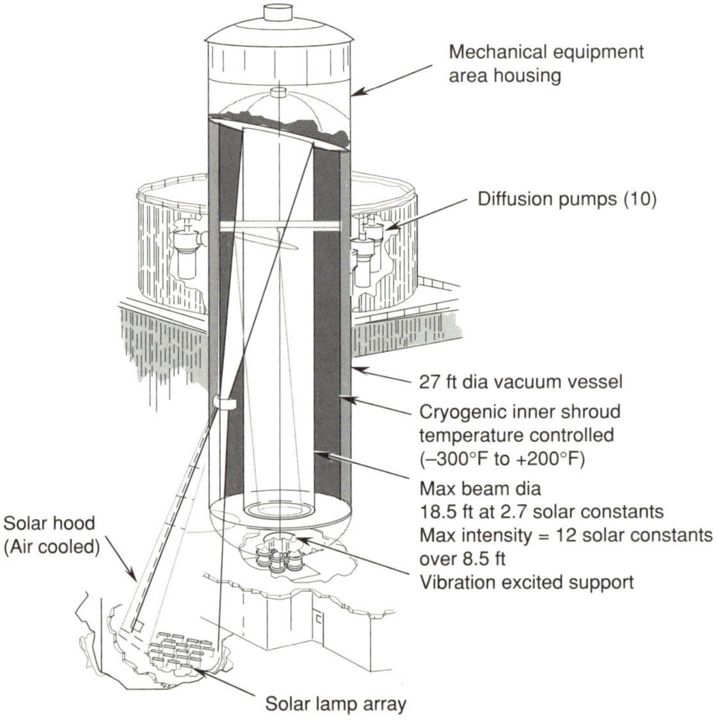

Figure 3. JPL thermal vacuum chamber

Pressure pumpdown is accomplished by mechanical pumps and diffusion pumps. Pressures as low as 10^{-5} to 10^{-6} Torr are readily obtainable. Further depressuration can be achieved with the use of cryopumps, sputter-ion pumps, or turbo-molecular pumps.

The cold environment of space is typically simulated by liquid-nitrogen-cooled internal walls. The Mark 1 facility has a 77-deg K capability, while the Delta chamber can hold a 90-deg K temperature. Because of the fourth-power dependence of radiant-energy exchange, a wall at these temperatures constitutes only a minor radiant-energy source for a room-temperature spacecraft.

Heating can be accomplished in a number of ways. Solar simulation is the most realistic of a space environment.

Solar simulation is accomplished using an array of modules, each containing a 1-kW quartz-iodine lamp and a water-cooled collimator tube. As the created spectrum approximates a 3000-deg K blackbody, with the sun more nearly like a 5800-deg K blackbody, augmenting xenon short-arc lamps can be used to improve spectral matching. Solar simulation is the preferred method of spacecraft heating, as this technique allows the natural blockage and cavity effects to occur, while imposing direct and reflected solar-like radiant heating. Infrared heating of the spacecraft by appendages or payloads, that cannot practically be deployed in the chamber can be simulated using quartz-envelope tungsten-filament lamps. Because of cost and complexity, spacecraft heating is often done by methods that do not simulate the spectral content and directionality of the sun, but do attempt to impose the proper intensity and distribution of heating. These will be discussed on subsequent pages.

Besides solar simulators, infrared-radiance simulators are also available. The Mark 1 facility uses T-3 quartz-envelope tungsten-filament lamps mounted in individual reflectors around the test article. The heat-flux simulator at the Delta chamber is not an infrared simulator, but rather uses wavelengths between infrared and solar. It consists of eighty-seven individually controlled lamp zones positioned around the test article.

FACTORY AND LAUNCH-SITE THERMAL TESTING

Checkout and functional tests are required at various stages during the buildup of a space vehicle. Such tests often are part of the formal developmental, qualification, and acceptance process. For example, these tests allow checkout at intermediate stages during the buildup process, can verify that a subsystem has not been damaged or degraded during shipment, and allow continuity, checkout, and limited functional tests during and after assembly at the launch site. Thermal control (i.e., gas or liquid cooling) often is required to ensure that components do not overheat during these tests. Compounding the difficulty of this requirement is the fact that the subsystem or space-vehicle configuration and surrounding environment can encumber the cooling process. The cold radiation sink for which the space vehicle is designed is lacking during these tests, and natural convection cooling is inefficient. Moreover, the subsystem or space-vehicle may be oriented so that heat pipes are inoperative and may be enveloped with contamination covers, shrouds, or the like, so that there is a limited accessibility to fluid cooling.

It is important to identify, early in a program, factory and launch-site cooling requirements for checkout and functional tests. This is especially important for sensitive components such as batteries. Such vehicle-design accommodations and auxiliary ground equipment that may be required to allow adequate cooling should be specified. This may include ducting and fans, piping and pumps, and leveling hardware and instrumentation.

TEST TECHNIQUES

Qualification by Similarity

The continued production and use of items designed for space vehicles of one program on space vehicles of another program is of interest because not only are design, tooling, and qualification costs eliminated for subsequent programs, but the continuing usage of the same item increases the confidence in the item's reliability. Of course, to accommodate the specific requirements of another program, it may not be possible to use the same exact item, so there may be changes required in the item or in its testing. If those changes are within reasonable bounds, then qualification of the revised item by similarity should be considered.

Although MIL-STD-1540B does not directly address criteria for the qualification of items by similarity, it does provide the standard test baselines for comparison. If component A is to be considered as a candidate for qualification by similarity to a component B that has already been qualified for space use, then all of the following conditions should apply:

a. Component A should be a minor variation of component B. Dissimilarities will require understanding and evaluation in terms of weight, mechanical configuration, thermal effects, and dynamic response. Minor design changes involving substitution of piece parts and materials with equivalent reliability items can generally be tolerated. Design dissimilarities resulting from addition or subtraction of piece parts and particularly moving parts, ceramic or glass parts, crystal, magnetic devices, and power conversion or distribution equipment should be given attention in this evaluation.

b. Components A and B should perform similar functions, with B having equivalent or greater operating life with variations only in terms of performance such as accuracy, sensitivity, formatting, and input-output characteristics.

c. Components A and B should be produced by the same manufacturer using identical tools and manufacturing processes.

d. The environments encountered by component B during its qualification or flight history should have been equal to or more severe then the qualification environments intended for component A.

e. Component B should successfully pass a post-environmental functional-test series indicating survival of the qualification stresses.

f. Component B should have been a representative flight article.

g. Component B should not have been qualified by similarity or analysis.

In some cases, the item to be qualified by similarity is not a component, but is another level of assembly, such as a subsystem. In this case, the criteria of the item to be qualified by similarity would be the same as those used if the item were a component

It is recognized that in some cases, where all the above criteria are not satisfied, qualification based on engineering analysis plus partial testing may be permissible. In this case, negotiation between the contracting agency and the contractor may result in an abbreviated testing program satisfactory for qualification of the component or item in question. The acceptability of qualification by similarity should be documented by test reports, drawings, and analyses. This justification or proof of qualification should be prepared in data packages and submitted to the contracting agency as required by the contract. The contracting agency usually has the final decision as to the acceptability of qualification by similarity, and the burden of proof of qualification is the responsibility of the contractor.

TESTING CHECKLIST

To prepare for thermal testing, coordination between the test facilities, equipment, and procedures must be complete. A checklist of handling/installation, instrumentation/data, test operations, thermal-vacuum-test procedures and thermal-vacuum-chamber procedures has been prepared by Hughes Aircraft Company as an aid to test readiness. The following pages present this checklist.

Test Article Handling/Installation

- Good housekeeping for all flight hardware areas.
- Handling and storage procedures should be in place for flight hardware.

- Special procedure should be defined for electrostatic discharge sensitive equipment.
- Proper contamination-control instructions shall be established.
- Requirements for electric grounding of test articles and test sets should be defined.
- Test cables should have end-to-end continuity checks.
- Test equipment should have readily retrievable maintenance and repair records.
- Connector pins should be visually inspected prior to mating and after demating.
- Component mounting requirements should be specified for interface filler and bolt torque.

Test Instrumentation/Data

- Test-recording and -measuring equipment and instrumentation should be clearly described on an interface diagram.
- Test-temperature limits and acceptable tolerances shall be clearly defined.
- Allowable test-temperature transition rates shall be defined.
- A clear definition of thermal stabilization (steady state) shall be established.
- Location of all temperature sensors and heaters (if applicable) shall be shown in pictures or sketches.
- Control-temperature sensors and primary backups shall be defined.
- A continuity check and response to temperature stimulus (where accessible) shall be performed on all temperature sensors prior to test start.
- Alarm setting shall be defined to give proper warning of temperature limits (plus tolerances) being exceeded.
- Audible alarms must not be left in "Silence Alarm" mode when heat exchangers are operating correctly. There shall be a signal to alert test personnel to manually reset the alarms.
- Use fuses with test heaters to prevent an overpower condition.
- Heaters should be powered at a low level and response verified by control temperature sensor.
- Manual data:
    - Types of data required and recording intervals shall be defined. Data sheets shall be provided with samples in the test procedure.
- Automated data:
    - Adequate consoles with data monitors should be available, including a dedicated one for the customer.
    - One set of all hardcopy results should be supplied to customer.
- Instrument test cables should be wrapped with low-ε surface material to minimize heat loss from the test article.

- Test-temperature-sensor locations should compliment and sometimes duplicate flight-sensor locations. Test sensors should be located so as to be meaningful when comparing to nodal computer models.
- Test sensors should be covered with thermal material similar to the surface to which they are attached. Lead wires should have low-emittance finish.
- When instrumenting a propellant line, run the sensor lead wire along the line for approximately 12 inches before exiting low-ε overwrap.
- Heaters used within the test article must be compatible with the slip-ring *current* capability for spinning tests. The number of thermal signatures that may be obtained are also limited by the slip-ring capability.
- When running a thermal-balance test where long heater lead-lengths are present, make certain that the voltage drop in the lines has been accounted for in setting the thermal dissipation. Also account for the heater-resistance change as function of temperature.

Test Operations

- Use test-facility requirements checklist prior to test.
- Pictorial timeline showing test phases and temperature transition should exist in procedures.
- Corona phenomena must be considered with proper safeguards.
- Automated test procedures should be validated prior to test start.
- Back-up power should be available.
- Performance measurements to be taken before, during, and after test should be defined.
- Personnel required during test should be clearly defined.

Component Thermal Vacuum Test

- Environmental test conditions and tolerances should be defined and should comply with the equipment specification.
- Pass/fail criteria should be specified.
- Test hazards and cautions shall be identified in the procedure at appropriate locations.
- A process for handling test discrepancies should be established.
- There should be regular test reviews while test is in progress.
- Quality Assurance (QA) and Engineering should review data prior to breaking configuration.
- A test chamber log-book should be used.
- A defined process should exist for test-shift overlaps.
- Independent QA monitoring of flight-hardware activities.
- Test personnel should be reasonably knowledgeable in thermal aspects of test.

- Test personnel should be trained and certified in operation of test equipment.
- Test personnel should be knowledgeable in safety circuitry.

System Thermal Vacuum Test

- Contamination monitoring and measuring instrumentation should be strategically placed.
- Test instrumentation should have current calibration and alignment data.
- Heat-flux calorimeters should be calibrated in-situ.
- Photographs should be taken of test article before and after installation in chamber.
- Power should be filtered and regulated.
- Safety procedures must be in effect in case of power outage or failure of the real-time data-acquisition system.
- Hydraulic lines and fluid-containing valves and components should be closely inspected before and after the test.
- Test heaters should be fused.
- Make periodic visual checks through port holes if possible.
- Assure that man-hole covers are bolted down prior to chamber closure.
- Critical test instrumentation and hardware should be redundant.
- Non-flight structure (i.e., spin fixture, spot-lamp supports, test hardware, etc.) should be hidden from spacecraft view with LN_2 shrouding.
- Reflections, even off black surfaces, should be accounted for when making predictions.
- Use small black-plate and Ag-Teflon radiometers to measure reflected solar energy where none is expected to occur.
- Test hardware in the chamber that is above the test article should be sufficiently out of the solar beam to insure no shadowing or reflections.
- Minimize the presence of warm test fixtures or pheripheral-equipment surfaces.
- Account for the effect of black surfaces (even if LN_2 cooled) that are close to the test article.
- Note the effective background temperature of the chamber.
- Use a small black-plate radiometer to measure the *local* effective background temperature.
- Heater planes that are used to produce IR to spacecraft surfaces should be suspended such that they will not sag under cold-wall/vacuum conditions. Make the stress group aware of the environmental condition to which these plates will be exposed.
- The test article should be isolated from the support fixture by a thermal isolator. A blocking heater should be placed on the fixture side of the interface to control the losses from the test article.

- Make certain that all materials used as last minute *fixes* (i.e., aluminum tape, Teflon, aluminum foil, paint, etc.) are space approved and also approved for your individual program. Also, material used should be compatible with expected temperature extremes (dental cement, epoxy, mylar/Teflon etc.).
- Do not use aluminum foil or tape on thermal isolation brackets used to conductively isolate components. The conduction of the foil or tape is too high and may short out the isolation bracket. Instead use aluminized Teflon or Kapton tape—aluminum side out—as the primary thermal control surface where it is not feasible to VDA the part.
- Perform a clearance check, at ambient conditions, around the periphery of all moving interfaces.
- Thermal surfaces on both the test article and test-chamber hardware should be cleaned prior to raising the end bell.
- Also, prior to raising the end bell, have a walk-around inspection of the test article, fixture, and chamber to (1) insure, as an example, the louver covers, Earth-sensor covers, sun-sensor covers, etc., have been removed and (2) last-minute thermal provisions have been implemented (do not assume that the above items have been done just because you have an understanding with the responsible manufacturing and test-facility engineer).
- When end bell has been raised, run a complete data scan to ensure that all instrumentation is working properly.

<u>Thermal Vacuum Chambers</u>

- Whenever diffusion pumps (DPs) are utilized, system interlocks and/or manual-control safety procedures shall be implemented, to include:
  - Automatic anti-oil migration and back-streaming controls.
  - Automatic DP oil high-temperature thermostat.
  - Prevention of roughing and high vacuum valve being open simultaneously.
  - Prevention of high vacuum valve being opened until roughing pressure is below a predetermined level.
  - Prevention of high vacuum valve being opened unless DP foreline valve is open.
  - Prevention of DP heaters being energized when DP foreline pressure is above a predetermined level.
  - Prevention of high vacuum valve being opened when DP heaters are off.
- Whenever a predetermined chamber pressure is exceeded,
  - An interlock should close the high vacuum valve.
  - An interlock should turn off power to corona-sensitive test articles.
- An interlock shall prevent cooling medium from being introduced until a predetermined chamber pressure is achieved.

- Independent pumping systems and valved systems are recommended. Whenever pumps performing multiple serial functions and/or valveless (semi-valveless) systems are utilized, interlocks or manual control with appropriate safety notes should be implemented to prevent out-of-sequence operations.

ONE-OF-A-KIND SPACECRAFT THERMAL TESTING

For high-priority, long-life complex space equipment, high reliability is usually achieved by strict compliance to specifications and standards that document the requirements and practices needed to achieve long life. Extensive design analyses, thorough screening processes, and special attention toward eliminating single-point failures in the design are implemented. To assure high reliability for these programs, a full qualification program is conducted on each component, each space experiment, and on each subsystem involved.

Not all space programs, however, are high-priority and long-life. Many are single missions of short duration and the equipment may be relatively simple. Although these programs present an opportunity for substantial time and cost savings, there is still a requirement for high reliability. Military Handbook 343 (MIL-HDBK-343) was written to identify cost-saving measures that are reasonable for one-of-a-kind space equipment or for the first of a series of space vehicles. The handbook is a basis for a consistent technical approach to achieve, in a cost-effective way, the high reliability required.

MIL-HDBK-343 defines the four classes of space programs, space vehicles, and space experiments as follows:

Class A — High Priority, Minimum Risk. Class A is defined as a high-priority, minimum-risk effort. The characteristics for Class A usually also involve some combination of the following features: high national prestige, long life, high complexity, high use of redundancy, soft failure modes, independent qualification items, complete flight spares, highest cost, and a critical launch time. Vehicle and experiment retrievability or in-orbit maintenance is usually not possible.

Class B — Risk with Cost Compromises. Class B is defined as a high-priority, medium-risk effort, with cost-saving compromises made primarily in areas other than design and construction. The characteristics for Class B usually involve some combination of the following features: high national prestige, medium life, high complexity, soft failure modes, protoflight qualification, limited flight spares, limited use of redundancy, high cost, short schedule, and a critical launch time.

Vehicle and experiment retrievability or in-orbit maintenance is usually not possible.

Class C — Economically Reflyable or Repeatable. Class C is defined as a medium- or higher-risk effort that is economically reflyable or repeatable. The characteristics for Class C usually involve some combination of the following features: medium to high national prestige, short life, low to medium complexity, small size, single-string designs, hard failure modes, very limited flight spares, medium cost, short schedule, and a noncritical launch time. Vehicle and experiment retrievability or in-orbit maintenance is usually possible, such as typified by Spacelab or Orbiter-attached payloads.

Class D — Minimum Acquisition Cost. Class D is defined as a higher-risk, minimum-cost effort. The characteristics for Class D usually involve some combination of the following features: medium to low national prestige, short life, low complexity, small size, single-string designs, spares, lowest cost, short schedule, and a noncritical launch schedule. Vehicle and experiment retrievability or in-orbit maintenance may or may not be possible.

Table 7 summarizes the thermal-test requirements for space vehicles or space experiments in each class. Terminology is consistent with that of MIL-STD-1540B. Analytic thermal computer modeling and the verification of the model through a thermal-balance test are required for all but Class D programs. As specified in MIL-STD-1540B, the test is to include maximum and minimum power-dissipation modes. If heat pipes are included, the attitude of the equipment shall not bias the test measurements. Test data shall be compared to thermal model predictions and correlation of the model will be performed.

Maximum operating environments and testing tolerances are in accordance with MIL-STD-1540B. Developmental testing is required for each class. Sufficient testing shall be performed to verify new designs, materials, and manufacturing processes necessary to achieve a high-reliability design. The required environmental design margins for Class A equipment are those specified in MIL-STD-1540B. Acceptance-test levels for components are -24-deg C to +61-deg C or 11-deg C beyond the nominal predicted temperature range. For experiments and vehicles, the acceptance temperature range is 11-deg C beyond the nominal predicted range. The qualification range uses an environmental design margin or qualification margin of 10-deg C.

Table 7. Comparison of Typical Test Requirements for Space Vehicles or Space Experiments of Each Class

Test Requirements	Class A	Class B	Class C	Class D
Computer thermal model	Required (software)	Required (software)	Required (software)	Not required
Thermal verification of computer model	Thermal vacuum test	Thermal vacuum test	Thermal test or thermal vacuum test	Not required
Maximum operating environments	MIL-STD-1540B definitions for each assembly level	MIL-STD-1540 definitions for each assembly level	MIL-STD-1540 definitions for each assembly	MIL-STD-1540 definitions for each assembly
Testing Tolerances	MIL-STD-1540	MIL-STD-1540	MIL-STD-1540	MIL-STD-1540
Development tests	As required	As required	As required	Not required
Component acceptance	MIL-STD-1540 (component acceptance)	Not required on 1st item; protoflight test only	MIL-STD-1540 (component acceptance)	Not required
Component qualification	MIL-STD-1540 (qual.) to design levels	MIL-STD-1540 (protoflight) to design levels	Not required (acceptance test only)	Not required
Qual. thermal margin	10 deg C	5 deg C	0 deg C	0 deg C
Experiment acceptance	MIL-STD-1540 (vehicle acceptance)	Not required on 1st item; protoflight test only	MIL-STD-1540 (vehicle acceptance)	MIL-STD-1540 (vehicle acceptance)
Experiment qualification	MIL-STD-1540 (vehicle qualification)	MIL-STD-1540 (protoflight) to design levels	Not required (acceptance test only)	Not required (acceptance test only)
Qual. margins (environ.)	10 deg C	10 deg C	0	0
Vehicle acceptance	MIL-STD-1540 (vehicle acceptance)	Not required on 1st item; protoflight test only	MIL-STD-1540 (vehicle acceptance)	MIL-STD-1540 (vehicle acceptance)
Vehicle qualification	MIL-STD-1540 (vehicle qualification)	MIL-STD-1540 (protoflight) 1st item to design levels	Not Required (acceptance test only)	Not Required (acceptance test only)
Qual. margins (environ.)	10 deg C	10 deg C	0	0

For Class A components or space vehicles, the first article manufactured of each type shall be acceptance tested and then qualification tested in accordance with MIL-STD-1540. In addition, for STS usage, it should be demonstrated that the component can operate in an explosive atmosphere: the component should not create an explosion in an explosive atmosphere; it should contain any explosion occurring inside the component, and the temperature of the component case and of all internal parts exposed to the atmosphere shall not exceed 178-deg C. Upon completion of the qualification-test program, the Class A qualification article is usually used as a development test article for extended margin evaluation tests and life tests. However, the qualification-article test history may be reviewed for excessive test time and potential fatigue-type failures to determine if the unit can be refurbished and used in the qualification-vehicle or experiment, or as a flight spare in a redundant flight set, but it should not otherwise be planned for flight. Subsequent Class A flight components after the first unit of each type shall be acceptance tested in accordance with MIL-STD-1540. Subsequent Class A space experiments or vehicles, after the first qualification unit, shall be acceptance tested in accordance with the space-vehicle-level acceptance test baseline of MIL-STD-1540.

For Class B equipment, qualification tests are slightly modified from the MIL-STD-1540B baseline for the first article manufactured of each type. These are called protoflight tests and they also serve as the acceptance test for that item. The required Class B component protoflight tests are the component-qualification tests specified in MIL-STD-1540B, with the two exceptions:

1) The environmental design margin is 5-deg C. Therefore, the thermal design range is 5-deg C beyond the minimum and maximum predicted temperatures (instead of 10-deg C for Class A components). Each component is to be designed to operate continuously within a temperature range of at least -29-deg C to +66-deg C.

2) The component burn-in acceptance shall be substituted for the component qualification life test.

For Class B space experiments and space vehicles protoflight tests, the following exceptions are made:

1) The environmental design margin is 5-deg C.

2) If the optional space vehicle thermal cycling test (6.2.9 in MIL-STD-1540) is adopted as baseline, the minimum space-vehicle temperature range shall be 60-deg C. The test should include 15 percent more thermal cycles than specified for the space vehicle thermal-cycling acceptance test (7.1.8.3 in MIL-STD-1540).

Following these Class B qualification tests, the qualification article may be used as a flight article and installed into the flight vehicle or experiment, or used as a flight spare, without further testing. Subsequent Class B flight components after the first unit of each type, if any, shall be acceptance tested in accordance with MIL-STD-1540.

Subsequent flight units of Class B space equipment after the first unit shall be acceptance tested in accordance with the space-vehicle-level acceptance test baseline of MIL-STD-1540.

For Class C and D equipment, no environmental design margins are required and no allowances for testing tolerances need to be made. The thermal design range is between the maximum and minimum predicted temperatures. Class C components and space vehicles only require acceptance testing in according with MIL-STD-1540B. Thermal testing of Class D components and space vehicles are optional.

MIL-HDBK-343 also specifies that to prevent generating a possible ignition source, the temperature of any part exposed to the atmosphere shall not exceed 178-deg C. Furthermore, for STS usage, it should be demonstrated that components can operate in an explosive atmosphere: the component should not create an explosion in an explosive atmosphere, and it should contain any explosion occurring inside the component.

REFERENCES

1. *Test Requirements for Space Vehicles,* MIL-STD-1540B (USAF) Military Standard, 10 October 1982.

2. *Application Guidelines for MIL-STD-1540B; Test Requirements for Space Vehicles,* MIL-STD-340 (USAF) Military Handbook, 01 July 1985.

3. *Design Construction and Testing Requirements from One of a Kind Space Equipment,* MIL-STD-343 (USAF) Military Handbook, 01 February 1986.

4. D. F. Gluck, *Space Vehicle Thermal Testing Environments, Related Design and Analysis, Requirements and Practice,* Aerospace Report TOR-0086(2902-08)-1, 27 September 1987.

5. O., Hamberg, *Space Division/Acquisition Logistics and Test Space Vehicle Test Training Course.*

6. D. F. Gluck, *Thermal Testing of Space Vehicle Electronic Components,* AIAA-86-1302, AIAA Thermophysics Conference, Boston, MA, 2-4 June 1986.

7. R. D. Stark, *Thermal Testing of Spacecraft,* Aerospace Report TOR-0172 (2441-01)-4, September 1971.

8. C. E. Mandel, *Environmental Stress Screening Guidelines for Assemblies,* Institute of Environmental Sciences, September 1984.

9. R. W. Burrows, Special Long-Life Assurance Studies, *Long Life Assurance Studies for Manned Spacecraft Long Life Hardware*, Martin-Marietta, Report MCR-72-169, Vol. IV, September 1972.

10. I. Quart and K. Wong, *Quantitative Screening Methods from Case Histories,* Environmental Stress Screening of Electronics Hardware Proceedings, Institute of Environmental Sciences, 1981.

11. J. W. Welch, *Thermal Testing of Small, Class C Space Vehicles: A Case Study,* AIAA-91-1302, AIAA Thermophysics Conference, Honolulu, HI, 24-26 June 1991.

Chapter X

Technology Projections

David G. Gilmore
The Aerospace Corporation

Thermal management technology for spacecraft 5 to 20 years in the future will experience dramatic changes as spacecraft evolution splits along two parallel paths. Large space systems will continue to grow as they have in the past and increase heat rejection requirements by an order of magnitude over that required for current systems. A second evolutionary path, however, is expected in the emergence of small satellites, some of which, referred to as "nano satellites," may eventually become small enough to fit in the palm of the hand.

While simple thermal-management technologies consisting of radiating areas on vehicle primary structure and low-power heaters are typical of today's spacecraft, future high-power spacecraft will require large deployable radiators, pumped fluid loops, rotating fluid joints, thermal storage, and high-performance heat exchangers to handle the large projected increase in waste heat levels. Thermal technologies that must be developed to meet these high-power requirements include high-capacity and flexible heat pipes, diode and variable-conductance heat pipes, capillary pumped loops, pumped two-phase heat acquisition and transport systems, large deployable radiators, and rotating radiative and fluid joints for steerable (sun avoiding) radiators. These will be enabling technologies for future high-power space systems such as space-based radars and space-station platforms.

The emergence of micro- and nano-engineering technology will allow many mechanical devices with dimensions on the order of microns to be fabricated in silicon or other materials using the same technology employed in the manufacture of micro circuits. Wafer-scale integration of microminiaturized mechanical components, such as gyroscopes and thrusters, and very-large-scale integrated circuits have the potential to reduce some satellites from the size and weight of an automobile to something that could easily fit in the palm of the hand and weigh less than a kilogram. This technology and its potential for reducing the size of satellite systems presents both advantages and challenges in the field of thermal control. Advantages include large reductions in waste heat from some satellite components and smaller satellite dimensions; this makes it easier to isothermalize the vehicle and thereby simplify the thermal design and analysis process. Challenges may develop, however, in removing waste heat from densely packaged high-power devices. Micromachining technology itself offers some potential solutions to thermal problems that may arise.

Moving heat from internal sources to the surface of microsat and isothermalizing the vehicle could be achieved through the use of a variety of technologies, depending on the amount of heat, the conduction distance, and the degree of heat concentration at the source. For small, low-power systems, simple conduction through aluminum structures or doubler plates would be sufficient. At higher power levels or somewhat larger sizes, advanced, highly conductive materials such as metal-matrix composites, carbon-carbon, or diamond may be useful. Alternatively, micro heat pipes or micro-pumped coolant loops may be employed. Heat pipes with diameters on the order of 1 mm and transport distances of several inches, such as the one shown in Figure 1, have been built and tested. Also, both pumps and heat exchangers constructed in silicon by means of micromachining technology have been demonstrated, and the integration of these technologies at a system level would be the next logical step.

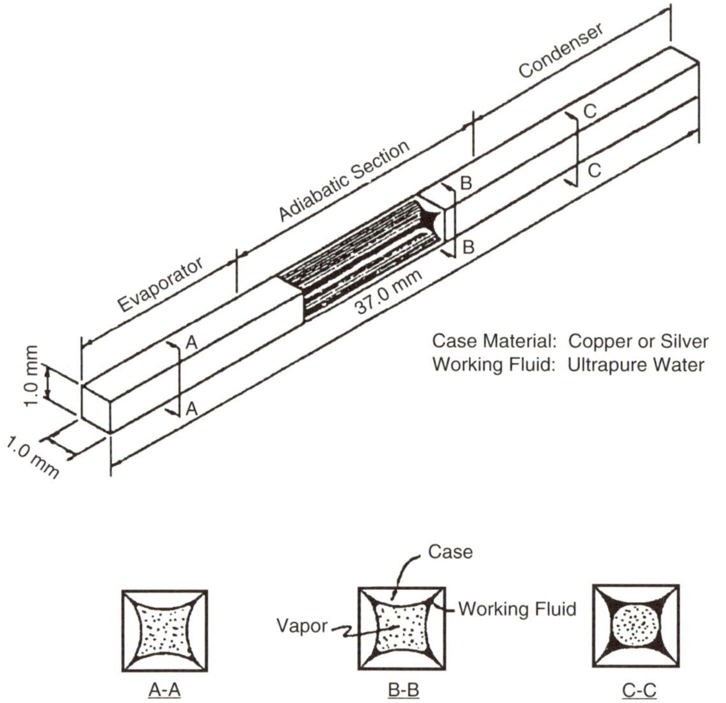

Figure 1. Trapezoidal micro heat pipe

Finding ways to spread heat out from very concentrated localized heat sources in micro satellites, such as high-power monolithic microwave integrated circuits (MMICs), high-throughput processors, and laser diodes is another potential challenge. In addition to existing multichip-module packaging technologies, more advanced systems able to handle very high power densities are achievable with the use of composite chips having highly conductive substrates (such as diamond), micropumped fluid-loop heat exchangers, and, in the long term, diamond semiconductors. The micro heat exchanger shown in Figure 2 has been built. Its coolant channels are each 25-μm wide and 200-μm deep, with a fin thickness of 15 μm. It is analytically predicted to be capable of handling local heat concentrations exceeding 1 k W/cm^2 with only a small temperature rise (10- to 20-deg C) between the mounting surface and the working fluid. A similar device has been built and demonstrated, cooling a laser-diode array at 3 kW/cm^2. Coupling such a micro heat exchanger with micromachined pumps would be a logical next step for this technology.

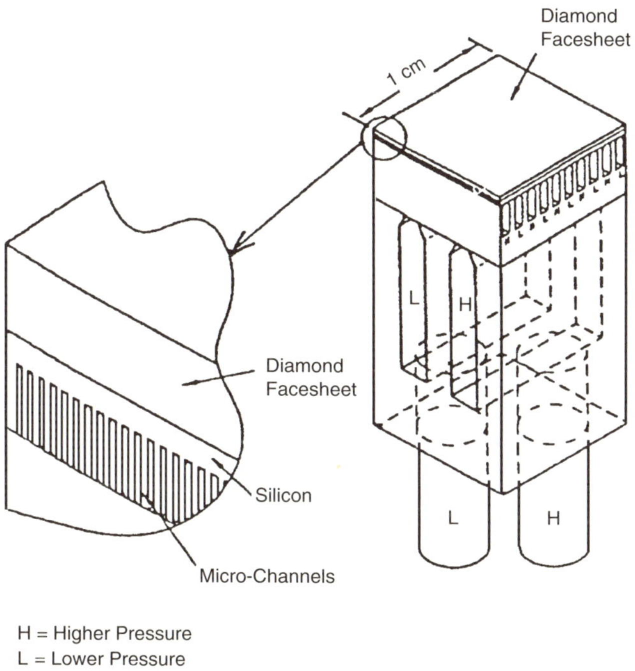

Figure 2. Microchannel heat exchanger with a diamond face sheet (TRW)

Although the development of mechanical cryorefrigerators and heat pumps based on micromachining technologies is theoretically possible, little has been demonstrated to date. A Joule-Thomson microrefrigerator with dimensions on the order of a few centimeters and capable of cooling an IR focal-plane array to 70-deg K is in commercial use today. However, such a refrigerator, which has no moving parts, is not desirable for long-term use on a nanosatellite because of its high consumption of coolant gas. Nevertheless, a closed-cycle version of the system, using micromechanical multistage compressors, could have potential use in long-term space systems. A number of small mechanical cryocoolers have been developed for missile guidance and night-vision applications, but these are still "large" by nanosatellite standards. Microminiature Stirling pulse-tube refrigerators based on silicon-etching technology are theoretically possible. Such coolers could be integrated directly into the focal plane of a microsatellite's IR sensor. This is, however, a far-term technology that would likely require more than ten years to develop. It should also be noted that room temperature mid- and long-wavelength IR sensors using microbolometers and requiring no cooling have been demonstrated. Such technologies could potentially eliminate the need for cryocoolers on satellite-based IR sensors.

Returning to move conventionally sized spacecraft, there are many subsystems that operate at or near cryogenic temperatures, such as IR sensors, low-noise amplifiers, and cryogenic fluid-storage tanks for orbit transfer-vehicle propulsion. Thermal control technologies for these systems include passive cryogenic radiators, insulation/low-conductance supports, and active cryo-refrigerators. For components cooled directly by a cryo-refrigerator, system life is directly dependent on refrigerator life and reliability. System weight may also be impacted by the need for a number of refrigerators to account for multiple failures. Cryogenic radiators provide higher reliability but may be very large and may require attitude restrictions on the spacecraft. Increasing service life of cryofluid systems can be accomplished by either increasing the quantity of cryogenic fluids stored on-board or reducing the amount of fluid vaporized and vented overboard due to heat "leaking" into the insulated storage tank. Increasing the effectiveness of the passive tank insulation and supports or cooling the storage-tank fluid with a radiator or an active refrigerator are two thermal-management approaches to reducing the amount of fluid. The optimum, minimum weight mix between the refrigerators or radiators and passive insulation depends on the relative performance of the technologies, mission duration, and tank size.

Passive insulation and radiators are attractive because of their inherent reliability and because they don't require additional electrical power. On the other hand, insulation can never completely eliminate heat leaks and fluid loss, and passive radiators may not be able to achieve very low temperatures. The cryocoolers are attractive because they can, theoretically, eliminate all fluid loss, but they have reliability/life issues and they add significantly to the power conversion and thermal-management-subsystem weights because of their extremely low cycle efficiency. Improvements in cryo-insulation, large cryo-radiators, low conductance or retractable tank supports, and cryo-refrigerators are enabling technologies for many spacecraft and launch-vehicle systems from military surveillance to manned missions to Mars.

In addition to the above, there are areas in which emerging technologies may themselves drive changes in the design of spacecraft systems. Such technologies that will have a significant impact on thermal management include development of high temperature superconductors and high-efficiency electronics, high-density electronics packaging, and constant-temperature spacecraft for reliability enhancement. Trade studies are required to evaluate the potential thermal-management-system weight (i.e., cost) savings to high-power spacecraft through the development of higher efficiency (lower waste heat) electronics and the use of high-temperature superconductors. Research to develop technologies for internal cooling of high-density electronic boxes is required to enable development of spacecraft with large on-board processing requirements. Statistical studies of spacecraft failures show a high degree of correlation between electronics failures and periods of temperature change. A cost/benefit analysis is needed to assess the savings due to increased spacecraft life or reduced ground testing requirements against the system cost of implementing constant-temperature thermal control techniques at both the spacecraft and electronics-box level. Depending on the results of such a study, additional investment in technologies such as variable-conductance heat pipes, variable-emittance / absorbance / conductance materials, and self-regulating heaters may be indicated. Materials that have very high thermal conductivities at cryogenic temperatures can significantly improve the design and lower the weight of cryogenic systems. A research program to identify and apply such materials is needed.

Appendix A

Thermal Finish α and ε Data

Note: Absorptance values are undegraded, beginning of life unless otherwise noted. See "Thermal Surface Finishes" section of Chapter IV for discussion of degradation effects.

BLACK COATINGS

	α† SOLAR	ε IR
ANODIZE BLACK	.88	.88
BLACK Z306 POLYURETHANE PAINT*, 3 MIL THICK BOL	.95	.87
BLACK Z306 POLYURETHANE PAINT*, 3 MIL THICK EOL 3 YEARS GEO	.93	.87
BLACK Z306 POLYURETHANE PAINT*, 3 MIL THICK EOL 5 YEARS GEO	.92	.87
CARBON BLACK PAINT NS-7	.96	.88
CATALAC BLACK PAINT	.96	.88
CHEMGLAZE BLACK PAINT Z306 BEGINNING OF LIFE	.96	.91
CHEMGLAZE BLACK PAINT* Z306 END OF LIFE	.96	.84
DELRIN BLACK PLASTIC	.96	.87
EBANOL C BLACK	.97	.73
EBANOL C BLACK-384 ESH*UV	.97	.75
GSFC BLACK PAINT 313-1	.96	.86
GSFC BLACK SILIFICATE MS-94	.96	.89
HUGHSON BLACK PAINT H322	.96	.86
HUGHSON BLACK PAINT L-300	.95	.84
MARTIN BLACK PAINT N-150-1	.94	.94
MARTIN BLACK VELVET PAINT	.91	.94
PALADIN BLACK LACQUER	.95	.75
PARSONS BLACK PAINT	.98	.91
POLYETHYLENE BLACK PLASTIC	.93	.92
PYRAMIL BLACK ON BERYLLIUM COPPER	.92	.72
ROUGH BLACK MATTE*, BLACK PAINT	.90	.90
TEDLAR BLACK PLASTIC	.94	.90
3M BLACK VELVET PAINT BEGINNING OF LIFE	.97	.91
3M BLACK VELVET PAINT* 2.5 YEARS	.97	.84
3M BLACK VELVET PAINT* END OF LIFE	.97	.84
VELESTAT BLACK PLASTIC	.96	.85

FILMS AND TAPES

	α SOLAR	ε IR
ACLAR FILM (ALUMINUM BACKING) 1 MIL	.12	.45
ACLAR FILM (ALUMINUM BACKING) 2 MIL	.11	.62
ACLAR FILM (ALUMINUM BACKING) 5 MIL	.11	.73
ASTROQUARTZ FABRIC*	.22	.80
BETA CLOTH*	.40	.86
GRAFOIL BOL	.65	.34
GRAFOIL EOL	.61	.34
INDIUM OXIDE COATED OPTICAL SOLAR REFLECTOR BOL	.07	.76
INDIUM OXIDE COATED OPTICAL SOLAR REFLECTOR EOL	.11	.76
ALUMINIZED KAPTON, FIRST SURFACE* BOL	.12	.03
KAPTON FILM (ALUMINUM BACKING) .08 MIL	.23	.24
KAPTON FILM (ALUMINUM BACKING) .15 MIL	.25	.34
KAPTON FILM (ALUMINUM BACKING) .25 MIL	.31	.45
KAPTON FILM (ALUMINUM BACKING) .50 MIL	.34	.55
ALUMINIZED KAPTON, SECOND SURFACE* (.5 MIL) REINFORCED WITH DACRON CLOTH	.35	.53
CHROMIZED KAPTON FILM*, .5 MIL THICK BOL	.70	.70
CHROMIZED KAPTON FILM*, .5 MIL THICK EOL 5 YEARS GEO	.70	.70

† Beginning of life, unless otherwise noted

FILMS AND TAPES (CONTINUED)

	α[†] SOLAR	ε IR
KAPTON FILM (ALUMINUM-SILICON OXIDE OVERCOATING) .5 MIL INITIAL	.12	.18
KAPTON FILM (ALUMINUM-SILICON OXIDE OVERCOATING) .5 MIL 4000 ESH UV	.28	.24
BLACK KAPTON FILM*, 1 MIL THICK BOL	.92	.88
BLACK KAPTON FILM*, 1 MIL THICK EOL 5 YEARS GEO	.92	.88
ALUMINIZED KAPTON*, 1 MIL THICK BOL	.36	.61
ALUMINIZED KAPTON*, 1 MIL THICK EOL 3 YEARS GEO	.54	.61
ALUMINIZED KAPTON*, 1 MIL THICK EOL 5 YEARS GEO	.66	.61
FRONT SURFACE ALUMINIZED KAPTON SHEET* (1. MIL)	.12	.04
SECOND SURFACE ALUMINIZED KAPTON SHEET* (1 MIL)	.37	.63
KAPTON FILM (GOLD BACKING)* 1.0 MIL	.19	.5
KAPTON FILM (CHROMIUM-SILICON OXIDE-ALUMINUM BACKING(GREEN)) 1. MIL	.79	.78
KAPTON FILM (ALUMINUM-ALUMINUM OXIDE OVERCOATING) 1. MIL INITIAL	.12	.20
KAPTON FILM (ALUMINUM-ALUMINUM OXIDE OVERCOATING) 1. MIL 1800 ESH UV	.12	.20
KAPTON FILM (ALUMINUM SILICON OXIDE OVERCOATING) 1. MIL INITIAL	.11	.33
KAPTON FILM (ALUMINUM-SILICON OXIDE OVERCOATING) 1 MIL 2400 ESH UV	.22	.33
KAPTON FILM (SILVER-ALUMINUM OXIDE OVERCOATING) 1. MIL INITIAL	.08	.19
KAPTON FILM (SILVER-ALUMINUM OXIDE OVERCOATING) 1. MIL 2400 ESH UV	.08	.21
KAPTON FILM (ALUMINUM BACKING) 1.5 MIL	.40	.71
ALUMINIZED KAPTON*, 2 MIL THICK BOL	.39	.73
ALUMINIZED KAPTON*, 2 MIL THICK EOL 3 YEARS	.55	.73
ALUMINIZED KAPTON*, 2 MIL THICK EOL 5 YEARS	.67	.73
ALUMINIZED KAPTON*, 2 MIL THICK WITH INDIUM OXIDE COATING BOL	.34	.75
ALUMINIZED KAPTON*, 2 MIL THICK WITH INDIUM OXIDE COATING EOL (3 YRS)	.47	.75
KAPTON FILM (ALUMINUM BACKING) 3.0 MIL	.45	.82
ALUMINIZED KAPTON*, 5 MIL THICK BOL	.49	.83
ALUMINIZED KAPTON*, 5 MIL THICK 2.5 YEARS	.61	.83
ALUMINIZED KAPTON*, 5 MIL THICK EOL	.70	.83
ALUMINIZED KAPTON SHEET (5 MIL) SECOND SURFACE	.34	.55
KIMFOIL-POLYCARBONATE FILM (ALUMINUM BACKING) .08 MIL	.19	.23
KIMFOIL-POLYCARBONATE FILM (ALUMINUM BACKING) .2 MIL	.20	.30
KIMFOIL-POLYCARBONATE FILM (ALUMINUM BACKING) .24 MIL	.17	.28
MYLAR FILM (ALUMINUM BACKING) .15 MIL	.14	.28
MYLAR FILM (ALUMINUM BACKING) .25 MIL	.15	.34
MYLAR FILM (ALUMINUM BACKING) 3.0 MIL	.17	.76
MYLAR FILM (ALUMINUM BACKING) 5.0 MIL	.19	.77
SILICA CLOTH*	.18	.86
SKYLAB SAIL INITIAL	.15	.35
SKYLAB SAIL 1900 ESH UV	.19	.36
SKYLAB PARASOL FABRIC (ORANGE) INITIAL	.51	.86
SKYLAB PARASOL FABRIC (ORANGE) 2400 ESH UV	.65	.86
TEDLAR (GOLD BACKING) .5 MIL	.30	.49

[†] Beginning of life, unless otherwise noted

FILMS AND TAPES (CONTINUED)

	α† SOLAR	ε IR
TEDLAR (GOLD BACKING) 1. MIL	.26	.58
TEFZEL (GOLD BACKING) .05 MIL	.29	.47
TEFLON (GOLD BACKING) .5 MIL	.24	.43
TEFLON (GOLD BACKING) 1.0 MIL	.22	.52
TEFLON (ALUMINUM BACKING) 2. MIL	.08	.66
ALUMINIZED TEFLON TAPE, SECOND SURFACE* 2 MIL	.17	.76
ALUMINIZED TEFLON SHEET, SECOND SURFACE*, 2 MIL	.16	.65
TEFLON (SILVER BACKING) 2 MIL BEGINNING OF LIFE	.08	.68
TEFLON (SILVER BACKING)* 2 MIL 2.5 YEARS	.32	.66
TEFLON (SILVER BACKING)* 2 MIL END OF LIFE	.54	.66
TEFLON (ALUMINUM BACKING) 5 MIL BOL	.13	.81
TEFLON (ALUMINUM BACKING)* 5 MIL EOL 3 YEARS GEO	.25	.80
TEFLON (ALUMINUM BACKING)* 5 MIL EOL 5 YEARS GEO	.32	.80
ALUMINIZED TEFLON SHEET, SECOND SURFACE*, 5 MIL	.17	.77
TEFLON (GOLD BACKING) 5 MIL	.22	.81
TEFLON (SILVER BACKING) 5 MIL BEGINNING OF LIFE	.08	.81
TEFLON (SILVER BACKING)* 5 MIL 2.5 YEARS	.32	.76
TEFLON (SILVER BACKING)* 5 MIL END OF LIFE	.54	.76
SILVERIZED FEP TEFLON FLEXIBLE, OPTICAL, SOLAR REFLECTOR* 5 MIL THICK BOL	.07	.80
SILVERIZED FEP TEFLON FLEXIBLE, OPTICAL, SOLAR REFLECTOR* 5 MIL THICK EOL 3 YEARS GEO	.20	.80
SILVERIZED FEP TEFLON FLEXIBLE, OPTICAL, SOLAR REFLECTOR* 5 MIL THICK EOL 5 YEARS GEO	.27	.80
TEFLON (ALUMINUM BACKING) 10 MIL	.13	.87
ALUMINIZED TEFLON SHEET, SECOND SURFACE*, 10 MIL	.17	.83
TEFLON (GOLD BACKING) 10 MIL	.23	.82
TEFLON (SILVER BACKING) 10 MIL	.09	.88
SILVER TEFLON (.01)	.09	.88
TEFZEL (GOLD BACKING) 1.0 MIL	.26	.61
TAPES 235-3M BLACK	.95	.90
ALUMINUM TAPE	.04	.10
TAPES 425-3M ALUMINUM FOIL	.20	.03
ALUMINUM FOIL TAPE, 2 MIL, 2 MIL ADHESIVE* BOL	.15	.04
ALUMINUM FOIL TAPE, 2 MIL, 2 MIL ADHESIVE* EOL 3 YEARS GEO	.16	.04
ALUMINUM FOIL TAPE, 2 MIL, 2 MIL ADHESIVE* EOL 5 YEARS GEO	.17	.04
TAPES 850-3M MYLAR-ALUMINUM BACKING	.15	.59
TAPES 7361 MYSTIC ALUMINIZED KAPTON	.09	.03
TAPES 7452 MYSTIC ALUMINUM FOIL	.14	.03
TAPES 7800 MYSTIC ALUMINUM FOIL	.21	.03
Y9360-3M ALUMINIZED MYLAR	.19	.03

WHITE COATINGS

	α SOLAR	ε IR
ANDREW BROWN CO. SKYSPAR	.22	.91
BARIUM SULPHATE WITH POLYVINYL ALCOHOL	.06	.88
BIPHENYL - WHITE SOLID	.23	.86
CATALAC WHITE PAINT	.24	.90
CHEMGLAZE A276 WHITE PAINT*	.23	.88

† Beginning of life, unless otherwise noted

WHITE COATINGS (CONTINUED)

	α† SOLAR	ε IR
CHEMGLAZE Z202 WHITE PAINT*	.58	.86
DUPONT LUCITE ACRYLIC LACQUER	.35	.90
DOW CORNING WHITE PAINT DC-007	.19	.88
FLAMEMASTER CORP. STM K797 WHITE PAINT BEGINNING OF LIFE	.22	.85
FLAMEMASTER CORP. STM K797 WHITE PAINT EOL FOR 4 YR SYNCH. ORBIT	.60	.85
GSFC WHITE PAINT NS43-C	.20	.92
GSFC WHITE PAINT NS44-B	.34	.91
GSFC WHITE PAINT NS74-B	.17	.92
GSFC WHITE PAINT NS-37	.36	.91
HUGHSON WHITE PAINT A-276	.26	.88
HUGHSON WHITE PAINT A-276+1036 ESH UV	.44	.88
HUGHSON WHITE PAINT V-200	.26	.89
HUGHSON WHITE PAINT Z-202	.25	.87
HUGHSON WHITE PAINT Z-202+1000 ESH UV	.40	.87
HUGHSON WHITE PAINT Z-255	.25	.89
MAUTZ WHITE HOUSE PAINT	.30	.90
MAGNESIUM OXIDE WHITE PAINT	.09	.90
MAGNESIUM OXIDE ALUMINUM OXIDE PAINT	.09	.92
OPAL GLASS	.28	.87
OSO-H WHITE PAINT 63W	.27	.83
P764-1A WHITE PAINT	.23	.92
POTASSIUM FLUOROTITANATE WHITE PAINT	.15	.88
SHERWIN WILLIAMS WHITE PAINT (A8W11)	.28	.87
SHERWIN WILLIAMS WHITE PAINT (F8W2030)	.36	.87
SHERWIN WILLIAMS F8W2030 WITH POLASOL V6V241	.36	.87
SHERWIN WILLIAMS KEMACRYL	.24	.86
SPEREX WHITE PAINT	.34	.85
TEDLAR WHITE PLASTIC	.39	.87
DOW CORNING THERMATROL* (DC-92-007) BEGINNING OF LIFE	.19	.82
DOW CORNING THERMATROL* (DC-92-007) EOL FOR 4 YR SYNCH. ORBIT	.57	.82
3M-401 WHITE PAINT	.25	.91
TITANIUM OXIDE WHITE PAINT WITH METHYL SILICONE	.20	.90
TITANIUM OXIDE WHITE PAINT WITH POTASSIUM SILICATE	.17	.92
VITA-VAR PV-100 WHITE PAINT	.22	.82
WHITE PAINT 293*	.19	.89
WHITE PAINT 513 GLO*	.19	.89
WHITE S13* BEGINNING OF LIFE	.21	.88
WHITE S13* EOL FOR 4 YR SYNCH. ORBIT	.56	.88
WHITE POLYURETHANE	.27	.84
WHITE S-13G-LO SILICONE PAINT*, 10 MIL THICK BOL	.22	.88
WHITE S-13G-LO SILICONE PAINT*, 10 MIL THICK 3 YEARS GEO	.39	.88
WHITE S-13G-LO SILICONE PAINT*, 10 MIL THICK EOL 5 YEARS GEO	.47	.88
WHITE VELVET 3M SERIES 400	.30	.87
WHITE ZOT (IITRI YB-71)* BEGINNING OF LIFE	.20	.91
WHITE ZOT (IITRI YB-71)* 2.5 YEARS	.45	.91
WHITE ZOT (IITRI YB-71)* END OF LIFE	.70	.91
ZERLAUTS S-13G WHITE PAINT BEGINNING OF LIFE	.20	.90
ZERLAUTS S-13G WHITE PAINT* 2.5 YEARS	.52	.85
ZERLAUTS S-13G WHITE PAINT* END OF LIFE	.70	.85
ZERLAUTS Z-93 WHITE PAINT	.17	.92
ZINC ORTHOTITANATE WITH POTASSIUM SILICATE	.13	.92

† Beginning of life, unless otherwise noted

WHITE COATINGS (CONTINUED)

	$\alpha^{\dagger}$ SOLAR	ε IR
ZINC OXIDE WITH SODIUM SILICATE	.15	.92
ZIRCONIUM OXIDE WITH 650 GLASS RESIN	.23.	.88

OTHER PAINTS

	α SOLAR	ε IR
BRILLIANT ALUMINUM PAINT	.30	.31
CHROMACOAT ALUMINUM BOL	.28	.05
CHROMACOAT ALUMINUM EOL (3 YRS)	.33	.05
CHROMERIC SILVER PAINT 586	.30	.30
DUPONT SILVER PAINT 4817	.43	.49
EPOXY ALUMINUM PAINT	.77	.81
FINCH ALUMINUM PAINT 643-1-1	.22	.23
GSFC YELLOW NS-43-G	.38	.90
GSFC GREEN NS-53-B	.52	.87
GSFC GREEN NS-43-E	.57	.89
GSFC WHITE NS-43-C	.20	.92
GSFC GREEN-NS-55-F	.57	.91
GSFC GREEN NS-79	.57	.91
LEAFING ALUMINUM IN EPON 828	.37	.36
LEAFING ALUMINUM 80-U	.29	.32
NRL LEAFING ALUMINUM PAINT	.24	.24
NRL LEAFING ALUMINUM PAINT	.28	.29
SILICONE ALUMINUM PAINT	.29	.30

ANODIZED ALUMINUM SAMPLES (COATING THICKNESS IS CRITICAL)

	α SOLAR	ε IR
BLACK		
BLACKANODIZED1	.67	.87
BLACKANODIZED2	.53	.82
BLUE		
BLUEANODIZED1	.67	.87
BLUEANODIZED2	.53	.82
BROWN		
BROWNANODIZED	.73	.86
CHROMIC		
CHROMANODIZED	.44	.56
CLEAR		
CLEARANODIZED1	.27	.76
CLEARANODIZED2	.35	.84
GOLD		

† Beginning of life, unless otherwise noted

ANODIZED ALUMINUM SAMPLES (COATING THICKNESS IS CRITICAL (CONTINUED)

	α† SOLAR	ε IR
GOLDANODIZED	.48	.82
GREEN		
GREENANODIZED	.66	.88
PLAIN		
PLAINANODIZED	.26	.04
RED		
REDANODIZED	.57	.88
SULPHURIC	.42	.87
YELLOW	.47	.87
BLUE ANODIZED TITANIUM FOIL*	.70	.13
ANODIZED TITANIUM FOIL C.P.*, 1 MIL THICK BOL	.70	.10
ANODIZED TITANIUM FOIL C.P.*, 1 MIL THICK EOL 5 YEARS GEO	.70	.10

METAL CONVERSION COATINGS

	α SOLAR	ε IR
BUFFED ALUMINUM	.16	.03
CLAD 7075 ALUMINUM* BOL	.25	.04
CLAD 7075 ALUMINUM* EOL 3 YEARS GEO	.26	.04
CLAD 7075 ALUMINUM* EOL 5 YEARS GEO	.27	.04
HEAVILY OXIDIZED ALUMINUM*	.13	.30
IRRIDITE ALUMINUM	.11	0.0
POLISHED ALUMINUM* BEGINNING OF LIFE	.15	.05
POLISHED ALUMINUM* END OF LIFE	.15	.05
ALZAC A-2	.16	.73
ALZAC A-5	.18	1.0
BLACK CHROME	.96	.62
BLACK COPPER	.98	.63
BUFFED COPPER	3	.03
BLACK IRRIDITE	.62	.17
BLACK NICKEL	.91	.66
CONSTANTAN-METAL STRIP	.37	.09
COPPER FOIL TAPE PLAIN	.32	.02
COPPER FOIL TAPE SANDED	.26	.04
COPPER FOIL TAPE TARNISHED	.55	.04
DOW 7 ON POLISHED MANGESIUM	1.0	.49
DOW 7 ON SANDED MAGNESIUM	1.0	.65
DOW 9 ON MAGNESIUM	1.0	.87
DOW 23 ON MAGNESIUM	.62	.67
EBANOL C BLACK	.97	.77
ELECTROPLATED GOLD	.23	.03
ELECTROLESS NICKEL	.39	.07
POLISHED GOLD* BEGINNING OF LIFE	.30	.05
POLISHED GOLD* END OF LIFE	.30	.05
SANDBLASTED GOLD*	.48	.14

† Beginning of life, unless otherwise noted

METAL CONVERSION COATINGS (CONTINUED)

	α[†] SOLAR	ε IR
INCONEL X FOIL (1 MIL)	.52	.10
KANNIGEN-NICKEL ALLOY	.45	.08
PLAIN BERYLLIUM COPPER	.31	.03
PLATINUM FOIL	.33	.04
STAINLESS STEEL POLISHED	.42	.11
STAINLESS STEEL SANDBLASTED	.58	.38
STAINLESS STEEL MACHINED	.47	.14
STAINLESS STEEL MACHINE ROLLED	.39	.11
STAINLESS STEEL BOOM-POLISHED	.44	.10
STAINLESS STEEL 1 MIL 304 FOIL	.4	.05
TANTALUM FOIL	.4	.05
TUNGSTEN POLISHED	.44	.03

VAPOR DEPOSITED COATINGS

	α SOLAR	ε IR
ALUMINUM	.08	.02
ALUMINUM ON FIBERGLASS	.15	.07
ALUMINUM ON STAINLESS STEEL	.08	.02
CHROMIUM	.56	.17
CHROMIUM ON 5-MIL KAPTON	.57	.24
GERMANIUM	.52	.09
GOLD	.19	.02
IRON OXIDE	.85	.56
MOLYBDENUM	.56	.21
NICKEL	.38	.04
RHODIUM	.18	.03
SILVER POLISHED	.04	.02
SILVER OXIDIZED*	.05	.03
DENTON SILVER*	.06	.03
TYPE A TEFLON X VACUUM DEPOSITED ALUMINUM X Y966 ACRYLIC PRESSURE SENSITIVE ADHESIVE*, .5 MIL	.14	.4
TYPE A TEFLON X VACUUM DEPOSITED ALUMINUM X Y966 ACRYLIC PRESSURE SENSITIVE ADHESIVE*, 1 MIL	.14	.48
TYPE A TEFLON X VACUUM DEPOSITED ALUMINUM X Y966 ACRYLIC PRESSURE C SENSITIVE ADHESIVE*, 2 MIL	.14	.6
TYPE A TEFLON X VACUUM DEPOSITED ALUMINUM X Y966 ACRYLIC PRESSURE C SENSITIVE ADHESIVE*, 5 MIL	.14	.75
TYPE A TEFLON X VACUUM DEPOSITED ALUMINUM*, .5 MIL	.14	.4
TYPE A TEFLON X VACUUM DEPOSITED ALUMINUM*, 1. MIL	.14	.48
TYPE A TEFLON X VACUUM DEPOSITED ALUMINUM*, 2. MIL	.14	.6
TYPE A TEFLON X VACUUM DEPOSITED ALUMINUM*, 5. MIL	.14	.75
TYPE A TEFLON X VACUUM DEPOSITED ALUMINUM*, 7.5 MIL	.15	.8

[†] Beginning of life, unless otherwise noted

VAPOR DEPOSITED COATINGS (CONTINUED)

	α† SOLAR	ε IR
TYPE A TEFLON X VACUUM DEPOSITED ALUMINUM*, 10 MIL	.15	.85
TITANIUM	.52	.12
BARE TITANIUM*	.40	.55
TUNGSTEN	.60	.27

COMPOSITE COATINGS

	α SOLAR	ε IR
ALUMINUM OXIDE (AL203)-(12 LAMBDA/4) ON BUFFED ALUMINUM INITIAL	.13	.23
ALUMINUM OXIDE (AL203)-(12 LAMBDA/4) ON BUFFED ALUMINUM 2560 ESH UV	.13	.23
ALUMINUM OXIDE (AL203)-(12 LAMBDA/4) ON FUSED SILICA	.12	.24
FIBERGLASS POLYIMIDE BOL	.75	.89
FIBERGLASS POLYIMIDE EOL	.80	.89
GLASS POLYIMIDE* BOL	.75	.89
GLASS POLYIMIDE* 2.5 YEARS	.78	.89
GLASS POLYIMIDE* EOL	.80	.89
GSFC DARK MIRROR COATING-SIO-CR-AL	.86	.04
GSFC COMPOSITE SIOX-AL2-AG	.07	.68
HELIOS SECOND SURFACE MIRROR/SILVER BACKING INITIAL	.07	.79
HELIOS SECOND SURFACE MIRROR/SILVER BACKING 24 HOURS AT 5 SUNS	.07	.08
HELIOS SECOND SURFACE MIROR/SILVER BACKING 48 HOURS AT 11 SUNS	.08	.79
INCONEL WITH TEFLON OVERCOATING- 1 MIL	.55	.46
SILVER BERYLLIUM COPPER	.19	.03
SILVER BERYLLIUM COPPER WITH KAPTON OVERCOATING	.31	.57
SILVER BERYLLIUM COPPER WITH PARYLENE C OVERCOATING	.22	.34
SILVER BERYLLIUM COPPER WITH TEFLON OVERCOATING	.12	.38
SILVERIZED FUSED SILICA OPTICAL, SOLAR REFLECTOR*, 8 MIL THICK BOL	.06	.8
SILVERIZED FUSED SILICA OPTICAL, SOLAR REFLECTOR* 8 MIL THICK EOL 3 YEARS GEO	.17	.8
SILVERIZED FUSED SILICA OPTICAL, SOLAR REFLECTOR*, C 8 MIL THICK EOL 5 YEARS GEO	.2	.8
VESPEL POLYIMIDE SP1	.89	.9

MISCELLANEOUS

	α SOLAR	ε IR
FIBERGLASS EPOXY* (BOL AND EOL ARE THE SAME)	.72	.89
GRAPHITE EPOXY* (BOL AND EOL ARE THE SAME)	.93	.85

† Beginning of life, unless otherwise noted

Appendix B

Thermal Vacuum Test Facilities

THERMAL VACUUM TESTING

FACILITY CAPABILITY

COMPANY, ADDRESS	CHAMBER CHARACTERISTICS				SOLAR BEAM			CHAMBER ORIENTATION		
	DIA (FT)	VERT HT (FT)	HORIZ LNG (FT)	WALL TEMP RANGE- (˚F)	SIZE (FT)	FROM	COLL (DEG)	LOAD FROM	OPEN DIA (FT)	COMMENTS
ARNOLD (AEDC) ARNOLD AFS. TENN.	12	25		-320	8	TOP	1.8	TOP	10	
BOEING KENT, WASH.	30	46		-320	10	TOP	1.5	TOP	30	
BOEING KENT, WASH.	30	46		-320	3.5	TOP	1.45	BOT	6	XENON
LORAL PALO ALTO, CA	10		22	-320	8	TOP		BOT	10	IN 10' CHAMB
GENERAL DYNAMICS, CONV. SAN DIEGO, CA	12		22	-320	5	SIDE	7	SIDE	12	CARB. ARC
MARTIN MARIETTA VALLEY FORGE, PA					15	TOP	3-5	TOP	21	XENON
MARTIN MARIETTA VALLEY FORGE, PA	10		12	-320 TO 170	3	SIDE	1.5	BOT	5	
GRUMMAN AEROSPACE BETHPAGE, NY	8		34	-320 TO AMB	4	END	.25		7	CARB ARC
HUGHES S&CG	13.5	35.5		-320 TO 250	8	TOP		BOT	13.5	XENON
IABG MUNICH, WEST GERMANY	11.5		21.3	-320	8	S	2	S	11.5	
JPL PASADENA, CA	27	85		-320 TO 25	15	TOP	1	SIDE	15X25	XENON
JPL PASADENA, CA	11.3	45		-320	8	TOP	1.5	BOT	10	XENON

THERMAL VACUUM TESTING

FACILITY CAPABILITY (CONTINUED)

COMPANY, ADDRESS	CHAMBER CHARACTERISTICS				SOLAR BEAM			CHAMBER ORIENTATION		
	DIA (FT)	VERT HT (FT)	HORIZ LNG (FT)	WALL TEMP RANGE- (˚F)	SIZE (FT)	FROM	COLL (DEG)	LOAD FROM	OPEN DIA (FT)	COMMENTS
LORAL DALLAS, TX	12		11	-320	6	SIDE	13	SIDE	10	HG-XENON
MARTIN MARIETTA DENVER, CO	24	45		-320 TO 125	16	TOP	2	TOP	24	
MARTIN MARIETTA DENVER, CO	6		15		3	SIDE	1.7	SIDE	6	
MARTIN MARIETTA DENVER, CO	6		15		2.5	SIDE	2.5	SIDE	6	
MCDONNELL DOUGLAS ST. LOUIS, MO	8		16		2.5 SQ	SIDE	2	SIDE	8	
MCDONNELL DOUGLAS HUNTINGTON BEACH, CA	10		13	-320	4	TOP	1.25		10	MOVEABLE
MCDONNELL DOUGLAS HUNTINGTON BEACH, CA	10		13	-320	4	TOP	1.25		5	MOVEABLE
NASA GODDARD GREENBELT, MD	12	15		-320 TO 300	10			TOP	12	
NASA JOHNSON HOUSTON, TX	65	120		-320 TO 100	13	TOP	1.5	SIDE	40	
NASA JOHNSON HOUSTON, TX	65	120		-320 TO 100	13X33	SIDE	1.5	SIDE	50	
NASA JOHNSON HOUSTON, TX	6	14		-320 TO 600	3.6	TOP		BOT	6	
NASA JOHNSON HOUSTON, TX	4	10		-320 TO 440	3.6	SIDE				
NASA LEWIS CLEVELAND, OH	100	120		-320 TO 200	15X30	TOP	1	SIDE	50X50	

THERMAL VACUUM TESTING

FACILITY CAPABILITY (CONTINUED)

COMPANY, ADDRESS	CHAMBER CHARACTERISTICS				SOLAR BEAM			CHAMBER ORIENTATION		
	DIA (FT)	VERT HT (FT)	HORIZ LNG (FT)	WALL TEMP RANGE- (˚F)	SIZE (FT)	FROM	COLL (DEG)	LOAD FROM	OPEN DIA (FT)	COMMENTS
NASA LEWIS CLEVELAND OH	100	120		-320 TO 200	15X30	SIDE	1	SIDE	50X50	
NASA LEWIS CLEVELAND, OH	6	10			2.5		2			
NASDA TSUKUBA, JAPAN	27	80		-320	13	TOP	1	SIDE	16	
ROCKWELL SPACE DIVISION DOWNEY, CA	5		12	-320 TO 400	4	END		SIDE	5	1.5 SUNS
SOPEMEA, TOULOUSE, FRANCE	19.7	23		-300 TO 370	10	TOP	1.5	BOT	13	XENON
TRW REDONDO BEACH, CA	22	46		-320	10X10	TOP	2	BOT	22	XENON
TRW REDONDO BEACH, CA	30			-320	8X8	SIDE		TOP	18	XENON
TRW REDONDO BEACH, CA	30			-320	4X4	SIDE		TOP	18	XENON
TRW REDONDO BEACH, CA	30			-320	1.6 SQ	SIDE		SIDE		2-CARB ARC
NASA MARSHALL HUNTSVILLE, AL	6	9		-320	3	TOP			3	HG-XENON
ESTEC NETHERLANDS	33	32		-320						18 KW IR SIM
ESTEC NETHERLANDS					8.5	BOT	4	TOP	9	XENON

THERMAL VACUUM TESTING

FACILITY CAPABILITY (CONTINUED)

COMPANY, ADDRESS	CHAMBER CHARACTERISTICS				SOLAR BEAM			CHAMBER ORIENTATION		
	DIA (FT)	VERT HT (FT)	HORIZ LNG (FT)	WALL TEMP RANGE- (°F)	SIZE (FT)	FROM	COLL (DEG)	LOAD FROM	OPEN DIA (FT)	COMMENTS
ARNOLD (AEDC) ARNOLD AFS, TN	42	82		-320				TOP	20	QUARTZ IODID
BOEING KENT, WA	30	46		-320						
EDWARDS AFB EDWARDS AFB, CA	30			-320 TO 400				TOP	19	
LORAL PALO ALTO, CA	38.5			-320						IR 250 KW
GEN. MOTORS-AC ELECT. MILWAUKEE, WI	25		33	-320 TO 300						
MARTIN MARIETTA VALLEY FORGE, PA	32	54		-320						
MARTIN MARIETTA VALLEY FORGE, PA	30	30		-320						3 CHAMBERS
JPL PASADENA, CA	27	85		-325 TO 25						
LOCKHEED SUNNYVALE, CA	30		78	-320						76-15 KW ZNS
MCDONNELL DOUGLAS ST. LOUIS, MO	30		36	-320 TO 275						MAN RATED
MCDONNELL DOUGLAS HUNTINGTON BEACH, CA	30			-320						
NASA GODDARD GREENBELT, MD	28		40	-320 TO 200	200					
NASA JOHNSON	65	120		-320 TO 100						

THERMAL VACUUM TESTING

FACILITY CAPABILITY (CONTINUED)

COMPANY, ADDRESS	CHAMBER CHARACTERISTICS				SOLAR BEAM			CHAMBER ORIENTATION		
	DIA (FT)	VERT HT (FT)	HORIZ LNG (FT)	WALL TEMP RANGE- (°F)	SIZE (FT)	FROM	COLL (DEG)	LOAD FROM	OPEN DIA (FT)	COMMENTS
NASA JOHNSON HOUSTON, TX	25	26		-320	13	TOP	1.5	TOP	35	
NASA LEWIS CLEVELAND, OH	100	120		-320 TO 200						
NASA LEWIS CLEVELAND, OH	30		100	-320						
NASA LEWIS CLEVELAND, OH	25		70	-320						
NASDA TSUKUBA, JAPAN	27	80		-320						
HUGHES DANBURY, CT	50			-320						
TRW REDONDO BEACH, CA	30			-320						
BENDIX ANN ARBOR, MI	20		27	-320	6.2	SIDE		SIDE	20	CARB. ARC
DAVID FLORIDA LAB OTTAWA, ONT	22	35		-320 TO 300						
MARTIN MARIETTA DENVER, CO	24	45		-320 TO 125						
MARTIN MARIETTA PRINCETON, NJ	24	20		-320 TO 257						TOP LOADER
TRW REDONDO BEACH, CA	22	46		-320						
USAF WRIGHT-PAT DAYTON, OH	23	27		-320						

THERMAL VACUUM TESTING

FACILITY CAPABILITY (CONTINUED)

COMPANY, ADDRESS	CHAMBER CHARACTERISTICS				SOLAR BEAM			CHAMBER ORIENTATION		
	DIA (FT)	VERT HT (FT)	HORIZ LNG (FT)	WALL TEMP RANGE- (˚F)	SIZE (FT)	FROM	COLL (DEG)	LOAD FROM	OPEN DIA (FT)	COMMENTS
GARRETT AIRESEARCH TORRANCE, CA	15	23								TOP LOADER
GRUMMAN AEROSPACE BETHPAGE, NY	15	20		-320 TO 300				TOP		
HUGHES S&CG	16	32		-320 TO 250						
LOCKHEED SUNNYVALE, CA	18	28		-320						53ZNS, 220 KW
MCDONNELL DOUGLAS ST. LOUIS, MO	18		30	-320						
NASA MARSHALL HUNTSVILLE, AL	15	20								
ROCKWELL SPACE DIV. DOWNEY, CA	17	18		-320						
SOPEMEA TOULOUSE, FRANCE	19.7	23		-300 TO 370						
ARNOLD (AEDC) ARNOLD AFS, TN	10	20								
ARNOLD (AEDC) ARNOLD AFS, TN	12	25		-320						
AVCO WILMINGTON, MA	10		12							AF FACILITY
BALL AEROSPACE BOULDER, CO	10		14	-240 TO 212						
DAVID FLORIDA LAB OTTAWA, ONT	10	30		-320						

THERMAL VACUUM TESTING

FACILITY CAPABILITY (CONTINUED)

COMPANY, ADDRESS	CHAMBER CHARACTERISTICS				SOLAR BEAM			CHAMBER ORIENTATION		
	DIA (FT)	VERT HT (FT)	HORIZ LNG (FT)	WALL TEMP RANGE- (°F)	SIZE (FT)	FROM	COLL (DEG)	LOAD FROM	OPEN DIA (FT)	COMMENTS
FAIRCHILD (REPUBLIC) FARMINGDALE, NY	13		20	AMB TO 400						30 K SHAKER
FAIRCHILD CAMERA SYOSSET, NY	15	14	13	-320 TO 550						INCL VIB 14 K
LORAL NEWPORT BEACH, CA	8		27							
LORAL PALO ALTO	10		22	-320						
GARRETT AIRESEARCH MINT CANYON, CA	11		12							
GENERAL DYNAMICS, CONV. SAN DIEGO, CA	12		19							
GENERAL DYNAMICS SAN DIEGO, CA	12		22	-320						
MARTIN MARIETTA VALLEY FORGE, PA	12		26	-32O TO 250						
MARTIN MARIETTA VALLEY FORGE, PA	10		12	-320 TO 170						
HAMILTON STD. WINDSOR LOCKS, CT	10		10	-320 TO 200	3X7	SIDE				QUARTZ LAMPS
HUGHES S&CG	13.5	35.5		-32O TO 250						
HUGHES S&CG	10	8		-32O TO 275						

THERMAL VACUUM TESTING

FACILITY CAPABILITY (CONTINUED)

COMPANY, ADDRESS	CHAMBER CHARACTERISTICS				SOLAR BEAM			CHAMBER ORIENTATION		
	DIA (FT)	VERT HT (FT)	HORIZ LNG (FT)	WALL TEMP RANGE- (˚F)	SIZE (FT)	FROM	COLL (DEG)	LOAD FROM	OPEN DIA (FT)	COMMENTS
HUGHES S&CG	10	19		-320 TO 275						
IABG MUNICH, WEST GERMANY	11.5		21.3	-320						
IABG MUNICH, WEST GERMANY	11	22		-280 TO 194						
JPL PASADENA, CA	11.3	45		-320						
JPL PASADENA, CA	10		10	-320 TO 300						
LOCKHEED SANTA CRUZ, CA	13	16		-320						
LORAL DALLAS, TX	12		11	-320						
MCDONNELL DOUGLAS HUNTINGTON BEACH, CA	10		13	-320						
NASA GODDARD GREENBELT, MD	12	15		-32O TO 300						
NASA LEWIS CLEVELAND, OH	15		60	-320 TO 200						
NASA MARSHALL HUNTSVILLE, AL	12	35								
NASA MARSHALL HUNTSVILLE, AL	12		16							

THERMAL VACUUM TESTING

FACILITY CAPABILITY (CONTINUED)

COMPANY, ADDRESS	CHAMBER CHARACTERISTICS				SOLAR BEAM			CHAMBER ORIENTATION		
	DIA (FT)	VERT HT (FT)	HORIZ LNG (FT)	WALL TEMP RANGE- (°F)	SIZE (FT)	FROM	COLL (DEG)	LOAD FROM	OPEN DIA (FT)	COMMENTS
NORTHROP HAWTHORNE, CA	12		15	-320	1.2					MAN RATED
US NAVY AVIONICS CTR INDIANAPOLIS, IN	10	11		-320						2 CHAMBERS
US NAVY MISSILE CTR FT. MUGU. CA	10	20								

Appendix C

Material Properties Data

	ρ(lb/in^3)	k(Btu/hr-ft-F)	c_p(Btu/lb-F)
ALUMINUM ALLOYS			
208.0	.10	70.	
222.0	.10	77.	
242.0	.10	87.	
295.0	.10	83.	
B295.0	.10	93.	
308.0	.10	82.	
319.0	.10	66.	
355.0	.10	87.	
C355.0	.10	82.	
356.0	.10	92.	
A356.	.10	92.	
A380.	.10	58.	
A413.0	.10	70.	
443.0	.10	84.	
B443.0	.10	85.	
514.0	.10	80.	
518.0	.10	56.	
520.0	.10	51.	
D712.0	.10	80.	
1060-0	.10	136.	
1060-H18	.10	133.	
1100-0	.10	128.	.22
1100-H18	.10	128.	
1350-0	.10	135.	
2011-0	.10	83.	.23
2011-T3	.10	87.	
2011-T8	.10	99.	
2014-0	.10	109.	.22
2014-T4	.10	77.	
2014-T6	.10	90.	
2017-0	.10	111.	.22
2024-0	.10	109.	.22
2024-T3	.10	70.	
2024-T36	.10	70.	
2024-T4	.10	70.	
2024-T6	.10	70.	
2025-T6	.10	90.	
2036-0	.10	92.	
2219-0	.10	100.	
3003-0	.10	102.	.22
3003-H18	.10	102.	
3004-0	.10	94.	.22
3004-H38	.10	94.	
4032-0	.10	90.	

	ρ(lb/in^3)	k(Btu/hr-ft-F)	c_p(Btu/lb-F)
4032-T6	.10	80.	
5005-0	.10	119.	.23
5005-H38	.10	119.	
5050-0	.10	111.	.22
5050-H38	.10	111.	
5052-0	.10	80.	.22
5052-H38	.10	.80	
5056-0	.10	67.	.22
5056-H38	.10	67.	
5083-0	.10	68.	.23
5083-H38	.10	68.	
5083-H113	.10	68.	
5086-0	.10	73.	.23
5086-H34	.10	73.	
5154-0	.10	73.	.23
5154-H38	.10	73.	
5252-0	.10	80.	
5254-0	.10	73.	
5254-H38	.10	73.	
5356-0	.10	68.	
5356-H38	.10	68.	
5357-0	.10	97.	
5357-H38	.10	97.	
5454-0	.10	78.	
5454-H38	.10	78.	
5456-0	.10	68.	.23
5456.H38	.10	68.	
5457-0	.10	102.	
5652.0	.10	80.	
5652-H38	.10	80.	
6009-0	.10	96.	
6053-0	.10	99.	
6053-T4	.10	90.	
6053-T5	.10	99.	
6053-T6	.10	90.	
6061-0	.10	104.	
6061-T4	.10	90.	.23
6061-T6	.10	97.	.23
6062-0	.10	99.	
6062-T4	.10	90.	
6062-T6	.10	90.	
6063-0	.10	126.	
6063-T42	.10	111.	
6063-T5	.10	116.	
6063-T6	.10	116.	
6063-T42	.10	111.	
6063-T5	.10	116.	

	ρ(lb/in³)	k(Btu/hr-ft-F)	c_p(Btu/lb-F)
6063-T6	.10	116.	
6262-T9	.10	99.	
6463-0	.10	126.	
6463-T42	.10	109.	
6463-T5	.10	121.	
6463-T6	.10	116.	
7075-T6-T7	.10	70.	.23
7079-T6	.10	73.	
7178-T6	.10	73.	
BERYLLIA	.11	157.	.25
BERYLLIUM			
Beryllium	.067	87.	.45
Be-38 Al	.075	123.	
Be-96A	.066	82.	.43
BERYLLIUM OXIDE	.083	80.	.25
COPPER			
C10200	.32	226.	.092
C10400, C10500, C10700	.32	224.	.092
C11000	.32	226.	.092
C11300, C11400, C11600	.32	224.	.092
C12200	.32	196.	.092
C14500	.32	205.	.092
C14700	.32	216.	.092
C15000	.32	212.	.092
C15500	.32	200.	.092
C17200	.32	62.-75.	.10
C17400	.32	120.-150.	
C18200	.32	187.	.092
C19400	.32	150.	.092
C21000	.32	.135.	.09
C22000	.32	109.	
C22600	.32	100.	
C23000	.32	92.	
C24000	.31	81.	
C26000	.31	70.	.09
C26800, C27000	.31	67.	.09
C28000	.30	71.	.09
C31400	.32	104.	.09
C33000; C33200, C33500	.31	.67	.09
C34000	.31	67.	.09
C34200, C35300	.31	67.	.09

	ρ(lb/in^3)	k(Btu/hr-ft-F)	c_p(Btu/lb-F)
C36000	.31	67.	.09
C36500, C36600, C36700, C36800	.30	71.	.09
C37000	.30	69.	.09
C37700	.31	71.	.09
C38500	.31	71.	.09
C40500, C42200, C42500	.32	69.-95.	.09
C44300, C44400, C44500	.31	64.	.09
C46400, C46500, C46600, C46700	.30	67.	.09
C48500	.31	67.	.09
C50500	.32	120.	.09
C51000	.32	40.	.09
C52100	.32	36.	.09
C52400	.32	29.	.09
C54400	.32	50.	.09
C61400	.29	39.	.09
C63800	.30	23.	.09
C64700	.32	102.	.09
C65100	.32	33.	.09
C65500	.31	21.	.09
C66700	.31	56.	.09
C67500	.30	61.	.09
C68700	.30	58.	.09
C68800	.30	23.	
C70600, C71000, C71500	.32	17.-26.	.09
C74500, C75200, C75400, C75700, C77000	.31	17.-26.	.09
C80100, C80300, C80500, C80700, C80900, C81100	.32	200.-226.	.09
C81400, C81500	.32	150.182.	.09
C82000	.31	150.	.10
C82200	.32	106.	.10
C82400	.30	77.	.10
C82500	.30	75.	.10
C82600	.30	73.	.10
C82800	.30	71.	.10
C83600	.32	42.	.09
C84400, C85200	.31	42.-49.	.09
C86200, C86300	.28	21.	.09
C87400	.3	16.	.09
C90300, C90500, C91600	.32	42.	.09
C92200, C92300	.31	42.	.09
C93200	.32	34.	.09
C93700	.32	27.	.09
C94700	.32	31.	.09
C94800	.32	22.	.09

	ρ(lb/in^3)	k(Btu/hr-ft-F)	c_p(Btu/lb-F)
C95200, C95300, C95400, C95500	.27	24.-36.	.09
C96200, C96400	.32	17.-26.	.09
C97300, C97400, C97600, C97800	.32	13.-17.	.09
C99300	.28	25.	.1
DELRIN	.047	.22	.35
FIBERGLASS			

Fiberglass properties are anisotropic and vary depending on lay-up. Manufacturer's data or testing recommended for critical applications.

GALLIUM ARSENIDE	.19	19.	.080
GERMANIUM	.19	35.	.077
GLASS			
Optical Fused Silica (Quartz)	.080	.87	.17
HYDRAZINE			
Liquid	.036	.30	.74
Solid	.042	.96	.48
INVAR	.29	7.7	.12
MAGNESIUM			
AZ31B-F, AZ318-H24	.065	44.	.25
AZ61A-F	.065	34.	.25
AZ80A-T5	.065	29.	.25
ZK60A-T5	.065	69.	.25
HK31A-H24	.065	66.	.25
HM21A-T8	.065	79.	.25
HM31A-T5	.065	60.	.25
AZ63A	.065	29.-39.	.25
AZ81A	.065	29.	.25
AZ91A, AZ91B	.065	31.	.25
AZ91C	.065	37.-21.	.25
AZ92A	.065	27.-34.	.25
AM100A	.065	24.-34.	.25
EZ33A-T5	.065	58.	.25

	ρ(lb/in^3)	k(Btu/hr-ft-F)	c_p(Btu/lb-F)
HK31A-T6	.065	52.	.25
KAPTON			
Standard Kapton	.051	.090	.24
Carbon Loaded (Black) Kapton	.047	.09	.26
KOVAR		9.5	
MOLYBDENUM			
Molybdenum	.37	85.	.065
TZM	.37	85.	.065
MYCALEX	.084	.24-.34	.12
NICKEL			
INCO Alloy HA-330	.29	7.1	.11
Incoloy 800HT	.29	6.6	.11
Inconel 825	.30	6.4	.11
Inconel 600	.30	8.6	.11
Inconel 601	.29	6.5	.11
Inconel 625	.31	5.7	.10
Beryllium Nickels	.29	7.3-18.3	.12
Monel 400	.29	12.6	.10
Monel 404	.32	12.2	.10
Monel R-405	.32	12.6	.10
Monel K-500	.31	10.1	.10
Monel 502	.31	10.1	.10
80 Ni	.30	.22	.11
75 Ni	.30	.22	.11
70 Ni	.30	.24	.11
60 Ni	.30	.22	.11
Hastelloy alloy B-2	.33	7.75	.09
Hastelloy alloy C, Uniloy HC	.32	6.5	.09
Alloy C-276	.32	7.5	.10
Alloy G	.30	89.	.09
IN102	.31	6.5	
Inconel 600	.29	6.5	.11
Inconel 617	.30	7.8	
Inconel 625	.29	6.8	.11
Inconel 690	.30	6.6	
Inconel 700	.29	7.2	.11
Inconel 706	.29	7.3	
Inconel 718	.30	6.5	
Inconel 722	.30	8.5	

	ρ(lb/in^3)	k(Btu/hr-ft-F)	c_p(Btu/lb-F)
Inconel X-750	.30	6.9	
901	.30	7.7	
B-1900	.30	6.8	
D-979	.30	7.3	
MAR-M-200	.31	12.2	.2
MAR-M-246	.31	14.5	
MAR-M-421	.29	15.6	
TD NI	.32	27.	.11
TD Ni Cr	.31	22.	
Udimet 500	.29	14.1	.10
Waspaloy	.30	9.7	
Nicrotung	.3	8.8	
Rene-41, R-41	.30	9.3	
GMR-235-D	.29	8.2	
Hastelloy alloy S	.32	8.2	
Hastelloy alloy X	.30	9.1	.11
Udimet HX	.30	9.1	.11
Unitemp HX	.30	9.1	.11
INCO alloy HX	.30	9.1	.11
RTV			
11	.040	.17	
21, 41	.043	.18	
31, 60, 88	.048	.18	
511	.039	.15	
560, 577	.046	.18	
615	.033	.11	
616	.040	.16	
630	.042	.18	
632	.041	.18	
634	.039	.18	
619	.032	.11	
627	.460	.18	
655, 670	.035	.11	
8111, 8112	.039	.17	
8262	.049	.18	
SILICON	.084	86.	.17
STAINLESS STEEL			
201, 202	.29	9.4	.12
203EX	.28	9.4	.12
211	.28		
216	.29		
301	.29	9.4	.12

	ρ(lb/in^3)	k(Btu/hr-ft-F)	c_p(Btu/lb-F)
302	.29	9.4	.12
302B	.29	9.2	.12
302HQ	.29	6.5	.12
303, 303 Se	.29	9.4	.12
303 PLUS-X	.29	9.4	12
304	.29	9.4	.12
304L, 304LN	.29	9.4	.12
305	.29	9.4	.12
308	.29	8.8	.12
309, 309S	.29	9.	.12
310, 310S	.29	8.2	.12
316	.29	9.4	.12
316L	.29	9.4	.12
317	.29	9.4	.12
321	.29	9.3	.12
347-348	.29	9.3	.12
384-385	.29	9.5	.12
403	.28	14.4	.11
405	.28	15.6	.11
410, 410CB	.28	14.4	.11
414	.28	14.4	.11
416, 416SE	.28	14.4	.11
420	.28	14.4	.11
420F	.28	14.4	.11
429	.28	14.8	.11
430	.28	15.1	.11
430F-430FSE	.28	15.1	.11
431	.28	11.7	.11
434	.28	15.2	.11
436	.28	13.8	.11
440A, B, C	.28	14.0	.11
446	.27	12.1	.12
501	.28	21.2	.11
502	.28	21.2	.11
Stainless W	.28	12.1	
17-4 PH	.28	10.4	
CB-7 Cu	.28	9.9	
17-7 PH	.28	9.7	
PH 15-7 MO	.28	9.3	
17-14 Cu Mo	.28	8.7	
AM-350	.28	8.9	
AM-355	.28	9.2	
JS700	.29	8.5	.12
Uniloy 326	.28	11.3	.10
Nitronic 40	.28	8.	
Nitronic 50	.28	9.	
CA-6NM	.28	14.5	.11

	ρ(lb/in^3)	k(Btu/hr-ft-F)	c_p(Btu/lb-F)
CA-15	.28	14.5	.11
CB-30	.27	12.8	.11
CC-50	.27	12.6	.12
CF-3M	.28	9.4	.12
CD-4M Cu	.28	8.8	.11
CE-30	.28	8.5	.14
CF-3	.28	9.2	.12
CF-8	.28	9.2	.12
CF-20	.28	9.2	.12
CF-8M, CF-12M	.28	9.4	.12
CF-8C	.28	9.3	.12
CF-16F	.28	9.4	.12
CG-8M	.28	9.4	.12
CH-20	.28	8.2	.12
CK-20	.28	7.9	.12
CN-7M	.28	12.1	.11
HA	.28	15.	.11
HC	.28	12.6	.12
HD	.28	12.6	.12
HE	.28	8.5	.14
HF	.28	8.3	.12
HH	.28	8.2	.12
HI	.28	8.2	.12
HK	.28	7.9	.12
HL	.28	8.2	.12
HN	.28	7.5	.11
HP	.28	7.5	.11
HT	.29	7.	.11
HU	.29	7.	.11
HW	.29	7.2	.11
HX	.29	7.2	.11

	ρ(lb/in^3)	k(Btu/hr-ft-F)	c_p(Btu/lb-F)
TANTALUM	.60	31.5	.036
TEFLON			
FEP	.080	.10	.24
TFE	.080	.13	.23
TITANIUM			
Titanium	.16	9.-10.	.125
Ti-0.15-0.2 Pd	.16	4.5	.125
Ti-5Al-2.5 Sn	.16	4.5	.125
Ti-5AL-6Sn-2Zr-1Mo	.16	3.8	.125
Ti-8AL-1Mo-1V	.16	4.2	.125
Ti-6Al-4V cast	.16	4.2	.137
Ti-8Mn	.17	6.3	.118
Ti-6Al-2Sn-4Zr-6Mo	.17	4.1	
Ti-6Al-6V-2Sn	.16	4.2	.155
Ti-6Al-2Sn-4Zr-2Mo	.16	3.5	.1
Ti-3Al-13V-11Cr		4.	.12
TUNGSTEN	.70	97.	.034

Note: All properties at room temperature.

HONEYCOMB CONDUCTION

Honeycomb composites of various types are commonly used on spacecraft as equipment shelves, solar-array substrates, etc. The following equations for calculating the effective conductivity through honeycomb core material in different directions were developed by Lee Hennis of Hughes Aircraft Company.

Due to its construction, honeycomb has directionally dependent conductivities. These are presented for each of the three directions for a general hexagonal honeycomb structure, as well as the typical regular hexagonal structure. The final "k" and "C" equations given at the end of each section are expressed in terms of variables that can be obtained from the face of an engineering drawing. It should be noted that this paper deals exclusively with the core material and does not in any way include the facesheets that will be bonded to the core. Also, radiation exchange between walls of hexagonal structure has been excluded from this discussion. [Note: radiation-heat transfer is small compared to conduction for aluminum honeycomb panels. Ed.]

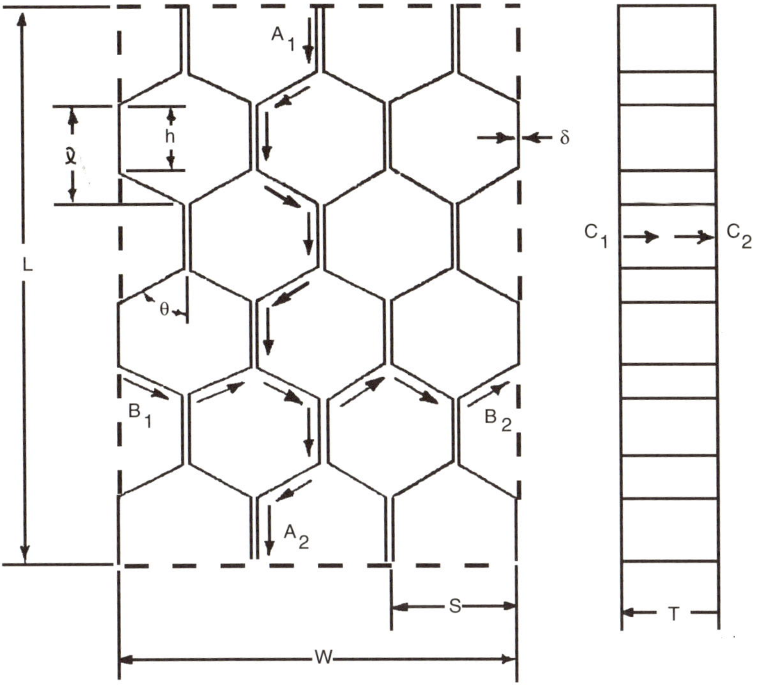

Nomenclature:

L = Overall honeycomb length (in the ribbon direction)
W – Overall honeycomb width (perpendicular to the ribbon direction)
T = Thickness of honeycomb
S = Cell size, face to face
δ = Ribbon thickness
h = Length of cell wall
θ = Cell angle

Conduction in the "L" direction:

For one ribbon:
$C_{A1\text{-}A2} = KA/x$
where k = conductivity of the ribbon material
A = cross sectional area of the ribbon
x = total length of the ribbon
A = δT
x = σL
where σ = an extension factor
$\sigma = 2h/\ell$

$\hat{\ell} = h + h\cos\theta$

therefore $\sigma = 2/(\hat{\ell} + \cos\theta)$

Substituting:

$C_{A1\text{-}A2} = k\delta T/\sigma L$

Now for n ribbons: (It is assumed that the net heat interchange between ribbons is negligible for this directional calculation.)

$n = \#$ of ribbons

$n = 2W/S$ ($W/S = \#$ of cells in the W direction)

$C_L = kA/x$

where k = conductivity of the honeycomb material

$A = \delta T(2W/S)$

$x = \sigma L$

Substituting:

$$C_L = (2k\delta/\sigma S)(WT/L)$$

or $\bar{k}_L$ = equivalent honeycomb conductivity in the "L" direction

$$\bar{k}_L = 2k\delta/\sigma S$$

For the normal hexagonal honeycomb structure:

$\theta = 60^\circ$

$\sigma = 2/(\hat{\ell}+\cos 60^\circ)$

$\sigma = 2/1.5 = 4/3$

Substituting:

$$\bar{k}_L = 3k\delta/2S$$

$$\boxed{C_L = (3k\,\delta/2S)(WT/L)}$$

__Conduction in the "W" direction__:

For one path B_1-B_2: (It is assumed that the net heat interchange between paths is negligible for this directional calculation. It can be shown that the contact resistance at the ribbon interfaces along the path is also negligible compared to the material resistance.)

$C_{B1\text{-}B2} = kA/x$

where k = path material conductivity

A = cross-sectional area of the path

x = total path length

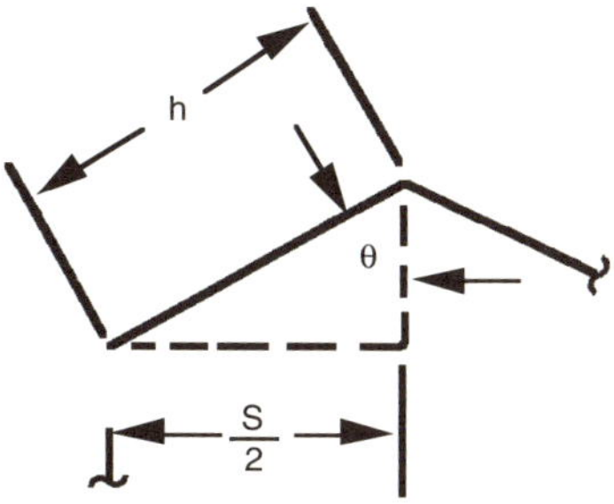

$A = \delta T$

$x = nh \ (n = 2W/S)$

$x = 2Wh/S$

$\sin\theta = S/2h$

$h = s/2 \sin\theta$

therefore $x = W/\sin\theta$

Substituting:

$C_{B1\text{-}B2} = k\delta T \sin\theta/W$

Now for m paths:

m = # of paths

$m = L/\ell = L/(h + h\cos\theta) = \sigma L/2h = \sigma L \sin\theta/S$

$C_W = kA/x$

where k = conductivity of the honeycomb material

$A = m\,\delta T = \delta T(\sigma L \sin\theta/S)$

$x = W/\sin\theta$

Substituting:

$C_W = (k\,\delta\sigma \sin^2\theta/S)(LT/W)$

or $\bar{k}_W$ = equivalent honeycomb conductivity in the "W" direction

$\bar{k}_W = k\,\delta\sigma \sin^2\theta/S$

For the normal hexagonal honeycomb structure:

$\theta = 60^\circ$

$\sin^2\theta = 3/4$

$\sigma = 4/3$

Substituting:

$\bar{k}_W = k\,\delta/S$

$$\boxed{C_W = (k\,\delta/S)(LT/W)}$$

<u>Conduction in the "T" direction</u>:

For one ribbon:

$C_{C1} - C_2 = kA/x$

where k = conductivity of the ribbon material

A = cross sectional area of the path

x = total path length

$A = \sigma L\delta$

$x = T$

Substituting:

$C_{C1} - C_2 = k\sigma\delta\, L/T$

Now for n ribbons: (It is assumed that the net heat interchange between ribbons is negligible for this directional calculation.)

$C_T = kA/x$

where k = conductivity of the honeycomb material

$A = n\,\sigma\delta L$

$n = 2W/S$

$A = 2\sigma\delta LW/S$

$x = T$

Substituting:

$C_T = 2k\sigma\delta LW/ST$

or $\bar{k}_T$ = equivalent honeycomb conductivity in the "T" direction

$\bar{k}_T = 2k\sigma\delta/S$

For the normal hexagonal honeycomb structure:

$\theta = 60^\circ$

$\sigma = 4/3$

Substituting:

$\bar{k}_T = 8k\delta/3S$

$$\boxed{C_T = (8k\delta/3S)\,(LW/T)}$$